Einar Hille, Professor Emeritus at Yale University, is currently Visiting Professor at the University of New Mexico. He was educated in Sweden, where he received the Ph.D. degree from the University of Stockholm. Before accepting an appointment at Yale, he served on the faculties of the University of Stockholm and Harvard and Princeton Universities. Dr. Hille has been a visiting professor at Stanford University, the University of Chicago, the University of California, Irvine, the University of Oregon, and at the Universities of Nancy, Mainz, Stockholm, Uppsala, Tata Institute, and Australian National University. He was also a Fulbright lecturer at the Sorbonne. Dr. Hille is the author of a number of mathematics texts, including *Lectures on Ordinary Differential Equations* (Addison-Wesley, 1969). He is past president of the American Academy of Arts and Sciences, the National Academy of Science, and the Royal Swedish Academy of Science. His field of specialization is mathematical analysis.

METHODS IN CLASSICAL AND FUNCTIONAL ANALYSIS

EINAR HILLE
Professor Emeritus
Yale University

ADDISON-WESLEY PUBLISHING COMPANY
Reading, Massachusetts · Menlo Park, California · London · Don Mills, Ontario

This book is in the

ADDISON - WESLEY SERIES IN MATHEMATICS

Consulting Editor:

LYNN H. LOOMIS

 Printed in the United States of America. Published simultaneously in Canada. Library of Congress Catalog Card No. 70-137840.

To KIRSTI

On revient toujours à son premier amour

PREFACE

Plus ca change, plus c'est la même chose

Modes come and go in mathematics as in most fields. During the half-century and more that I have worked in the vineyard I have heard many dire predictions for the fate of my ideas and interests. Abstraction has been in the saddle during most of the time and has ridden us mercilessly. In a modest way I have taken part in this development. I did not believe in abstraction *per se*; one should know what one is trying to generalize and one should show that the generalization is significant. I have tried to keep at least one foot on the ground while craning my neck to look into Heaven. Was it Heaven? There are some doubts, and the more extravagant claims of the abstract mathematicians to be the sole dispensers of the true faith and the arbiters of values are received with a healthy scepticism. A recent letter to the editor of the Notices of the American Mathematical Society headed "Can Mathematics be Saved?," understood to be from the modern mathematicians, is more than a straw in the wind. "Applicable Analysis" is becoming sufficiently popular to sport its own journal.

This book may be regarded as part of the backlash. If the book has a thesis, it is that a functional analyst is an analyst, first and foremost, and not a degenerate species of a topologist. His problems come from analysis and his results should throw light on analysis. The book was originally planned as an elementary introduction to functional analysis, but in working out the guiding ideas I soon found that my program took in only part of what nowadays goes under the title of functional analysis. On the other hand, applications to analysis were stressed and unvarnished classical analysis was handed out in significant doses. My interests have always been toward the concrete sides of analysis, a tendency that has not become less pronounced with the years.

It seemed to me that I could do some useful work in giving the student a historical perspective and in showing how the multitude of abstract concepts have arisen and are present *in nuce* already in Euclidean spaces. This required a rather detailed discussion of the space C^n and applying the general ideas to illustrative material taken from the basic infinite-dimensional spaces such as continuous functions, functions of bounded variation, sequence, and Lebesgue spaces. The further development of the book reflects my own interest and research during the last twenty-five years. I stress complex analysis in Banach spaces and Banach algebras because it has meant so much to me and may benefit a budding analyst even in the seventies. Fixed point theorems and functional inequalities are more recent interests and are eminently applicable. So are functional equations and mean values and, even if they come last, they are not least. A mean value may be an abstract notion, but it has applications to extremal problems, to set functions, to potential theory, and probably to more to come. Linear transformations and inner product spaces are more canonical items and need no justification.

Practically the whole book has been lecture material given from Bombay to Kingston, R.I., under varying titles. The book is not polished; there are loose ends and many unsolved problems waiting for the craftsman. Such research openings occur chiefly in the last four chapters. There are over 850 exercises, a fair part of which are byproducts of the author's own research. While problem solving should not become an obsession, it provides good training in research and stimulates an inquisitive mind to venture into new directions.

The material in the book can serve several different purposes. Chapters 1 to 4 and 7 to 11 can serve as text for an introductory course in functional analysis. In the author's judgment it offers a fairly easy introduction and supplement to the more advanced books such as Dunford–Schwartz, Hille–Phillips, and Yosida. Chapters 5, 6, and 12 through 15 stand closer to classical analysis than does the rest of the book and can serve as a text for a course in aspects of analysis.

The book has been written at the University of New Mexico, where the Department of Mathematics and Statistics has given friendly help and support. Many knotty questions have been clarified in discussions with colleagues here. B. Epstein, R. Hersh and R. Metzler provided many illuminating comments. I am particularly indebted to Ih-Ching Hsu, who read the whole manuscript and has made elaborate suggestions for improvements. †J. B. Miller of Monash University, Victoria, has also read the manuscript and given me the benefit of his advice and detailed comments. The last two chapters owe much to the criticism of J. Aczél and C. T. Ng of the University of Waterloo, Ontario. Comments have also been received from the referees. To all these helpful friends I express my sincere gratitude.

I also wish to thank the staff of Addison-Wesley Publishing Company for consideration, kindness, and understanding during the writing phase and for a work well done in presenting the book to the public.

Albuquerque, New Mexico E. H.
April, 1971

† J. Donaldson and I.-C. Hsu have also helped with the proof-reading.

CONTENTS

1 COMPLEX EUCLIDEAN SPACES

Modern analysis, especially functional analysis, is concerned with analysis in abstract, usually linear, spaces. The elements of such a space are called "points" or "vectors" in analogy with the usage in Euclidean spaces. The latter form the prototype of all abstract spaces and the theory of Euclidean spaces is a necessary tool for the study of the generalizations. Hence we must start our discourse with a study of Euclidean spaces. Much of the theory is just linear algebra and thus supposedly familiar to the student, but our emphasis is directed by the needs for the generalizations. The reader will encounter a large number of concepts which will play a basic role in later chapters. They will largely be just labels at this early stage because we are not ready for formal definitions. The reader may be assured, however, that there is a purpose behind the introduction of this multitude of concepts. They serve as prototypes for later generalizations, and one has to have a fairly good idea of the elementary models before fruitful extensions can be attempted.

This chapter is divided into seven sections: Euclidean three space; The space C^n; Linear transformations; Matrices; The resolvent; Invariantive and metric properties; and Hermitian forms.

1.1 EUCLIDEAN THREE SPACE

The ordinary Euclidean three space will be denoted by R^3. To simplify the exposition we introduce *Cartesian coordinates* in the familiar manner. We choose an *origin* O, a *unit of length*, and three perpendicular lines through O referred to as the *axes*. If P is any point in space other than O, the directed line segment from O to P is called the *vector* $\overrightarrow{OP}$ or simply $\mathbf{x}$, which is used both to denote the vector and its endpoint P. Thus we speak equivalently of the vector $\mathbf{x}$ and the point $\mathbf{x}$. With $\mathbf{x}$ is associated an ordered triple (x_1, x_2, x_3) of three real numbers, the *coordinates* of $\mathbf{x}$. We use roman bold face type for vectors, italics for their coordinates. The coordinates are obtained by the following construction. We take orthogonal projections of $\overrightarrow{OP}$ on the three axes and obtain three vectors, $\overrightarrow{OX}_1, \overrightarrow{OX}_2, \overrightarrow{OX}_3$ as a result. Each of the three axes is given an orientation so that for that axis one direction from the origin is said to be *positive*, the opposite *negative*. In this way it makes sense to speak of a positive vector $\overrightarrow{OX}_j$, $j = 1, 2, 3$. The orientation and numbering of the axes is chosen in such a manner that three positive vectors $\overrightarrow{OX}_1$, $\overrightarrow{OX}_2, \overrightarrow{OX}_3$ correspond to a right hand screw: if the axes are built into the screw with the third axis being the axis of the screw, then a rotation from $\overrightarrow{OX}_1$ to $\overrightarrow{OX}_2$ carries the screw upwards in the direction of $\overrightarrow{OX}_3$.

We denote the Euclidean length of the vector by

$$\|\overrightarrow{OP}\| \quad \text{or} \quad \|\mathbf{x}\|. \tag{1.1.1}$$

The symbol $\|\mathbf{x}\|$ is read "norm of $\mathbf{x}$" and will be used throughout this treatise as notation for various generalizations of the elementary Euclidean length. We now set

$$x_j = \pm\|\overrightarrow{OX}_j\|, \qquad j = 1, 2, 3. \tag{1.1.2}$$

Here the plus sign is chosen if $\overrightarrow{OX}_j$ has the positive direction, otherwise the minus sign.

In this manner we obtain a one-to-one correspondence between the points P of the space and the ordered number triples

$$P \leftrightarrow (x_1, x_2, x_3) = \mathbf{x}. \tag{1.1.3}$$

Here x_1, x_2, x_3 are known as the coordinates of $\mathbf{x}$ in the chosen coordinate system.

We now proceed to define vector addition and scalar multiplication.

Definition 1.1.1. *If* $\mathbf{x} = (x_1, x_2, x_3)$, $\mathbf{y} = (y_1, y_2, y_3)$ *and if* α *is a real number, then* $\mathbf{x} + \mathbf{y}$ *and* $\alpha\mathbf{x}$ *are the vectors*

$$\begin{aligned} \mathbf{x} + \mathbf{y} &= (x_1 + y_1, x_2 + y_2, x_3 + y_3), \\ \alpha\mathbf{x} &= (\alpha x_1, \alpha x_2, \alpha x_3). \end{aligned} \tag{1.1.4}$$

In terms of the unit vectors

$$\begin{aligned} \mathbf{u}_1 &= (1, 0, 0), \\ \mathbf{u}_2 &= (0, 1, 0), \\ \mathbf{u}_3 &= (0, 0, 1), \end{aligned} \tag{1.1.5}$$

we have now the representation

$$\mathbf{x} = x_1\mathbf{u}_1 + x_2\mathbf{u}_2 + x_3\mathbf{u}_3 \tag{1.1.6}$$

which is unique since it implies and is implied by

$$\mathbf{x} = (x_1, x_2, x_3)$$

together with Definition 1.1.1.

Formula (1.1.6) suggests the following question. Given three vectors $\mathbf{v}_1, \mathbf{v}_2, \mathbf{v}_3$ in R^3, when is it possible to represent an arbitrary vector $\mathbf{x}$ of R^3 as a linear combination of the given vectors with constant coefficients, say

$$\mathbf{x} = s_1\mathbf{v}_1 + s_2\mathbf{v}_2 + s_3\mathbf{v}_3? \tag{1.1.7}$$

if $u_1, \ldots, u_k$ are linearly independent and span V then this set of vectors is called a basis of V.

A necessary and sufficient condition for the existence of such a representation is clearly that the unit vectors $\mathbf{u}_1, \mathbf{u}_2, \mathbf{u}_3$ can be so represented. Since all vectors are to be representable, $\mathbf{u}_1, \mathbf{u}_2, \mathbf{u}_3$ must have this property, so the condition is necessary. On the other hand, if

$$\mathbf{u}_j = \sum_{k=1}^{3} s_{jk} \mathbf{v}_k, \quad j = 1, 2, 3, \tag{1.1.8}$$

then combining this with (1.1.6) leads to (1.1.7) with

$$s_j = \sum_{k=1}^{3} x_k s_{kj}. \tag{1.1.9}$$

Now we know that the **v**'s are linearly representable in terms of the **u**'s, say

$$\mathbf{v}_j = \sum_{t=1}^{3} a_{jk} \mathbf{u}_k, \quad j = 1, 2, 3. \tag{1.1.10}$$

This is a system of three linear equations for the **u**'s, so our problem is solvable iff (read "if and only if") the system (1.1.10) is solvable for the **u**'s for any given set of the **v**'s and this is the case iff the *determinant* of the system

$$\det(a_{jk}) \neq 0. \tag{1.1.11}$$

If this is the case, we can express the **u**'s linearly in terms of the **v**'s, for instance by Cramer's rule.

Now if the determinant is not zero, then the **v**'s defined by (1.1.10) are *linearly independent*, i.e. a relation

$$a_1 \mathbf{v}_1 + a_2 \mathbf{v}_2 + a_3 \mathbf{v}_3 = 0 \tag{1.1.12}$$

with real numbers a_1, a_2, a_3 implies that all three numbers are zero. Conversely, given any three linearly independent vectors $\mathbf{v}_1, \mathbf{v}_2, \mathbf{v}_3$, we can then find numbers a_{jk} such that (1.1.10) holds and the linear independence implies that $\det(a_{jk}) \neq 0$ so the **u**'s can be expressed linearly in the **v**'s. We refer to such a set of three vectors $\mathbf{v}_1, \mathbf{v}_2, \mathbf{v}_3$ as a *basis* of R^3.

In geometrical language condition (1.1.12) expresses that the three vectors lie in the same plane through the origin. They are then said to be *coplanar*. Two vectors are said to be *collinear* if one is a constant multiple of the other so that they lie on the same straight line through the origin.

Suppose that we start with any two non-collinear vectors $\mathbf{v}_1$ and $\mathbf{v}_2$. They determine uniquely a plane through the origin. Now take any third vector $\mathbf{v}_3$ not coplanar with $\mathbf{v}_1$ and $\mathbf{v}_2$. Then the three vectors $\mathbf{v}_1, \mathbf{v}_2, \mathbf{v}_3$ form a basis for R^3 since they are linearly independent.

Moreover, given any basis $\mathbf{v}_1, \mathbf{v}_2, \mathbf{v}_3$ we can construct an *orthogonal basis*. We may impose the additional restriction that the new vectors be of unit length, in which case one speaks of an *orthonormal basis*. The set $\mathbf{u}_1, \mathbf{u}_2, \mathbf{u}_3$ introduced above is an example of an orthonormal basis. The following construction is a simple special case of what is known as the *Gram-Schmidt orthogonalization process* after the Danish actuary Jörgen Pedersen Gram (1850–1916) and the German mathematician Erhard Schmidt (1876–1959).

We start with $\mathbf{v}_1$, which may be assumed to be of length one; if not, divide by its length. This gives a new vector of the same direction as $\mathbf{v}_1$ and of length one. Consider now the plane Π determined by $\mathbf{v}_1$ and $\mathbf{v}_2$. All vectors in this plane are linear combinations of the form

$$\mathbf{v} = \alpha\mathbf{v}_1 + \beta\mathbf{v}_2.$$

Geometrical intuition asserts that among these vectors there are two and only two vectors which are of length one and perpendicular to $\mathbf{v}_1$. Let one of them be denoted by $\mathbf{u}_2$ and write $\mathbf{u}_1$ for $\mathbf{v}_1$. Then there are constants α_{21} and α_{22} such that

$$\mathbf{u}_2 = \alpha_{21}\mathbf{v}_1 + \alpha_{22}\mathbf{v}_2.$$

Here $\alpha_{22} \neq 0$ for $\mathbf{u}_2$ may conceivably be a multiple of $\mathbf{v}_2$ but it cannot be a multiple of $\mathbf{u}_1$. It follows that $\mathbf{u}_1$ and $\mathbf{u}_2$ form a basis for the vectors in Π. Now consider all vectors in R^3 which are not coplanar with $\mathbf{u}_1$ and $\mathbf{u}_2$. Again, geometrical intuition asserts that there are two and only two such vectors which are of length one and are perpendicular to $\mathbf{u}_1$ and $\mathbf{u}_2$. We simply erect the normal to Π at the origin and lay off the distance one to either side of the origin. One of these vectors is taken as $\mathbf{u}_3$ and it is clear that there are three numbers $\alpha_{31}, \alpha_{32}, \alpha_{33}$ such that

$$\mathbf{u}_3 = \alpha_{31}\mathbf{v}_1 + \alpha_{32}\mathbf{v}_2 + \alpha_{33}\mathbf{v}_3,$$

where $\alpha_{33} \neq 0$ since the vectors $\mathbf{v}_1, \mathbf{v}_2, \mathbf{v}_3$ are not coplanar. This set of vectors $\mathbf{u}_1, \mathbf{u}_2, \mathbf{u}_3$ forms an orthonormal basis of R^3. In choosing $\mathbf{u}_2$ and $\mathbf{u}_3$ we may make the choice in such a manner that the new orthonormal basis is similarly oriented as the vectors of (1.1.5). If this is done, there exists a uniquely defined rotation which keeps the origin fixed and carries the old orthonormal system into the new one.

Let us get rid of the "geometric intuition" which is too feeble to support generalizations. What we need is the notion of an *inner product* which in R^3 coincides with the *dot product* of classical vector analysis. Any pair of vectors $\mathbf{x}$ and $\mathbf{y}$ in R^3, distinct or not, has an inner product which is a real number denoted by $(\mathbf{x}, \mathbf{y})$. We shall specify the desired properties of $(\mathbf{x}, \mathbf{y})$ which will enable us to give explicit formulas for this number. The following definition makes sense in any *inner-product space over the reals*, whatever this may mean, and requires little change in the complex case.

Definition 1.1.2. *For any pair of vectors* $\mathbf{x}, \mathbf{y}$ *of the space, distinct or not, there is a uniquely defined number* $(\mathbf{x}, \mathbf{y})$ *called the inner product of* $\mathbf{x}$ *with* $\mathbf{y}$ *such that*

1) $(\mathbf{x}, \mathbf{y})$ *is real;*
2) $(\mathbf{y}, \mathbf{x}) = (\mathbf{x}, \mathbf{y})$*;*
3) $(\alpha\mathbf{x}_1 + \beta\mathbf{x}_2, \mathbf{y}) = \alpha(\mathbf{x}_1, \mathbf{y}) + \beta(\mathbf{x}_2, \mathbf{y})$ *(linearity);*
4) $(\mathbf{x}, \mathbf{x}) \geqslant 0$ *for all* $\mathbf{x}$, $(\mathbf{x}, \mathbf{x}) = 0$ *iff* $\mathbf{x} = \mathbf{0}$*;*
5) $(\mathbf{x}, \mathbf{x}) = \|\mathbf{x}\|^2$.

Here (3) is to hold for all real numbers α, β and all vectors $\mathbf{x}_1, \mathbf{x}_2, \mathbf{y}$. Let us remark in passing that in the complex case, considered already in the next section, the word "real" is to be replaced by "complex" and in (2) the right member is to be replaced by its complex conjugate.

Combining (2) and (3) we see that

$$(\mathbf{x}, \alpha\mathbf{y}_1 + \beta\mathbf{y}_2) = \alpha(\mathbf{x}, \mathbf{y}_1) + \beta(\mathbf{x}, \mathbf{y}_2). \tag{1.1.13}$$

The two relations (3) and (1.1.13) express that the inner product is *bilinear*.

Lemma 1.1.1. *If* $\mathbf{x}, \mathbf{y} \in R^3$ *and make the angle* θ *with each other, then*

$$(\mathbf{x}, \mathbf{y}) = \|\mathbf{x}\| \, \|\mathbf{y}\| \cos \theta. \tag{1.1.14}$$

Proof. We use the Law of Cosines on the triangle two sides of which are $\mathbf{x}$ and $\mathbf{y}$. The third side, oriented from $\mathbf{x}$ to $\mathbf{y}$, is parallel to and has the same length and orientation as the vector $\mathbf{y} - \mathbf{x}$. Its length is $\|\mathbf{y} - \mathbf{x}\|$ and the Law of Cosines gives

$$\|\mathbf{y} - \mathbf{x}\|^2 = \|\mathbf{x}\|^2 + \|\mathbf{y}\|^2 - 2\,\|\mathbf{x}\|\,\|\mathbf{y}\| \cos \theta.$$

Here the left member equals

$$(\mathbf{y} - \mathbf{x}, \mathbf{y} - \mathbf{x}) = (\mathbf{y}, \mathbf{y}) - (\mathbf{y}, \mathbf{x}) - (\mathbf{x}, \mathbf{y}) + (\mathbf{x}, \mathbf{x})$$

by properties (2) and (3). Using (4) we see that this reduces to

$$\|\mathbf{x}\|^2 + \|\mathbf{y}\|^2 - 2\,(\mathbf{x}, \mathbf{y}).$$

Comparison of the two expressions for the length of the third side gives (1.1.14). ∎

This lemma shows that the inner product $(\mathbf{x}, \mathbf{y})$ is unique and expressible in terms of intrinsic properties of the vectors $\mathbf{x}, \mathbf{y}$, namely their lengths and the angle between them. We also note

Corollary 1. *The vectors* $\mathbf{x}$ *and* $\mathbf{y}$ *are orthogonal iff*

$$(\mathbf{x}, \mathbf{y}) = 0. \tag{1.1.15}$$

Corollary 2. *We have*

$$|(\mathbf{x}, \mathbf{y})| \leqslant \|\mathbf{x}\| \, \|\mathbf{y}\| \tag{1.1.16}$$

with equality iff $\mathbf{y}$ *is a constant multiple of* $\mathbf{x}$.

This is the first instance of an inequality listed in this treatise together with a statement of when the inequality becomes an equality. Such considerations will occur again and again in the following.

Lemma 1.1.2. *Three vectors* $\mathbf{v}_1, \mathbf{v}_2, \mathbf{v}_3$ *are linearly independent iff the determinant of their inner products is different from zero, i.e.*

$$G \equiv \det\big((\mathbf{v}_j, \mathbf{v}_k)\big) \neq 0. \tag{1.1.17}$$

Remark. This determinant is known as the *Gramian*. It has a geometric interpretation as the volume of the parallelepiped formed by the three vectors and is non-negative. If the lengths of the vectors are held fixed, the volume is a maximum when the vectors are pairwise orthogonal, i.e. the solid is a rectangular parallelepiped. The volume decreases to zero when the vectors become coplanar.

Proof. We form the vector

$$\mathbf{v} = \alpha\mathbf{v}_1 + \beta\mathbf{v}_2 + \gamma\mathbf{v}_3 \tag{1.1.18}$$

and the inner products

$$\begin{aligned}
(\mathbf{v}_1, \mathbf{v}) &= \alpha(\mathbf{v}_1, \mathbf{v}_1) + \beta(\mathbf{v}_1, \mathbf{v}_2) + \gamma(\mathbf{v}_1, \mathbf{v}_3),\\
(\mathbf{v}_2, \mathbf{v}) &= \alpha(\mathbf{v}_2, \mathbf{v}_1) + \beta(\mathbf{v}_2, \mathbf{v}_2) + \gamma(\mathbf{v}_2, \mathbf{v}_3),\\
(\mathbf{v}_3, \mathbf{v}) &= \alpha(\mathbf{v}_3, \mathbf{v}_1) + \beta(\mathbf{v}_3, \mathbf{v}_2) + \gamma(\mathbf{v}_3, \mathbf{v}_3).
\end{aligned} \tag{1.1.19}$$

The determinant of this system (regarded as linear equations satisfied by α, β, γ) is G. If $G \neq 0$ and $\mathbf{v}$ is any given vector in R^3, then we can determine α, β, γ uniquely from the system (1.1.19). This means that $\mathbf{v}_1, \mathbf{v}_2, \mathbf{v}_3$ is a basis for R^3 and they are linearly independent vectors. Conversely, if the vectors are linearly independent, then the representation (1.1.18) is unique. This implies the existence of a unique solution (α, β, γ) of the system (1.1.19) for any given vector $\mathbf{v}$, i.e. for arbitrary values of the left hand side in the system. This is possible iff the determinant $G \neq 0$. ∎

The orthogonalization process can now be formulated as follows. Given three linearly independent vectors $\mathbf{v}_1, \mathbf{v}_2, \mathbf{v}_3$. Find six constants a_{jk}, $1 \leqslant k \leqslant j \leqslant 3$, such that the three vectors

$$\begin{aligned}
\mathbf{u}_1 &= a_{11}\mathbf{v}_1,\\
\mathbf{u}_2 &= a_{21}\mathbf{v}_1 + a_{22}\mathbf{v}_2,\\
\mathbf{u}_3 &= a_{31}\mathbf{v}_1 + a_{32}\mathbf{v}_2 + a_{33}\mathbf{v}_3
\end{aligned} \tag{1.1.20}$$

satisfy the six conditions

$$(\mathbf{u}_j, \mathbf{u}_k) = \delta_{jk},$$

where δ_{jk} is the Kronecker delta, i.e. 0 if $j \neq k$ and 1 if $j = k$. The solution is easily found except for the normalization factors. We have

$$\begin{gathered}
\mathbf{u}_1 = c_1\mathbf{v}_1, \qquad \mathbf{u}_2 = c_2\begin{vmatrix} \mathbf{v}_1 & \mathbf{v}_2 \\ (\mathbf{v}_1, \mathbf{v}_1) & (\mathbf{v}_1, \mathbf{v}_2) \end{vmatrix},\\
\mathbf{u}_3 = c_3\begin{vmatrix} \mathbf{v}_1 & \mathbf{v}_2 & \mathbf{v}_3 \\ (\mathbf{v}_1, \mathbf{v}_1) & (\mathbf{v}_1, \mathbf{v}_2) & (\mathbf{v}_1, \mathbf{v}_3) \\ (\mathbf{v}_2, \mathbf{v}_1) & (\mathbf{v}_2, \mathbf{v}_2) & (\mathbf{v}_2, \mathbf{v}_3) \end{vmatrix}.
\end{gathered} \tag{1.1.21}$$

Here the vector-valued determinants are to be understood as follows. Expand each determinant according to the elements of the first row. The result is clearly of the form (1.1.20). We can form inner products of the form $(\mathbf{w}, \det(\cdot))$ by taking inner products of the first row. Thus

$$(\mathbf{w}, \mathbf{u}_2) = c_2 \begin{vmatrix} (\mathbf{w}, \mathbf{v}_1) & (\mathbf{w}, \mathbf{v}_2) \\ (\mathbf{v}_1, \mathbf{v}_1) & (\mathbf{v}_1, \mathbf{v}_2) \end{vmatrix}.$$

In particular, we see that $(\mathbf{v}_1, \mathbf{u}_2)$ involves a determinant with two equal rows which is zero for any choice of c_2. This gives $(\mathbf{u}_1, \mathbf{u}_2) = 0$. Similarly, we see that

$$(\mathbf{u}_3, \mathbf{v}_1) = 0, \qquad (\mathbf{u}_3, \mathbf{v}_2) = 0$$

for any choice of c_3, and this gives two more of the conditions (1.1.20), namely

$$(\mathbf{u}_1, \mathbf{u}_3) = 0, \qquad (\mathbf{u}_2, \mathbf{u}_3) = 0.$$

Thus our three vectors are orthogonal, and by choosing the factors c_1, c_2, c_3 properly we obtain an orthonormal system. The constants are expressible in terms of minors of G, but we omit further details. Here the "geometric intuition" has been eliminated and we have a formulation of the Gram-Schmidt process which extends to higher dimensions as well as to more general linear spaces.

Some further comments should be attached to the notion of the inner product. If $(\mathbf{u}_1, \mathbf{u}_2, \mathbf{u}_3)$ is an ordered orthonormal basis of R^3 and if

$$\mathbf{x} = (x_1, x_2, x_3), \qquad \mathbf{y} = (y_1, y_2, y_3) \tag{1.1.22}$$

in terms of this basis, then

$$(\mathbf{x}, \mathbf{y}) = x_1 y_1 + x_2 y_2 + x_3 y_3. \tag{1.1.23}$$

This is the classical formula of vector analysis for the dot product.

The inner product $(\mathbf{x}, \mathbf{y})$ is a function on ordered pairs of vectors to numbers. Such a function is usually referred to as a *functional.* Definition 1.1.2, (2), (3) shows that $(\mathbf{x}, \mathbf{y})$ is linear in both arguments. Hence it is called a *bilinear functional.*

Definition 1.1.3. *A linear functional on R^3 is a function $f(\mathbf{x})$ from vectors to numbers such that for all vectors $\mathbf{x}, \mathbf{y}$ and all real numbers α*

$$f(\mathbf{x} + \mathbf{y}) = f(\mathbf{x}) + f(\mathbf{y}), \qquad f(\alpha\mathbf{x}) = \alpha f(\mathbf{x}). \tag{1.1.24}$$

EXERCISE 1.1

1. Verify (1.1.23).
2. In the proof of Lemma 1.1.1 it is stated that the distance from the point $\mathbf{x}$ to the point $\mathbf{y}$ is $\|\mathbf{x} - \mathbf{y}\|$. Use (1.1.23) to verify that this is indeed the Euclidean distance.
3. Formula (1.1.16) states that $(\mathbf{x}, \mathbf{x})(\mathbf{y}, \mathbf{y}) - (\mathbf{x}, \mathbf{y})^2 \geqslant 0$. If $\mathbf{x}$ and $\mathbf{y}$ are given by (1.1.22) show that this is equivalent to

$$\left(\sum_{j=1}^{3} x_j y_j\right)^2 \leqslant \sum_{j=1}^{3} x_j^2 \sum_{j=1}^{3} y_j^2.$$

This is a special case of the so-called *Cauchy inequality* [after Augustin Louis, Baron de Cauchy (1789–1857)]. What happens to the condition for equality?

4. Show that the principal minors of the Gramian G are positive. A principal minor of a determinant is obtained by striking out rows and columns with the same subscripts.
5. In formula (1.1.21) the coefficients c_j are expressible as square roots of principal minors. Find these expressions.
6. Try to prove that G is the volume of the parallelepiped defined by the vectors $\mathbf{v}_1, \mathbf{v}_2, \mathbf{v}_3$.
7. Justify the method for forming inner products when one of the vectors is a vector determinant like (1.1.21).
8. While the dot product is a function from vectors to numbers the so-called *vector* or *cross product* of vector analysis is a vector. With the usual notation

$$\mathbf{x} \times \mathbf{y} = \begin{vmatrix} \mathbf{u}_1 & \mathbf{u}_2 & \mathbf{u}_3 \\ x_1 & x_2 & x_3 \\ y_1 & y_2 & y_3 \end{vmatrix}.$$

In vector analysis it is customary to denote the unit vectors by $\mathbf{i}$, $\mathbf{j}$, $\mathbf{k}$ instead of $\mathbf{u}_1, \mathbf{u}_2, \mathbf{u}_3$. Show that $\mathbf{x} \times \mathbf{y} = -\mathbf{y} \times \mathbf{x}$ so that this type of multiplication is *non-commutative* in the sense that $\mathbf{x} \times \mathbf{y} \neq \mathbf{y} \times \mathbf{x}$. Note also that $\mathbf{x} \times \mathbf{x} = \mathbf{0}$ for all $\mathbf{x}$, so a product can vanish without either factor being zero.

9. Verify that $\mathbf{i} \times \mathbf{j} = \mathbf{k}$, $\mathbf{j} \times \mathbf{k} = \mathbf{i}$, $\mathbf{k} \times \mathbf{i} = \mathbf{j}$.
10. A linear functional on R^3 was introduced in Definition 1.1.3. Such a functional is said to be bounded if there is a fixed positive number M such that $|f(\mathbf{x})| \leqslant M\|\mathbf{x}\|$ for all $\mathbf{x}$. Show that there exists a fixed uniquely determined vector $\mathbf{y}$ such that $f(\mathbf{x}) = (\mathbf{x}, \mathbf{y})$. [*Hint:* Specify the action of f on a basis and use linearity.]

1.2 THE SPACE C^n

We have given a fairly thorough discussion of the space R^3 except for linear transformations, which are postponed until Section 1.3. Here we take up the passage from three dimensions to n and from real coordinates to complex.

We say that R^3 is a *three-dimensional vector space* because there are sets of three linearly independent vectors, but any set of four or more vectors is necessarily linearly dependent. We now proceed to define C^n.

Definition 1.2.1. *The space C^n is the set of all ordered n-tuples of complex numbers $(z_1, z_2, \ldots, z_n) \equiv \mathbf{z}$ with the following conventions concerning equality, algebraic operations, and metric properties:*

1) $\mathbf{z} = \mathbf{w}$ *iff* $z_j = w_j, \quad j = 1, 2, \ldots, n;$
2) $\mathbf{z} + \mathbf{w} = (z_1 + w_1, z_2 + w_2, \ldots, z_n + w_n);$
3) $\alpha\mathbf{z} = (\alpha z_1, \alpha z_2, \ldots, \alpha z_n)$, α *any complex number;*

4) *There is an inner product*

$$(\mathbf{z}, \mathbf{w}) = \sum_{j=1}^{n} z_j \bar{w}_j$$

where the bar denotes the complex conjugate;

5) $(\mathbf{z}, \mathbf{z}) = \|\mathbf{z}\|^2$*; and*

6) *The Euclidean distance between* $\mathbf{z}$ *and* $\mathbf{w}$ *is* $\|\mathbf{z} - \mathbf{w}\|$.

The space R^n is obtained from C^n by restricting the coordinates z_j to be real numbers. This also means restricting the multipliers α to be real numbers and the bar can be omitted in (4).

The n-tuples are still regarded as vectors and are also spoken of as points or elements of C^n. In C^n we have a definition of addition, *vector addition*, given by (2). There is also a notion of *scalar multiplication* defined by (3). The symbol $\|\mathbf{z}\|$ (read "the norm of $\mathbf{z}$") is the natural generalization of the length of a vector or of the distance from the origin to the point $\mathbf{z}$. We speak of Euclidean distance in (6) because that is what it reduces to for $n = 2$ or 3 and also because, as we shall see later, there are other possibilities of defining a useful notion of distance in the space C^n.

From (4) we get

$$(\mathbf{w}, \mathbf{z}) = \overline{(\mathbf{z}, \mathbf{w})}. \tag{1.2.1}$$

The definition of the inner product is in agreement with Definition 1.1.2, where we replace "real" by "complex." The bilinearity of the inner product is expressed by the formulas

$$(\alpha\mathbf{z}_1 + \beta\mathbf{z}_2, \mathbf{w}) = \alpha(\mathbf{z}_1, \mathbf{w}) + \beta(\mathbf{z}_2, \mathbf{w}), \tag{1.2.2}$$

$$(\mathbf{z}, \alpha\mathbf{w}_1 + \beta\mathbf{w}_2) = \bar{\alpha}(\mathbf{z}, \mathbf{w}_1) + \bar{\beta}(\mathbf{z}, \mathbf{w}_2). \tag{1.2.3}$$

Note that the multipliers α and β are replaced by their complex conjugates in the right member of (1.2.3).

Lemma 1.1.1 remains valid in R^n, for two vectors $\mathbf{x}$ and $\mathbf{y}$ determine a two-dimensional plane formed by all the vectors

$$\alpha\mathbf{x} + \beta\mathbf{y}$$

with real α and β, and the Law of Cosines may be applied to the triangle with sides of lengths $\|\mathbf{x}\|$, $\|\mathbf{y}\|$, and $\|\mathbf{y} - \mathbf{x}\|$. On the other hand, the lemma normally does not make sense in C^n since $(\mathbf{z}, \mathbf{w})$ is complex valued.

Corollary 1 of Lemma 1.1.1 becomes

Definition 1.2.2. *The vectors* $\mathbf{z}$ *and* $\mathbf{w}$ *are said to be orthogonal if*

$$(\mathbf{z}, \mathbf{w}) = 0. \tag{1.2.4}$$

Corollary 2, on the other hand, becomes a theorem to be proved.

Theorem 1.2.1. *For any two vectors* $\mathbf{z}$ *and* $\mathbf{w}$ *in* C^n

$$|(\mathbf{z}, \mathbf{w})| \leqslant \|\mathbf{z}\| \, \|\mathbf{w}\| \tag{1.2.5}$$

with equality iff $\mathbf{w} = \gamma\mathbf{z}$ *for some complex number* γ.

Proof. Let α be a complex number to be disposed of later and consider $\|\mathbf{w} - \alpha\mathbf{z}\|^2$. This is a non-negative real number and zero iff $\mathbf{w} = \alpha\mathbf{z}$. Then

$$\begin{aligned}\|\mathbf{w} - \alpha\mathbf{z}\|^2 &= (\mathbf{w} - \alpha\mathbf{z}, \mathbf{w} - \alpha\mathbf{z}) \\ &= (\mathbf{w}, \mathbf{w}) - \bar{\alpha}(\mathbf{w}, \mathbf{z}) - \alpha(\mathbf{z}, \mathbf{w}) + |\alpha|^2 (\mathbf{z}, \mathbf{z}).\end{aligned}$$

Note that the second and the third terms of the last member are complex conjugates. If now $(\mathbf{z}, \mathbf{z}) = 0$, then $\mathbf{z} = \mathbf{0}$ and $(\mathbf{z}, \mathbf{w}) = 0$ so that (1.2.5) is trivially true in this case. If $(\mathbf{z}, \mathbf{z}) \neq 0$, we can take

$$\alpha = \frac{(\mathbf{w}, \mathbf{z})}{(\mathbf{z}, \mathbf{z})}$$

and obtain

$$0 \leqslant (\mathbf{w}, \mathbf{w}) - \frac{|(\mathbf{w}, \mathbf{z})|^2}{(\mathbf{z}, \mathbf{z})},$$

and this implies (1.2.5). ∎

Using Definition 1.2.1 we see that (1.2.5) may be written

$$\left| \sum_{j=1}^{n} z_j \bar{w}_j \right|^2 \leqslant \sum_{j=1}^{n} |z_j|^2 \sum_{j=1}^{n} |w_j|^2 \tag{1.2.6}$$

with equality iff $w_j = \gamma z_j$, $j = 1, 2, \ldots, n$ for some fixed γ. This is the general form of *Cauchy's inequality*.

Definition 1.2.3. *A set of k vectors* $\mathbf{z}_1, \mathbf{z}_2, \ldots, \mathbf{z}_k$ *in* C^n *is linearly independent over the complex field C if*

$$c_1\mathbf{z}_1 + c_2\mathbf{z}_2 + \cdots + c_k\mathbf{z}_k = \mathbf{0} \tag{1.2.7}$$

for complex numbers $c_1, c_2, \ldots, c_k$ *implies*

$$c_1 = c_2 = \cdots = c_k = 0.$$

The set is linearly dependent if k numbers $c_1, c_2, \ldots, c_k$, *not all zero, can be found for which* (1.2.7) *holds.*

Note that we are here considering linear independence over the complex field while in Section 1.1 the real field R was used. The two vectors $(1, 0)$ and $(i, 0)$ belong to C^2. They are independent over R but not over C.

Theorem 1.2.2. *There exist sets of n vectors in C^n which are linearly independent over C but any set of $n + 1$ vectors is linearly dependent.*

Proof. Let $\mathbf{u}_j$ denote the unit vector which has a one in the jth place and zeros elsewhere. Then

$$\sum_{j=1}^{n} c_j \mathbf{u}_j = (c_1, c_2, \ldots, c_n) \equiv \mathbf{c}.$$

This is the zero vector iff all the c_j's are zero. Thus the set $\{\mathbf{u}_k\}$ is a set of n vectors in C^n linearly independent over C. On the other hand, given $n + 1$ vectors $\mathbf{v}_1, \mathbf{v}_2, \ldots, \mathbf{v}_{n+1}$, say

$$\mathbf{v}_k = (c_{1k}, c_{2k}, \ldots, c_{nk}), \qquad k = 1, 2, \ldots, n + 1,$$

we can find $n + 1$ numbers $c_1, c_2, \ldots, c_{n+1}$, not all zero, such that if we multiply the kth vector by c_k and add the results, we obtain the zero vector. The conditions that have to be satisfied are

$$\sum_{k=1}^{n+1} c_{jk} c_k = 0, \qquad j = 1, 2, \ldots, n.$$

This is a system of n linear equations in $n + 1$ unknowns $c_1, c_2, \ldots, c_{n+1}$. Since the system is homogeneous, there exist non-trivial solutions, i.e. there is at least one set $\{c_k\}$ where not all c_k's are zero, such that

$$\sum_{k=1}^{n+1} c_k \mathbf{v}_k = \mathbf{0}$$

and the vectors $\mathbf{v}_1, \mathbf{v}_2, \ldots, \mathbf{v}_{n+1}$ are linearly dependent over C. ∎

We say that C^n is *n-dimensional* over C since we can find n linearly independent vectors and any larger set is linearly dependent.

A set of n linearly independent vectors $\mathbf{v}_1, \mathbf{v}_2, \ldots, \mathbf{v}_n$ forms a *basis* of C^n, i.e. every vector $\mathbf{v} \in C^n$ can be written in a unique manner as a linear combination of the $\mathbf{v}_k$'s, say

$$\mathbf{v} = \sum_{k=1}^{n} c_k \mathbf{v}_k \tag{1.2.8}$$

with complex coefficients c_k. There is such a representation, for the $n + 1$ vectors $\mathbf{v}, \mathbf{v}_1, \mathbf{v}_2, \ldots, \mathbf{v}_n$ are linearly dependent while the set $\mathbf{v}_1, \mathbf{v}_2, \ldots, \mathbf{v}_n$ is made up of linearly independent vectors by assumption. This gives a relation of the type (1.2.8). There is one and only one such representation, for if there were two, then there would be a linear relation between the basis vectors and this is absurd. Hence every vector $\mathbf{v} \in C^n$ has a unique representation in terms of the given basis.

Just as in R^3, we can replace a given basis by an orthonormal one using the Gram–Schmidt process. Formula (1.1.21) generalizes right away and we find for $k = 1, 2, \ldots, n$ that

$$\mathbf{u}_k = c_k \begin{vmatrix} \mathbf{v}_1 & \mathbf{v}_2 & \cdots & \mathbf{v}_k \\ (\mathbf{v}_1, \mathbf{v}_1) & (\mathbf{v}_1, \mathbf{v}_2) & \cdots & (\mathbf{v}_1, \mathbf{v}_k) \\ (\mathbf{v}_2, \mathbf{v}_1) & (\mathbf{v}_2, \mathbf{v}_2) & \cdots & (\mathbf{v}_2, \mathbf{v}_k) \\ \cdots & \cdots & \cdots & \cdots \\ (\mathbf{v}_{k-1}, \mathbf{v}_1) & (\mathbf{v}_{k-1}, \mathbf{v}_2) & \cdots & (\mathbf{v}_{k-1}, \mathbf{v}_k) \end{vmatrix} \tag{1.2.9}$$

where c_k is a real normalization factor. The verification is left to the reader.

Finally, a definition of an important concept which we shall meet over and over again.

Definition 1.2.4. *A subset K of C^n is said to be convex if $\mathbf{x}, \mathbf{y} \in K$ implies that the points $t\mathbf{x} + (1 - t)\mathbf{y} \in K$ where $0 < t < 1$. These points form a line segment joining* $\mathbf{x}$ *and* $\mathbf{y}$.

EXERCISE 1.2

1. Prove the *subadditive* property of the norm, namely the inequality

$$\|\mathbf{z} + \mathbf{w}\| \leqslant \|\mathbf{z}\| + \|\mathbf{w}\|,$$

and decide when equality holds. [*Hint:* Use the inner product and Theorem 1.2.1.]

2. Prove the *triangle inequality* for distances in C^n

$$\|\mathbf{z} - \mathbf{w}\| \leqslant \|\mathbf{z} - \mathbf{v}\| + \|\mathbf{v} - \mathbf{w}\|$$

and decide when equality holds. Why the name?

3. State and prove the analogue of Lemma 1.1.2.
4. For arbitrary vectors $\mathbf{z}$ and $\mathbf{w}$ in C^n prove the *Parallelogram Law*

$$\|\mathbf{z} + \mathbf{w}\|^2 + \|\mathbf{z} - \mathbf{w}\|^2 = 2[\|\mathbf{z}\|^2 + \|\mathbf{w}\|^2].$$

Why the name?

5. For any three vectors $\mathbf{z}_1, \mathbf{z}_2, \mathbf{z}_3$, prove that

$$\|\mathbf{z}_1 - \mathbf{z}_2\|^2 + \|\mathbf{z}_2 - \mathbf{z}_3\|^2 + \|\mathbf{z}_3 - \mathbf{z}_1\|^2 + \|\mathbf{z}_1 + \mathbf{z}_2 + \mathbf{z}_3\|^2 = 3[\|\mathbf{z}_1\|^2 + \|\mathbf{z}_2\|^2 + \|\mathbf{z}_3\|^2].$$

6. In the preceding problem, suppose that $\mathbf{z}_1, \mathbf{z}_2, \mathbf{z}_3$ are points on the unit circle in the complex plane. Use the formula to discuss the following problem in maxima and minima with a side condition. A triangle is inscribed in a circle of radius one. What triangle would maximize the sum of the squares of the lengths of its sides?
7. Extend the result of Problem 10, Exercise 1.1, to C^n, i.e. prove that a linear bounded functional $f(\mathbf{z})$ on C^n is necessarily an inner product, $f(\mathbf{z}) = (\mathbf{z}, \mathbf{w})$ for some fixed $\mathbf{w}$.

8. Given a set of n linearly independent vectors in C^n, show that any non-void subset is also linearly independent.

9. Convergence in the complex plane is reduced to convergence of positive numbers by the convention that $\lim_{j\to\infty} \mathbf{z}_j = \mathbf{z}_0$ iff $\lim_{j\to\infty} |\mathbf{z}_j - \mathbf{z}_0| = 0$. Similarly for vectors in C^n. Here $\lim_{j\to\infty} \mathbf{z}_j = \mathbf{z}_0$ iff $\lim_{j\to\infty} \|\mathbf{z}_j - \mathbf{z}_0\| = 0$. Suppose $\mathbf{z}_j = (z_{j1}, z_{j2}, \ldots, z_{jn})$ and $\mathbf{z}_0 = (z_{01}, z_{02}, \ldots, z_{0n})$. Show that $\lim_{j\to\infty} \mathbf{z}_j = \mathbf{z}_0$ iff $\lim_{j\to\infty} z_{jk} = z_{0k}$ for $k = 1, 2, \ldots, n$.

10. In R^n the *unit hypercube* is the set of points of coordinates between 0 and 1, limits included. Show that this is a convex set.

11. Show that the set of points $(\mathbf{z}; \|\mathbf{z}\| \leqslant 1)$, known as the *closed unit sphere* or the *unit ball*, is a convex set.

12. If K is a convex set in C^n, show that for any choice of m points $\mathbf{z}_1, \mathbf{z}_2, \ldots, \mathbf{z}_m$ in K and m numbers α_j such that $0 \leqslant \alpha_j \leqslant 1$, $j = 1, 2, \ldots, m$, and $\sum_{j=1}^m \alpha_j = 1$, then $\sum_{j=1}^m \alpha_j \mathbf{z}_j \in K$.

13. Prove that the intersection of two convex sets is convex or void.

14. The *convex hull* H of a set S in C^n is the intersection of all convex sets containing S. Show that H is the set of all finite sums of the form $\sum_{j=1}^m \alpha_j \mathbf{z}_j$ with $\mathbf{z}_j \in S$, $0 \leqslant \alpha_j \leqslant 1$, $\sum_{j=1}^m \alpha_j = 1$.

15. A set $\mathbf{C}_0 \subset C^n$ is called a *cone* if $\mathbf{z} \in \mathbf{C}_0$, $\alpha \geqslant 0$, implies that $\alpha\mathbf{z} \in \mathbf{C}_0$. A cone contains the origin. It is said to be proper if $\mathbf{z} \in \mathbf{C}_0$, $-\mathbf{z} \in \mathbf{C}_0$ implies $\mathbf{z} = \mathbf{0}$. Find when a cone is convex.

16. Verify (1.2.9).

17. The Gramian of n vectors $\mathbf{v}_k$ in C^n is by definition

$$G = \det [(\mathbf{v}_j, \mathbf{v}_k)] \equiv \det (g_{jk}).$$

Show that $g_{kj} = \overline{g_{kj}}$ and prove that G and all its principal minors are real and different from zero. Are they positive?

1.3 LINEAR TRANSFORMATIONS

We shall have to use some of the conventions and notations of *intuitive set theory* some of which have already figured implicitly in the preceding. A *set* S is a *collection of objects* denoted by some symbol like x. Here the nature of the objects is immaterial and only the membership counts. We demand that one and only one of the statements "*x is a member of S*" and "*x is not a member of S*" be true. If the first one is true, we write $x \in S$; if the second, $x \notin S$. This implies that two sets are equal iff they have the same members. A set S_1 is a *subset* of S, written

$$S_1 \subseteq S, \tag{1.3.1}$$

if all members of S_1 are also members of S. It is a *proper subset*, written $S_1 \subset S$, if (1.3.1) holds but there are some elements (= members) of S not in S_1.

Consider now two sets S_1 and S_2 and a collection T of ordered pairs $[x, y]$ such that (1) $x \in S_1$, $y \in S_2$, (2) every $x_0 \in S_1$ is the first element of at most one pair $[x, y]$. Such a collection T of ordered pairs is a *mapping* from S_1 into S_2. If $[x_0, y_0] \in T$, we call y_0 the *image* of x_0 under the mapping T and x_0 is the *inverse image* or *pre-image* of y_0. The set of all elements $x \in S_1$ which actually occur as first elements of pairs $[x, y]$ in T forms the *domain* of T while the elements $y \in S_2$ which are present as second elements of pairs $[x, y]$ form the *range* of T. In symbols

$$\mathfrak{D}[T] = D = \{x; x \in S_1, [x, y] \in T \text{ for some } y\},$$
$$\mathfrak{R}[T] = R = \{y; y \in S_2, [x, y] \in T \text{ for some } x\}. \tag{1.3.2}$$

We shall now be concerned with a class of mappings of C^n into itself, known as *linear transformations*. Here $S_1 = S_2 = C^n$, the domain will be all of C^n while the range may be a proper subset.

Definition 1.3.1. *A is a linear transformation from C^n into itself, if* (1) *to each vector $\mathbf{x} \in C^n$ corresponds a unique vector $\mathbf{y} = A(\mathbf{x})$ in C^n, and* (2) *for each pair of complex numbers α, β and each pair of vectors $\mathbf{x}_1, \mathbf{x}_2$*

$$A(\alpha\mathbf{x}_1 + \beta\mathbf{x}_2) = \alpha A(\mathbf{x}_1) + \beta A(\mathbf{x}_2). \tag{1.3.3}$$

The range of A is the set of all vectors $A(\mathbf{x})$ and (1.3.3) shows that $\mathfrak{R}[A]$ *is a linear subspace of C^n*. Verify!

We define the *null space* of A, also called the *kernel* of A, as the set of all vectors $\mathbf{x}$ which are mapped into the zero vector

$$\mathfrak{N}[A] = [\mathbf{x}; \mathbf{x} \in C^n, A(\mathbf{x}) = \mathbf{0}]. \tag{1.3.4}$$

Since

$$A(\mathbf{0}) = A(\mathbf{0} + \mathbf{0}) = 2\,A(\mathbf{0})$$

we see that

$$A(\mathbf{0}) = \mathbf{0} \tag{1.3.5}$$

so that the zero element always belongs to $\mathfrak{N}[A]$.

Formula (1.3.3) shows that $\mathfrak{N}[A]$ is a linear subspace of C^n. (Why?) The transformation A turns out to be particularly simple when the null space reduces to the zero element, $\mathfrak{N}[A] = \{\mathbf{0}\}$. Then $\mathfrak{R}[A] = C^n$ and the mapping is one-to-one so that each $\mathbf{y}$ is the image of one and only one point $\mathbf{x}$. In the collection of ordered pairs $[\mathbf{x}, \mathbf{y}]$ which defines A, of course, each $\mathbf{x}$ occurs once and only once, but now each $\mathbf{y}$ occurs once and only once. Thus the collection of pairs $[\mathbf{y}, \mathbf{x}]$ is also a mapping of C^n into itself, the so-called *inverse mapping* A^{-1}, and this mapping is also a linear transformation. All this will be proved below.

It should be pointed out that $\mathfrak{N}[A]$ does not have to reduce to the zero element. If $\mathfrak{N}[A] \neq \{\mathbf{0}\}$, we shall see that $\mathfrak{R}[A]$ is a linear subspace and the mapping is no longer one-to-one. In an extreme case, the range may reduce to the zero element, $\mathfrak{N}[A] = C^n$ and A "annihilates" the whole space.

We start by proving

Theorem 1.3.1. *The linear mapping defined by A is one-to-one iff* $\mathfrak{N}[A] = \{\mathbf{0}\}$.

Remark. We write "1–1" for "one-to-one." This property expresses that $\mathbf{x}_1 \neq \mathbf{x}_2$ implies $A(\mathbf{x}_1) \neq A(\mathbf{x}_2)$.

Proof. If for some pair of distinct vectors $\mathbf{x}_1, \mathbf{x}_2$ we should have $A(\mathbf{x}_1) = A(\mathbf{x}_2)$, then by the linearity of A

$$A(\mathbf{x}_1 - \mathbf{x}_2) = \mathbf{0}$$

and $\mathbf{x}_1 - \mathbf{x}_2 \in \mathfrak{N}[A]$. Conversely, if $\mathfrak{N}[A] = \{\mathbf{0}\}$, then $A(\mathbf{x}_1) = A(\mathbf{x}_2)$ for $\mathbf{x}_1 \neq \mathbf{x}_2$ leads to a contradiction so the condition $\mathfrak{N}[A] = \{\mathbf{0}\}$ is necessary and sufficient for A to be 1–1. ∎

In the modern terminology of N. Bourbaki the term *injective* is used to designate a 1–1 transformation (an *injection*) and *surjective* (*surjection*) for a transformation which is *onto*, i.e. whose range is the whole space.

Theorem 1.3.2. $\mathfrak{R}[A] = C^n$ *iff* $\mathfrak{N}[A] = \{\mathbf{0}\}$.

Proof. For this fact we shall give two proofs. The first, based on a personal communication by Bertram Yood, is abstract and proves the statement in a few lines. The second is computational and leads to the representation of A by a matrix which is basic for the following.

I. Let $\mathbf{u}_1, \mathbf{u}_2, \ldots, \mathbf{u}_n$ be a basis for C^n. Suppose that $\mathfrak{N}[A] = \{\mathbf{0}\}$. Then the n vectors $A(\mathbf{u}_1), A(\mathbf{u}_2), \ldots, A(\mathbf{u}_n)$ are linearly independent. For if

$$\sum_{j=1}^{n} c_j A(\mathbf{u}_j) = \mathbf{0}, \quad \text{then} \quad A\left(\sum_{j=1}^{n} c_j \mathbf{u}_j\right) = \mathbf{0} \quad \text{and} \quad \sum_{j=1}^{n} c_j \mathbf{u}_j = \mathbf{0}$$

and all the c_j's are zero since $\{\mathbf{u}_j\}$ is a basis. But then $\mathfrak{R}[A]$, a linear subspace of C^n, contains a set of n linearly independent vectors, the $A(\mathbf{u}_j)$, and is thus of dimension n. Hence $\mathfrak{R}[A] = C^n$. On the other hand, suppose that $\mathfrak{R}[A] = C^n$. Then $\mathfrak{R}[A]$ contains a set of n linearly independent vectors $\mathbf{v}_1, \mathbf{v}_2, \ldots, \mathbf{v}_n$. These are images of vectors $\mathbf{u}_j$ in C^n under the mapping A, say $\mathbf{v}_j = A(\mathbf{u}_j)$. Then

$$\sum_{j=1}^{n} c_j \mathbf{v}_j = \sum_{j=1}^{n} c_j A(\mathbf{u}_j) = A\left(\sum_{j=1}^{n} c_j \mathbf{u}_j\right).$$

The first member can be the zero vector iff all $c_j = 0$. Hence $A(\mathbf{v}) = 0$ iff $\mathbf{v} = \mathbf{0}$ or $\mathfrak{N}[A] = \{\mathbf{0}\}$.

II. In the second proof we determine the structure of A. We take as the basis for C^n the unit vectors $\mathbf{u}_j$, where $\mathbf{u}_j$ has a 1 in the jth place, all other entries being 0. The mapping A is uniquely determined by its effect on the unit vectors plus the assumption that A is linear. Since $A(\mathbf{u}_k)$ is a vector in C^n, it is a linear combination of the basis vectors. This implies the existence of n^2 numbers a_{jk} such that

$$A(\mathbf{u}_k) = \sum_{j=1}^{n} a_{jk} \mathbf{u}_j, \qquad k = 1, 2, \ldots, n. \tag{1.3.6}$$

It is clear that this set exists and is uniquely determined by A.

Suppose now that

$$\mathbf{v} = \sum_{k=1}^{n} \hat{v}_k \mathbf{u}_k \tag{1.3.7}$$

is the representation of the vector $\mathbf{v}$. By the linearity of A

$$\mathbf{w} = A(\mathbf{v}) = \sum_{k=1}^{n} \hat{v}_k A(\mathbf{u}_k) = \sum_{k=1}^{n} \hat{v}_k \left\{ \sum_{j=1}^{n} a_{jk} \mathbf{u}_j \right\} = \sum_{j=1}^{n} \left\{ \sum_{k=1}^{n} a_{jk} \hat{v}_k \right\} \mathbf{u}_j. \tag{1.3.8}$$

This means that in the chosen coordinate system the mapping A is completely determined by the array $a_{jk}, j, k = 1, 2, \dots, n$, or n^2 complex numbers. We introduce the symbol

$$\mathcal{A} = \begin{bmatrix} a_{11} & a_{12} \cdots a_{1n} \\ a_{21} & a_{22} \cdots a_{2n} \\ \cdots & \cdots \\ a_{n1} & a_{n2} \cdots a_{nn} \end{bmatrix} \tag{1.3.9}$$

and refer to $\mathcal{A}$ as the *matrix* whose entry in the place (j, k) is $(\mathcal{A})_{jk} = a_{jk}$. We also write $\mathcal{A} = (a_{jk})$ and speak of the jth row and the kth column of $\mathcal{A}$. We write symbolically

$$\mathbf{w} = \mathcal{A} \cdot \mathbf{v} \tag{1.3.10}$$

and think of the right hand side as a "product" of a matrix with a vector. The result is again a vector, namely $\mathbf{w} = (\mathbf{w}_1, \mathbf{w}_2, \dots, \mathbf{w}_n)$,

$$\mathbf{w}_j = \sum_{k=1}^{n} a_{jk} \hat{v}_k. \tag{1.3.11}$$

Let us now again examine the implications of the hypothesis $\mathfrak{N}[A] = \{\mathbf{0}\}$. This says that the homogeneous system of n linear equations in n unknowns

$$\sum_{k=1}^{n} a_{jk} \hat{v}_k = 0, \qquad j = 1, 2, \dots, n, \tag{1.3.12}$$

has the unique solution

$$\hat{v}_1 = \hat{v}_2 = \cdots = \hat{v}_n = 0. \tag{1.3.13}$$

For this to be the case, it is necessary and sufficient that the determinant of the system is different from zero

$$\det (a_{jk}) \neq 0. \tag{1.3.14}$$

By definition this is the *determinant of the matrix* $\mathcal{A}$.

Now condition (1.3.14) implies that the non-homogeneous system

$$\sum_{k=1}^{n} a_{jk} \hat{v}_k = w_j, \qquad j = 1, 2, \dots, n, \tag{1.3.15}$$

has a unique solution $(\hat{v}_1, \hat{v}_2, \dots, \hat{v}_n) = \mathbf{v}$ for each given vector $\mathbf{w} = (w_1, w_2, \dots, w_n)$ of C^n. Hence $\mathfrak{R}[A] = C^n$.

On the other hand, if $\mathfrak{R}[A] = C^n$, then the system (1.3.15) must have a solution vector $\mathbf{v}$ for each given $\mathbf{w}$. This implies that (1.3.14) holds and $\mathbf{v} = \mathbf{0}$ is the only solution of (1.3.12) or $\mathfrak{N}[A] = \{\mathbf{0}\}$. ∎

The matrix $\mathcal{A}$ is the representation of A under the chosen coordinate system. Changing the basis changes the representation matrix. The passage from one basis to another also involves a matrix transformation, so we shall have to consider the composition of linear transformations and of matrices in the next section.

We refer to $\mathcal{A}$ as an n by n matrix and denote the set of all such matrices by $\mathfrak{M}_n$. The structure of this set will also be examined in the next section.

If det $(a_{jk}) \neq 0$ we say that the matrix $\mathcal{A}$ is *regular*, otherwise *singular*. The singular case calls for more detailed comments. A square *submatrix* of $\mathcal{A}$ is one obtained by crossing out the same number of rows and columns of $\mathcal{A}$. We say that $\mathcal{A}$ is of *rank* $n - q$ if there is a non-singular submatrix of $\mathcal{A}$ with $n - q$ rows and columns but all submatrices with more than $n - q$ rows and columns are singular. The number q is known as the *nullity* of the matrix (and of the determinant and the system of equations).

Using dim $(\mathfrak{X})$ as notation for the dimension of a linear space $\mathfrak{X}$ we have

$$\dim\{\mathfrak{N}[A]\} = q, \tag{1.3.16}$$

for now the system (1.3.12) has q and exactly q linearly independent solution vectors. The nullity also crops up in the discussion of the system (1.3.15) where now the right hand sides can no longer be chosen arbitrarily, but must satisfy linear relations of the form

$$c_1 w_1 + c_2 w_2 + \cdots + c_n w_n = 0, \tag{1.3.17}$$

q in number. These are conditions which the image $\mathbf{w} = A(\mathbf{v})$ must satisfy, i.e. restrictions imposed on $\mathfrak{R}[A]$. The range space is now of dimension $n - q$, $\dim\{\mathfrak{R}[A]\} = n - q$. Various interpretations may be put on the conditional equations (1.3.17). Each of them is the equation of a *hyperplane* through the origin in C^n. Equivalently (1.3.17) figures as the null space of a linear functional

$$f(\mathbf{w}) = (\mathbf{w}, \bar{\mathbf{c}}), \qquad \bar{\mathbf{c}} = (\overline{c_1}, \overline{c_2}, \ldots, \overline{c_n}). \tag{1.3.18}$$

Thus it is seen that there is a profound difference between regular and singular mapping matrices. If $\mathcal{A}$ is regular, the mapping is 1–1 and onto and there is an inverse with the same properties. If $\mathcal{A}$ is singular, then all of C^n collapses onto a proper linear subspace. Only points of this space are images under A, i.e. occur as second elements of pairs $[\mathbf{x}, \mathbf{y}]$ but each such $\mathbf{y}$ figures as second element in infinitely many pairs. Finally, we restate one of the main results of this section as

Theorem 1.3.3. *Once a basis has been chosen for C^n, there is a* 1–1 *correspondence between transformations A of $\mathfrak{E}(C^n)$ and matrices $\mathcal{A}$ of $\mathfrak{M}_n$. Here $\mathfrak{E}(C^n)$ denotes the set of all linear bounded transformations from C^n to itself.*

EXERCISE 1.3

1. Give an example of a 3 by 3 matrix of rank 2. Find its range and nullspace.
2. If $K \subset C^n$ is a convex set, show that its image $A(K)$ under the mapping $A \in \mathfrak{E}(C^n)$ is also convex.
3. Let $E = \{\mathbf{z};\ \|\mathbf{z}\| \leqslant 1\}$ be the unit ball of C^n. Show that it is convex and prove that no point of $A(E)$, $A \in \mathfrak{E}(C^n)$, can have a distance from the origin exceeding $\{\sum_{j=1}^{n} \sum_{k=1}^{n} |a_{jk}|^2\}^{\frac{1}{2}}$.
4. A linear transformation $U \in \mathfrak{E}(C^n)$ which preserves inner products, $(U\mathbf{z}, U\mathbf{w}) = (\mathbf{z}, \mathbf{w})$, is called *unitary*. Prove that distances are preserved. Take an orthonormal basis for C^n and let $\mathfrak{U}$ be the matrix representing U in this system. $\mathfrak{U}$ is called a *unitary* matrix. Show that its column vectors form an orthonormal system. Take $\mathfrak{U} = (u_{jk})$ and prove that $\mathfrak{U}$ is regular.
5. What can be said about the row vectors in the unitary case? The jth row vector is the vector $(u_{j1}, u_{j2}, \ldots, u_{jn})$.
6. What is the value of det (u_{jk})?
7. In the unitary case the unit ball is obviously mapped in a 1–1 manner onto itself. Why? What is the bound given in Problem 3 when $a_{jk} = u_{jk}$?
8. In the notation of Problem 17, Exercise 1.2, let $\mathfrak{G}$ be the matrix corresponding to the Gramian G. When is such a matrix unitary?
9. What is the argument that leads to formula (1.3.17) when the determinant of the matrix $\mathcal{A}$ is zero? Justify the assertion that the number of independent such relations equals q.
10. How is formula (1.3.18) obtained?
11. If $\mathcal{A}$ is singular, prove that if $\mathbf{y}$ is second element of a pair $[\mathbf{x}, \mathbf{y}]$, then $\mathbf{y}$ is second element of infinitely many such pairs.

1.4 MATRICES

Let us consider the set $\mathfrak{M}_n$ of all n by n matrices over the complex field. Since $\mathfrak{M}_1 = C^1 = C$, we assume $n > 1$. Does the set $\mathfrak{M}_n$ have any discernible structure, algebraic or geometrical? To study this question we consider simultaneously the set $\mathfrak{E}(C^n)$ of all linear bounded transformations from C^n to C^n. This set is in 1–1 correspondence with the set $\mathfrak{M}_n$ once a basis has been chosen for C^n.

Now the set $\mathfrak{E}(C^n)$ may be endowed with an algebraic structure in a natural manner. We can define addition, scalar multiplication, and element multiplication for linear transformations in $\mathfrak{E}(C^n)$ in a way which extends immediately to more general cases.

Definition 1.4.1. *For any vector* $\mathbf{x} \in C^n$, *any transformations* $A, B \in \mathfrak{E}(C^n)$ *and any complex number* α *set*

$$(A + B)(\mathbf{x}) = A(\mathbf{x}) + B(\mathbf{x}), \tag{1.4.1}$$

$$(\alpha A)(\mathbf{x}) = \alpha A(\mathbf{x}), \tag{1.4.2}$$

$$(AB)(\mathbf{x}) = A[B(\mathbf{x})], \tag{1.4.3}$$

Here the right hand sides are meaningful and serve to define the transformation in the left member: $A + B$, αA, AB, respectively. The meaning of the first two transformations is obvious. In the third case we first let B act on $\mathbf{x}$ and obtain a vector $\mathbf{y} = B(\mathbf{x})$. We then let A act on $\mathbf{y}$. Since the domain of A is all of C^n, we can form $A(\mathbf{y}) = \mathbf{z}$. Now the passage from $\mathbf{x}$ to $\mathbf{z}$ is clearly a linear transformation from C^n to C^n, i.e. an element of $\mathfrak{E}(C^n)$. We define this transformation as *the product of A and B in this order*. It is important to observe the order for normally $A[B(\mathbf{x})] \neq B[A(\mathbf{x})]$ so that

$$AB \neq BA. \tag{1.4.4}$$

Thus we are dealing with *non-commutative multiplication* in $\mathfrak{E}(C^n)$. If, exceptionally, $AB = BA$, these two transformations are said to *commute*. Note that *A commutes with its own powers* and that the *law of exponents* is valid

$$A^j A^k = A^{j+k}. \tag{1.4.5}$$

The successive powers are defined recursively by

$$A^{j+1}(\mathbf{x}) = A[A^j(\mathbf{x})]. \tag{1.4.6}$$

We observed above that there is a 1–1 correspondence between linear transformations A and matrices $\mathcal{A}$ once a particular, say orthonormal basis $\{\mathbf{u}_k;\ k = 1, 2, \ldots, n\}$, has been chosen. We can now define the algebraic operations for matrices in such a way that the correspondence preserves the algebraic operations, i.e. *sums go into sums, scalar products into scalar products, and products of elements into products of elements.* Such a correspondence is known as an *isomorphism.* What is required is that

$$A \leftrightarrow \mathcal{A} = (a_{jk}) \quad \text{and} \quad B \leftrightarrow \mathcal{B} = (b_{jk}) \tag{1.4.7}$$

shall imply that

$$A + B \leftrightarrow \mathcal{A} + \mathcal{B}, \quad \alpha A \leftrightarrow \alpha\mathcal{A}, \quad AB \leftrightarrow \mathcal{A}\mathcal{B}. \tag{1.4.8}$$

This leads to the following definition of the algebraic operations on matrices:

$$\mathcal{A} + \mathcal{B} = (a_{jk} + b_{jk}), \tag{1.4.9}$$

$$\alpha\mathcal{A} = (\alpha a_{jk}), \tag{1.4.10}$$

$$\mathcal{A}\mathcal{B} = (c_{jk}) \quad \text{with} \quad c_{jk} = \sum_{m=1}^{n} a_{jm} b_{mk}. \tag{1.4.11}$$

Only the last convention needs justification. We have

$$\mathbf{y} = \mathcal{B}\mathbf{x}, \qquad \mathbf{z} = \mathcal{A}\mathbf{y}$$

or, in terms of components (= coordinates),

$$y_m = \sum_{k=1}^{n} b_{mk} x_k,$$

$$z_j = \sum_{m=1}^{n} a_{jm} y_m = \sum_{m=1}^{n} a_{jm} \sum_{k=1}^{n} b_{mk} x_k = \sum_{k=1}^{n} \left\{ \sum_{m=1}^{n} a_{jm} b_{mk} \right\} x_k,$$

so that c_{jk} has the value stated in (1.4.11).

A set in which the operations of addition, multiplication by scalars, and multiplication of elements may be performed is known as an *algebra* provided the operations satisfy certain simple conditions which will be specified later (see Sections 2.1 and 2.6). It is an *algebra over the complex* (real) *field* if the scalars are complex (real) numbers. It is a *non-commutative algebra* if element multiplication is non-commutative.

The reader undoubtedly realizes that he has encountered a number of algebras in his earlier study of mathematics. Thus functions of a real variable which are continuous at a fixed point form an algebra. So do the differentiable functions. Functions of a complex variable which are holomorphic in a given domain form an algebra. All these function algebras are *commutative* in the sense that $fg = gf$ for all elements f and g of the algebra. On the other hand, as shown in Exercise 1.1, the vectors in R^3 form a non-commutative algebra under the operations of vector addition, multiplication by real scalars, and the forming of cross products. Here we have now two new examples of non-commutative algebras, namely, $\mathfrak{E}(C^n)$ and $\mathfrak{M}_n$.

The algebra $\mathfrak{E}(C^n)$ is *associative* in the sense that

$$(A + B) + C = A + (B + C), \tag{1.4.12}$$

$$(AB)C = A(BC). \tag{1.4.13}$$

It is also *distributive,* meaning that

$$(A + B)C = AC + BC, \qquad A(B + C) = AB + AC. \tag{1.4.14}$$

The same properties hold for the corresponding elements of $\mathfrak{M}_n$. Verification is left to the reader. These properties should not be treated as trivial and self-evident. Thus, for example, the vector algebra in R^3 mentioned above is non-associative since the relation

$$(\mathbf{x} \times \mathbf{y}) \times \mathbf{z} \neq \mathbf{x} \times (\mathbf{y} \times \mathbf{z}) \tag{1.4.15}$$

may very well hold. A case in point is given by $\mathbf{x} = \mathbf{i}$, $\mathbf{y} = \mathbf{z} = \mathbf{j}$, where the left member is $-\mathbf{i}$ while the right is $\mathbf{0}$.

An algebra may have a *unit element*, i.e. an element which is *neutral* for multiplication. If there is such an element, it is unique. Both $\mathfrak{E}(C^n)$ and $\mathfrak{M}_n$ have unit elements. There is a linear transformation which leaves C^n pointwise invariant,

$$\mathbf{y} = \mathbf{x} \quad \text{or} \quad \mathbf{y} = I(\mathbf{x}), \tag{1.4.16}$$

known as the *identity mapping*. We have clearly

$$IA = AI = A \tag{1.4.17}$$

for all A. There is a corresponding matrix $\mathcal{E}$, known as the *unit matrix*, such that

$$\mathcal{A}\mathcal{E} = \mathcal{E}\mathcal{A} = \mathcal{A} \tag{1.4.18}$$

for all $\mathcal{A}$. $\mathcal{E}$ has ones along the main diagonal, all other entries are zero, i.e. $(\mathcal{E})_{jk} = \delta_{jk}$.

In an algebra there may be *inverses* and *quotients*, but it is to be expected that the case where all elements have inverses, the zero element excepted, is of rare occurrence. We say that A has an *inverse* B if there exists an element $B \in \mathfrak{E}(C^n)$ such that

$$AB = BA = I. \tag{1.4.19}$$

We then write $B = A^{-1}$. Note that B is a linear transformation by definition since the only inverses accepted at this stage must be members of $\mathfrak{E}(C^n)$. If B exists, then it is unique and, as will be shown later, B is 1–1 and onto.

To the inverse transformation B corresponds a matrix $\mathcal{B}$ which is the inverse of the matrix $\mathcal{A}$ so that

$$\mathcal{A}\mathcal{B} = \mathcal{B}\mathcal{A} = \mathcal{E}. \tag{1.4.20}$$

We then write $\mathcal{B} = \mathcal{A}^{-1}$.

Lemma 1.4.1. *If B and $\mathcal{B}$ exist, they are unique.*

Proof. It is enough to prove the case when $B \in \mathfrak{E}(C^n)$. The matrix case is handled in the same manner. Suppose that B exists as an element of $\mathfrak{E}(C^n)$ and satisfies (1.4.19). Suppose there is a transformation $C \in \mathfrak{E}(C^n)$ such that

$$AC = CA = I.$$

We have then

$$B(AC) = (BA)C = IC = C = B(I) = B \quad \text{or} \quad C = B$$

as asserted. ∎

Lemma 1.4.2. *If B exists in $\mathfrak{E}(C^n)$, then B is* 1–1 *and onto.*

Proof. By assumption B is linear. The domain of B in C^n is the range of A: $\mathfrak{D}[B] = \mathfrak{R}[A]$. Now a necessary and sufficient condition for A to have an inverse is that $\mathfrak{N}[A] = \{\mathbf{0}\}$. By Theorem 1.3.2 this implies and is implied by $\mathfrak{R}[A] = C^n$ so that $\mathfrak{D}[B] = \mathfrak{R}[A] = C^n$. We must have $\mathfrak{N}[B] = \{\mathbf{0}\}$, for if

$$B(\mathbf{z}) = \mathbf{0}, \quad \text{then} \quad (AB)(\mathbf{z}) = I(\mathbf{z}) = \mathbf{z} = \mathbf{0}.$$

This implies $\mathfrak{R}[B] = C^n$, so that B is onto as well as 1–1. ∎

Now consider

$$\mathcal{A}_1 \cdot \mathbf{v}_j = [\mathcal{B}\mathcal{A}\mathcal{B}^{-1}]\,\mathbf{v}_j = [\mathcal{B}\mathcal{A}][\mathcal{B}^{-1}\,\mathbf{v}_j] = [\mathcal{B}\mathcal{A}]\,\mathbf{u}_j$$

$$= \mathcal{B}\left[\sum_{k=1}^{n} a_{jk}\mathbf{u}_k\right] = \sum_{k=1}^{n} a_{jk}\mathcal{B}\cdot\mathbf{u}_k = \sum_{k=1}^{n} a_{jk}\mathbf{v}_k.$$

Thus to any given ordered basis F of C^n and any pair of similar matrices $\mathcal{A}$ and $\mathcal{A}_1$ we have constructed an ordered basis F_1 such that the action of $\mathcal{A}_1$ on F_1 is the same as that of $\mathcal{A}$ on F. ∎

We have, of course, similar relations between bases and similar transformations A and A_1.

Let us now consider some operations that may be performed on matrices. To a given n by n matrix $\mathcal{A} = (a_{jk})$ we order two other matrices $\mathcal{A}^t$ and $\mathcal{A}^*$ known as the *transpose* and the *conjugate transpose* of $\mathcal{A}$. Here

$$(\mathcal{A}^t)_{jk} = a_{kj}, \qquad (\mathcal{A}^*)_{jk} = \overline{a_{kj}}, \tag{1.4.36}$$

where the bar as usual indicates the conjugate complex. If the a_{jk} are real, $\mathcal{A}^t = \mathcal{A}^*$. We say that $\mathcal{A}$ is *symmetric*, if

$$\mathcal{A}^t = \mathcal{A}, \tag{1.4.37}$$

Hermitian if

$$\mathcal{A}^* = \mathcal{A}. \tag{1.4.38}$$

The latter case is named after the famous French mathematician Charles Hermite (1822–1901).

The Gramian defined above is an example of a Hermitian matrix. As we shall see later (Section 1.7), Hermitian matrices have many important properties. The transformation

$$\mathcal{A} \to \mathcal{A}^* \tag{1.4.39}$$

is called a *conjugation operator*. It is an *involution* in the sense that

$$(\mathcal{A}^*)^* = \mathcal{A}. \tag{1.4.40}$$

Let us consider some special singular matrices. A matrix $\mathfrak{J}$ such that

$$\mathfrak{J}^2 = \mathfrak{J} \tag{1.4.41}$$

is called an *idempotent*. This equation is clearly satisfied by the unit matrix and by the zero matrix which are the trivial idempotents. In $\mathfrak{M}^2$

$$\begin{bmatrix} 1 & 0 \\ 0 & 0 \end{bmatrix}, \quad \begin{bmatrix} 0 & 0 \\ 0 & 1 \end{bmatrix}, \quad \begin{bmatrix} 1 & 1 \\ 0 & 0 \end{bmatrix}$$

are examples of non-trivial idempotents. All idempotents, the unit matrix excepted, are singular. To an idempotent matrix corresponds an idempotent transformation, often called a *projection*.

The product of two matrices may very well be the zero matrix without either factor being zero. Since

$$\begin{bmatrix} 0 & 0 \\ 1 & 0 \end{bmatrix} \begin{bmatrix} 0 & 0 \\ 0 & 1 \end{bmatrix} = \begin{bmatrix} 0 & 0 \\ 0 & 0 \end{bmatrix}$$

this phenomenon can occur already in $\mathfrak{M}_2$. Such a matrix is called a *divisor of zero*. It is clearly singular.

Another type of a singularity is shown by

$$\mathcal{A} = \begin{bmatrix} 0 & 1 & 1 \\ 0 & 0 & 1 \\ 0 & 0 & 0 \end{bmatrix}.$$

Here $\mathcal{A}^3 = \mathcal{O}$. This is a *nilpotent* matrix. The lowest exponent k such that $\mathcal{A}^k = 0$ is the *degree of nilpotency* of $\mathcal{A}$.

At the beginning of this section the question was raised if the sets $\mathfrak{E}(C^n)$ and $\mathfrak{M}_n$ have some structure, algebraic or geometric. We have endowed both spaces with algebraic structures. Can we also provide them with geometric structures, preferably metric ones? The answer is in the affirmative for both spaces. We postpone the discussion of $\mathfrak{E}(C^n)$, but observe that there are several ways of defining a norm and a distance in $\mathfrak{M}_n$.

We may regard $\mathfrak{M}_n$ as a linear vector space of n^2 dimensions over the complex field with addition and scalar multiplication defined by (1.4.9) and (1.4.10). In this context it is natural to endow $\mathfrak{M}_n$ with a Euclidean metric by setting

$$\|\mathcal{A}\|_2 = \left\{ \sum_{j=1}^{n} \sum_{k=1}^{n} |a_{jk}|^2 \right\}^{1/2}. \tag{1.4.42}$$

This is known as the *Frobenius-Wedderburn norm* [G. Frobenius (1849–1917) and J. H. Maclagan Wedderburn (1882–1948)] and is commonly used in matrix theory. It has the disadvantage of assigning the norm $\sqrt{n}$ to the unit matrix. Since this is undesirable, we shall use instead

$$\|\mathcal{A}\|_1 = \max_j \sum_{k=1}^{n} |a_{jk}|, \tag{1.4.43}$$

which gives $\|\mathcal{E}\|_1 = 1$. This notion of length will be used below and we define the distance between two matrices in $\mathfrak{M}_n$ as

$$\|\mathcal{A} - \mathcal{B}\|_1 = \|\mathcal{A} + (-1)\mathcal{B}\|_1. \tag{1.4.44}$$

A third alternative is based on the definition of the norm in $\mathfrak{E}(C^n)$ which will be given later. See (1.6.23). This amounts to setting $\|A\| = \|\mathcal{A}\|_0$.

Lemma 1.4.3. *The norm of the product of two matrices is at most equal to the product of the norms of the factors.*

Remark. This holds for all three norms proposed. We shall give the proof for the norm defined by (1.4.43).

Proof. We have

$$\|\mathcal{A}\mathcal{B}\|_1 = \max_j \sum_{k=1}^n \left| \sum_{m=1}^n a_{jm} b_{mk} \right| \leqslant \max_j \sum_{m=1}^n |a_{jm}| \sum_{k=1}^n |b_{mk}|$$

$$\leqslant \max_j \sum_{m=1}^n |a_{jm}| \, \|\mathcal{B}\|_1 = \|\mathcal{A}\|_1 \, \|\mathcal{B}\|_1,$$

as asserted. ∎

The existence of alternate definitions of the distance between matrices raises the question of equivalence of the definitions in the case of convergence. Suppose we have an infinite set of matrices

$$\mathcal{A}_m = (a_{jk}^m), \qquad m = 0, 1, 2, \ldots .$$

What should be meant by the statement

$$\lim_{m \to \infty} \mathcal{A}_m = \mathcal{A}_0? \tag{1.4.45}$$

We encountered a similar problem for vectors in C^n (see Problem 9, Exercise 1.2). There we introduced convergence in the sense of the metric of the space but found that this type of convergence is equivalent to convergence for each coordinate separately. The same situation holds here. We have then at least four possibilities of defining convergence for matrices. One would be

$$\lim_{m \to \infty} a_{jk}^m = a_{jk}{}^0, \qquad j, k = 1, 2, 3, \ldots, n, \tag{1.4.46}$$

the others are of the type

$$\lim_{m \to \infty} \|\mathcal{A}_m - \mathcal{A}_0\| = 0, \tag{1.4.47}$$

where for the norm we can take that defined by (1.4.42) or (1.4.43) or (1.6.20). Fortunately all four possibilities are equivalent.

Lemma 1.4.4. *Formula* (1.4.47) *is valid for one of the stated norms iff* (1.4.46) *holds and then* (1.4.47) *holds also for the other norms.*

Proof. Using the symbol ⇒ to denote implication, we shall prove (1.4.46) ⇒ (1.4.47) ⇒ (1.4.46).

I. Suppose that (1.4.46) holds and consider the norm defined by (1.4.43). Then, as $m \to \infty$,

$$\|\mathcal{A}_m - \mathcal{A}_0\|_1 = \max_j \sum_{k=1}^n |a_{jk}^m - a_{jk}{}^0| \to 0. \tag{1.4.48}$$

Similarly,

$$\|\mathcal{A}_m - \mathcal{A}_0\|_2 = \left\{ \sum_{j=1}^n \sum_{k=1}^n |a_{jk}^m - a_{jk}{}^0|^2 \right\}^{1/2} \to 0.$$

We omit the operator norm, which has not been defined at this stage.

II. Suppose that $\|\mathcal{A}_m - \mathcal{A}_0\|_1 \to 0$. Then (1.4.48) shows that this implies (1.4.46). ∎

EXERCISE 1.4

1. C^n is a linear vector space over the complex numbers and vectors may be added and multiplied by complex numbers. Make a list of the properties of addition and scalar multiplication which you have encountered so far. In particular, what types of associativity, commutativity, and distributivity have you noticed?
2. Make a similar list of properties of an algebra using $\mathfrak{E}(C^n)$ as a concrete illustration.
3. Verify (1.4.12) and (1.4.13).
4. Verify (1.4.14).
5. Verify (1.4.15).
6. Find two operators A and B in $\mathfrak{E}(C^2)$ which do not commute.
7. Show that a polynomial in a matrix $\mathcal{A}$, say

$$P(\mathcal{A}) = a_0\mathcal{E} + \sum_{j=1}^{p} a_j\mathcal{A}^j, \qquad a_j \in C,$$

commutes with any other polynomial in $\mathcal{A}$.

8. A finite set of matrices $\{\mathcal{A}_j; j = 1, 2, \ldots, p, \mathcal{A}_j \in \mathfrak{M}_n\}$ is linearly independent over the complex field if

$$\sum_{j=1}^{p} c_j\mathcal{A}_j = \mathcal{O}, \qquad c_j \in C,$$

implies $c_1 = c_2 = \cdots = c_p = 0$. Show that p cannot exceed n^2.

9. Show that this implies that the matrix $\mathcal{A}$ satisfies an algebraic equation with coefficients in C.
10. Write down the equation satisfied by (1) an idempotent, (2) a nilpotent. Note that both equations are without constant term.
11. If $\mathcal{A}$ and $\mathcal{B}$ are regular, show that so is $\mathcal{A}\mathcal{B}$ and find its inverse. Conversely, if $\mathcal{C} = \mathcal{A}\mathcal{B}$ is regular, show that $\mathcal{A}$ and $\mathcal{B}$ must be regular.
12. Use this to show that an algebraic equation satisfied by a singular matrix cannot involve a constant term.
13. If the equation $P(x) = 0$ with $P(0) \neq 0$ is satisfied by a regular matrix, find the equation satisfied by its inverse.
14. Let $\mathcal{E}_{jk}$ be the matrix with a one in the place (j, k) and a zero everywhere else. Show that the set $\{\mathcal{E}_{jk}; j, k = 1, 2, \ldots, n\}$ is a basis for $\mathfrak{M}_n$ regarded as a linear vector space of dimension n^2.
15. Take $n = 3$ and let $\mathcal{E}_{11}$ be defined as in the preceding problem. Find all matrices in $\mathfrak{M}_3$ which commute with $\mathcal{E}_{11}$.

16. Prove that the degree of nilpotency of a matrix in $\mathfrak{M}_n$ cannot exceed n. Can the degree n be reached?
17. Show that the unit elements of $\mathfrak{E}(C^n)$ and $\mathfrak{M}_n$ are unique.
18. If $\mathbf{x}$ and $\mathbf{y}$ are vectors in C^n, find a matrix $\mathcal{A}$ in $\mathfrak{M}_n$ such that $\mathbf{y} = \mathcal{A} \cdot \mathbf{x}$. Is the solution unique? Does the problem always have a solution?
19. A diagonal matrix is one in which all elements off the main diagonal are zero, i.e. $j \neq k$ implies $a_{jk} = 0$. When is such a matrix singular? Show that the set of all diagonal matrices in $\mathfrak{M}_n$ forms a commutative subalgebra.
20. Characterize the idempotents of this subalgebra and determine the divisors of zero. Are there any non-trivial nilpotents?
21. Prove that an idempotent in $\mathfrak{M}_n$ which is regular must coincide with the unit matrix and hence all other idempotents are singular.
22. If $\mathcal{A}$ and $\mathcal{A}_1$ are similar matrices show that $\det(\mathcal{A})$ and $\det(\mathcal{A}_1)$ have the same numerical value. Is the converse true?

1.5 THE RESOLVENT

The linear transformation $A \in \mathfrak{E}(C^n)$ whose representation in terms of a chosen basis $\{\mathbf{u}_j\}$ is the matrix $\mathcal{A}$ maps C^n into the linear subspace $\mathfrak{R}[A]$ of C^n made up of all vectors of the form

$$\mathbf{y} = \mathcal{A} \cdot \mathbf{x}. \tag{1.5.1}$$

The zero element is left invariant by $\mathcal{A}$. Are there other invariant elements? As a rule not, but if we relax the demands and require only that

$$\mathcal{A}\mathbf{x} = \alpha \mathbf{x} \tag{1.5.2}$$

for some constant α and some vector $\mathbf{x}$ with $\|\mathbf{x}\| = 1$, then a more favorable situation arises. Here (1.5.2) may be written

$$(\alpha \mathcal{E} - \mathcal{A}) \cdot \mathbf{x} = 0. \tag{1.5.3}$$

Equivalently we require $(\alpha I - A)\,\mathbf{x} = 0$. Thus we find that the matrix $\alpha\mathcal{E} - \mathcal{A}$ is required to annihilate a non-zero vector $\mathbf{x}$ and this can happen iff the matrix is singular. Now if $\alpha\mathcal{E} - \mathcal{A}$ is singular, then its determinant is zero,

$$\det(\alpha \mathcal{E} - \mathcal{A}) = 0. \tag{1.5.4}$$

Consider the equation in λ

$$\Delta(\lambda) \equiv \det(\lambda \mathcal{E} - \mathcal{A}) = 0, \tag{1.5.5}$$

or, in more detail,

$$\Delta(\lambda) = \begin{vmatrix} \lambda - a_{11} & -a_{12} & \cdots & -a_{1n} \\ -a_{21} & \lambda - a_{22} & \cdots & -a_{2n} \\ \cdots & \cdots & \cdots & \cdots \\ -a_{n1} & -a_{n2} & \cdots & \lambda - a_{nn} \end{vmatrix} = 0. \tag{1.5.6}$$

Expansion gives an algebraic equation in λ of degree n which has a total number of n roots in the complex field. Suppose that there are p distinct roots $\lambda_1, \lambda_2, \ldots, \lambda_p$ and that the multiplicity of λ_j is k_j. Here

$$p \leqslant n, \qquad \sum_{j=1}^{p} k_j = n, \tag{1.5.7}$$

and we have

$$\Delta(\lambda) = \prod_{j=1}^{p} (\lambda - \lambda_j)^{k_j}. \tag{1.5.8}$$

It follows that the number α in (1.5.2) must be one of the roots λ_j. Here (1.5.6) is known as the *characteristic equation* of the matrix $\mathcal{A}$. Incidentally, as will be shown later, it is also satisfied by $\mathcal{A}$ itself, i.e. $\Delta(\mathcal{A}) = 0$. Note that we are concerned with matrices and vectors over the complex field. Over the real field and for n even (1.5.6) may have no roots at all.

The roots are known under many names: *characteristic roots* or *values, eigen values, latent roots* are the most common. The set of roots for which (1.5.3) has a non-trivial solution is called the *spectrum* of $\mathcal{A}$ and will be denoted by $\sigma(\mathcal{A})$ in the following. If operators are held in the foreground, then $\sigma(A)$ will be corresponding notation.

If λ has a value distinct from all the roots, then $\lambda\mathcal{E} - \mathcal{A}$ is regular, i.e. has an inverse and we set

$$(\lambda\mathcal{E} - \mathcal{A})^{-1} = \mathfrak{R}(\lambda, \mathcal{A}), \tag{1.5.9}$$

known as the *resolvent* of $\mathcal{A}$. We know how to form the inverse of a regular matrix, see formula (1.4.25). Applied to the present case this gives as the element of the resolvent in the place (j, k)

$$\frac{A_{kj}(\lambda)}{\Delta(\lambda)} \equiv r_{jk}(\lambda), \tag{1.5.10}$$

where $A_{jk}(\lambda)$ is the cofactor of $\lambda\delta_{jk} - a_{jk}$ in the determinant $\Delta(\lambda)$. Thus $A_{kj}(\lambda)$ is a polynomial in λ with complex coefficients and of degree $\leqslant n - 1$. It follows that $r_{jk}(\lambda)$ is a rational function of λ with poles included in the set of zeros of $\Delta(\lambda)$ and $r_{jk}(\lambda) \to 0$ as $\lambda \to \infty$. Hence

$$\mathfrak{R}(\lambda, \mathcal{A}) = \frac{\mathcal{S}(\lambda, \mathcal{A})}{\Delta(\lambda)}, \tag{1.5.11}$$

where $\mathcal{S}(\lambda, \mathcal{A})$ is a polynomial in λ of degree $\leqslant n - 1$ whose coefficients are constant matrices in $\mathfrak{M}_n$.

It was stated above that the poles of $r_{jk}(\lambda)$ are included in the set of zeros of $\Delta(\lambda)$. That no stronger statement can be made is illustrated by the case of a diagonal matrix $\mathcal{A}$ with $a_{jj} = a_j$, $a_{jk} = 0$ if $j \neq k$. If the a_j are distinct, then

$$r_{jj}(\lambda) = (\lambda - a_j)^{-1}, \qquad r_{jk}(\lambda) = 0, \qquad j \neq k. \tag{1.5.12}$$

Here each diagonal element has a single simple pole and the non-diagonal elements have none.

Formula (1.5.11) shows that $\mathfrak{R}(\lambda, \mathfrak{A})$ is a rational function of λ with matrix coefficients. Note that the matrices enter only in the numerator of this rational function. The denominator is a scalar polynomial. Since each $r_{jk}(\lambda) \to 0$ as $\lambda \to \infty$, $\mathfrak{R}(\lambda, \mathfrak{A})$ has the same property. Here we can apply classical results on scalar rational functions to show that each $r_{jk}(\lambda)$ has an expansion in partial fractions of the form

$$r_{jk}(\lambda) = \sum_{l} \sum_{m} \alpha_{jklm} (\lambda - \lambda_m)^{-l}. \tag{1.5.13}$$

Here the α_{jklm} are complex numbers, the λ_m's run through the p distinct roots of $\Delta(\lambda)$ and l is a positive integer taking on values from 1 to k_m, the multiplicity of the root λ_m.

Passing from elements to matrices we obtain an expansion

$$\mathfrak{R}(\lambda, \mathfrak{A}) = \sum_{l} \sum_{m} (\lambda - \lambda_m)^{-l} \mathfrak{C}_{lm}, \tag{1.5.14}$$

where now the $\mathfrak{C}_{lm}$ are constant matrices in $\mathfrak{M}_n$. These matrices have very special properties. They commute with each other and with $\mathfrak{A}$. They are all singular matrices, divisors of zero, idempotent for $l = 1$, nilpotent for $l > 1$. Moreover, matrices $\mathfrak{C}_{lm}$ with different second subscript are orthogonal to each other, i.e. their product is the zero matrix.

It is possible to prove these properties by fairly simple means. Let us take the expansion of $\mathfrak{R}(\lambda, \mathfrak{A})$ in the neighborhood of one of the zeros of $\Delta(\lambda)$, say $\lambda = \alpha$, to simplify the notation. We have then

$$\mathfrak{R}(\lambda, \mathfrak{A}) = \frac{\mathfrak{C}_1}{\lambda - \alpha} + \frac{\mathfrak{C}_2}{(\lambda - \alpha)^2} + \cdots + \frac{\mathfrak{C}_s}{(\lambda - \alpha)^s} + \mathfrak{B}(\lambda). \tag{1.5.15}$$

Here s is at most equal to the multiplicity of the root α and $\mathfrak{B}(\lambda)$ is a power series

$$\mathfrak{B}(\lambda) = \sum_{k=0}^{\infty} \mathfrak{B}_k (\lambda - \alpha)^k, \tag{1.5.16}$$

convergent in norm for $|\lambda - \alpha| < \rho$, the distance from α to the rest of the spectrum. The $\mathfrak{B}$'s and $\mathfrak{C}$'s are constant matrices, elements of $\mathfrak{M}_n$. To see that the remainder $\mathfrak{B}(\lambda)$ in (1.5.15) really has a representation of type (1.5.16), we revert to (1.5.13). Here we picked out the terms where $\lambda_m = \alpha$. The rest should be $\mathfrak{B}(\lambda)$, i.e.

$$\mathfrak{B}(\lambda) = \sum_{l} \sum_{m}{}' (\lambda - \lambda_m)^{-l} \mathfrak{C}_{lm}, \tag{1.5.17}$$

where the prime after the summation sign indicates that terms with $\lambda_m = \alpha$ are to be omitted. Now if $\lambda_m = \beta \neq \alpha$

$$(\lambda - \beta)^{-l} = [(\alpha - \beta) + (\lambda - \alpha)]^{-l} = (\alpha - \beta)^{-l} \left\{1 - \frac{\lambda - \alpha}{\beta - \alpha}\right\}^{-l}$$

$$= \sum_{k=0}^{\infty} (-1)^{k+l} \binom{-l}{k} (\beta - \alpha)^{-k-l} (\lambda - \alpha)^k,$$

a binomial series in $\lambda - \alpha$ convergent for $|\lambda - \alpha| < |\beta - \alpha|$ which is at least ρ. We repeat this for each $\beta \neq \alpha$ in the spectrum and add the results. It follows that for $|\lambda - \alpha| < \rho$ the rational function $\mathcal{B}(\lambda)$ is the sum of a finite number of matrices $\mathcal{C}_{lm}$ each multiplied by an absolutely convergent binomial series. Collecting terms, we obtain a series of type (1.5.16).

Thus the representation (1.5.15) with $\mathcal{B}(\lambda)$ given by (1.5.16) is valid in the punctured disk $0 < |\lambda - \alpha| < \rho$ and in this domain

$$(\lambda\mathcal{E} - \mathcal{A})\,\mathcal{R}(\lambda, \mathcal{A}) = \mathcal{R}(\lambda, \mathcal{A})\,(\lambda\mathcal{E} - \mathcal{A}) = \mathcal{E}. \tag{1.5.18}$$

Here we substitute

$$(\lambda - \alpha)\,\mathcal{E} + (\alpha\mathcal{E} - \mathcal{A})$$

for $\lambda\mathcal{E} - \mathcal{A}$ and the expansion (1.5.15) for $\mathcal{R}(\lambda, \mathcal{A})$. We multiply out, collect terms, and note that the result should be identically $\mathcal{E}$. This gives the conditions that the coefficient matrices must satisfy. To force the negative powers of $\lambda - \alpha$ to drop out we must have

$$\begin{aligned} &\mathcal{C}_s\,(\mathcal{A} - \alpha\mathcal{E}) = \mathcal{O}, \\ &\mathcal{C}_s = \mathcal{C}_{s-1}\,(\mathcal{A} - \alpha\mathcal{E}), \\ &\cdots\cdots\cdots\cdots\cdots\cdots \\ &\mathcal{C}_2 = \mathcal{C}_1\,(\mathcal{A} - \alpha\mathcal{E}), \end{aligned} \tag{1.5.19}$$

as well as the relations obtained by permuting the factors in the products. The constant term gives

$$\mathcal{C}_1 = \mathcal{B}_0\,(\mathcal{A} - \alpha\mathcal{E}) + \mathcal{E} \tag{1.5.20}$$

and the positive powers will drop out provided for each $k > 0$

$$\mathcal{B}_{k-1} = \mathcal{B}_k\,(\mathcal{A} - \alpha\mathcal{E}). \tag{1.5.21}$$

Again the factors commute.

The next step is to examine $\mathcal{B}(\lambda)$. The last two relations imply that

$$(\lambda\mathcal{E} - \mathcal{A})\,\mathcal{B}(\lambda) = \mathcal{B}(\lambda)\,(\lambda\mathcal{E} - \mathcal{A}) = -\mathcal{B}_0(\mathcal{A} - \alpha\mathcal{E}) \equiv \mathcal{T}_0. \tag{1.5.22}$$

Hence, for λ not in the spectrum of $\mathcal{A}$,

$$\mathcal{B}(\lambda) = \mathcal{T}_0\,\mathcal{R}(\lambda, \mathcal{A}) = \mathcal{R}(\lambda, \mathcal{A})\,\mathcal{T}_0. \tag{1.5.23}$$

Here we compare the expansions for $\mathcal{B}(\lambda)$ and for $\mathcal{R}(\lambda, \mathcal{A})$ with the relations implied by (1.5.23). It is clear that the matrix $\mathcal{T}_0$ must annihilate all the negative powers in (1.5.15) and, since this must hold identically in λ, the coefficient matrices must be annihilated, i.e.

$$\mathcal{C}_1\,\mathcal{T}_0 = \mathcal{C}_2\,\mathcal{T}_0 = \cdots = \mathcal{C}_s\,\mathcal{T}_0 = 0, \tag{1.5.24}$$

and here we can also permute the factors.

We now return to (1.5.20), which may be written

$$\mathcal{C}_1 = \mathcal{E} - \mathcal{T}_0.$$

Multiplying both sides by $\mathcal{C}_1$ and using (1.5.24) we get

$$\mathcal{C}_1^{\,2} = \mathcal{C}_1 \equiv \mathcal{J},$$

i.e. an idempotent as asserted. Recall that (1.5.15) is the expansion of $\mathcal{R}(\lambda, \mathcal{A})$ in the neighborhood of $\lambda = \alpha$, one of the zeros of $\Delta(\lambda)$. If α is actually in the spectrum, then $\mathcal{J}$ cannot be the zero matrix. For if $\mathcal{J} = \mathcal{C}_1 = \mathcal{O}$, then by (1.5.19) $\mathcal{C}_1 = \mathcal{C}_2 = \cdots = \mathcal{C}_s = \mathcal{O}$ and $\mathcal{R}(\alpha, \mathcal{A}) = \mathcal{B}(\alpha)$ would exist, contradicting the fact that α is in the spectrum. $\mathcal{J}$ may be the unit matrix but only in an exceptional case, for which see below.

We now set

$$\mathcal{C}_2 = \mathcal{Q} \tag{1.5.25}$$

and find that

$$\mathcal{J}\mathcal{Q} = \mathcal{Q}\mathcal{J} = \mathcal{Q}, \tag{1.5.26}$$

of interest only if $\mathcal{Q}$ is not the zero matrix. Equations (1.5.19) now give

$$\mathcal{C}_j = \mathcal{Q}^{j-1}, \qquad j = 2, 3, \ldots, s \tag{1.5.27}$$

and

$$\mathcal{Q}^s = \mathcal{O}, \tag{1.5.28}$$

so that $\mathcal{Q}$ is nilpotent as asserted above.

From (1.5.23) we obtain

$$\mathcal{B}(\lambda) = \mathcal{J}_0\,\mathcal{B}(\lambda) = \mathcal{B}(\lambda)\mathcal{J}_0. \tag{1.5.29}$$

Since $\mathcal{J}_0\,\mathcal{J} = \mathcal{J}\,\mathcal{J}_0 = \mathcal{O}$, we get

$$\mathcal{J}\,\mathcal{B}(\lambda) = \mathcal{B}(\lambda)\,\mathcal{J} = \mathcal{O}, \qquad \mathcal{Q}\,\mathcal{B}(\lambda) = \mathcal{B}(\lambda)\,\mathcal{Q} = \mathcal{O}. \tag{1.5.30}$$

We have now obtained the principal part of $\mathcal{R}(\lambda, \mathcal{A})$ at the spectral singularity $\lambda = \alpha$ and the same type of expansion holds at the other points of the spectrum. Returning to (1.5.17) and using (1.5.30), we see that $\mathcal{J}$ and $\mathcal{Q}$ annihilate all the matrices $\mathcal{C}_{lm}$ with m such that $\lambda_m \neq \alpha$. Now $\mathcal{C}_{1m}$ is the idempotent associated with the spectral value λ_m. It follows that all these idempotents are annihilated by $\mathcal{J}$ and hence also by $\mathcal{Q}$. In other words, the idempotents associated with distinct spectral values of $\mathcal{A}$ are mutually orthogonal. We denote now the idempotent corresponding to $\lambda = \lambda_m$ by $\mathcal{J}_m$ and the nilpotent by $\mathcal{Q}_m$. We have then proved

Theorem 1.5.1. *The resolvent of the matrix $\mathcal{A}$ has a partial fraction expansion of the form*

$$\mathcal{R}(\lambda, \mathcal{A}) = \sum_{m=1}^{p} \left\{ \frac{\mathcal{J}_m}{\lambda - \lambda_m} + \frac{\mathcal{Q}_m}{(\lambda - \lambda_m)^2} + \cdots + \frac{\mathcal{Q}_m^{\,s_m - 1}}{(\lambda - \lambda_m)^{s_m}} \right\}. \tag{1.5.31}$$

Here the summation extends over the p distinct characteristic values of $\mathcal{A}$ and

$$\mathcal{J}_j\mathcal{J}_k = \delta_{jk}\mathcal{J}_j, \qquad \mathcal{J}_m\mathcal{Q}_m = \mathcal{Q}_m\mathcal{J}_m = \mathcal{Q}_m, \tag{1.5.32}$$

$$\mathcal{J}_j\mathcal{Q}_k = \mathcal{Q}_k\mathcal{J}_j = \mathcal{O}, \qquad \mathcal{Q}_j\mathcal{Q}_k = \mathcal{Q}_k\mathcal{Q}_j = \mathcal{O}, \qquad j \neq k. \tag{1.5.33}$$

Further, s_m is the least integer k such that

$$\mathcal{Q}_m{}^k = \mathcal{O} \tag{1.5.34}$$

and s_m is at most equal to the multiplicity k_m of λ_m as a root of $\Delta(\lambda) = 0$. Finally,

$$\mathcal{Q}_m = \mathcal{J}_m(\mathcal{A} - \lambda_m \mathcal{E}). \tag{1.5.35}$$

There are still various issues to be faced. First, can $\mathcal{J}_m$ be the unit matrix? Since the idempotents are mutually orthogonal and $\mathcal{E}$ is orthogonal only to $\mathcal{O}$, there is a single spectral value, say $\lambda = \alpha$, and

$$\mathcal{R}(\lambda, \mathcal{A}) = (\lambda - \alpha)^{-1}\mathcal{E} + (\lambda - \alpha)^{-2}\mathcal{Q} + \cdots + (\lambda - \alpha)^{-s}\mathcal{Q}^{s-1}, \tag{1.5.36}$$

so that

$$\mathcal{A} = \alpha\mathcal{E} + \mathcal{Q}$$

and $\mathcal{A}$ differs from $\alpha\mathcal{E}$ only by a nilpotent. The conclusion also follows from (1.5.35).

A second question is: Can s_m be less than the multiplicity of λ_m as a root of $\Delta(\lambda) = 0$? An affirmative answer to this question can be read off from (1.5.12). This formula is valid whether or not the a_j's are distinct. Suppose that $a_j = 0$ with multiplicity k and $a_j = 1$ with multiplicity $n - k$. There are only two characteristic values, namely 0 and 1 of stated multiplicities. Here $\mathcal{A}$ itself is idempotent and the expansion of its resolvent is simply

$$\mathcal{R}(\lambda, \mathcal{A}) = \lambda^{-1}(\mathcal{E} - \mathcal{A}) + (\lambda - 1)^{-1}\mathcal{A}. \tag{1.5.37}$$

Thus the resolvent has only two simple poles. If $k > 1$, $n - k > 1$ we have $s_j < k_j, j = 1, 2$.

Our third question is: Can it happen that a zero of $\Delta(\lambda)$ is not a pole of $\mathcal{R}(\lambda, \mathcal{A})$? This presupposes that the corresponding idempotent is the zero matrix. Suppose that this takes place at $\lambda = \alpha$. Note that $\alpha\mathcal{E} - \mathcal{A}$ is definitely a singular matrix since $\Delta(\alpha) = 0$. This means that $\mathcal{R}(\lambda, \mathcal{A})$ does not exist for $\lambda = \alpha$, so this point must be a singularity of $\mathcal{R}(\lambda, \mathcal{A})$, a function of theoretical singularity. Now formula (1.5.11) shows that $\mathcal{R}(\lambda, \mathcal{A})$ is a rational function of λ, the quotient of a matrix polynomial of degree $\leqslant n - 1$ in λ and a scalar polynomial in λ of degree n. Such a function can have no singularities save for poles in the complex λ-plane. Hence $\lambda = \alpha$ is a pole and the idempotent cannot be the zero matrix. Thus the third question is answered in the negative.

For the further development we need

Lemma 1.5.1. *For large values of $|\lambda|$*

$$\mathcal{R}(\lambda, \mathcal{A}) = \lambda^{-1}\mathcal{E} + \lambda^{-2}\mathcal{A} + \cdots + \lambda^{-k-1}\mathcal{A}^k + \cdots. \tag{1.5.38}$$

In particular, the series converges and the expansion is valid for $|\lambda| > \|\mathcal{A}\|$.

Proof. By Lemma 1.4.3 we have

$$\|\mathcal{A}^k\| \leqslant \|\mathcal{A}\|^k, \tag{1.5.39}$$

valid for the norm (1.4.42) as well as for (1.4.43). This proves the assertion concerning convergence of the series for $|\lambda| > \|\mathcal{A}\|$. For such values we can multiply the convergent series left or right by $\lambda\mathcal{E} - \mathcal{A}$ and collect terms. They all cancel except for the constant term, which is $\mathcal{E}$ as it should be. ∎

The reader should note that the lemma does not assert that the radius of convergence of the series is $\|\mathcal{A}\|$, an expression which has at least two different meanings. Actually the series converges for

$$|\lambda| > \max_j |\lambda_j|, \tag{1.5.40}$$

as we can deduce from (1.5.31), and this is always the exact radius of convergence.

Among the consequences of Lemma 1.5.1 we list

Lemma 1.5.2. *If $\mathcal{A}$ has p distinct characteristic roots and if $\mathcal{J}_1, \mathcal{J}_2, \ldots, \mathcal{J}_p$ are the corresponding idempotents, then*

$$\mathcal{J}_1 + \mathcal{J}_2 + \cdots + \mathcal{J}_p = \mathcal{E}. \tag{1.5.41}$$

Proof. We compare formulas (1.5.31) and (1.5.38), both valid for large values of $|\lambda|$. We can expand the partial fractions in descending powers of λ. Only the first order fractions can give a term in $1/\lambda$ and it will have the coefficient $\mathcal{J}_m$. Adding terms and comparing the result with (1.5.38) we obtain (1.5.41). ∎

We call this formula a *resolution of the identity*. It is a decomposition of the unit matrix into a sum of mutually orthogonal idempotents belonging to the matrix $\mathcal{A}$.

We shall use this result to prove the *Cayley–Hamilton Theorem* [after Arthur Cayley (1821–95) and Sir William Rowen Hamilton (1805–65)]. With the notation of Theorem 1.5.1 set

$$H(\lambda) = \prod_{j=1}^{p} (\lambda - \lambda_j)^{s_j}. \tag{1.5.42}$$

This polynomial in λ is a divisor of $\Delta(\lambda)$ and will coincide with the latter iff each $s_j = k_j$, the multiplicity of λ_j as a characteristic root. We now form

$$\mathcal{H}(\mathcal{A}) = \prod_{j=1}^{p} (\mathcal{A} - \lambda_j\mathcal{E})^{s_j} \tag{1.5.43}$$

and note that the factors on the right commute. The Cayley–Hamilton theorem now reads:

Theorem 1.5.2. *$\mathcal{H}(\mathcal{A})$ is the zero matrix.*

Proof. Lemma 1.5.3 gives

$$H(\mathcal{A}) = H(\mathcal{A})[\mathcal{J}_1 + \mathcal{J}_2 + \cdots + \mathcal{J}_p]. \tag{1.5.44}$$

Here

$$H(\mathcal{A})\mathcal{T}_m = \prod_{j=1}^{p} (\mathcal{A} - \lambda_j\mathcal{E})^{s_j}\mathcal{T}_m = H_m(\mathcal{A})\,(\mathcal{A} - \lambda_m\mathcal{E})^{s_m}\mathcal{T}_m,$$

where

$$H_m(\mathcal{A}) = \prod_{j\neq m} (\mathcal{A} - \lambda_j\mathcal{E})^{s_j}.$$

We now refer to formula (1.5.35), according to which

$$(\mathcal{A} - \lambda_m\mathcal{E})^{s_m}\mathcal{T}_m = \mathcal{Q}_m^{\ s_m} = \mathcal{O}$$

Hence each of the p products in (1.5.44) is zero and so is their sum. ∎

Thus $\mathcal{A}$ satisfies

$$H(\lambda) = 0, \tag{1.5.45}$$

known as the *minimal equation*. "Satisfies" means here that if λ is replaced by $\mathcal{A}$ and 1 by $\mathcal{E}$ in the polynomial $H(\lambda)$, the result is the zero matrix. It is clear that $\mathcal{A}$ also satisfies the characteristic equation. The term minimal refers to the fact that if $\mathcal{A}$ satisfies an algebraic equation $G(\lambda) = 0$ in the sense just mentioned, then $G(\lambda)$ must be divisible by $H(\lambda)$. See Problems 2 (Exercise 1.5) and 3 and 4 (Exercise 1.6).

Let us return for a moment to formula (1.5.31), which expresses $\mathcal{R}(\lambda, \mathcal{A})$ as the sum of terms of the form

$$\mathcal{B}(\lambda - \alpha)^{-k}, \tag{1.5.46}$$

where $\mathcal{B}$ is a matrix and the other factor is a rational function of λ. Each of these factors has derivatives of all orders with respect to λ. From this we may conclude that $\mathcal{R}(\lambda, \mathcal{A})$ also has derivatives of all orders with respect to λ. In particular we can find $\mathcal{R}'(\lambda, \mathcal{A})$ by termwise differentiation in formula (1.5.31).

EXERCISE 1.5

1. If $\mathcal{E}_{jj}$ is the matrix with a 1 in the place (j, j) and zeros elsewhere and if $n > 2$ find the characteristic and the minimal equations of $\mathcal{E}_{jj}$.
2. Let $G(\lambda)$ be a scalar polynomial in λ with no root in common with the minimal equation $H(\lambda) = 0$ satisfied by a given matrix $\mathcal{A}$. Show that $G(\mathcal{A})$ is a regular matrix and hence not zero.
3. Let $G(\lambda)$ be a scalar polynomial in λ of degree $m \leqslant n$. How is the spectrum of $G(\mathcal{A})$ related to that of $\mathcal{A}$?
4. Assuming the spectral values of $\mathcal{A}$ to be distinct, find $\mathcal{R}(\lambda, \mathcal{A}^2)$.
5. If a matrix $\mathcal{A}$ satisfies the equation $\mathcal{A}^3 - 4\mathcal{A} = \mathcal{O}$, find the spectrum and the resolvent.
6. Show that formula (1.5.37) is valid for any idempotent matrix $\mathcal{A}$ and not merely for diagonal idempotent matrices.

7. What is the minimal equation of a nilpotent element? The characteristic equation?
8. Assuming $\mathcal{A}$ to be a non-singular diagonal matrix, in how many ways can a diagonal matrix $\mathcal{B}$ be found such that $\mathcal{B}^2 = \mathcal{A}$?
9. Show that the characteristic roots of a Hermitian matrix are real.
10. Show that the spectrum of a unitary matrix (see Problem 4, Exercise 1.3) lies on the unit circle in the complex plane and is symmetric with respect to the real axis.
11. What is the inverse of a unitary matrix? Show that the minimal equation has real coefficients and is a so-called reciprocal equation, i.e. if α is a root, so is α^{-1}.
12. What is the relation between the spectrum of $\mathcal{A}$ and that of its conjugate transpose $\mathcal{A}^*$?
13. If λ and μ are not in the spectrum of A, show that
$$(\lambda - \mu)\,\mathcal{R}(\lambda, \mathcal{A})\,\mathcal{R}(\mu, \mathcal{A}) = \mathcal{R}(\mu, \mathcal{A}) - \mathcal{R}(\lambda, \mathcal{A}).$$
This functional equation is known as the *first resolvent equation.*
14. At the end of Section 1.5 it was shown that $\mathcal{R}(\lambda, \mathcal{A})$ has derivatives of all orders with respect to λ. Use the result of the preceding problem to show that $\mathcal{R}'(\lambda, \mathcal{A}) = -[\mathcal{R}(\lambda, \mathcal{A})]^2$ and obtain from this $\mathcal{R}^{(m)}(\lambda, \mathcal{A}) = (-1)^m m!\,[\mathcal{R}(\lambda, \mathcal{A})]^{m+1}$, $\forall\, m$.

1.6 INVARIANTIVE AND METRIC PROPERTIES

Let us write $C^n = \mathfrak{X}$ to abbreviate and consider a linear transformation A on $\mathfrak{X}$ to itself which in terms of a given ordered basis $\mathsf{F} = (\mathbf{u}_1, \ldots, \mathbf{u}_n)$ is represented by the matrix $\mathcal{A}$ of $\mathfrak{M}_n$. Let the distinct characteristic values of $\mathcal{A}$ be $\lambda_1, \lambda_2, \ldots, \lambda_p$, where the multiplicity of λ_j is k_j. As usual, $\mathcal{J}_j$ shall be the idempotent, $\mathcal{Q}_j$ the nilpotent corresponding to λ_j. Further, s_j is the least integer such that

$$\mathcal{Q}_j^{s_j} = \mathcal{O}. \tag{1.6.1}$$

We know that $s_j \leqslant k_j$.

In this section we shall consider two disconnected topics. The first is the *invariant subspaces* connected with the transformation A. The second is concerned with *norm* and *distance* for the spaces $\mathfrak{X}$ and $\mathfrak{E}(\mathfrak{X}) = \mathfrak{E}(C^n)$.

The matrix $\mathcal{J}_j$ defines a projection of $\mathfrak{X}$ onto a linear subspace $\mathfrak{X}_j \subset \mathfrak{X}$. The subspaces $\mathfrak{X}_i$ and $\mathfrak{X}_j$, $i \neq j$, have only the zero element in common since the corresponding matrices $\mathcal{J}_i$ and $\mathcal{J}_j$ are orthogonal to each other. Hence any element $\mathbf{x} \in \mathfrak{X}$, $\mathbf{x} \neq \mathbf{0}$, has a unique decomposition obtained by applying (1.5.41),

$$\mathbf{x} = \mathbf{x}_1 + \mathbf{x}_2 + \cdots + \mathbf{x}_p, \qquad \mathbf{x}_j \in \mathfrak{X}_j. \tag{1.6.2}$$

Here $\mathfrak{X}_j$ is known as the jth *root space* of $\mathfrak{X}$ and *it is left pointwise invariant under* $\mathcal{J}_j$,

$$\mathcal{J}_j(\mathbf{x}) = \mathbf{x} \quad \text{if} \quad \mathbf{x} \in \mathfrak{X}_j. \tag{1.6.3}$$

Note also that $\mathcal{J}_j$ maps $\mathfrak{X}_k$, $k \neq j$, into the zero vector.

Some vectors of $\mathfrak{X}$ are annihilated by the matrix $\mathcal{A} - \lambda_j \mathcal{E}$. We start with vectors in $\mathfrak{X}_j$. As we shall see, no vector in $\mathfrak{X}_k$, $\mathbf{x} \neq \mathbf{0}$, can have this property for a $k \neq j$.

Suppose that $\mathbf{x}$ is a *characteristic vector*, i.e.

$$(\mathcal{A} - \lambda_j \mathcal{E})\,\mathbf{x} = \mathbf{0}, \qquad \mathbf{x} \neq \mathbf{0}, \qquad \mathbf{x} \in \mathfrak{X}_j. \tag{1.6.4}$$

The set of such vectors along with the zero vector is a linear subspace Ξ_j of $\mathfrak{X}_j$, known as the *characteristic* or *eigen space* corresponding to λ_j. Here Ξ_j may coincide with $\mathfrak{X}_j$. If it is a proper subspace, however, then there are vectors in $\mathfrak{X}_j \ominus \Xi_j$ which are annihilated by some power of $\mathcal{A} - \lambda_j \mathcal{E}$ higher than the first. In fact, all of $\mathfrak{X}_j$ is annihilated by the s_jth power, i.e.

$$(\mathcal{A} - \lambda_j \mathcal{E})^{s_j}\,\mathbf{x} = \mathbf{0}, \qquad \mathbf{x} \in X_j. \tag{1.6.5}$$

This follows from the definitions of $\mathcal{Q}_j$ and of s_j. We have

$$\mathcal{Q}_j = (\mathcal{A} - \lambda_j \mathcal{E})\,\mathcal{T}_j,$$

so that on $\mathfrak{X}_j$ the matrices $\mathcal{A} - \lambda_j \mathcal{E}$ and $\mathcal{Q}_j$ exert the same action. Hence $\mathbf{x} \in \mathfrak{X}_j$ implies that

$$(\mathcal{A} - \lambda_j \mathcal{E})^{s_j}\mathbf{x} = \mathcal{Q}_j^{\,s_j}\mathbf{x} = \mathbf{0}. \tag{1.6.6}$$

In particular, this argument shows that Ξ_j cannot reduce to the zero vector. It was remarked above that no vector in $\mathfrak{X}_k$, different from the zero vector, can be annihilated by $\mathcal{A} - \lambda_j \mathcal{E}$ for a $j \neq k$. This will now be proved.

Suppose that $\mathbf{x} \in \mathfrak{X}_k$, $\mathbf{x} \neq \mathbf{0}$, $j \neq k$, and $\mathcal{A}\mathbf{x} = \lambda_j \mathbf{x}$. Then

$$(\mathcal{A} - \lambda_k \mathcal{E})\,\mathbf{x} = (\lambda_j - \lambda_k)\,\mathbf{x}$$

and, generally,

$$(\mathcal{A} - \lambda_k \mathcal{E})^m\,\mathbf{x} = (\lambda_j - \lambda_k)\,(\mathcal{A} - \lambda_k \mathcal{E})^{m-1}\,\mathbf{x} \tag{1.6.7}$$

for any positive integer m. Since $\mathbf{x} \in \mathfrak{X}_k$, there exists an m such that the left member of (1.6.7) is zero. Since $j \neq k$, this implies that $(\mathcal{A} - \lambda_k \mathcal{E})^{m-1}\,\mathbf{x}$ is also zero and by complete induction that $\mathbf{x} = \mathbf{0}$. The same argument shows that no power of $\mathcal{A} - \lambda_j \mathcal{E}$ can annihilate an $\mathbf{x} \in \mathfrak{X}_k$, $\mathbf{x} \neq \mathbf{0}$, $j \neq k$.

We shall now discuss the dimensions of Ξ_j and $\mathfrak{X}_j$. These are linear vector spaces and the notion of dimension was defined in Section 1.2 as the largest number of linearly independent vectors belonging to the space in question. We shall need the following:

Lemma 1.6.1. *Similar matrices have identical characteristic equations and thus have the same characteristic values with the same multiplicities.*

Proof. Suppose that $\mathcal{A}_1 = \mathcal{B}\mathcal{A}\mathcal{B}^{-1}$, where $\mathcal{B}$ is a regular matrix. We recall that

$$\det(\mathcal{B})\det(\mathcal{B}^{-1}) = \det(\mathcal{E}) = 1.$$

It follows that

$$\begin{aligned}
\det(\lambda\mathcal{E} - \mathcal{A}_1) &= \det(\lambda\mathcal{E} - \mathcal{B}\mathcal{A}\mathcal{B}^{-1})\\
&= \det[\mathcal{B}(\lambda\mathcal{E} - \mathcal{A})\,\mathcal{B}^{-1}]\\
&= \det(\mathcal{B})\det(\lambda\mathcal{E} - \mathcal{A})\det(\mathcal{B}^{-1})\\
&= \det(\lambda\mathcal{E} - \mathcal{A})
\end{aligned}$$

as asserted. From the identity of the characteristic equations the other assertions follow. ▮

We denote the dimension of a linear vector space $\mathfrak{X}_0$ by dim $(\mathfrak{X}_0)$ and prove

Theorem 1.6.1. $1 \leqslant \dim(\Xi_j) \leqslant k_j$.

Proof. It was noted above that there is at least one $\mathbf{x}$ with $\mathbf{x} \neq 0$ belonging to Ξ_j. Since all constant multiples of this vector also belong to Ξ_j, it follows that $\dim(\Xi_j) \geqslant 1$. To prove that $\dim(\Xi_j) \leqslant k_j$, the multiplicity of λ_j, we argue as follows. Suppose that $\dim(\Xi_j) = m$. We can now find a new ordered basis for $\mathfrak{X}$, say $\mathsf{F}_1 = (\mathbf{v}_1, \ldots, \mathbf{v}_m, \mathbf{v}_{m+1}, \ldots, \mathbf{v}_n)$, such that the first m vectors belong to Ξ_j. Now the action of transformation A on F_1 is represented by a matrix $\mathcal{A}_1$ similar to $\mathcal{A}$. Here the first m columns of $\mathcal{A}_1$ are made up of zeros except for the elements of the main diagonal, which are all equal to λ_j. It follows that $\det(\lambda\mathcal{E} - \mathcal{A}_1)$ has $(\lambda - \lambda_j)^m$ as a factor so that $\lambda = \lambda_j$ is a characteristic root of $\mathcal{A}_1$ of multiplicity $\geqslant m$. But by Lemma 1.6.1, $\mathcal{A}$ and $\mathcal{A}_1$, being similar, have the same characteristics value with the same multiplicities. Thus $m \leqslant k_j$ as asserted. ▮

We come now to the question of the dimension of $\mathfrak{X}_j$. We shall prove that $\dim(\mathfrak{X}_j) = k_j$. For this purpose we need

Lemma 1.6.2. *If the matrix $\mathcal{B}$ has $\lambda = 0$ as a characteristic value of multiplicity k, then the same is true for $\mathcal{B}^r$, where r is any positive integer.*

Proof. Let ω be a primitive rth root of unity and consider

$$\lambda^r\mathcal{E} - \mathcal{B}^r = \prod_{j=0}^{r-1} (\lambda\omega^j\mathcal{E} - \mathcal{B}). \tag{1.6.8}$$

Hence, if one of the factors on the right is singular, so is the product. Now

$$\det(\lambda^r\mathcal{E} - \mathcal{B}^r) = \prod_{j=0}^{r-1} \det(\lambda\omega^j\mathcal{E} - \mathcal{B}). \tag{1.6.9}$$

Here the first factor on the right with $j = 0$ is divisible by λ^k and by no higher power of λ. The various factors on the right are simply permuted under the mapping $\lambda \to \omega\lambda$. It follows that each factor is divisible by λ^k and by no higher power of λ. This shows that the left member is divisible by λ^{kr} and

$$\det(\alpha\mathcal{E} - \mathcal{B}^r)$$

is divisible by α^k and by no higher power. Hence $\alpha = 0$ is a characteristic root of $\mathcal{B}^r$ of multiplicity k. ▮

We apply this lemma to the case

$$\mathcal{B} = \mathcal{A} - \lambda_j\mathcal{E}, \qquad r = s_j$$

to get

Theorem 1.6.2. *The dimension of the root space $\mathfrak{X}_j$ equals the multiplicity of the corresponding characteristic root λ_j.*

Proof. We consider the matrix $(\mathcal{A} - \lambda_j \mathcal{E})^{s_j}$. It has the characteristic value 0 with the multiplicity k_j by Lemma 1.6.2 since $\mathcal{A} - \lambda_j \mathcal{E}$ has this property. We now appeal to (1.6.6) which says that all of $\mathfrak{X}_j$ is annihilated by $(\mathcal{A} - \lambda_j \mathcal{E})^{s_j}$. Moreover, it was shown that a vector $\mathbf{x}$, $\mathbf{x} \neq 0$, is annihilated by this matrix iff $\mathbf{x} \in \mathfrak{X}_j$. This shows that $(\mathcal{A} - \lambda_j \mathcal{E})^{s_j}$ admits $\mathfrak{X}_j$ as the characteristic space for the characteristic value zero. By Theorem 1.6.1 the dimension of a characteristic space is at most equal to the multiplicity of the corresponding characteristic value. In the present case this gives $\dim(\mathfrak{X}_j) \leqslant k_j$. Hence, for $j = 1, 2, \ldots, p$,

$$\dim(\mathfrak{X}_j) \leqslant k_j. \tag{1.6.10}$$

Since $\mathfrak{X}_i \cap \mathfrak{X}_j = \{\mathbf{0}\}$ if $i \neq j$,

$$\sum_{j=1}^{p} \dim(\mathfrak{X}_j) = \dim(\mathfrak{X}) = n. \tag{1.6.11}$$

On the other hand, the total number of characteristic values with attention paid to multiplicity is also n so that

$$\sum_{j=1}^{n} k_j = n. \tag{1.6.12}$$

Combining the last three relations we see that equality must hold for each j in (1.6.10). ∎

The rest of this section will be occupied by questions concerning the metrics of C^n and $\mathfrak{E}(C^n)$. Originally we had an orthonormal basis $\{\mathbf{u}_j\}$, and if

$$\mathbf{x} = \sum_{j=1}^{n} x_j \mathbf{u}_j \tag{1.6.13}$$

we defined the Euclidean norm by

$$\|\mathbf{x}\| = \left\{ \sum_{j=1}^{n} |x_j|^2 \right\}^{1/2}. \tag{1.6.14}$$

This is a real number associated with the vector $\mathbf{x}$ having the following properties:

(N_1) $\|\mathbf{x}\| \geqslant 0$ *for all* $\mathbf{x}$*;* $\|\mathbf{x}\| = 0$ *iff* $\mathbf{x} = \mathbf{0}$.

(N_2) $\|\alpha \mathbf{x}\| = |\alpha| \, \|\mathbf{x}\|$, $\alpha \in C$.

(N_3) $\|\mathbf{x} + \mathbf{y}\| \leqslant \|\mathbf{x}\| + \|\mathbf{y}\|$.

For the last property see Problem 1, Exercise 1.2.
Once we had a norm in C^n we could define a distance simply by setting

$$d(\mathbf{x}, \mathbf{y}) = \|\mathbf{x} - \mathbf{y}\|. \tag{1.6.15}$$

This notion of distance satisfies the conditions:

(D_1) $d(\mathbf{x}, \mathbf{y}) \geqslant 0$ *for all* $\mathbf{x}, \mathbf{y}$; $d(\mathbf{x}, \mathbf{y}) = 0$ *iff* $\mathbf{x} = \mathbf{y}$.

(D_2) $d(\mathbf{x}, \mathbf{y}) = d(\mathbf{y}, \mathbf{x})$.

(D_3) $d(\mathbf{x}, \mathbf{y}) \leqslant d(\mathbf{x}, \mathbf{z}) + d(\mathbf{z}, \mathbf{y})$.

The last condition is known as the *triangle inequality* for obvious reasons. See Problem 2, Exercise 1.2.

In investigations of what he called the "geometry of numbers" the German mathematician Hermann Minkowski (1864–1909) found it desirable to introduce alternate metrics in R^n. In particular, he introduced what was later called the l_p-metric. Here one sets

$$\|\mathbf{x}\|_p = \left[\sum_{j=1}^{n} |x_j|^p\right]^{1/p}, \tag{1.6.16}$$

where p is any real number $\geqslant 1$. He also considered the l_∞-norm obtained by letting $p \to \infty$ in (1.6.16) so that

$$\|\mathbf{x}\|_\infty = \max_{1 \leqslant j \leqslant n} |x_j|. \tag{1.6.17}$$

The Euclidean norm is the case $p = 2$.

In addition, Minkowski did two other noteworthy things. He formulated conditions (D_1), (D_2), (D_3) as properties that a notion of distance ought to possess. As a matter of fact, he did not always insist on (D_2), so that the distance from $\mathbf{x}$ to $\mathbf{y}$ could be different from the distance from $\mathbf{y}$ to $\mathbf{x}$.

He also introduced a class of functions from vectors to reals which could serve for defining a distance. In recent years such functionals have become known as *semi-norms*. The two basic properties of a semi-norm are *subadditivity* and *homogeneity*:

$$\rho(\mathbf{x} + \mathbf{y}) \leqslant \rho(\mathbf{x}) + \rho(\mathbf{y}), \tag{1.6.18}$$

$$\rho(\alpha\mathbf{x}) = \alpha\rho(\mathbf{x}). \tag{1.6.19}$$

Minkowski, who was concerned essentially only with R^n, required the second property only for real positive values of α. The distance from $\mathbf{x}$ to $\mathbf{y}$ is then taken as

$$d(\mathbf{x}, \mathbf{y}) = \rho(\mathbf{y} - \mathbf{x}) \tag{1.6.20}$$

and the distance from $\mathbf{y}$ to $\mathbf{x}$ as $\rho(\mathbf{x} - \mathbf{y})$. If $\mathbf{x} \in R^n$ and

$$\rho(-\mathbf{x}) = \rho(\mathbf{x}), \tag{1.6.21}$$

then these two distances are equal, otherwise not.

Minkowski verified that the functions defined by (1.6.16) and (1.6.17) satisfy the stated conditions. The homogeneity is obvious but the subadditivity requires an elaborate argument which becomes simple only for $p = 1$, 2, and ∞. This discussion is postponed until Section 3.1.

In Exercise 1.3 the unit ball in C^n was considered as well as its image under a linear transformation A. In these problems the underlying metric of C^n was the Euclidean one, but the same problems are meaningful and significant for any $p \geqslant 1$. The unit ball changes its shape with p, as the reader will find by considering the situation in R^2 and R^3.

In Problem 3, Exercise 1.3, the reader had to show an inequality which may be written in the form

$$\|A\mathbf{x}\|_2 \leqslant \|\mathcal{A}\|_2 \, \|\mathbf{x}\|_2, \tag{1.6.22}$$

where the Euclidean norm for the corresponding matrix $\mathcal{A}$ is defined by (1.4.42). This is usually a fairly crude estimate, but at least it shows that the norm of the linear transformation A when defined by

$$\|A\|_2 = \sup_{\mathbf{x}} \{\|A\mathbf{x}\|_2; \; \|\mathbf{x}\|_2 \leqslant 1\} \tag{1.6.23}$$

is a finite number depending only upon A. This is the norm of A as an element of the algebra $\mathfrak{E}(C^n)$ in the l_2-metric. Similarly, we can define an l_p-norm

$$\|A\|_p = \sup_{\mathbf{x}} \{\|A\mathbf{x}\|_p; \; \|\mathbf{x}\|_p \leqslant 1\} \tag{1.6.24}$$

which depends only upon A and p. In Section 1.4 it was mentioned that the norm of the linear transformation A could also be used as norm of the corresponding representative matrix $\mathcal{A}$. All we have to do is to replace A by $\mathcal{A}$ in the last two formulas.

EXERCISE 1.6

1. Why is $\Xi_j \neq \{\mathbf{0}\}$?
2. Let $\mathbf{x}_0 \in \mathfrak{X}_j$, $\mathbf{x}_0 \neq \mathbf{0}$, and form $\mathbf{x}_m = (\mathcal{A} - \lambda_j \mathcal{E})^m \mathbf{x}_0$, $m = 1, 2, \ldots, k$. If $\mathbf{x}_k \neq \mathbf{0}$, show that the vectors $\mathbf{x}_0, \mathbf{x}_1, \ldots, \mathbf{x}_k$ are linearly independent vectors of $\mathfrak{X}_j$.
3. It is desired to justify the name minimal equation given to (1.5.45). In Problem 2, Exercise 1.5, the reader proved that if $G(\lambda)$ and $H(\lambda)$ have no zeros in common, then $G(\mathcal{A})$ is regular and hence cannot be the zero matrix. Suppose now that $G(\lambda) = 0$ has the root $\lambda = \lambda_j$ but with a lower multiplicity than for $H(\lambda) = 0$. Show that there is a vector $\mathbf{x}_0$, $\mathbf{x}_0 \neq \mathbf{0}$, $\mathbf{x}_0 \in \mathfrak{X}_j$, such that $G(\mathcal{A})\,\mathbf{x}_0 \neq \mathbf{0}$, so that $G(\mathcal{A})$ cannot be the zero matrix.
4. Use the preceding result to show that $G(\lambda)$ must be divisible by $H(\lambda)$ for $G(\mathcal{A})$ to be the zero matrix.
5. If $\|\mathbf{x}\|_p$ is defined by (1.6.16) and (1.6.17) show that this functional is subadditive in the extreme cases $p = 1$ and $p = \infty$.
6. The shortest distance between two places A and B is in daily speech often given as so and so many hours' ride. Leaving aside the lack of precision in this notion of distance, show that it satisfies (D_1) and (D_3). Is (D_3) necessarily valid?

7. Using the l_p-metric construct the "unit circle" in R^2. How do the curves $\|\mathbf{x}\|_p = 1$ vary when p increases from $p = 1$ to $p = \infty$? What points are common to all these curves?
8. Consider R^n and $p = 1$. Show that the unit ball is bounded by hyperplanes, 2^n in number, and give their equations. Find the minimum of the Euclidean distance of a point P on the l_1-unit sphere from the origin. Is the unit ball a convex set?
9. The unit ball in any l_p-metric is a convex subset of C^n. What properties of the metric do you need for the proof? (Generalization of Problem 3, Exercise 1.3.)
10. Suppose that a functional $\rho(\mathbf{x})$ on C^n to reals satisfies the conditions (1.6.18) and (1.6.19). Suppose that $\rho(\mathbf{x}) \geqslant 0$ and $\rho(\mathbf{x}) = 0$ iff $\mathbf{x} = \mathbf{0}$. Suppose also that $\rho(-\mathbf{x}) = \rho(\mathbf{x})$. Show that $d(\mathbf{x}, \mathbf{y}) = \rho(\mathbf{x} - \mathbf{y})$ defines a distance in C^n satisfying conditions (D_1), (D_2), (D_3).

1.7 HERMITIAN FORMS

In the study of the properties of linear transformations $A \in \mathfrak{E}(C^n)$ valuable information may be obtained from the inner product

$$(A\mathbf{x}, \mathbf{x}). \tag{1.7.1}$$

Let $\{\mathbf{u}_j\}$ be an ordered orthonormal basis of C^n in terms of which

$$\mathbf{x} = \sum_{j=1}^{n} x_j \mathbf{u}_j.$$

There is a matrix $\mathcal{A} \in \mathfrak{M}_n$ such that

$$A\mathbf{x} = \mathcal{A} \cdot x = \sum_{j=1}^{n} \left[\sum_{k=1}^{n} a_{jk} x_k \right] \mathbf{u}_j.$$

We have then

$$(A\mathbf{x}, \mathbf{x}) = \sum_{j=1}^{n} \sum_{k=1}^{n} a_{jk} x_k \bar{x}_j, \tag{1.7.2}$$

$$(\mathbf{x}, A\mathbf{x}) = \sum_{j=1}^{n} \sum_{k=1}^{n} \bar{a}_{jk} \bar{x}_k x_j. \tag{1.7.3}$$

For the case of R^n and a real symmetric matrix $\mathcal{A}$, $\mathcal{A}^t = \mathcal{A}$, we have then

$$(\mathcal{A}\mathbf{x}, \mathbf{x}) = (\mathbf{x}, \mathcal{A}\mathbf{x}) = \sum_{j=1}^{n} \sum_{k=1}^{n} a_{jk} x_j x_k, \tag{1.7.4}$$

where the last member is known as a *quadratic form*. Such forms arose at an early stage in analytic geometry. In R^3 the locus

$$\sum_{j=1}^{3} \sum_{k=1}^{3} a_{jk} x_j x_k = C \tag{1.7.5}$$

is known as a *central quadric surface*, "central" because there is a center, in this case the origin. An important problem in analytic geometry is the *reduction of a general quadric to principal axes* and to the corresponding normal form of the equation of the surface. This reduction involves a discussion of the characteristic roots of associated symmetric matrices and of certain minimax problems. These facts may serve as a motivation for the following discussion, which applies to the more general case where $\mathcal{A} = \mathcal{A}^*$ is a Hermitian matrix. In Chapter 11 we shall encounter extensions to Hilbert space.

Besides quadratic and Hermitian forms we shall also encounter the associated *bilinear polar forms*

$$(\mathcal{A}\mathbf{x}, \mathbf{y}) = \sum_{j=1}^{n} \sum_{k=1}^{n} a_{jk} x_k \bar{y}_j, \tag{1.7.6}$$

$$(\mathbf{x}, \mathcal{A}\mathbf{y}) = \sum_{j=1}^{n} \sum_{k=1}^{n} \bar{a}_{jk} \bar{y}_k x_j. \tag{1.7.7}$$

The name polar form comes from analytic geometry where $(A\mathbf{x}, \mathbf{y}) = C$ is the *polar plane* of the quadric $(A\mathbf{x}, \mathbf{x}) = C$ with respect to the *pole* $\mathbf{y}$. If $\mathbf{y}$ is on the quadric, the equation of the *tangent plane* at $\mathbf{x} = \mathbf{y}$ is obtained. The following theorems play an important role for various uniqueness theorems.

Theorem 1.7.1. *A necessary and sufficient condition that $\mathcal{A}$ be the zero matrix is that $(\mathcal{A}\mathbf{x}, \mathbf{y}) = 0$ for all $\mathbf{x}$ and $\mathbf{y}$ in C^n.*

Proof. The condition is obviously necessary. It is also sufficient for if $(\mathcal{A}\mathbf{x}, \mathbf{y}) = 0$ for all $\mathbf{x}$ and $\mathbf{y}$, we can take $\mathbf{y} = \mathcal{A}\mathbf{x}$ and obtain $(\mathcal{A}\mathbf{x}, \mathcal{A}\mathbf{x}) = 0$ or $\mathcal{A}\mathbf{x} = \mathbf{0}$ for all $\mathbf{x}$. This forces $\mathcal{A}$ to be the zero matrix. ∎

For the next theorem we need an important identity which holds for all complex numbers s and t and all vectors $\mathbf{x}$ and $\mathbf{y}$:

$$s\bar{t}(\mathcal{A}\mathbf{x}, \mathbf{y}) + \bar{s}t(\mathcal{A}\mathbf{y}, \mathbf{x}) = [\mathcal{A}(s\mathbf{x} + t\mathbf{y}), s\mathbf{x} + t\mathbf{y}] - |s|^2(\mathcal{A}\mathbf{x}, \mathbf{x}) - |t|^2(\mathcal{A}\mathbf{y}, \mathbf{y}). \tag{1.7.8}$$

The verification is left to the reader.

Theorem 1.7.2. *A necessary and sufficient condition that $\mathcal{A}$ be the zero matrix in $\mathfrak{M}_n$ is that $(\mathcal{A}\mathbf{x}, \mathbf{x}) = 0$ for all $\mathbf{x}$ in C^n.*

Proof. The necessity is again obvious. If the condition is satisfied, then the right hand side of (1.7.8) is zero for any choice of s, t, $\mathbf{x}$, $\mathbf{y}$. We leave $\mathbf{x}$ and $\mathbf{y}$ arbitrary and take first $s = t = 1$, obtaining

$$(\mathcal{A}\mathbf{x}, \mathbf{y}) + (\mathcal{A}\mathbf{y}, \mathbf{x}) = 0.$$

Next we take $s = i$, $t = 1$, the factor i cancels and we obtain

$$(\mathcal{A}\mathbf{x}, \mathbf{y}) - (\mathcal{A}\mathbf{y}, \mathbf{x}) = 0.$$

Addition gives

$$(\mathcal{A}\mathbf{x}, \mathbf{y}) = 0$$

for all $\mathbf{x}$ and $\mathbf{y}$, and by Theorem 1.7.1 this implies that $\mathcal{A}$ is the zero matrix. ∎

In these theorems we can replace matrices by linear transformations. We may also replace $(A\mathbf{x}, \mathbf{y})$ by $(\mathbf{x}, A\mathbf{y})$, etc.

Suppose now that $A \in \mathfrak{E}(C^n)$. We can then find a transformation $B \in \mathfrak{E}(C^n)$ such that

$$(A\mathbf{x}, \mathbf{x}) = (\mathbf{x}, B\mathbf{x}) \tag{1.7.9}$$

for all $\mathbf{x}$. There is at most one such transformation B, for if there were two, B_1 and B_2, then

$$[\mathbf{x}, (B_1 - B_2)\,\mathbf{x}] = 0$$

for all $\mathbf{x}$ and by the alternative form of Theorem 1.7.2 this implies that $B_1 - B_2$ is the zero operator. To prove that there is at least one B we use the matrix formulation. If $\mathcal{A} = (a_{jk})$, $\mathcal{B} = (b_{jk})$ are corresponding matrices, then (1.7.9) implies the identity

$$\sum_{j=1}^{n} \sum_{k=1}^{n} a_{jk} x_k \bar{x}_j = \sum_{j=1}^{n} \sum_{k=1}^{n} \bar{b}_{jk} \bar{x}_k x_j$$

in the $2n$ variables $x_1, x_2, \ldots, x_n, \bar{x}_1, \bar{x}_2, \ldots, \bar{x}_n$. Hence

$$b_{jk} = \bar{a}_{kj} \tag{1.7.10}$$

so that

$$\mathcal{B} = \mathcal{A}^*. \tag{1.7.11}$$

As a consequence we write

$$B = A^* \tag{1.7.12}$$

and refer to A^* as the *adjoint transformation* of A. If

$$A = A^*$$

we say that A is *self-adjoint* or *Hermitian*. This is the case to be studied in the following. The corresponding form (1.7.2) is known as a *Hermitian form* and the matrix is also Hermitian.

Theorem 1.7.3. *The characteristic roots of a Hermitian matrix are real.*

Proof. Let λ_0 be a characteristic root of a Hermitian matrix $\mathcal{H}$ and $\mathbf{x}_0$, $\|\mathbf{x}_0\| = 1$, an associated characteristic vector. Then

$$(\mathcal{H}\mathbf{x}_0, \mathbf{x}_0) = (\lambda_0 \mathbf{x}_0, \mathbf{x}_0) = \lambda_0(\mathbf{x}_0, \mathbf{x}_0) = \lambda_0.$$

Since $\mathcal{H}$ is Hermitian, the first member equals

$$(\mathbf{x}_0, \mathcal{H}\mathbf{x}_0) = \overline{(\mathcal{H}\mathbf{x}_0, \mathbf{x}_0)},$$

and hence real so that λ_0 is real. ∎

The corresponding Hermitian form is said to be *positive definite*, *indefinite*, or *negative definite* according as the characteristic roots of $\mathcal{H}$ are all positive, some positive and some negative, or all negative, respectively.

Theorem 1.7.4. *Characteristic vectors corresponding to distinct characteristic roots of a Hermitian matrix are orthogonal.*

Proof. Suppose that λ_1 and λ_2 are distinct characteristic roots of the Hermitian matrix $\mathcal{H}$ and that $\mathbf{x}_1$ and $\mathbf{x}_2$ are corresponding normalized characteristic vectors. Since λ_1 and λ_2 are real,

$$\lambda_1(\mathbf{x}_1, \mathbf{x}_2) = (\mathcal{H}\mathbf{x}_1, \mathbf{x}_2) = (\mathbf{x}_1, \mathcal{H}\mathbf{x}_2) = \lambda_2(\mathbf{x}_1, \mathbf{x}_2),$$

so that $(\mathbf{x}_1, \mathbf{x}_2) = 0$ as asserted. ∎

Here we have used the

Lemma 1.7.1. *If $\mathcal{H}$ is Hermitian, then for all $\mathbf{x}$ and $\mathbf{y}$*

$$(\mathcal{H}\mathbf{x}, \mathbf{y}) = (\mathbf{x}, \mathcal{H}\mathbf{y}). \tag{1.7.13}$$

The verification is left to the reader.

Let us get some idea of what operations may be performed in the set of Hermitian matrices in $\mathfrak{M}_n$. We state without proof

Theorem 1.7.5. *The sum of two Hermitian matrices is Hermitian. The scalar product of a real number and a Hermitian matrix is Hermitian. Any positive power of a Hermitian matrix is Hermitian but in general the product of two Hermitian matrices need not be Hermitian.*

The last point is illustrated by the example of

$$\begin{bmatrix} 1 & 2 \\ 2 & 1 \end{bmatrix} \begin{bmatrix} 2 & i \\ -i & 2 \end{bmatrix} = \begin{bmatrix} 2-2i & 4+i \\ 4-i & 2+2i \end{bmatrix}$$

where the factors are Hermitian but the product is not.

Another operation which preserves the Hermitian character of a matrix is transforming it by a unitary matrix. Unitary transformations U and unitary matrices $\mathcal{U}$ were introduced in Exercise 1.3. We recall that distances and inner products are preserved under unitary transformations $(\mathbf{z}, \mathbf{w}) = (U\mathbf{z}, U\mathbf{w})$. Moreover, a unitary transformation has an inverse. With $\mathbf{w} = \mathcal{U}^{-1}\mathbf{z}$ we get

$$(\mathbf{z}, \mathcal{U}^{-1}\mathbf{z}) = (\mathcal{U}\mathbf{z}, \mathcal{U}\mathcal{U}^{-1}\mathbf{z}) = (\mathcal{U}\mathbf{z}, \mathbf{z}) = (\mathbf{z}, \mathcal{U}^*\mathbf{z}).$$

This implies

$$\mathcal{U}^{-1} = \mathcal{U}^* \tag{1.7.14}$$

by Theorem 1.7.2. Transforming a Hermitian matrix $\mathcal{H}$ by a unitary matrix $\mathcal{U}$ we obtain a similar matrix $\mathcal{H}_1$. Here

$$(\mathcal{H}_1)^* = (\mathcal{U}\mathcal{H}\mathcal{U}^{-1})^* = (\mathcal{U}^{-1})^*\mathcal{H}^*\mathcal{U}^* = \mathcal{U}\mathcal{H}\mathcal{U}^{-1} = \mathcal{H}_1.$$

Compare Problem 3, Exercise 1.7. Thus $\mathcal{H}_1$ is also Hermitian. It has the same characteristic values as $\mathcal{H}$. Actually this device may be used to construct a similar Hermitian matrix which is also diagonal.

Theorem 1.7.6. *If H is a Hermitian linear transformation on C^n, there exists an ordered orthonormal basis $\{\mathbf{u}_k\}$ such that*

$$H\mathbf{u}_k = \lambda_k \mathbf{u}_k, \qquad k = 1, 2, \ldots, n, \tag{1.7.15}$$

where $\{\lambda_k\}$ is some arrangement of the characteristic roots of $\mathfrak{H}$, each root repeated as often as its multiplicity indicates.

Proof. We use induction on n. The theorem is obviously true for $n = 1$. Suppose that it holds for $n = m$ and consider a Hermitian linear transformation on C^{m+1}. It has $m + 1$ characteristic values, all real. For simplicity, let λ_1 be the largest such value and let $\mathbf{u}_1$ be the corresponding normalized characteristic vector so that

$$H\mathbf{u}_1 = \lambda_1 \mathbf{u}_1.$$

The vectors in C^{m+1} which are orthogonal to $\mathbf{u}_1$ form a linear subspace $\mathfrak{X}_1$ of dimension m. We note that $H(\mathfrak{X}_1) \subseteq \mathfrak{X}_1$. For if $\mathbf{x}_0 \in \mathfrak{X}_1$ and $\mathbf{x}_0 \neq \mathbf{0}$, then, by (1.7.13),

$$(H\mathbf{x}_0, \mathbf{u}_1) = (\mathbf{x}_0, H\mathbf{u}_1) = (\mathbf{x}_0, \lambda_1 \mathbf{u}_1) = \lambda_1(\mathbf{x}_0, \mathbf{u}_1) = 0,$$

so $H\mathbf{x}_0$ is also orthogonal to $\mathbf{u}_1$ and hence confined to $\mathfrak{X}_1$. We now restrict H to $\mathfrak{X}_1$. It is still Hermitian (why?) and its characteristic roots are $\lambda_2, \lambda_3, \ldots, \lambda_{m+1}$, i.e. the characteristic values of H on C^{m+1} except for λ_1 (why?). By the induction hypothesis we can find an orthonormal basis for $\mathfrak{X}_1$, say $\mathbf{u}_2, \mathbf{u}_3, \ldots, \mathbf{u}_{m+1}$, in terms of which

$$H\mathbf{u}_k = \lambda_k \mathbf{u}_k, \qquad k = 2, 3, \ldots, m + 1.$$

Adjoining $\mathbf{u}_1$ to this system, we obtain an orthonormal basis for C^{m+1} in terms of which (1.7.15) holds. Hence the theorem holds for all n. ∎

Theorem 1.7.7. *For a Hermitian matrix each root space coincides with the corresponding characteristic space.*

Proof. Take a characteristic value λ_0 of multiplicity k and let $\mathfrak{X}_0$ and Ξ_0 be the corresponding root space and characteristic space. Consider the orthonormal basis defined in the preceding theorem. To simplify the notation we assume that $\lambda_j = \lambda_0$ for $j = 1, 2, \ldots, k$. Consider the corresponding elements of the basis $\mathbf{u}_1, \mathbf{u}_2, \ldots, \mathbf{u}_k$. They are all annihilated by the first power of $(H - \lambda_0 I)$ so they belong to Ξ_0 and hence also to $\mathfrak{X}_0$. These elements are linearly independent since they are mutually orthogonal by construction. Hence $\dim(\Xi_0) \geqslant k$ and by Theorem 1.6.2 $\dim(\mathfrak{X}_0) = k$. It follows that $\dim(\Xi_0) = \dim(\mathfrak{X}_0) = k$ and since $\Xi_0 \subseteq \mathfrak{X}_0$ we must have $\Xi_0 = \mathfrak{X}_0$ as asserted. ∎

Corollary 1. *For a Hermitian matrix the associated nilpotents reduce to the zero matrix.*

Proof. In terms of linear transformations rather than matrices we have

$$Q_j = (H - \lambda_j I) P_j.$$

Here P_j leaves the jth root space invariant and annihilates the rest of C^n. Since the root space is also the characteristic space, it is annihilated by $H - \lambda_j I$. It follows that Q_j annihilates all of C^n and is hence the zero operator. ∎

Corollary 2. *The Hermitian matrix* $\mathcal{H}$ *is similar to the diagonal matrix* $(\lambda_j \delta_{jk})$.

Proof. Interpret (1.7.15). ∎

Corollary 3. *If* $\mathcal{H}$ *is Hermitian and the similar matrix* $\mathcal{H}_1 = \mathcal{A}\mathcal{H}\mathcal{A}^{-1}$ *is also Hermitian, then* $\mathcal{A}$ *is unitary.*

Proof. If

$$(\mathcal{A}\mathcal{H}\mathcal{A}^{-1})^* = (\mathcal{A}^{-1})^* \mathcal{H}^* \mathcal{A}^* = \mathcal{A}\mathcal{H}\mathcal{A}^{-1},$$

then $\mathcal{A}^{-1} = \mathcal{A}^*$ and $\mathcal{A}$ is unitary. ∎

Thus *the Hermitian character of a matrix is preserved under a similarity transformation iff the transforming matrix is unitary.*

The rest of this section is devoted to the relations between the Hermitian form $(H\mathbf{x}, \mathbf{x})$ and the spectrum of H. Consider the orthonormal basis $\{\mathbf{u}_k\}$ defined in Theorem 1.7.6. Suppose that

$$\mathbf{x} = c_1\mathbf{u}_1 + c_2\mathbf{u}_2 + \cdots + c_n\mathbf{u}_n, \qquad c_j = (\mathbf{x}, \mathbf{u}_j). \tag{1.7.16}$$

Then

$$H\mathbf{x} = \lambda_1 c_1 \mathbf{u}_1 + \lambda_2 c_2 \mathbf{u}_2 + \cdots + \lambda_n c_n \mathbf{u}_n, \tag{1.7.17}$$

so that

$$(H\mathbf{x}, \mathbf{x}) = \lambda_1 |c_1|^2 + \lambda_2 |c_2|^2 + \cdots + \lambda_n |c_n|^2. \tag{1.7.18}$$

We have

$$\|\mathbf{x}\|^2 = |c_1|^2 + |c_2|^2 + \cdots + |c_n|^2. \tag{1.7.19}$$

Suppose now that the characteristic values are numbered so that

$$\lambda_1 \geqslant \lambda_2 \geqslant \cdots \geqslant \lambda_n.$$

We have then

$$(H\mathbf{x}, \mathbf{x}) \leqslant \lambda_1 \|\mathbf{x}\|^2. \tag{1.7.20}$$

If all the λ's are equal to λ_1, we have equality in (1.7.20). If $\lambda_1 > \lambda_2$, we have equality iff $|c_1| = 1$ and $c_2 = c_3 = \ldots = c_n = 0$. Since $(H\mathbf{x}, \mathbf{x})$ is continuous and bounded on the closed and bounded unit sphere $\|\mathbf{x}\| = 1$, $(H\mathbf{x}, \mathbf{x})$ takes its maximum value at some point. This gives

Theorem 1.7.8. *The maximum value of the Hermitian form* $(H\mathbf{x}, \mathbf{x})$ *on the unit sphere* $\|\mathbf{x}\| = 1$ *equals* λ_1, *the numerically largest of the characteristic values.*

The other characteristic values are also obtainable as solutions of maxima-minima problems, in this case with side conditions.

Theorem 1.7.9. *The maximum value of* $(H\mathbf{x}, \mathbf{x})$ *is* λ_j, *if* $\mathbf{x}$ *is restricted to the set of vectors which satisfy*

$$\|\mathbf{x}\| = 1,\ (\mathbf{x}, \mathbf{u}_1) = (\mathbf{x}, \mathbf{u}_2) = \cdots = (\mathbf{x}, \mathbf{u}_{j-1}) = 0. \tag{1.7.21}$$

Proof. We have $c_k = 0$ for $1 \leqslant k \leqslant j-1$. Hence (1.7.18) reduces to

$$(H\mathbf{x}, \mathbf{x}) = \lambda_j|c_j|^2 + \lambda_{j+1}|c_{j+1}|^2 + \cdots + \lambda_n|c_n|^2 \leqslant \lambda_j.$$

If $\lambda_j > \lambda_{j+1}$, we have $(H\mathbf{x}, \mathbf{x}) = \lambda_j$, iff $|(\mathbf{x}, \mathbf{u}_j)| = 1$ and $(\mathbf{x}, \mathbf{u}_k) = 0$ for $k \neq j$. If $\lambda_j = \lambda_{j+1}$, the maximum is still λ_j, but the maximum is reached on a larger subset of the sphere. ▮

Thus $(H\mathbf{x}, \mathbf{x})$ takes on all spectral values of H for unit vectors. We call attention to the fact that intermediary values are also assumed (why?) so that *the range of* $(H\mathbf{x}, \mathbf{x})$ *is the closed interval bounded by the least and the largest characteristic roots.*

The proofs of the extremal properties of the characteristic values of a Hermitian linear transformation given above presuppose a knowledge of these values and corresponding vectors. Actually it is possible to formulate a *minimax problem* which leads directly to the jth characteristic value.

Theorem 1.7.10. *Consider any set of* $j-1$ *linearly independent vectors* $\mathbf{v}_1, \mathbf{v}_2, \ldots, \mathbf{v}_{j-1}$ *in* C^n. *Let* S *be the set of all vectors* $\mathbf{x}$ *such that*

$$\|\mathbf{x}\| = 1,\ (\mathbf{x}, \mathbf{v}_1) = (\mathbf{x}, \mathbf{v}_2) = \cdots = (\mathbf{x}, \mathbf{v}_{j-1}) = 0, \tag{1.7.22}$$

and let $\lambda(S)$ *denote the maximum of* $(H\mathbf{x}, \mathbf{x})$ *for* $\mathbf{x}$ *in* S. *Then for any admissible set* S *we have* $\lambda(S) \geqslant \lambda_j$.

Proof. We know the existence of a set $\{\mathbf{v}_k\}$ for which $\lambda(S) = \lambda_j$, namely the vectors $\mathbf{u}_1, \mathbf{u}_2, \ldots, \mathbf{u}_{j-1}$ defined by (1.7.15). We shall prove that

$$\min_S \max_{\mathbf{x}\in S} (H\mathbf{x}, \mathbf{x}) = \lambda_j, \tag{1.7.23}$$

which is a stronger assertion than that made in the theorem. To show this, it is enough to show that any admissible set S contains an element $\mathbf{x}$ for which $(H\mathbf{x}, \mathbf{x}) \geqslant \lambda_j$. Since S is a closed bounded set $\max_{\mathbf{x}\in S} (H\mathbf{x}, \mathbf{x})$ exists and will not be less than λ_j for this value is reached for some choice of $\mathbf{x}$. From $\lambda(S) \geqslant \lambda_j$ and Theorem 1.7.9, which asserts that the infimum of $\lambda(S)$ is assumed for at least one choice of S, formula (1.7.23) would follow.

Now take a fixed set of $j-1$ linearly independent vectors $\{\mathbf{v}_k\}$ and consider all vectors of the form

$$\mathbf{x} = \alpha_1\mathbf{u}_1 + \alpha_2\mathbf{u}_2 + \cdots + \alpha_j\mathbf{u}_j, \tag{1.7.24}$$

where the $\mathbf{u}$'s are defined as in Theorem 1.7.6 and the α's are complex numbers. It is possible to choose these numbers so that $\mathbf{x} \in S$. The conditions $(\mathbf{x}, \mathbf{v}_k) = 0$. $k = 1, 2, \ldots, j-1$, lead to the system

$$\alpha_1(\mathbf{u}_1, \mathbf{v}_k) + \alpha_2(\mathbf{u}_2, \mathbf{v}_k) + \cdots + \alpha_j(\mathbf{u}_j, \mathbf{v}_k) = 0,$$

where $k = 1, 2, \ldots, j-1$. This is a set of $j-1$ linear homogeneous equations in j unknowns and as such always has a non-trivial solution. Thus, for any S there is a vector of the form (1.7.24) which is contained in S after normalization. We have now

$$(H\mathbf{x}, \mathbf{x}) = \lambda_1|\alpha_1|^2 + \lambda_2|\alpha_2|^2 + \cdots + \lambda_j|\alpha_j|^2 \geqslant \lambda_j,$$

if, as we may assume, $\lambda_1 \geqslant \lambda_2 \geqslant \ldots \geqslant \lambda_j$, and we use that $\sum_{k=1}^{j} |\alpha_k|^2 = 1$. This proves the assertion. ∎

Since we are dealing with finite forms $(H\mathbf{x}, \mathbf{x})$, bounded and continuous on $\|\mathbf{x}\| = 1$, it makes sense to speak of maxima and minima. We could, however, just as well consider suprema and infima in (1.7.23). This formulation of the problem is easier to carry over to infinite-dimensional situations.

EXERCISE 1.7

1. Verify (1.7.8).
2. Prove Lemma 1.7.1.
3. Prove that $(AB)^* = B^*A^*$.
4. Let $\mathcal{A}$ be a real symmetric matrix, $\mathcal{A} \in \mathfrak{M}_n$, and consider the quadratic form $(\mathcal{A}\mathbf{x}, \mathbf{x})$. Prove the analogues of Theorems 1.7.8 and 1.7.9 by the methods of the Calculus (e.g. by Lagrange's methods of multipliers). *Hint:* In the case of the absolute maximum consider

$$G(x) = \sum_{j=1}^{n} \sum_{k=1}^{n} a_{jk} x_j x_k - \tfrac{1}{2}\lambda \left[\sum_{j=1}^{n} x_j^2 - 1\right].$$

5. If A is a linear transformation, $A \in \mathfrak{E}(C^n]$, show the existence of a unique decomposition $A = H + iK$, where H and K are Hermitian.
6. If H and K are Hermitian and α is a real number, show that $H + K$ and αH are Hermitian.
7. Show that HK is Hermitian iff H and K commute.
8. If H is Hermitian, show that H^p is Hermitian for any positive integer p.
   The preceding three problems cover Theorem 1.7.5.
9. How are the spectra of the Hermitian operators H and H^p related?
10. If the Hermitian operator H is positive definite, show that among the various square roots of H there is one which is positive definite Hermitian. T is a square root of H if $T^2(\mathbf{x}) = H(\mathbf{x})$, $\forall\, \mathbf{x}$.
11. If H is Hermitian, show that its Euclidean operator norm $\|H\| = \max_{\|\mathbf{x}\|=1} \|H\mathbf{x}\|$ equals $\max |\lambda_j|$.

12. Define $\exp(H) = I + \sum_{m=1}^{\infty} 1/m!\, H^m$. Show that the series converges in norm, i.e. the sum of the norms is convergent. Is this an element of $\mathfrak{E}(C^n)$, and if so, why? Show that $\exp(H)$ is Hermitian, if H has this property. Show that the spectrum of $\exp(H)$ is the set $\{\exp(\lambda_j)\}$.
13. Show that if H and K are Hermitian, then HK and KH have the same spectra. [*Hint:* Reduce the discussion to the case where one of the factors is regular.]
14. Is $\exp(H + K)$ necessarily equal to $\exp(H)\exp(K)$? Show that this is true if H and K commute.
15. Prove that the range of $(H\mathbf{x}, \mathbf{x})$ with H Hermitian for $\|\mathbf{x}\| = 1$ is an interval and determine the latter.
16. Let $\lambda_1, \lambda_2, \ldots, \lambda_m$ be the distinct characteristic values of a Hermitian linear transformation $H \in \mathfrak{E}(C^n)$ and let P_j be the projection corresponding to λ_j. Prove the spectral representation
$$H = \lambda_1 P_1 + \lambda_2 P_2 + \cdots + \lambda_m P_m.$$
17. Find the spectral representation of H^k, if H is Hermitian and k is a positive integer.
18. Find the spectral representation of $\exp(H)$, H Hermitian.
19. If H is Hermitian, find the spectral representation of its resolvent. [*Hint:* Remember there are no nilpotents.]

COLLATERAL READING

For a general review of the field consult, for instance:

STOLL, R. R. and E. T. WONG, *Linear Algebra*, Academic Press, New York (1968).

The following books have a stronger slant in the general direction toward which we are aiming:

BACHMAN, G. and L. NARICI, *Functional Analysis*, Chapter 1, Academic Press, New York (1966).

HALMOS, P., *Finite-dimensional Vector Spaces*, 2nd ed., Van Nostrand, New York (1958).

2 LINEAR VECTOR SPACES

In the preceding chapter it was found that the discussion of complex Euclidean spaces leads to a large number of concepts which would seem to have a bearing on more general situations. Among such concepts we list in alphabetic order: algebra, convexity, distance, functional, inner product, invariant subspace, linear transformation, matrix, norm, orthonormal system, projection, resolvent, space, subspace, vector, and vector space. There are others, but this sample will indicate the profusion of ideas that we have to play with. In this chapter we take a first step in this direction. We bring a number of definitions centering around the notion of a *linear vector space*. Many theorems will be stated, some without proofs which are postponed to later chapters. Examples of linear vector spaces are given in Chapters 3 and 4. Analysis in linear vector spaces is discussed in Chapters 7 and 8; Banach algebras in Chapter 9; linear transformations in Chapter 10; inner product spaces in Chapter 11. Omitted proofs will be found in various places in these later chapters.

The present chapter has six sections: Banach spaces; Linear transformations; Linear functionals; Linear operator spaces; Inner product spaces; and Banach algebras.

2.1 BANACH SPACES

In analysis we have often to consider classes of mathematical systems where the elements of the class are important as well as the structures of the system. Such a structure may be algebraic or topological or possibly both, in which case the algebra and the topology must be adjusted to each other. It is customary to refer to such a system as an *abstract space*. We use $\mathfrak{X}$ as symbol for the space; $\mathbf{x}, \mathbf{y}, \ldots$, for the elements which are usually called *points* or *vectors*. We say that $\mathfrak{X}$ *has an algebraic structure if one or more algebraic operations may be performed on the elements or if a notion of order is meaningful.*

At this stage we shall not be concerned with general topological spaces. The only type of topology that will be considered in the first six chapters of this treatise is one based on a notion of *distance*. This leads to a *metric space*.

In the preceding chapter we have encountered a number of abstract spaces with an algebraic as well as a metric structure. We recall the spaces C^n, $\mathfrak{M}_n$, and $\mathfrak{E}(C^n)$. In all three we had notions of vector addition, scalar multiplication and in the latter two cases also multiplication of elements. There were also various notions of distance. On the other hand, so far order between elements has played no role. The notion of *partial ordering* will come in Chapter 5.

We now proceed to some general definitions.

Definition 2.1.1. *An abstract space $\mathfrak{X}$ is called a linear vector space if the operations of addition and scalar multiplication may be performed on the elements of $\mathfrak{X}$ subject to the following postulates:*

(A_1) *Any two elements* $\mathbf{x}$ *and* $\mathbf{y}$ *of* $\mathfrak{X}$, *distinct or not, have a unique sum* $\mathbf{x} + \mathbf{y}$ *which is an element of* $\mathfrak{X}$.

(A_2) *Addition is commutative:* $\mathbf{x} + \mathbf{y} = \mathbf{y} + \mathbf{x}$.

(A_3) *Addition is associative:* $\mathbf{x} + (\mathbf{y} + \mathbf{z}) = (\mathbf{x} + \mathbf{y}) + \mathbf{z}$.

(A_4) *There is a zero element* $\mathbf{0}$ *such that* $\mathbf{x} + \mathbf{0} = \mathbf{0} + \mathbf{x} = \mathbf{x}$ *for all* $\mathbf{x}$.

(A_5) *For every* $\mathbf{x}$ *there is a negative, written* $-\mathbf{x}$, *such that* $\mathbf{x} + (-\mathbf{x}) = \mathbf{0}$.

There is a field F of scalars, normally taken to be either the real field R or the complex field C. Scalar multiplication is subjected to the following postulates:

(S_1) *To every number* $\alpha \in F$ *and every element* $\mathbf{x} \in \mathfrak{X}$ *there is a unique scalar product* $\alpha\mathbf{x}$ *in* $\mathfrak{X}$.

(S_2) $(\alpha + \beta)\,\mathbf{x} = \alpha\mathbf{x} + \beta\mathbf{x}$.

(S_3) $\alpha(\mathbf{x} + \mathbf{y}) = \alpha\mathbf{x} + \alpha\mathbf{y}$.

(S_4) $\alpha(\beta\mathbf{x}) = (\alpha\beta)\,\mathbf{x}$.

(S_5) $1 \cdot \mathbf{x} = \mathbf{x}$.

A "postulate" is an "assumption" or a "rule of the game" or a "working hypothesis." It is an undefined term, not subject to proof. The only requirements that a system of postulates should satisfy are freedom from contradiction and being satisfied by one or more systems of interest to mathematics.

These postulates involve another undefined term, that of *equality*, which in the applications differs from space to space. Whatever the definition is, we assume the following conditions to be satisfied:

(E_1) *Equality is reflexive,* $\mathbf{x} = \mathbf{x}$.

(E_2) *Equality is symmetric,* $\mathbf{x} = \mathbf{y}$ *implies* $\mathbf{y} = \mathbf{x}$.

(E_3) *Equality is transitive,* $\mathbf{x} = \mathbf{y}$ *and* $\mathbf{y} = \mathbf{z}$ *implies* $\mathbf{x} = \mathbf{z}$.

(E_4) *Equality is preserved under addition,* $\mathbf{x} = \mathbf{y}$ *implies* $\mathbf{x} + \mathbf{z} = \mathbf{y} + \mathbf{z}$ *for all* $\mathbf{z}$ *in* $\mathfrak{X}$.

(E_5) *Equality is preserved under scalar multiplication,* $\mathbf{x} = \mathbf{y}$ *implies* $\alpha\mathbf{x} = \alpha\mathbf{y}$ *for all* $\alpha \in F$.

The postulates have a number of implications which will be stated as lemmas.

Lemma 2.1.1. *There is one and only one zero element.*

Proof. If $\mathbf{0}^*$ satisfies (A_4), then $\mathbf{0}^* = \mathbf{0}^* + \mathbf{0} = \mathbf{0}$ and $\mathbf{0}^* = \mathbf{0}$. ▮

Lemma 2.1.2. *Every* $\mathbf{x}$ *has one and only one negative.*

Proof. Suppose that $\mathbf{x} + \mathbf{y} = \mathbf{0}$. Then

$$(-\mathbf{x}) + \mathbf{x} + \mathbf{y} = -\mathbf{x} + \mathbf{0} = -\mathbf{x} \quad \text{or} \quad \mathbf{0} + \mathbf{y} = -\mathbf{x} \quad \text{or} \quad \mathbf{y} = -\mathbf{x}$$

as asserted. Here use is made of (A_4) and (E_4). ▮

Lemma 2.1.3. *The negative of* $-\mathbf{x}$ *is* $\mathbf{x}$.

Proof. $(-\mathbf{x}) + \mathbf{x} = \mathbf{0}$ for all $\mathbf{x}$. This says that $\mathbf{x}$ is the negative of $-\mathbf{x}$ and, since there is a unique negative, the assertion follows. ▮

Lemma 2.1.4. *There is a Law of Cancellation:*

$$\mathbf{x} + \mathbf{z} = \mathbf{y} + \mathbf{z} \quad \textit{implies} \quad \mathbf{x} = \mathbf{y}.$$

Proof. Add $-\mathbf{z}$ to both sides and use (A_4) and (E_4). ▮

Postulates (S_2) and (S_3) state that scalar multiplication and element addition are *distributive operations.* In (S_5) the 1 is the unit element of F.

Lemma 2.1.5. *We have* $0 \cdot \mathbf{x} = \mathbf{0}$ *and* $-\mathbf{x} = (-1)\,\mathbf{x}$ *for all* $\mathbf{x}$.

Proof. Note that $\mathbf{x} = 1 \cdot \mathbf{x} = (1 + 0) \cdot \mathbf{x} = \mathbf{x} + 0 \cdot \mathbf{x}$ so that $0 \cdot \mathbf{x}$ is a zero element. Since the latter is unique, $0 \cdot \mathbf{x} = \mathbf{0}$. Next

$$\mathbf{0} = 0 \cdot \mathbf{x} = [1 + (-1)] \cdot \mathbf{x} = \mathbf{x} + (-1) \cdot \mathbf{x} \quad \text{while} \quad \mathbf{0} = \mathbf{x} + (-\mathbf{x}).$$

Since the negative is unique, $(-1) \cdot \mathbf{x} = -\mathbf{x}$. ▮

We say that $\mathfrak{X}$ is a *real* linear vector space if $F = R$, a *complex* linear vector space if $F = C$.

Let $\mathbf{x}_j \in \mathfrak{X}$, $j = 1, 2, \ldots, n$. These vectors are *linearly independent over* F if

$$\alpha_1 \mathbf{x}_1 + \alpha_2 \mathbf{x}_2 + \cdots + \alpha_n \mathbf{x}_n = \mathbf{0}, \qquad \alpha_j \in F, \tag{2.1.1}$$

implies $\alpha_1 = \alpha_2 = \cdots = \alpha_n = 0$. They are *linearly dependent* if there exist α's not all zero such that (2.1.1) holds. The space $\mathfrak{X}$ is said to be of *dimension* n, if there are n linearly independent vectors in $\mathfrak{X}$ but any system of $n + 1$ vectors is linearly dependent. The space is of *infinite dimension*, if n linearly independent vectors may be found for any n.

We assume the possibility of introducing a *metric* in our space.

Definition 2.1.2. *An abstract space* $\mathfrak{X}$ *is said to be a metric space if for any ordered pair of points* $\mathbf{x}$, $\mathbf{y}$ *a number* $d(\mathbf{x}, \mathbf{y})$ *is defined, called the distance from* $\mathbf{x}$ *to* $\mathbf{y}$ *and satisfying the following conditions:*

(D_1) $d(\mathbf{x}, \mathbf{y}) \geqslant 0$ *for all* $\mathbf{x}$, $\mathbf{y}$ *and* $d(\mathbf{x}, \mathbf{y}) = 0$ *iff* $\mathbf{x} = \mathbf{y}$.

(D_2) $d(\mathbf{x}, \mathbf{y}) = d(\mathbf{y}, \mathbf{x})$.

(D_3) $d(\mathbf{x}, \mathbf{y}) \leqslant d(\mathbf{x}, \mathbf{z}) + d(\mathbf{z}, \mathbf{y})$, *the triangle inequality.*

We recognize the postulates given in Section 1.6. We shall have more to say about metric spaces in Chapter 5. Here it will be necessary to introduce some ideas connected with metric spaces before we turn over to the special case of a normed metric.

In metric spaces many of the concepts of real analysis become meaningful if Euclidean distance is replaced by the notion of distance assumed to hold in the space. Suppose that $\mathbf{x}_0 \in \mathfrak{X}$. Then the set

$$\{\mathbf{x};\ d(\mathbf{x}_0, \mathbf{x}) < \varepsilon, \quad \mathbf{x} \in \mathfrak{X}\} \tag{2.1.2}$$

is called an *ε-neighborhood* of $\mathbf{x}_0$. It is also known as an *open ε-sphere with center at* $\mathbf{x}_0$. Let $S \subset \mathfrak{X}$ be a subset of $\mathfrak{X}$. The point $\mathbf{x}_0 \in \mathfrak{X}$ (but not necessarily to S) is called a *cluster point* of S, if every ε-neighborhood of $\mathbf{x}_0$ contains two distinct points of S. The union of S with its set of cluster points is called the *closure* of S and is denoted by $\bar{S}$. The set S is *closed* if $\bar{S} = S$, *dense* in $\mathfrak{X}$ if $\bar{S} = \mathfrak{X}$. The set S is *open* in $\mathfrak{X}$ if $\mathbf{x}_0 \in S$ implies the existence of an ε-neighborhood of $\mathbf{x}_0$ the points of which are also in S. The set is *nowhere dense* in $\mathfrak{X}$ if every open set in $\mathfrak{X}$ contains an open subset, no points of which are in S.

We also need the notion of a *sequence*

$$\{\mathbf{x}_k;\ \mathbf{x}_k \in \mathfrak{X}, \qquad k = 1, 2, \ldots\}. \tag{2.1.3}$$

The reader will find a discussion of sequences in Section 3.1. In real analysis sequences play a basic role in the discussion of convergence. Since there is a notion of distance in a metric space, we are able to discuss convergence in such spaces. We shall see, however, that the situation is not always so favorable in general metric spaces as it is in R or for that matter in any of the metric spaces considered in Chapter 1.

Definition 2.1.3. *The sequence* (2.1.3) *is said to be convergent if, for each* $\varepsilon > 0$, *there is an integer* $N = N_\varepsilon$ *such that*

$$d(\mathbf{x}_m, \mathbf{x}_n) < \varepsilon \quad \text{for all } m, n > N_\varepsilon. \tag{2.1.4}$$

The sequence converges to $\mathbf{x}_0 \in \mathfrak{X}$, *or* $\lim \mathbf{x}_m = \mathbf{x}_0$, *if for each* $\varepsilon > 0$

$$d(\mathbf{x}_m, \mathbf{x}_0) < \varepsilon \quad \text{for } m > N_\varepsilon. \tag{2.1.5}$$

A sequence satisfying (2.1.4) for each $\varepsilon > 0$ is called a *Cauchy sequence.* The metric space $\mathfrak{X}$ is said to be *complete* if every Cauchy sequence converges to an element of $\mathfrak{X}$.

The spaces R^n and C^n are complete for all n, the space of rational numbers is not, both in terms of the Euclidean metric. See further Section 5.2.

We now restrict ourselves to linear spaces and a metric based on a norm. The latter concept has figured repeatedly in Chapter 1 but the definition is repeated for the sake of completeness.

Definition 2.1.4. *A linear vector space $\mathfrak{X}$ over C (over R) is said to be normed if for each $\mathbf{x} \in \mathfrak{X}$ there is defined a real number $\|\mathbf{x}\|$ called the norm of $\mathbf{x}$ such that*

(N_1) $\|\mathbf{x}\| \geqslant 0$ *and* $\|\mathbf{x}\| = 0$ *iff* $\mathbf{x} = \mathbf{0}$.

(N_2) $\|\alpha\mathbf{x}\| = |\alpha|\,\|\mathbf{x}\|$ *for all* α *in* C *(or* R*).*

(N_3) $\|\mathbf{x} + \mathbf{y}\| \leqslant \|\mathbf{x}\| + \|\mathbf{y}\|$.

The distance in the linear normed space $\mathfrak{X}$ is defined by

$$d(\mathbf{x}, \mathbf{y}) = \|\mathbf{x} - \mathbf{y}\|. \tag{2.1.6}$$

It is left to the reader to verify that this definition of distance satisfies Definition 2.1.2.

In Chapter 1 the spaces C^n, R^n, $\mathfrak{M}_n$, $\mathfrak{E}(C^n)$ were considered and shown to be normed linear vector spaces, complete in their normed metrics.

Definition 2.1.5. *A normed linear vector space which is complete under the normed metric is called a Banach space, B-space for short.*

The name is in honor of the Polish mathematician Stefan Banach (1892–1945), one of the founders of modern functional analysis. He introduced this range of ideas in his dissertation of 1922. Among his forerunners should be mentioned the Frenchman Maurice Fréchet (1878) and the Hungarian Frigyes Riesz (1880–1956).

We give one more definition.

Definition 2.1.6. *A metric space is said to be separable if it contains a countably infinite set of points dense in the space.*

EXERCISE 2.1

1. Prove that C^n is complete in the Euclidean metric.
2. How should the argument be modified for the l_p-metric?
3. Show that C^n is separable in either metric.
4. Why is the space of rational numbers not complete under the Euclidean distance?
5. Verify that (2.1.6) defines a distance.
6. Is $\mathfrak{M}_n$ complete in each of the metrics defined in Section 1.4? Is it separable?

7. On the real line R, can the expression

$$\frac{|x - y|}{1 + |x - y|}$$

be used as a distance between x and y? If so, what is the upper bound of the distances? Can the expression be used as a norm?

8. Prove that a convergent sequence in a B-space is necessarily bounded. Is a bounded sequence always convergent?

9. In a B-space addition and scalar multiplication are continuous operations. Explain in some detail what the assertions mean and give a proof.

10. Prove that in a B-space the norm of $\mathbf{x}$ is a continuous function of $\mathbf{x}$ so that $\lim_{m\to\infty} \mathbf{x}_m = \mathbf{x}_0$ implies $\lim_{m\to\infty} \|\mathbf{x}_m\| = \|\mathbf{x}_0\|$.

11. Prove that the Bolzano–Weierstrass theorem holds in $\mathfrak{M}_n$. In other words, a bounded infinite set of matrices in $\mathfrak{M}_n$ admits of at least one cluster point.

2.2 LINEAR TRANSFORMATIONS

Consider two abstract spaces $\mathfrak{X}$ and $\mathfrak{Y}$ where $\mathfrak{Y}$ may conceivably be a second copy of $\mathfrak{X}$. The *product space* $\mathfrak{X} \times \mathfrak{Y}$ (or *Cartesian product space*) is by definition the set of all ordered pairs $(\mathbf{x}, \mathbf{y})$ with $\mathbf{x} \in \mathfrak{X}$, $\mathbf{y} \in \mathfrak{Y}$.

Definition 2.2.1. *A mapping T from $\mathfrak{X}$ into $\mathfrak{Y}$ is a collection of ordered pairs $(\mathbf{x}, \mathbf{y})$, $\mathbf{x} \in \mathfrak{X}$, $\mathbf{y} \in \mathfrak{Y}$, such that every $\mathbf{x}$ of $\mathfrak{X}$ belongs to one and only one pair $(\mathbf{x}, \mathbf{y})$. Here $\mathbf{y} = T(\mathbf{x})$ is called the image of $\mathbf{x}$ induced by T and $\mathbf{x}$ is the pre-image of $\mathbf{y}$.*

Note that $\mathbf{y}$ may be the image of several points $\mathbf{x}$ and it is not excluded that all of $\mathfrak{X}$ may be mapped on a single point $\mathbf{y}$. Such instances were encountered in Section 1.4.

So far the spaces are arbitrary. We now specialize: $\mathfrak{X}$ and $\mathfrak{Y}$ shall be linear vector spaces and T a linear mapping.

Definition 2.2.2. *Let $\mathfrak{X}$ and $\mathfrak{Y}$ be linear vector spaces over the same scalar field F. T is said to be a linear mapping (equivalently, a linear transformation or a linear operator) from $\mathfrak{X}$ into $\mathfrak{Y}$ if for all $\mathbf{x}_1, \mathbf{x}_2 \in \mathfrak{X}$ and all $\alpha, \beta \in F$*

$$T(\alpha\mathbf{x}_1 + \beta\mathbf{x}_2) = \alpha T(\mathbf{x}_1) + \beta T(\mathbf{x}_2). \tag{2.2.1}$$

Note that here we are dealing with equality in $\mathfrak{Y}$, which may be different from the notion of equality in $\mathfrak{X}$. The same observation applies to the algebraic operations.

If T is a linear transformation, then

$$T(\mathbf{0}) = \mathbf{0}. \tag{2.2.2}$$

Here on the left we have the zero element of $\mathfrak{X}$ and the assertion is that its image is the zero element of $\mathfrak{Y}$. It is not customary to distinguish between zero elements in different spaces by differences in notation. The context shows which zero element is meant. To prove (2.2.2) note that

$$T(\mathbf{0}) = T(\mathbf{0} + \mathbf{0}) = T(\mathbf{0}) + T(\mathbf{0}),$$

which implies (2.2.2).

In general, the zero element of $\mathfrak{X}$ is not the only element of $\mathfrak{X}$ mapped into the zero element of $\mathfrak{Y}$. The set of all such vectors is the *null space* or *kernel* of T. Thus

$$\mathfrak{N}[T] = \{\mathbf{x};\ T(\mathbf{x}) = \mathbf{0}\}. \tag{2.2.3}$$

Just as in C^n, there is a profound difference between the two cases $\mathfrak{N}[T] = \{\mathbf{0}\}$ and $\mathfrak{N}[T] \neq \{\mathbf{0}\}$. This will be taken up below at least for B-spaces.

We now specialize still further: $\mathfrak{X}$ and $\mathfrak{Y}$ shall be B-spaces.

Definition 2.2.3. *If $\mathfrak{X}$ and $\mathfrak{Y}$ are Banach spaces and T is a linear transformation from $\mathfrak{X}$ to $\mathfrak{Y}$, then T is said to be bounded if there exists a constant M such that for all* $\mathbf{x}$

$$\|T(\mathbf{x})\| \leqslant M\|\mathbf{x}\|. \tag{2.2.4}$$

Here again we are dealing with two different versions of the same concept: on the left is the norm as defined in $\mathfrak{Y}$, on the right as defined in $\mathfrak{X}$. If confusion is likely, we mark the norms by affixing suitable subscripts.

If there is an M satisfying (2.2.4), then any larger M will also do. The greatest lower bound of the set of admissible M's is called the *norm* of T, written $\|T\|$, and it may be defined directly as follows:

$$\|T\| = \sup_{\|\mathbf{x}\|=1} \|T(\mathbf{x})\|. \tag{2.2.5}$$

Note that to define the norm it is enough to consider the unit sphere of $\mathfrak{X}$, for any $\mathbf{x}, \mathbf{x} \neq \mathbf{0}$, not on the sphere, may be written $\mathbf{x} = r\mathbf{x}_0$, where $\|\mathbf{x}_0\| = 1$ and $r = \|\mathbf{x}\|$.

Actually $\|T\|$ is the norm of T in still another space, that of all *linear bounded transformations* from $\mathfrak{X}$ to $\mathfrak{Y}$, denoted by $\mathfrak{E}(\mathfrak{X}, \mathfrak{Y})$. If $\mathfrak{Y} = \mathfrak{X}$, we write simply $\mathfrak{E}(\mathfrak{X})$. Here "$\mathfrak{E}$" stands for *endomorphism*. For these operator spaces see further Sections 2.4 and 10.1. There it will be shown that $\mathfrak{E}(\mathfrak{X}, \mathfrak{Y})$ and $\mathfrak{E}(\mathfrak{X})$ are linear vector spaces for a suitable definition of addition and scalar multiplication. Moreover, (2.2.5) defines a norm in $\mathfrak{E}(\mathfrak{X}, \mathfrak{Y})$ and a normed topology in terms of which the space is complete. This applies, in particular, to the space $\mathfrak{E}(\mathfrak{X})$, but here we have also a definition of element multiplication and $\mathfrak{E}(\mathfrak{X})$ is a normed algebra complete in its metric, a so-called *Banach algebra.*

The reader will recall the space $\mathfrak{E}(C^n)$ defined in Section 1.4 as the set of all linear transformations from C^n to itself. In this case "boundedness" is a consequence of "linearity", but in general for infinite dimensional B-spaces $\mathfrak{X}$ this is not the case. All the properties stated for $\mathfrak{E}(\mathfrak{X})$ are obviously true for $\mathfrak{E}(C^n)$, so we know in advance that the proposed discussion is neither vacuous nor lacking in interest.

In the rest of this section we make some observations on individual transformations T in $\mathfrak{E}(\mathfrak{X}, \mathfrak{Y})$. Our first remark is attached to (2.2.3).

Theorem 2.2.1. *If* $\mathfrak{N}[T] = \{\mathbf{0}\}$, *then the linear transformation* T *is* 1–1, *i.e.*

$$\mathbf{x}_1 \neq \mathbf{x}_2 \quad \textit{implies} \quad T(\mathbf{x}_1) \neq T(\mathbf{x}_2). \tag{2.2.6}$$

Proof. If $\mathbf{x}_1 \neq \mathbf{x}_2$ and $T(\mathbf{x}_1) = T(\mathbf{x}_2)$, then

$$T(\mathbf{x}_1 - \mathbf{x}_2) = \mathbf{0},$$

so that $\mathbf{x}_1 - \mathbf{x}_2 \in \mathfrak{N}[T]$ against the assumption. ∎

Let $\mathfrak{R}[T]$ be the range of T, i.e.

$$\mathfrak{R}[T] = \{\mathbf{y};\ \mathbf{y} \in \mathfrak{Y},\ T(\mathbf{x}) = \mathbf{y} \quad \text{for some} \quad \mathbf{x} \in \mathfrak{X}\}. \tag{2.2.7}$$

If $\mathfrak{N}[T] = \{\mathbf{0}\}$, then each $\mathbf{y} \in \mathfrak{R}[T]$ is the image of one and only one $\mathbf{x} \in \mathfrak{X}$. This says that the set of ordered pairs

$$\{(\mathbf{y}, \mathbf{x});\ \mathbf{y} \in \mathfrak{R}[T],\ \mathbf{y} = T(\mathbf{x}),\ \mathbf{x} \in \mathfrak{X}\} \tag{2.2.8}$$

is also a mapping, in this case, of $\mathfrak{R}[T]$ onto $\mathfrak{X}$ and this mapping is also 1–1. We recall that "onto $\mathfrak{X}$" means that every $\mathbf{x} \in \mathfrak{X}$ is second coordinate of one of the pairs $(\mathbf{y}, \mathbf{x})$, in fact of one and only one pair since the mapping is 1–1. We denote this second mapping by T^{-1} and call it the *inverse* of T. We have

$$T^{-1}[T(\mathbf{x})] = \mathbf{x}, \quad \forall\, \mathbf{x} \in \mathfrak{X}, \tag{2.2.9}$$

$$T[T^{-1}(\mathbf{y})] = \mathbf{y}, \quad \forall\, \mathbf{y} \in \mathfrak{R}[T]. \tag{2.2.10}$$

Theorem 2.2.2. *Let* $T \in \mathfrak{E}(\mathfrak{X}, \mathfrak{Y})$ *define a* 1–1 *mapping. Then* T^{-1} *defines a linear mapping of* $\mathfrak{R}[T]$ *onto* $\mathfrak{X}$. *Further, in order that* T^{-1} *belong to* $\mathfrak{E}(\mathfrak{Y}, \mathfrak{X})$ *it is necessary and sufficient that* (1) $\mathfrak{R}[T] = \mathfrak{Y}$ *and* (2) *there exists a positive number* m *such that*

$$\|T(\mathbf{x})\| \geqslant m\|\mathbf{x}\|, \quad \forall\, \mathbf{x}. \tag{2.2.11}$$

Proof. That T^{-1} is linear on $\mathfrak{R}[T]$ is shown as follows. If

$$\mathbf{y}_1 = T(\mathbf{x}_1), \quad \mathbf{y}_2 = T(\mathbf{x}_2), \quad \text{then} \quad \mathbf{x}_1 = T^{-1}(\mathbf{y}_1), \quad \mathbf{x}_2 = T^{-1}(\mathbf{y}_2)$$

and the set $\mathfrak{R}[T]$ is linear by definition so that

$$T(\alpha\mathbf{x}_1 + \beta\mathbf{x}_2) = \alpha\mathbf{y}_1 + \beta\mathbf{y}_2$$

implies that

$$\begin{aligned} T^{-1}(\alpha\mathbf{y}_1 + \beta\mathbf{y}_2) &= \alpha\mathbf{x}_1 + \beta\mathbf{x}_2 \\ &= \alpha T^{-1}(\mathbf{y}_1) + \beta T^{-1}(\mathbf{y}_2) \end{aligned} \tag{2.2.12}$$

as asserted.

The assumption that $T^{-1} \in \mathfrak{E}(\mathfrak{Y}, \mathfrak{X})$ implies that T^{-1} is defined on all of $\mathfrak{Y}$, i.e. $\mathfrak{R}[T] = \mathfrak{Y}$, and T^{-1} is linear and bounded on $\mathfrak{Y}$. Now the boundedness implies the existence of a constant $M > 0$ such that for all $\mathbf{y}$

$$\|\mathbf{x}\| = \|T^{-1}(\mathbf{y})\| \leqslant M\|\mathbf{y}\| = M\|T(\mathbf{x})\|. \tag{2.2.13}$$

This implies (2.2.11) with $m = 1/M$. Thus the conditions are necessary.

In order to prove the sufficiency we note first that (2.2.11) implies that T is 1–1. For

$$\|T(\mathbf{x}_1 - \mathbf{x}_2)\| \geqslant m\|\mathbf{x}_1 - \mathbf{x}_2\|$$

shows that the left member can be zero iff the right member is zero. Thus T^{-1} exists and its domain of definition is $\mathfrak{R}[T] = \mathfrak{Y}$. Then from (2.2.10) and (2.2.12) for all $\mathbf{y}$

$$\|\mathbf{y}\| = \|T[T^{-1}(\mathbf{y})]\| \geqslant m\|T^{-1}(\mathbf{y})\|$$

or

$$\|T^{-1}(\mathbf{y})\| \leqslant (1/m)\|\mathbf{y}\|,$$

so that T^{-1} is bounded. This shows that $T^{-1} \in \mathfrak{E}(\mathfrak{Y}, \mathfrak{X})$. ∎

It should be observed that even if T is 1–1, its range need not be all of $\mathfrak{Y}$ nor is T^{-1} necessarily a bounded transformation on $\mathfrak{R}[T]$. If $\mathfrak{N}[T] \neq \{\mathbf{0}\}$, then we lose the existence of T^{-1} altogether and T is no longer 1–1. We saw in Chapter 1 how this could happen for mappings in $\mathfrak{E}(C^n)$. For the first type of pathology we have to go to linear vector spaces of infinite dimension. Examples will be given later.

EXERCISE 2.2

1. The set $\mathfrak{M}_n$ of all n by n matrices is a B-space (in fact, a B-algebra) under the definitions of algebraic operations and (alternative) norms given in Section 1.4. What is the general form of a transformation in $\mathfrak{E}(\mathfrak{M}_n)$?

2. Find necessary and sufficient conditions for such a transformation to be 1–1.

3. $C[0, 1]$ denotes the space of all functions $t \to f(t)$ defined and continuous for $0 \leqslant t \leqslant 1$. It will be shown in Section 3.2 that this is a B-algebra under the obvious definition of the algebraic operations and with $\|f\| = \sup_{0 \leqslant t \leqslant 1} |f(t)|$. Define

$$T[f](t) = t(1 + t^2)^{-1} f(t).$$

Find $\|T\|$.

4. The transformation T of the preceding problem is obviously linear. Is it 1–1? Find $\mathfrak{N}[T]$ and $\mathfrak{R}[T]$.

5. With the same setting, is T^{-1} bounded on $\mathfrak{R}[T]$?

6. With the same space define $U[f](t) = 1 + tf(t)$. Is this a linear transformation? Could you state some sense in which U could be said to be bounded?

7. Prove that a linear transformation from a B-space $\mathfrak{X}$ to a B-space $\mathfrak{Y}$ is continuous iff it is bounded.

2.3 LINEAR FUNCTIONALS

In this section we shall be concerned with the space $\mathfrak{E}(\mathfrak{X}, C)$, assuming $\mathfrak{X}$ to be a complex B-space. This is the *space of all linear bounded functionals defined on* $\mathfrak{X}$. It is referred to as the *adjoint space* of $\mathfrak{X}$ and commonly denoted by $\mathfrak{X}^*$. For its elements symbols like x^* are used.

We note that $\mathfrak{X}^*$ becomes a linear vector space if addition and scalar multiplication are defined by

$$(x^*_1 + x^*_2)(\mathbf{x}) = x^*_1(\mathbf{x}) + x^*_2(\mathbf{x}), \quad \forall\, \mathbf{x}, \tag{2.3.1}$$

$$(\alpha x^*)(\mathbf{x}) = \alpha x^*(\mathbf{x}), \quad \forall\, \alpha. \tag{2.3.2}$$

Since the norm in C is the absolute value we define

$$\|x^*\| = \sup_{\|\mathbf{x}\|=1} |x^*(\mathbf{x})|. \tag{2.3.3}$$

In this manner $\mathfrak{X}^*$ becomes a normed linear vector space.

Theorem 2.3.1. *$\mathfrak{X}^*$ is complete in the normed metric.*

Proof. Suppose that $\{x^*_k\}$ is a Cauchy sequence in $\mathfrak{X}^*$. Then, for each $\varepsilon > 0$, there is an $N = N_\varepsilon$ such that

$$\|x^*_m - x^*_n\| < \varepsilon \quad \text{for} \quad m, n > N_\varepsilon. \tag{2.3.4}$$

Moreover, by Problem 7, Exercise 2.1, there is a finite M with

$$\|x^*_k\| \leqslant M, \quad \forall\, k. \tag{2.3.5}$$

Then for each $\mathbf{x} \in \mathfrak{X}$

$$\begin{aligned} |x^*_m(\mathbf{x}) - x^*_n(\mathbf{x})| &= |(x^*_n - x^*_m)(\mathbf{x})| \\ &\leqslant \|(x^*_n - x^*_m)\| \; \|\mathbf{x}\| \leqslant \varepsilon\|\mathbf{x}\|. \end{aligned} \tag{2.3.6}$$

Now for each fixed $\mathbf{x}$ the sequence $\{x^*_n(\mathbf{x})\}$ is a sequence of complex numbers and (2.3.6) shows that this is a Cauchy sequence. Since C is complete

$$\lim_{k\to\infty} x^*_k(\mathbf{x}) = f(\mathbf{x}) \tag{2.3.7}$$

exists as a linear complex valued function on $\mathfrak{X}$. Moreover,

$$|f(\mathbf{x})| \leqslant \sup_k \|x^*_k\| \; \|\mathbf{x}\| \leqslant M\|\mathbf{x}\|, \tag{2.3.8}$$

so that $f \in \mathfrak{X}^*$. By (2.3.6)

$$\lim_{m\to\infty} |x^*_m(\mathbf{x}) - x^*_n(\mathbf{x})| = |f(\mathbf{x}) - x^*_n(\mathbf{x})| \leqslant \varepsilon\|\mathbf{x}\|.$$

Now for $\|\mathbf{x}\| = 1$

$$|f(\mathbf{x}) - x^*_n(\mathbf{x})| \leqslant \varepsilon$$

uniformly in $\mathbf{x}$. Since

$$\sup_{\|\mathbf{x}\|=1} |f(\mathbf{x}) - x^*_n(\mathbf{x})| = \|f - x^*_n\| \leqslant \varepsilon$$

by (2.3.3). This shows that

$$\|f - x^*_n\| \leqslant \varepsilon, \quad n > N_\varepsilon,$$

and

$$\lim_{n\to\infty} \|f - x^*_n\| = 0. \tag{2.3.9}$$

Thus every Cauchy sequence converges to a limit in the space and $\mathfrak{X}^*$ is complete and hence a B-space. ▮

In Problem 4, Exercise 1.2, the reader showed that a linear functional on the space C^n is necessarily an inner product. Thus to each $x^* \in \mathfrak{E}(C^n, C)$ there is a $\mathbf{y} \in C^n$ such that

$$x^*(\mathbf{x}) = (\mathbf{x}, \mathbf{y}), \quad \forall\, \mathbf{x}, \quad \text{and} \quad \|x^*\| = \|\mathbf{y}\|. \tag{2.3.10}$$

Since inner products are not defined in general B-spaces, we cannot expect this result to hold except in inner product spaces. Nevertheless, it is instructive to examine some special cases involving functionals with particular properties which, as we shall see later, exist for all B-spaces even if the construction that is used here breaks down.

We note first that $(\mathbf{x}, \mathbf{y}) = 0$ for all $\mathbf{x}$ iff $\mathbf{y} = \mathbf{0}$, i.e. the zero functional is the only element of $\mathfrak{X}^*$ which annihilates all of $\mathfrak{X}$. Next set

$$\mathbf{y} = \frac{\mathbf{x}_0}{\|\mathbf{x}_0\|}, \quad \mathbf{x}_0 \text{ fixed } \neq \mathbf{0}, \quad x^*(\mathbf{x}) = (\mathbf{x}, \mathbf{y}). \tag{2.3.11}$$

Here

$$(\mathbf{x}_0, \mathbf{y}) = \|\mathbf{x}_0\| \quad \text{and} \quad \|\mathbf{x}^*\| = 1.$$

Suppose that $\mathfrak{X}_0$ is a linear proper subspace of C^n of dimension $q < n$. It is desired to find a linear functional on C^n whose null space contains $\mathfrak{X}_0$ and which is of norm 1. We can find an orthonormal basis $\mathbf{u}_1, \mathbf{u}_2, \ldots, \mathbf{u}_q, \mathbf{u}_{q+1}, \ldots, \mathbf{u}_n$ such that the first q vectors form a basis of $\mathfrak{X}_0$. Now set

$$\begin{aligned} &\mathbf{y} = \alpha_1\mathbf{u}_{q+1} + \alpha_2\mathbf{u}_{q+2} + \cdots + \alpha_{n-q}\mathbf{u}_n, \\ &\sum_{j=1}^{n-q} |\alpha_j|^2 = 1, \quad x^*(\mathbf{x}) = (\mathbf{x}, \mathbf{y}). \end{aligned} \tag{2.3.12}$$

Then

$$x^*(\mathbf{x}) = 0, \quad \forall\, \mathbf{x} \in \mathfrak{X}_0 \quad \text{and} \quad \|x^*\| = 1.$$

Thus the null space of x^* contains $\mathfrak{X}_0$ but may conceivably be a larger linear subspace. It cannot be all of $\mathfrak{X}$, however, since, for example, $x^*(\alpha\mathbf{y}) = \alpha \neq 0$ unless $\alpha = 0$.

As the last problem, we show how to construct a linear functional on C^n which distinguishes between two given elements $\mathbf{x}_1$ and $\mathbf{x}_2$ where $\mathbf{x}_1 \neq \mathbf{x}_2$. It is desired to find $x^* \in \mathfrak{E}(C^n, C)$ such that $x^*(\mathbf{x}_1) \neq x^*(\mathbf{x}_2)$. We may assume $\mathbf{x}_1$ and $\mathbf{x}_2$ different from zero since the case where one of them is zero is trivial. If $\mathbf{x}_1$ and $\mathbf{x}_2$ are orthogonal to each other, then the functional $(\mathbf{x}, \mathbf{x}_1)$ will give a solution since $(\mathbf{x}_1, \mathbf{x}_2) = 0$ while $(\mathbf{x}_1, \mathbf{x}_1) = \|\mathbf{x}_1\|^2 > 0$. If $(\mathbf{x}_1, \mathbf{x}_2) \neq 0$ and $\mathbf{x}_1 \neq \gamma\mathbf{x}_2$, then there exists a vector $\mathbf{z}$ orthogonal to $\mathbf{x}_2$ but not to $\mathbf{x}_1$. Hence $(\mathbf{x}, \mathbf{z}) \neq 0$ for $\mathbf{x} = \mathbf{x}_1$ but is equal to zero for $\mathbf{x} = \mathbf{x}_2$. Thus $(\mathbf{x}, \mathbf{z})$ solves the problem in this case. Finally, if $\mathbf{x}_1 = \gamma\mathbf{x}_2$ we have $\gamma \neq 1$ by assumption and $(\mathbf{x}, \mathbf{x}_2)$ has the desired properties.

We have carried out this discussion in C^n in some detail because functionals with such properties exist in all B-spaces. This will be proved in Chapter 10, but we state the result here for future reference.

Theorem 2.3.2. *Every B-space has infinitely many linear bounded functionals. In particular, there exist functionals with one of the following properties:*

1) *If* $\mathbf{x}_0 \in \mathfrak{X}$, $\mathbf{x}_0 \neq \mathbf{0}$, *then there exists an* $x_0^* \in \mathfrak{X}^*$ *such that* $x_0^*(\mathbf{x}_0) = \|\mathbf{x}_0\|$ *and* $\|x_0^*\| = 1$.
2) *If* $\mathfrak{X}_0$ *is a linear subspace of* $\mathfrak{X}$, *not dense in* $\mathfrak{X}$, *then there is an* $x_1^* \in \mathfrak{X}^*$ *such that* $x_1^*(\mathbf{x}) = 0$, $\forall\, \mathbf{x} \in \mathfrak{X}_0$, *and* $\|x_1^*\| = 1$.
3) *If* $\mathbf{x}_1$ *and* $\mathbf{x}_2 \in \mathfrak{X}$, $\mathbf{x}_1 \neq \mathbf{x}_2$, *then there exists an* $x_2^* \in \mathfrak{X}^*$ *with* $x_2^*(\mathbf{x}_1) \neq x_2^*(\mathbf{x}_2)$.

From (3) we conclude that *the only element of* $\mathfrak{X}$ *that annihilates all functionals is the zero element*, while from (1) it follows that *the only functional which annihilates all of* $\mathfrak{X}$ *is the zero functional.* B-spaces have a profusion of linear bounded functionals. Not all linear vector spaces are so richly endowed. In fact, there are some where the zero functional is the only linear bounded functional.

EXERCISE 2.3

1. Determine the linear functionals on the space $\mathfrak{M}_n$.
2. Discuss the validity of Theorem 2.3.2 in this space.
3. Let $C[0, 1]$ be defined as in Exercise 2.2, i.e. as the linear vector space formed by functions $t \to f(t)$ defined and continuous on the closed interval $[0, 1]$ and normed by the sup-norm. Let t_0 be fixed, $0 \leqslant t_0 \leqslant 1$, and define $L[f] = f(t_0)$ for all $f \in C[0, 1]$. Show that this is a linear bounded functional on the space. Find the norm of L.
4. Let $\mathfrak{N}[L]$ denote the null space of this functional. Show that it is a convex subset of $\mathfrak{X} = C[0, 1]$.

5. Let f, g, h be elements of $\mathfrak{X}$ where f and $g \in \mathfrak{N}[L]$ and h is arbitrary. Show that $f + g$ and $fh \in \mathfrak{N}[L]$. In algebraic parlance: $\mathfrak{N}[L]$ *is an ideal of the B-algebra* $C[0, 1]$.
6. Show that $\int_0^1 f(t)\,dt = L_1[f]$ is also a linear bounded functional on $C[0, 1]$. Find the norm of L_1.
7. Characterize $\mathfrak{N}[L_1]$. Is the null space an ideal in the sense of Problem 5 above?
8. Show that $L[f]$ is multiplicative as well as linear, i.e. $L[fg] = L[f]\,L[g]$, and verify that $L_1[f]$ lacks this property.
9. Construct two functionals which distinguish between two given elements f and g of $C[0, 1]$.

2.4 LINEAR OPERATOR SPACES

Let $\mathfrak{X}$ and $\mathfrak{Y}$ be two B-spaces over the complex field. The spaces $\mathfrak{E}(\mathfrak{X}, \mathfrak{Y})$ and $\mathfrak{E}(\mathfrak{X})$ were defined in Section 2.2. A closer examination is now in order.

Here $\mathfrak{E}(\mathfrak{X}, \mathfrak{Y})$ is the set of all linear bounded transformations T from $\mathfrak{X}$ to $\mathfrak{Y}$. We have also defined the norm of T by

$$\|T\| = \sup_{\|\mathbf{x}\|=1} \|T(\mathbf{x})\|. \tag{2.4.1}$$

Our first order of business is to show how the algebraic operations can be defined in $\mathfrak{E}(\mathfrak{X}, \mathfrak{Y})$ so that it becomes a linear vector space. To this end we define sums and scalar multiples by

$$(T_1 + T_2)(\mathbf{x}) = T_1(\mathbf{x}) + T_2(\mathbf{x}), \tag{2.4.2}$$

$$(\alpha T)(\mathbf{x}) = \alpha T(\mathbf{x}). \tag{2.4.3}$$

Here α is arbitrary in C while T, T_1, T_2 are arbitrary in $\mathfrak{E}(\mathfrak{X}, \mathfrak{Y})$. These conventions are natural and involve no difficulties.

Next we observe that

$$\begin{aligned}\|T_1 + T_2\| &= \sup_{\|\mathbf{x}\|=1} \|T_1(\mathbf{x}) + T_2(\mathbf{x})\| \\ &\leqslant \sup_{\|\mathbf{x}\|=1} \|T_1(\mathbf{x})\| + \sup_{\|\mathbf{x}\|=1} \|T_2(\mathbf{x})\| = \|T_1\| + \|T_2\|\end{aligned} \tag{2.4.4}$$

and $\|\alpha T\| = |\alpha|\,\|T\|$. Thus the norm already introduced satisfies conditions N_1 to N_3. Further, we define

$$d(T_1, T_2) = \|T_1 - T_2\|. \tag{2.4.5}$$

We have now to show that $\mathfrak{E}(\mathfrak{X}, \mathfrak{Y})$ is actually a B-space under the normed metric. This has already been shown for the special case $\mathfrak{Y} = C$ and the argument given in Theorem 2.3.1 may be followed step by step.

Theorem 2.4.1. *The space* $\mathfrak{E}(\mathfrak{X}, \mathfrak{Y})$ *is complete under the normed metric.*

Proof. Suppose that $\{T_k\}$ is a Cauchy sequence in $\mathfrak{E}(\mathfrak{X}, \mathfrak{Y})$. Then for each $\varepsilon > 0$ there is an $N = N_\varepsilon$ such that

$$\|T_m - T_n\| \leqslant \varepsilon, \quad m, n > N_\varepsilon, \tag{2.4.6}$$

and for suitable M

$$\|T_k\| \leqslant M, \quad \forall k. \tag{2.4.7}$$

For each $\mathbf{x} \in \mathfrak{X}$

$$\|T_m(\mathbf{x}) - T_n(\mathbf{x})\| = \|(T_m - T_n)(\mathbf{x})\| \leqslant \|T_m - T_n\| \, \|\mathbf{x}\| \leqslant \varepsilon\|\mathbf{x}\|. \tag{2.4.8}$$

This shows that $\{T_k(\mathbf{x})\}$ is a Cauchy sequence in the complete space $\mathfrak{Y}$ for each $\mathbf{x}$. Hence

$$\lim_{k\to\infty} T_k(\mathbf{x}) = T(\mathbf{x}) \tag{2.4.9}$$

exists as an element of $\mathfrak{Y}$. Moreover,

$$\|T(\mathbf{x})\| \leqslant \sup_k \|T_k\| \, \|\mathbf{x}\| \leqslant M\|\mathbf{x}\|. \tag{2.4.10}$$

Thus (2.4.9) defines a mapping from $\mathfrak{X}$ into $\mathfrak{Y}$ which is obviously linear and (2.4.10) shows that it is also bounded. Hence $T \in \mathfrak{E}(\mathfrak{X}, \mathfrak{Y})$. We have further

$$\lim_{m\to\infty} \|T_m(\mathbf{x}) - T_n(\mathbf{x})\| = \|T(\mathbf{x}) - T_n(\mathbf{x})\| \leqslant \varepsilon\|\mathbf{x}\|.$$

Now for $\|\mathbf{x}\| = 1$

$$\|T(\mathbf{x}) - T_n(\mathbf{x})\| \leqslant \varepsilon$$

uniformly in $\mathbf{x}$. Since

$$\sup_{\|\mathbf{x}\|=1} \|T(\mathbf{x}) - T_n(\mathbf{x})\| = \|T - T_n\|,$$

we see that

$$\|T - T_n\| \leqslant \varepsilon, \quad n > N_\varepsilon,$$

and

$$\lim_{n\to\infty} \|T - T_n\| = 0. \tag{2.4.11}$$

Thus every Cauchy sequence in $\mathfrak{E}(\mathfrak{X}, \mathfrak{Y})$ converges to a limit in the space, which is hence complete in the normed metric and a B-space. ∎

If $\mathfrak{Y} = \mathfrak{X}$ we are dealing with the space $\mathfrak{E}(\mathfrak{X})$. Here products of elements may also be defined. For if T_1, $T_2 \in \mathfrak{E}(\mathfrak{X})$, then $T_2(\mathbf{x})$ is defined and is an element of $\mathfrak{X}$ for each $\mathbf{x} \in \mathfrak{X}$. But then $T_1[T_2(\mathbf{x})]$ is also defined and is in $\mathfrak{X}$. Thus $T_1[T_2(\mathbf{x})]$ defines a linear mapping from $\mathfrak{X}$ into itself. Moreover, it is bounded for

$$\sup_{\|\mathbf{x}\|=1} \|T_1[T_2(\mathbf{x})]\| \leqslant \|T_1\| \sup_{\|\mathbf{x}\|=1} \|T_2(\mathbf{x})\| = \|T_1\| \, \|T_2\|. \tag{2.4.12}$$

We write

$$T_1[T_2(\mathbf{x})] = [T_1 T_2](\mathbf{x}) \tag{2.4.13}$$

and define this transformation as the *product of the operators* T_1 *and* T_2 *in this order.*

Note that we operate first with T_2 on $\mathbf{x}$ and then apply T_1 to the result. This type of multiplication is normally *non-commutative*, i.e.

$$T_1 T_2 \neq T_2 T_1. \tag{2.4.14}$$

From (2.4.11) we get

$$\|T_1 T_2\| \leqslant \|T_1\| \, \|T_2\|. \tag{2.4.15}$$

Multiplication is associative

$$(T_1 T_2) T_3 = T_1 (T_2 T_3) \tag{2.4.16}$$

and distributive with respect to addition

$$(T_1 + T_2) T_3 = T_1 T_3 + T_2 T_3, \tag{2.4.17}$$

$$T_1 (T_2 + T_3) = T_1 T_2 + T_1 T_3. \tag{2.4.18}$$

We say that $\mathfrak{E}(\mathfrak{X})$ is a *Banach algebra* in the sense to be defined in Section 2.6. $\mathfrak{E}(\mathfrak{X})$ has a *unit element*, the *identity operator* I with the properties

$$I(\mathbf{x}) = \mathbf{x}, \quad \forall\, \mathbf{x}, \tag{2.4.19}$$

$$IT = TI = T, \quad \forall\, T. \tag{2.4.20}$$

In an algebra some elements have inverses. We say that $S \in \mathfrak{E}(\mathfrak{X})$ is the *inverse* of $T \in \mathfrak{E}(\mathfrak{X})$ if

$$(ST)(\mathbf{x}) = (TS)(\mathbf{x}) = \mathbf{x}, \quad \forall\, \mathbf{x}. \tag{2.4.21}$$

Here S is T^{-1} provided this transformation exists and is in $\mathfrak{E}(\mathfrak{X})$. For the existence of T^{-1} see Theorem 2.2.2. The inverse is unique if it exists.

Let λ be a complex number and consider the operator

$$T_\lambda = \lambda I - T, \tag{2.4.22}$$

where $T \in \mathfrak{E}(\mathfrak{X})$ is fixed. The values of λ fall into two mutually exclusive classes according as T_λ^{-1} exists as an element of $\mathfrak{E}(\mathfrak{X})$ or not. In the first case we say that λ belongs to the *resolvent set* $\rho(T)$ of T, in the second to the *spectrum* $\sigma(T)$. In the first case we write

$$T_\lambda^{-1} \equiv R(\lambda, T) \tag{2.4.23}$$

and call it the *resolvent* of T. The set $\sigma(T)$ is bounded and closed, the set $\rho(T)$ is unbounded, open and need not be connected. The special case $\mathfrak{X} = C^n$ was considered in Section 1.5. Here $\sigma(T)$ or $\sigma(A)$ in our previous notation was a finite point set containing at most n distinct points. Things are evidently much more complicated in the general B-space case. We note, however, that a considerable part of the analytical properties of the resolvent carries over to the general case. See Chapter 9. Here we shall merely carry over Lemma 1.5.1.

Lemma 2.4.1. *The domain $|\lambda| > \|T\|$ in the complex plane is a subset of $\rho(T)$ and for such values of λ*

$$R(\lambda, T) = \lambda^{-1}I + \lambda^{-2}T + \cdots + \lambda^{-n-1}T^n + \cdots \tag{2.4.24}$$

is a valid representation where the series converges in norm.

Proof. Since the powers of T are defined for all n by recurrence the individual terms of the series have a meaning. The partial sums of the series form a Cauchy sequence in $\mathfrak{E}(\mathfrak{X})$ for $|\lambda| > \|T\|$ since the series

$$\Sigma\, |\lambda|^{-n-1}\, \|T\|^n$$

converges for such values. Thus the sum of the series is a well-defined element of $\mathfrak{E}(\mathfrak{X})$. If the series is multiplied by $\lambda I - T$, either on the left or on the right, and the powers of λ are collected, then all terms cancel except the coefficient of λ^0, which is I. Hence

$$(\lambda I - T)\, R(\lambda, T) = R(\lambda, T)(\lambda I - T) = I$$

as asserted. ∎

EXERCISE 2.4

1. Show that $T(\mathbf{x})$ is a continuous function of $\mathbf{x}$ on $\mathfrak{X}$ if $T \in \mathfrak{E}(\mathfrak{X}, \mathfrak{Y})$.
2. Show that addition and scalar multiplication are continuous operations in $\mathfrak{E}(\mathfrak{X}, \mathfrak{Y})$.
3. Show that element multiplication is continuous in $\mathfrak{E}(\mathfrak{X})$.
4. Verify (2.4.16)–(2.4.18).
5. If $T \in \mathfrak{E}(\mathfrak{X})$, how are the positive integral powers of T defined? Verify the law of exponents.
6. If $T \in \mathfrak{E}(\mathfrak{X})$ and T^{-1} exists and is in $\mathfrak{E}(\mathfrak{X})$, show that
$$(T^{-1})^n = (T^n)^{-1}.$$
7. If T_1, T_2, T_1^{-1}, $T_2^{-1} \in \mathfrak{E}(\mathfrak{X})$, show that T_1T_2 has an inverse and
$$(T_1T_2)^{-1} = T_2^{-1}\, T_1^{-1}.$$
8. If T_1, $T_2 \in \mathfrak{E}(\mathfrak{X})$ and T_1T_2 has an inverse in $\mathfrak{E}(\mathfrak{X})$, show that T_1 and T_2 must have inverses in $\mathfrak{E}(\mathfrak{X})$.
9. A bounded linear operator $P \in \mathfrak{E}(\mathfrak{X})$ is called a *projection* if $P^2 = P$. Suppose $P \neq I$ and determine $\sigma(P)$. Find $R(\lambda, P)$. [Generalization of (1.5.37).]
10. Same question for an operator $T \in \mathfrak{E}(\mathfrak{X})$ such that $T^3 = 4T$. [Generalization of Problem 5, Exercise 1.5.]
11. Prove the first resolvent equation for $R(\lambda, T)$. [Generalization of Problem 13, Exercise 1.5.]

12. An operator $T \in \mathfrak{E}\{C[0, 1]\}$ is defined by $T[f](t) = tf(t)$. Find its resolvent and spectrum.
13. Same question for $T[f](t) = t(1 + t^2)^{-1} f(t)$.
14. If (2.4.21) holds for $S, T \in \mathfrak{E}(\mathfrak{X})$, why is S unique?

2.5 INNER PRODUCT SPACES

Let $\mathfrak{X}$ be a linear vector space over the complex numbers. Inner products were introduced for C^n in Section 1.2. Here follows the generalization to more general spaces.

Definition 2.5.1. *$\mathfrak{X}$ is an inner product space (also called a pre-Hilbert space), if, for any ordered pair of vectors $\mathbf{x}, \mathbf{y} \in \mathfrak{X}$, an inner product $(\mathbf{x}, \mathbf{y})$ is defined subject to the following conditions:*

1) $(\mathbf{x}, \mathbf{y})$ *is a complex number.*
2) $(\mathbf{x}, \mathbf{x})$ *is* $\geqslant 0$ *and* $(\mathbf{x}, \mathbf{x}) = 0$ *iff* $\mathbf{x} = \mathbf{0}$.
3) $(\mathbf{y}, \mathbf{x}) = \overline{(\mathbf{x}, \mathbf{y})}$.
4) $(\alpha \mathbf{x}_1 + \beta \mathbf{x}_2, \mathbf{y}) = \alpha(\mathbf{x}_1, \mathbf{y}) + \beta(\mathbf{x}_2, \mathbf{y})$.

Various facts are direct consequences of these postulates. Thus (3) plus (4) gives

$$(\mathbf{x}, \alpha \mathbf{y}_1 + \beta \mathbf{y}_2) = \bar{\alpha}(\mathbf{x}, \mathbf{y}_1) + \bar{\beta}(\mathbf{x}, \mathbf{y}_2). \tag{2.5.1}$$

This relation, together with (4), expresses the *bilinearity* of the inner product.

Since

$$(\mathbf{x}, \mathbf{0}) = (\mathbf{x}, \mathbf{0} + \mathbf{0}) = (\mathbf{x}, \mathbf{0}) + (\mathbf{x}, \mathbf{0}),$$

we see that

$$(\mathbf{x}, \mathbf{0}) = 0, \quad (\mathbf{0}, \mathbf{x}) = 0, \quad \forall\, \mathbf{x}. \tag{2.5.2}$$

Lemma 2.5.1. *If* $(\mathbf{x}, \mathbf{y}) = 0$ *for a fixed* $\mathbf{y}$ *and all* $\mathbf{x}$, *then* $\mathbf{y} = \mathbf{0}$.

Proof. We have, in particular, $(\mathbf{y}, \mathbf{y}) = 0$, so by (2) $\mathbf{y} = \mathbf{0}$. ∎

We have the following extension of Theorem 1.2.1.

Theorem 2.5.1. *For all* $\mathbf{x}, \mathbf{y} \in \mathfrak{X}$

$$|(\mathbf{x}, \mathbf{y})|^2 \leqslant (\mathbf{x}, \mathbf{x})(\mathbf{y}, \mathbf{y}) \tag{2.5.3}$$

with equality iff $\mathbf{y} = \gamma \mathbf{x}$ *for some number* γ.

Proof. We can follow the pattern of Theorem 1.2.1. Let α be a complex number to be disposed of later. Then

$$0 \leqslant (\mathbf{y} - \alpha \mathbf{x}, \mathbf{y} - \alpha \mathbf{x}) = (\mathbf{y}, \mathbf{y}) - \alpha(\mathbf{x}, \mathbf{y}) - \bar{\alpha}(\mathbf{y}, \mathbf{x}) + |\alpha|^2 (\mathbf{x}, \mathbf{x}).$$

We may assume $\mathbf{x} \neq \mathbf{0}$ since otherwise there is nothing to prove. Under this assumption set

$$\alpha = \frac{(\mathbf{y}, \mathbf{x})}{(\mathbf{x}, \mathbf{x})}$$

to obtain

$$0 \leqslant (\mathbf{y}, \mathbf{y}) - \frac{|(\mathbf{y}, \mathbf{x})|^2}{(\mathbf{x}, \mathbf{x})},$$

which is (2.5.3). It is clear that equality holds iff $\mathbf{y}$ is a constant multiple of $\mathbf{x}$. ∎

This inequality is a generalization of the classical inequalities of Cauchy and of Bounyakovsky-Schwarz.

Properties (4) and (2.5.3) imply that for fixed $\mathbf{y}$ the inner product $(\mathbf{x}, \mathbf{y})$ is a linear bounded functional on $\mathfrak{X}$. We introduce a metric in $\mathfrak{X}$ by

Theorem 2.5.2. *The convention*

$$\|\mathbf{x}\| = (\mathbf{x}, \mathbf{x})^{1/2} \tag{2.5.4}$$

defines a norm in $\mathfrak{X}$ and a normed topology by

$$d(\mathbf{x}, \mathbf{y}) = \|\mathbf{x} - \mathbf{y}\|. \tag{2.5.5}$$

Proof. It is clear that postulates (N_1) and (N_2) are satisfied by virtue of properties (2) and (4). To prove (N_3) consider

$$\begin{aligned}\|\mathbf{x} + \mathbf{y}\|^2 = (\mathbf{x} + \mathbf{y}, \mathbf{x} + \mathbf{y}) &= (\mathbf{x}, \mathbf{x}) + (\mathbf{x}, \mathbf{y}) + (\mathbf{y}, \mathbf{x}) + (\mathbf{y}, \mathbf{y}) \\ &\leqslant \|\mathbf{x}\|^2 + 2\,|(\mathbf{x}, \mathbf{y})| + \|\mathbf{y}\|^2 \\ &\leqslant \|\mathbf{x}\|^2 + 2\,\|\mathbf{x}\|\,\|\mathbf{y}\| + \|\mathbf{y}\|^2 \\ &= (\|\mathbf{x}\| + \|\mathbf{y}\|)^2,\end{aligned}$$

and this implies (N_3). ∎

Corollary. *The norm of the linear functional $x^*(\mathbf{x}) = (\mathbf{x}, \mathbf{y})$, where $\mathbf{y}$ is a fixed element of $\mathfrak{X}$, equals $\|\mathbf{y}\|$.*

Proof. Definition of the norm plus (2.5.3). ∎

The verification of the following theorems is left to the reader. The first is known as the *Parallelogram Law*, the second as the *Extended Parallelogram Law*, the third as the *Pythagorean Theorem*. Cf. Problems 4 and 5, Exercise 1.2. The "original" Pythagorean theorem was familiar to the Babylonians a good 1000 years before Pythagoras.

Theorem 2.5.3. *For any two vectors $\mathbf{x}_1, \mathbf{x}_2 \in \mathfrak{X}$*

$$\|\mathbf{x}_1 + \mathbf{x}_2\|^2 + \|\mathbf{x}_1 - \mathbf{x}_2\|^2 = 2\,[\|\mathbf{x}_1\|^2 + \|\mathbf{x}_2\|^2]. \tag{2.5.6}$$

Theorem 2.5.4. *For any n vectors $\mathbf{x}_1, \mathbf{x}_2, \ldots, \mathbf{x}_n$ in $\mathfrak{X}$*

$$\sum_{1 \leq j < k \leq n} \|\mathbf{x}_j - \mathbf{x}_k\|^2 + \left\| \sum_{j=1}^{n} \mathbf{x}_j \right\|^2 = n \sum_{j=1}^{n} \|\mathbf{x}_j\|^2. \tag{2.5.7}$$

Definition 2.5.2. *The vectors $\mathbf{x}$ and $\mathbf{y}$ are orthogonal if*

$$(\mathbf{x}, \mathbf{y}) = 0. \tag{2.5.8}$$

We also say that $\mathbf{x}$ and $\mathbf{y}$ are perpendicular, written $\mathbf{x} \perp \mathbf{y}$.

The Pythagorean theorem is a corollary of Theorem 2.5.3 and the definition of orthogonality.

Theorem 2.5.5. *If $\mathbf{x} \perp \mathbf{y}$, then*

$$\|\mathbf{x} + \mathbf{y}\|^2 = \|\mathbf{x}\|^2 + \|\mathbf{y}\|^2. \tag{2.5.9}$$

Definition 2.5.3. *An inner product space is called a Hilbert space if it is complete in the normed metric.*

We use a German capital $\mathfrak{H}$ as a generic notation for a Hilbert space. The name is in honor of the German mathematician David Hilbert (1861–1943), one of the last giants in our science.

The various spaces C^n are elementary instances of a Hilbert space. More sophisticated examples will be found in Chapters 3 and 4. Since $\mathfrak{H}$ is a linear vector space over the complex numbers, the notions of linear independence of vectors and dimension of the space are meaningful. If $\mathfrak{H}$ is of finite dimension n, then there is a 1–1 correspondence between $\mathfrak{H}$ and C^n which preserves distance and the algebraic operations. The first property is known as an *isometry*, the second as an *isomorphism*. We shall assume here that $\mathfrak{H}$ is infinitely dimensional.

This assumption implies that for every n there is a set of n linearly independent vectors. Without restricting the generality we may assume that in passing from n to $n + 1$ we simply add a vector $\mathbf{v}_{n+1}$ to the set $\mathbf{v}_1, \mathbf{v}_2, \ldots, \mathbf{v}_n$ already obtained. This means that there is an infinite system of vectors $\mathbf{v}_k$ in $\mathfrak{H}$ any n of which are linearly independent over C. From this system we can pass to an *orthonormal system* (OS) $\{\mathbf{u}_k\}$ by the Gram–Schmidt orthogonalization process. See Sections 1.1 and 1.2. The process applies to $\mathfrak{H}$ just as easily as to C^n and we obtain

$$\mathbf{u}_k = c_k \begin{vmatrix} \mathbf{v}_1 & \mathbf{v}_2 & \cdots & \mathbf{v}_k \\ (\mathbf{v}_1, \mathbf{v}_1) & (\mathbf{v}_1, \mathbf{v}_2) & \cdots & (\mathbf{v}_1, \mathbf{v}_k) \\ (\mathbf{v}_2, \mathbf{v}_1) & (\mathbf{v}_2, \mathbf{v}_2) & \cdots & (\mathbf{v}_2, \mathbf{v}_k) \\ \cdots & \cdots & \cdots & \cdots \\ (\mathbf{v}_{k-1}, \mathbf{v}_1) & (\mathbf{v}_{k-1}, \mathbf{v}_2) & \cdots & (\mathbf{v}_{k-1}, \mathbf{v}_k) \end{vmatrix} \tag{2.5.10}$$

where c_k is a real normalization factor. This formula is (1.2.9) in an obvious extension. We have

$$(\mathbf{u}_j, \mathbf{u}_k) = \delta_{jk} \tag{2.5.11}$$

where δ_{jk} is the Kronecker delta.

Suppose now that $\mathbf{x}$ is any vector in $\mathfrak{H}$ and form the numbers

$$\hat{x}_k = (\mathbf{x}, \mathbf{u}_k), \quad k = 1, 2, \ldots . \tag{2.5.12}$$

The circumflex in the left member is known as a "hat" in the vernacular. These numbers are by definition the *Fourier coefficients* of the vector $\mathbf{x}$ with respect to the OS $\{\mathbf{u}_k\}$. With these coefficients we form the corresponding *Fourier series*

$$\sum_{k=1}^{\infty} \hat{x}_k \mathbf{u}_k. \tag{2.5.13}$$

These objects are named after the colorful French mathematician Joseph Fourier (1768–1830), who participated in Napoleon's expedition to Egypt in 1798–99 where he served in the French military government as chief of jurisdiction and secretary of the Egyptian Institute in Cairo. He directed the first archaeological survey of Egypt. After his return to France he was prefect of the department of Isère for 14 years. His most famous work in mathematics is his *Théorie Analytique de la Chaleur* (1822), where a profusion of orthogonal series occur.

Let us return to the series (2.5.13). What is the meaning of the series? We shall show that the series converges, though normally not in norm, but the partial sums form a Cauchy sequence in $\mathfrak{H}$. This raises a further question: do the partial sums converge to the element $\mathbf{x}$ with which the series is associated through (2.5.12)? This does not follow and one of our tasks is to find conditions under which this holds.

We start by proving

Theorem 2.5.6. *The partial sums of* (2.5.13) *form a sequence of best approximation to* $\mathbf{x}$ *in* $\mathfrak{H}$ *in the sense that for any n and any choice of n numbers* $c_1, c_2, \ldots, c_n$

$$\left\| \mathbf{x} - \sum_{k=1}^{n} c_k \mathbf{u}_k \right\|^2 \geqslant \|\mathbf{x}\|^2 - \sum_{k=1}^{n} |\hat{x}_k|^2 \tag{2.5.14}$$

with equality iff $c_k = \hat{x}_k$ *for all k.*

Proof. To simplify the formulas we use c^* for $\bar{c}$, the conjugate of c, in this proof and the following. We have

$$\left\| \mathbf{x} - \sum_{k=1}^{n} c_k \mathbf{u}_k \right\|^2 = \|\mathbf{x}\|^2 - \sum_{k=1}^{n} c_k^* (\mathbf{x}, \mathbf{u}_k) - \sum_{k=1}^{n} c_k (\mathbf{u}_k, \mathbf{x})$$

$$+ \sum_{j=1}^{n} \sum_{k=1}^{n} c_j c_k^* (\mathbf{u}_j, \mathbf{u}_k).$$

This equals

$$\|\mathbf{x}\|^2 - \sum_{k=1}^{n} c_k{}^*\hat{x}_k - \sum_{k=1}^{n} c_k\hat{x}_k{}^* + \sum_{j=1}^{n} |c_j|^2.$$

To this we add and subtract

$$\sum_{k=1}^{n} |\hat{x}_k|^2$$

and obtain

$$\|\mathbf{x}\|^2 - \sum_{k=1}^{n} |\hat{x}_k|^2 + \sum_{k=1}^{n} |c_k - \hat{x}_k|^2.$$

This gives (2.5.14) with equality iff $c_k = \hat{x}_k$ for all k. ∎

Corollary [*Bessel's inequality*]. *We have*

$$\sum_{k=1}^{\infty} |\hat{x}_k|^2 \leqslant \|\mathbf{x}\|^2. \tag{2.5.15}$$

This is named after the German astronomer Friedrich Wilhelm Bessel (1784–1846), who noted a special case involving trigonometric Fourier series. He also gave his name to the Bessel functions.

Theorem 2.5.7. *The partial sums of the series* (2.5.13) *form a Cauchy sequence.*

Proof. For $1 \leqslant m < n < \infty$ we have the identity

$$\left\| \sum_{k=m}^{n} \hat{x}_k \mathbf{u}_k \right\|^2 = \sum_{k=m}^{n} |\hat{x}_k|^2. \tag{2.5.16}$$

Here the right member goes to zero as $m \to \infty$ by virtue of (2.5.15) and the assertion follows. ∎

We write temporarily $\tilde{\mathbf{x}}$ for the limit of the Cauchy sequence. The problem is now to find the relation between $\mathbf{x}$ and $\tilde{\mathbf{x}}$. The solution is furnished by the following theorem.

Theorem 2.5.8. *The following assertions are equivalent:*

1) *The set of all finite linear combinations of the* $\mathbf{u}_k$*'s is dense in* $\mathfrak{H}$.
2) $\sum_{k=1}^{\infty} |\hat{x}_k|^2 = \|\mathbf{x}\|^2$ *holds for all* $\mathbf{x}$.
3) $(\mathbf{x}, \mathbf{y}) = \sum_{k=1}^{\infty} \hat{x}_k (\hat{y}_k)^*$ *holds for all* $\mathbf{x}$ *and* $\mathbf{y}$.
4) $\hat{x}_k = 0$ *for all* k *iff* $\mathbf{x} = \mathbf{0}$.
5) $\tilde{\mathbf{x}} = \mathbf{x}$ *for all* $\mathbf{x}$.

Proof. We plan to show that (1) ⇒ (2) ⇒ (3) ⇒ (4) ⇒ (5) ⇒ (1). Here the symbol ⇒ is read "implies."

(1) ⇒ (2). If (1) holds and $\varepsilon > 0$ is given, then there is an n and n numbers

$c_1, c_2, \ldots, c_n$ can be found such that

$$\left\| \mathbf{x} - \sum_{k=1}^{n} c_k \mathbf{u}_k \right\|^2 < \varepsilon^2.$$

By (2.5.14) the left member is at most equal to

$$\|\mathbf{x}\|^2 - \sum_{k=1}^{n} |\hat{x}_k|^2,$$

so that the difference is at most equal to ε^2. Here ε is arbitrary and n goes to infinity when ε goes to zero. In other words, equality holds in (2.5.15) and this is the assertion of (2), known as *Parseval's identity*.

(2) $\Rightarrow$ (3). If α is an arbitrary complex number and $\mathbf{x}$, $\mathbf{y}$ two arbitrary vectors in $\mathfrak{H}$, then (2) implies that

$$\|\mathbf{x} + \alpha\mathbf{y}\|^2 = \sum_{k=1}^{\infty} |\hat{x}_k + \alpha\hat{y}_k|^2$$

with obvious notation. Here the left member equals

$$\|\mathbf{x}\|^2 + \alpha(\mathbf{y}, \mathbf{x}) + \alpha^*(\mathbf{x}, \mathbf{y}) + |\alpha|^2 \|\mathbf{y}\|^2$$

while the right member equals

$$\sum_{k=1}^{\infty} |\hat{x}_k|^2 + \alpha \sum_{k=1}^{\infty} (\hat{x}_k)^* \hat{y}_k + \alpha^* \sum_{k=1}^{\infty} \hat{x}_k(\hat{y}_k)^* + |\alpha|^2 \sum_{k=1}^{\infty} |\hat{y}_k|^2.$$

These two expressions are equal for all values of α, which means that the coefficients on both sides of the identity must be equal. Equating the coefficients of α^* we get (3).

(3) $\Rightarrow$ (4). If for some $\mathbf{x} \in \mathfrak{H}$ all Fourier coefficients $\hat{x}_k$ are 0, then by (3) the inner product $(\mathbf{x}, \mathbf{y}) = 0$ for all $\mathbf{y}$ and by Lemma 2.5.1 this implies $\mathbf{x} = \mathbf{0}$, which is (4).

(4) $\Rightarrow$ (5). Let $1 \leqslant m < n$ and consider

$$\left(\sum_{k=1}^{n} \hat{x}_k \mathbf{u}_k, \mathbf{u}_m \right) = \sum_{k=1}^{n} \hat{x}_k (\mathbf{u}_k, \mathbf{u}_m) = \hat{x}_m.$$

Now let $n \to \infty$. The limit of the first member is

$$(\tilde{\mathbf{x}}, \mathbf{u}_m),$$

(why?), so that $\tilde{\mathbf{x}}$ and $\mathbf{x}$ have the same Fourier coefficients. This means that all the Fourier coefficients of $\tilde{\mathbf{x}} - \mathbf{x}$ are zero. By (4) this requires that $\tilde{\mathbf{x}} - \mathbf{x} = \mathbf{0}$, which is (5).

(5) $\Rightarrow$ (1). Let $\mathbf{x} \in \mathfrak{H}$ be arbitrary and suppose that $\tilde{\mathbf{x}} = \mathbf{x}$. This implies

$$\lim_{n \to \infty} \left\| x - \sum_{k=1}^{n} \hat{x}_k \mathbf{u}_k \right\| = \|\mathbf{x} - \tilde{\mathbf{x}}\| = 0.$$

The partial sums of the Fourier series of $\mathbf{x}$ are finite linear combinations of the $\mathbf{u}_k$'s converging to the given element $\mathbf{x}$. Since $\mathbf{x}$ is arbitrary, this says that every element $\mathbf{x}$ of $\mathfrak{H}$ can be approximated arbitrarily closely by linear combinations of the $\mathbf{u}_k$'s, i.e. (1) holds. ∎

Another way of formulating these results is to say that the *OS can serve as a basis for* $\mathfrak{H}$ *iff the system is maximal.* Here the statement that OS is a *basis* for $\mathfrak{H}$ means (5): for every $\mathbf{x} \in \mathfrak{H}$

$$\mathbf{x} = \sum_{k=1}^{\infty} \hat{x}_k \mathbf{u}_k, \tag{2.5.17}$$

where the series converges in the sense of the metric. This is occasionally still referred to as *mean square convergence*. On the other hand, the statement that OS is *maximal* or *complete* means that OS cannot be a proper subset of another orthonormal system. For this would violate (4) since it would imply the existence of a non-zero vector orthogonal to all vectors of OS.

The next notion to be studied is that of *linear closed subspaces* of $\mathfrak{H}$ and their *orthogonal complements*. Given a set S, finite or not, of vectors in $\mathfrak{H}$, there is a closed linear subspace $\mathfrak{M} \subset \mathfrak{H}$ *spanned* or *generated* by S. By definition $\mathfrak{M}$ contains all finite linear combinations of vectors in S as well as all limit points of Cauchy sequences whose elements are such linear combinations. $\mathfrak{M}$ is the smallest closed linear subspace of $\mathfrak{H}$ with these properties.

Now let $\mathfrak{M}$ be a closed linear subspace of $\mathfrak{H}$ and $\mathfrak{M} \neq \mathfrak{H}$. Let $\mathbf{x}$ be a point of $\mathfrak{H}$ not in $\mathfrak{M}$. The distance $d(\mathbf{x}, \mathfrak{M})$ of $\mathbf{x}$ from $\mathfrak{M}$ is by definition $\inf \|\mathbf{y} - \mathbf{x}\| \equiv \delta$, where $\mathbf{y}$ ranges over $\mathfrak{M}$. Here $\delta > 0$ for if $\delta = 0$, then $\mathbf{x}$ would be limit point of elements in $\mathfrak{M}$ and hence in $\mathfrak{M}$ since $\mathfrak{M}$ is closed. Geometric intuition is not very reliable in a space of infinite dimension, but it does suggest that there is a point $\mathbf{z}_0 \in \mathfrak{M}$ such that $\|\mathbf{x} - \mathbf{z}_0\| = \delta$. Moreover, the vector $\mathbf{x} - \mathbf{z}_0$ should be perpendicular to $\mathbf{z}_0$. This will now be proved.

Theorem 2.5.9. *If* $\mathfrak{M}$ *is a closed linear proper subspace of* $\mathfrak{H}$ *and if* $\mathbf{x}$ *is a point of* $\mathfrak{H}$ *not in* $\mathfrak{M}$*, then there exists a unique point* $\mathbf{z}_0 \in \mathfrak{M}$ *such that* $d(\mathbf{x}, \mathfrak{M}) = \|\mathbf{x} - \mathbf{z}_0\|$. *The vector* $\mathbf{x} - z_0$ *is orthogonal to all of* $\mathfrak{M}$.

Proof. We have remarked that $d(\mathbf{x}, \mathfrak{M}) = \delta > 0$. Then there exists a sequence of points $\{\mathbf{z}_n\} \in \mathfrak{M}$ such that

$$\|\mathbf{x} - \mathbf{z}_n\| \to \delta$$

by the definition of the infimum. It is to be shown that $\{\mathbf{z}_n\}$ is a Cauchy sequence. To this end consider the vectors $\mathbf{x} - \mathbf{z}_m$ and $\mathbf{x} - \mathbf{z}_n$. Their sum is $2\mathbf{x} - \mathbf{z}_m - \mathbf{z}_n$, their difference $\mathbf{z}_n - \mathbf{z}_m$. By the Parallelogram Law (Theorem 2.5.3)

$$\|\mathbf{z}_n - \mathbf{z}_m\|^2 + \|2\mathbf{x} - \mathbf{z}_m - \mathbf{z}_n\|^2 = 2[\|\mathbf{x} - \mathbf{z}_m\|^2 + \|\mathbf{x} - \mathbf{z}_n\|^2],$$

and this may be written

$$\|\mathbf{z}_n - \mathbf{z}_m\|^2 = 2\|\mathbf{x} - \mathbf{z}_m\|^2 + 2\|\mathbf{x} - \mathbf{z}_n\|^2 - 4\|\mathbf{x} - \tfrac{1}{2}(\mathbf{z}_m + \mathbf{z}_n)\|^2.$$

Since $\mathfrak{M}$ is linear and convex, $\frac{1}{2}(\mathbf{z}_m + \mathbf{z}_n) \in \mathfrak{M}$. In the last display the first term on the right goes to the limit $2\delta^2$ and so does the second term, while the third may not have a limit as m and $n \to \infty$ but in any case stays $\leqslant -4\delta^2$. It follows that the superior limit of the right member is $\leqslant 0$ and this requires that it has the limit zero. Hence $\{\mathbf{z}_k\}$ is a Cauchy sequence and we denote its limit by $\mathbf{z}_0$. It belongs to $\mathfrak{M}$ since $\mathfrak{M}$ is closed. Further,

$$\|\mathbf{x} - \mathbf{z}_0\| = \lim_{n\to\infty} \|\mathbf{x} - \mathbf{z}_n\| = \delta \tag{2.5.18}$$

by the continuity of the norm. Thus $\mathbf{z}_0$ is a point in $\mathfrak{M}$ where $d(\mathbf{x}, \mathbf{y})$ reaches its minimum for $\mathbf{y} \in \mathfrak{M}$.

There is one and only one such point. For if there were two, $\mathbf{z}_0$ and $\mathbf{w}_0$, both at the distance δ from $\mathbf{x}$, then we consider the vectors $\mathbf{x} - \mathbf{z}_0$ and $\mathbf{x} - \mathbf{w}_0$ and use the Parallelogram Law again to obtain

$$\begin{aligned} 4\delta^2 &= 2\|\mathbf{x} - \mathbf{z}_0\|^2 + 2\|\mathbf{x} - \mathbf{w}_0\|^2 \\ &= \|\mathbf{z}_0 - \mathbf{w}_0\|^2 + 4\|\mathbf{x} - \tfrac{1}{2}(\mathbf{z}_0 + \mathbf{w}_0)\|^2 \geqslant \|\mathbf{z}_0 - \mathbf{w}_0\|^2 + 4\delta^2 \end{aligned}$$

or $\|\mathbf{z}_0 - \mathbf{w}_0\|^2 \leqslant 0$. Here equality must hold so that $\mathbf{w}_0 = \mathbf{z}_0$ and the point of $\mathfrak{M}$ nearest to $\mathbf{x}$ is unique.

To prove the orthogonality of $\mathbf{x} - \mathbf{z}_0$ to $\mathfrak{M}$, we consider any point $\mathbf{y}$ of $\mathfrak{M}$ and any complex number α. Form

$$\|\mathbf{z}_0 + \alpha\mathbf{y} - \mathbf{x}\|^2 = \|\mathbf{z}_0 - \mathbf{x}\|^2 + \alpha^*(\mathbf{z}_0 - \mathbf{x}, \mathbf{y}) + \alpha(\mathbf{y}, \mathbf{z}_0 - \mathbf{x}) + |\alpha|^2 \|\mathbf{y}\|^2.$$

The term on the left is $\geqslant \delta^2$ since $\mathbf{z}_0 + \alpha\mathbf{y} \in \mathfrak{M}$ and the first term on the right equals δ^2. It follows that

$$2\mathfrak{R}[\alpha^*(\mathbf{z}_0 - \mathbf{x}, \mathbf{y})] + |\alpha|^2 \|\mathbf{y}\|^2 \geqslant 0, \tag{2.5.19}$$

where $\mathfrak{R}$ denotes "the real part of." Here there are two possibilities: $(\mathbf{z}_0 - \mathbf{x}, \mathbf{y}) = 0$ or is $\neq 0$. In the first case we are through. In the second case we note that the real part may be positive or negative and the sign is at our disposal since α is arbitrary. Moreover, making $|\alpha|$ small we can make the second term in (2.5.19) small in comparison with the first. This implies that the left member of (2.5.19) is capable of taking on negative values. Thus the second alternative leads to a contradiction and the first one must hold. ∎

Consider again the closed linear subspace $\mathfrak{M} \neq \mathfrak{H}$. If $\mathbf{x}$ is a point of $\mathfrak{H} \ominus \mathfrak{M}$, then, as we have seen, there is a unique point $\mathbf{z}_0$ of $\mathfrak{M}$ nearest to $\mathbf{x}$ and $\mathbf{z}_0 - \mathbf{x}$ is orthogonal to all of $\mathfrak{M}$. Since $d(\mathbf{x}, \mathfrak{M}) = \delta > 0$, there are points $\mathbf{y} \in \mathfrak{H} \ominus \mathfrak{M}$ in a δ-neighborhood of $\mathbf{x}$. Each such point $\mathbf{y}$ has a nearest point $\mathbf{w}_0 \in \mathfrak{M}$ and $\mathbf{w}_0 - \mathbf{y}$ is orthogonal to all of $\mathfrak{M}$. Normally $\mathbf{z}_0 - \mathbf{x}$ and $\mathbf{w}_0 - \mathbf{y}$ are linearly independent.

Denote the set of vectors orthogonal to $\mathfrak{M}$ by $\mathfrak{M}^\perp$, colloquially if not elegantly known as "$\mathfrak{M}$-perp." This is a linear closed subspace of $\mathfrak{H}$. For if $\mathbf{v}_1$ and $\mathbf{v}_2$ are orthogonal to $\mathfrak{M}$, so is $\alpha\mathbf{v}_1 + \beta\mathbf{v}_2$ by the linearity of the inner product. Further, if $\{\mathbf{v}_n\}$ is a Cauchy sequence in $\mathfrak{M}^\perp$ which converges to $\mathbf{v}_0$, then

$$0 = (\mathbf{z}, \mathbf{v}_n) \quad \text{implies} \quad (\mathbf{z}, \mathbf{v}_0) = 0$$

by the continuity of the inner product which is implied by the continuity of the norm. Thus $\mathbf{v}_0 \in M^\perp$.

It is clear that $\mathfrak{M} \cap \mathfrak{M}^\perp = \{\mathbf{0}\}$. Further,

$$\mathfrak{M}_1 \subset \mathfrak{M}_2 \Rightarrow \mathfrak{M}_2^\perp \subset \mathfrak{M}_1^\perp. \tag{2.5.20}$$

Somewhat less obvious is

$$(\mathfrak{M}^\perp)^\perp = \mathfrak{M}. \tag{2.5.21}$$

The proof is left to the reader.

If $\mathfrak{A}$ and $\mathfrak{B}$ are two linear subspaces of a linear vector space $\mathfrak{X}$ and $\mathfrak{A} \cap \mathfrak{B} = \{\mathbf{0}\}$, then the set

$$\mathfrak{V} = \{\mathbf{x} + \mathbf{y};\, \mathbf{x} \in \mathfrak{A},\, \mathbf{y} \in \mathfrak{B}\} \tag{2.5.22}$$

is known as the *vector sum* of $\mathfrak{A}$ and $\mathfrak{B}$.

Theorem 2.5.10. *If $\mathfrak{M}$ is a closed linear subspace of $\mathfrak{H}$ and $\mathfrak{M}^\perp$ is the set of all vectors of $\mathfrak{H}$ orthogonal to $\mathfrak{M}$, then the vector sum of $\mathfrak{M}$ and $\mathfrak{M}^\perp$ is all of $\mathfrak{H}$. If $\mathbf{w}$ is any vector of $\mathfrak{H}$ there exists a unique representation of $\mathbf{w}$ of the form*

$$\mathbf{w} = \mathbf{u} + \mathbf{v}, \qquad \mathbf{u} \in \mathfrak{M}, \qquad \mathbf{v} \in \mathfrak{M}^\perp. \tag{2.5.23}$$

Proof. Denote temporarily the vector sum of $\mathfrak{M}$ and $\mathfrak{M}^\perp$ by $\mathfrak{V}$. It is clear that $\mathfrak{V}$ is a linear subspace of $\mathfrak{H}$. It is also closed. For suppose that

$$\mathbf{w}_n = \mathbf{u}_n + \mathbf{v}_n, \qquad \mathbf{u}_n \in \mathfrak{M}, \qquad \mathbf{v}_n \in \mathfrak{M}^\perp$$

and $\{\mathbf{w}_n\}$ is a Cauchy sequence in $\mathfrak{V}$. Then

$$\begin{aligned}(\mathbf{u}_m &+ \mathbf{v}_m - \mathbf{u}_n - \mathbf{v}_n,\, \mathbf{u}_m + \mathbf{v}_m - \mathbf{u}_n - \mathbf{v}_n)\\ &= \|\mathbf{u}_m - \mathbf{u}_n\|^2 + (\mathbf{u}_m - \mathbf{u}_n, \mathbf{v}_m - \mathbf{v}_n) + (\mathbf{v}_m - \mathbf{v}_n, \mathbf{u}_m - \mathbf{u}_n) + \|\mathbf{v}_m - \mathbf{v}_n\|^2.\end{aligned}$$

Here $\mathbf{u}_m - \mathbf{u}_n \in \mathfrak{M}$, $\mathbf{v}_m - \mathbf{v}_n \in \mathfrak{M}^\perp$ and they are orthogonal. By the Pythagorean theorem

$$\|(\mathbf{u}_m + \mathbf{v}_m) - (\mathbf{u}_n + \mathbf{v}_n)\|^2 = \|\mathbf{u}_m - \mathbf{u}_n\|^2 + \|\mathbf{v}_m - \mathbf{v}_n\|^2.$$

If the left member converges to zero as m and $n \to \infty$, each of the terms on the right converges to zero. Hence $\{\mathbf{u}_k\}$ and $\{\mathbf{v}_k\}$ are Cauchy sequences. If they converge to $\mathbf{u}_0$ and $\mathbf{v}_0$, respectively, then $\mathbf{u}_k + \mathbf{v}_k \to \mathbf{u}_0 + \mathbf{v}_0 \in \mathfrak{V}$ by the continuity of vector addition. It follows that the vector sum $\mathfrak{V}$ is a closed linear subspace of $\mathfrak{H}$.

Hence $\mathfrak{V}^\perp$ is well defined. Now if $\mathbf{z} \in \mathfrak{V}^\perp$, then $\mathbf{z}$ is orthogonal to all of $\mathfrak{M}$ as well as to all of $\mathfrak{M}^\perp$. For $\mathbf{z}$ is orthogonal to every vector of the form

$$\mathbf{x} + \mathbf{y}, \qquad \mathbf{x} \in \mathfrak{M}, \qquad \mathbf{y} \in \mathfrak{M}^\perp.$$

In particular, for $\mathbf{y} = \mathbf{0}$ we see that $(\mathbf{x}, \mathbf{z}) = 0$, $\forall\, \mathbf{x} \in \mathfrak{M}$. Similarly for $\mathbf{x} = \mathbf{0}$, $(\mathbf{y}, \mathbf{z}) = 0$, $\forall\, \mathbf{y} \in \mathfrak{M}^\perp$. But if $\mathbf{z}$ is orthogonal to all of $\mathfrak{M}$, then $\mathbf{z} \in \mathfrak{M}^\perp$ and if $\mathbf{z}$ is orthogonal to all of $\mathfrak{M}^\perp$, then $\mathbf{z} \in \mathfrak{M}$ by (2.5.21). Since $\mathfrak{M} \cap \mathfrak{M}^\perp = \{\mathbf{0}\}$ we have $\mathbf{z} = \mathbf{0}$ and $\mathfrak{V}^\perp = \{\mathbf{0}\}$. Hence $\mathfrak{V} = \mathfrak{H}$.

This asserts that every element $\mathbf{w}$ of $\mathfrak{H}$ admits at least one representation of the form (2.5.23). But if one such representation exists, it is clearly unique. For if

$$\mathbf{w} = \mathbf{u}_1 + \mathbf{v}_1 = \mathbf{u}_2 + \mathbf{v}_2,$$

then $\mathbf{u}_1 - \mathbf{u}_2 = \mathbf{v}_2 - \mathbf{v}_1$. Here the left member belongs to $\mathfrak{M}$, the right to $\mathfrak{M}^{\perp}$, and these subspaces of $\mathfrak{H}$ have only the zero element in common. ∎

We have proved that

$$\mathfrak{M}^{\perp} = \mathfrak{H} \ominus \mathfrak{M} \tag{2.5.24}$$

and we can express Theorem 2.5.10 by the formula

$$\mathfrak{H} = \mathfrak{M} \oplus \mathfrak{M}^{\perp}. \tag{2.5.25}$$

This is a decomposition of $\mathfrak{H}$ into the sum of two orthogonal subspaces having only the zero element in common. In view of this, $\mathfrak{M}^{\perp}$ is called the *orthogonal complement* of $\mathfrak{M}$, and vice versa.

Next we give an application of linear closed subspaces to the theory of linear bounded functionals of $\mathfrak{H}$. The set of all such functionals forms the adjoint space $\mathfrak{H}^*$. Actually there is a 1–1 mapping of $\mathfrak{H}$ into $\mathfrak{H}^*$ which is an *isometry* and a *skew isomorphism* in the sense that sums go into sums but scalar products into their conjugates so that $\mathbf{y} \to x^*$ implies

$$\lambda \mathbf{y} \to \bar{\lambda} x^*. \tag{2.5.26}$$

Here x^* denotes a generic element of $\mathfrak{H}^*$ and the bar indicates the conjugate complex number. We express this correspondence by saying that $\mathfrak{H}$ is *skew self-adjoint*. We shall prove

Theorem 2.5.11. *If $x^* \in \mathfrak{H}^*$, then there exists a unique element $\mathbf{y} \in \mathfrak{H}$ such that $x^*(\mathbf{x}) = (\mathbf{x}, \mathbf{y})$ for all $\mathbf{x}$ in $\mathfrak{H}$.*

Proof. Consider the nullspace of x^*

$$\mathfrak{N} = \mathfrak{N}[x^*] = \{\mathbf{x};\ x^*(\mathbf{x}) = 0\}.$$

Here $\mathfrak{N}$ is a linear closed subspace of $\mathfrak{H}$ (why?), so there is an orthogonal complement $\mathfrak{N}^{\perp}$. If now $\mathfrak{N}^{\perp}$ should reduce to the zero element, then $\mathfrak{N} = \mathfrak{H}$ and $x^*(\mathbf{x}) = (\mathbf{x}, \mathbf{0})$ is the trivial representation of the functional as an inner product. Suppose that $\mathfrak{N}^{\perp}$ contains an element $\mathbf{z} \neq \mathbf{0}$. We shall show that there is a constant γ such that

$$x^*(\mathbf{x}) = (\mathbf{x}, \gamma \mathbf{z}). \tag{2.5.27}$$

To this end consider the decomposition

$$\mathbf{x} = (\mathbf{x} - \beta \mathbf{z}) + \beta \mathbf{z},$$

where β is a number that will depend upon $\mathbf{x}$. We choose β so that $\mathbf{x} - \beta \mathbf{z} \in \mathfrak{N}$. This calls for

$$\beta = \frac{x^*(\mathbf{x})}{x^*(\mathbf{z})}.$$

Note that $x^*(\mathbf{z}) \neq 0$ since $\mathbf{z} \in \mathfrak{N}^{\perp}$. We have then

$$(\mathbf{x} - \beta\mathbf{z}, \gamma\mathbf{z}) = 0$$

for any choice of γ. Now γ is disposed of so that

$$x^*(\beta\mathbf{z}) = (\beta\mathbf{z}, \gamma\mathbf{z}) = \beta\bar{\gamma}\|\mathbf{z}\|^2. \tag{2.5.28}$$

This gives the condition

$$\bar{\gamma}\|\mathbf{z}\|^2 = x^*(\mathbf{z}), \tag{2.5.29}$$

which determines γ uniquely. We have then

$$x^*(\mathbf{x}) = x^*(\mathbf{x} - \beta\mathbf{z}) + \beta x^*(\mathbf{z}) = (\mathbf{x} - \beta\mathbf{z}, \gamma\mathbf{z}) + \beta(\mathbf{z}, \gamma\mathbf{z}) = (\mathbf{x}, \gamma\mathbf{z})$$

with γ determined by (2.5.29). The uniqueness follows from

$$(\mathbf{x}, \gamma\mathbf{z}) = (\mathbf{x}, \mathbf{y})$$

for all $\mathbf{x}$ implies $\mathbf{y} = \gamma\mathbf{z}$. The mapping $\mathbf{y} \to x^*$ clearly has the properties of isometry and skew isomorphism. ∎

As the last topic of this section we shall give a brief discussion of the analogues of quadratic forms, polar forms, and Hermitian forms for Hilbert space. It is just an introduction, a preview of Chapter 11.

Suppose that $T \in \mathfrak{E}(\mathfrak{H})$, the space of linear bounded transformations from $\mathfrak{H}$ into itself, and consider the linear bounded functional

$$(T\mathbf{x}, \mathbf{y}) \tag{2.5.30}$$

obtained by fixing $\mathbf{y}$ and letting $\mathbf{x}$ vary. This is the *polar form* analogous to (1.7.6) for the case $\mathfrak{H} = C^n$. For $\mathbf{y} = \mathbf{x}$ we get the *quadratic form* analogous to (1.7.1). We have also an *adjoint transformation* T^* such that for all $\mathbf{x}$ and $\mathbf{y}$ in $\mathfrak{H}$

$$(T\mathbf{x}, \mathbf{y}) = (\mathbf{x}, T^*\mathbf{y}). \tag{2.5.31}$$

Here T^* is a uniquely defined element of $\mathfrak{E}(\mathfrak{H})$. If

$$T^* = T, \tag{2.5.32}$$

the operator T is said to be *self-adjoint* or *Hermitian*, and the corresponding forms $(T\mathbf{x}, \mathbf{y})$ and $(T\mathbf{x}, \mathbf{x})$ are *Hermitian*.

Theorem 2.5.12. *If T is Hermitian, then $(T\mathbf{x}, \mathbf{x})$ is real-valued for all $\mathbf{x}$ and for $\|\mathbf{x}\| = 1$ its value belongs to the interval $[-\|T\|, \|T\|]$, where $\|T\|$ is the norm of T in $\mathfrak{E}(\mathfrak{H})$.*

Proof. We have

$$(T\mathbf{x}, \mathbf{x}) = \overline{(\mathbf{x}, T\mathbf{x})}$$

by property (3) of the inner product. Since

$$(T\mathbf{x}, \mathbf{x}) = (\mathbf{x}, T\mathbf{x})$$

for Hermitian forms, $(T\mathbf{x}, \mathbf{x})$ must be real. We have

$$|(T\mathbf{x}, \mathbf{x})| \leqslant \|T\mathbf{x}\| \, \|\mathbf{x}\| \leqslant \|T\|$$

for $\|\mathbf{x}\| = 1$. ∎

Actually at least one of the values $\|T\|$ and $-\|T\|$ is assumed by $(T\mathbf{x}, \mathbf{x})$ on the unit sphere. Compare Theorem 1.7.8 and subsequent discussion.

Just as in C^n, we have

Theorem 2.5.13. *If S and T are Hermitian operators in $\mathfrak{E}(\mathfrak{H})$, then so are $S + T$ and αS, where α is real, while ST is Hermitian iff S and T commute.*

Proof. The first two assertions are immediate consequences of the properties of Hermitian forms. Since $(ST)^* = T^*S^* = TS$ it follows that ST is Hermitian iff $ST = TS$. ∎

We also note that for any complex number α

$$(\alpha S)^* = \bar{\alpha} S^*. \tag{2.5.33}$$

Furthermore, an arbitrary operator $A \in \mathfrak{E}(\mathfrak{H})$ admits of a unique representation in terms of Hermitian operators

$$A = B + iC, \tag{2.5.34}$$

where

$$B = \tfrac{1}{2}(A + A^*), \qquad C = \frac{1}{2i}(A - A^*). \tag{2.5.35}$$

The operators B and C are known as the *real* and the *imaginary* parts of A, respectively. It should be noted that B and C are self-adjoint. This is obvious in the case of B; to prove the statement for C we have to use (2.5.33). The reader will observe that the *involution* defined by $T \to T^*$ is a generalization of the elementary operation of *conjugation*, $\alpha \to \bar{\alpha}$, in the theory of complex numbers, with self-adjoint operators playing the part of real numbers.

Definition 2.5.4. *$A \in \mathfrak{E}(\mathfrak{H})$ is said to be normal if A and A^* commute.*

A necessary and sufficient condition for A to be normal is that its real and imaginary parts commute for

$$A^* = (B + iC)^* = B^* - iC^* = B - iC. \tag{2.5.36}$$

Unitary operators are particularly important instances of normal operators. We can define U to be unitary if

$$UU^* = U^*U = I,$$

or, equivalently, by requiring that U have an inverse and

$$U^{-1} = U^*. \tag{2.5.37}$$

Since

$$(\mathbf{x}, \mathbf{y}) = (U^*U\mathbf{x}, \mathbf{y}) = (U\mathbf{x}, U^{**}\mathbf{y}) = (U\mathbf{x}, U\mathbf{y}), \tag{2.5.38}$$

unitary operators preserve inner products and, in particular, *distances.* Thus U defines a mapping of $\mathfrak{H}$ onto itself which is an isometry as well as an isomorphism. Such a mapping is called an *automorphism.* Conversely, *every automorphism of a Hilbert space is defined by a unitary transformation* (why?).

The last topic featured in this preview of Hermitian forms is *projections.*

Definition 2.5.5. *A mapping P from $\mathfrak{H}$ into itself is called a projection of $\mathfrak{H}$ onto the linear closed subspace $\mathfrak{M}$ of $\mathfrak{H}$ if $\mathbf{x} = \mathbf{u} + \mathbf{v}$, $\mathbf{u} \in \mathfrak{M}$, $\mathbf{v} \in \mathfrak{M}^{\perp}$, implies $P\mathbf{x} = \mathbf{u}$.*

This convention makes the range of P

$$\mathfrak{R}[P] = \mathfrak{M} = \{\mathbf{x}; P\mathbf{x} = \mathbf{x}\}. \tag{2.5.39}$$

Theorem 2.5.14. *The projection P onto $\mathfrak{M}$ is an idempotent Hermitian operator and, if $\mathfrak{M} \neq \{\mathbf{0}\}$, then $\|P\| = 1$.*

Proof. It is clear that P is a linear bounded operator (why?) and thus belongs to $\mathfrak{E}(\mathfrak{H})$. Since

$$P^2\mathbf{x} = P(P\mathbf{x}) = P\mathbf{u} = \mathbf{u} = P\mathbf{x}$$

we have $P^2 = P$ and P is idempotent. Next, since $(\mathbf{u}, \mathbf{v}) = (\mathbf{v}, \mathbf{u}) = 0$ we have

$$\begin{aligned}(P\mathbf{x}, \mathbf{x}) &= (\mathbf{u}, \mathbf{u} + \mathbf{v}) = \|\mathbf{u}\|^2 + (\mathbf{u}, \mathbf{v}) = \|\mathbf{u}\|^2 \\ &= (\mathbf{u} + \mathbf{v}, \mathbf{u}) = (\mathbf{x}, P\mathbf{x})\end{aligned}$$

or $P^* = P$ and P is Hermitian. By the Pythagorean theorem

$$\|\mathbf{x}\|^2 = \|\mathbf{u} + \mathbf{v}\|^2 = \|\mathbf{u}\|^2 + \|\mathbf{v}\|^2$$

or

$$\|\mathbf{u}\|^2 \leqslant \|\mathbf{x}\|^2 \quad \text{implies} \quad \|P\mathbf{x}\| \leqslant \|\mathbf{x}\|,$$

so that $\|P\| \leqslant 1$. If $\mathfrak{M} \neq \{\mathbf{0}\}$, then there is an $\mathbf{x} \neq \mathbf{0}$ in $\mathfrak{M}$ and for such an $\mathbf{x}$ we have $P\mathbf{x} = \mathbf{x}$ so that $\|P\| = 1$.

It should be noted that $I - P$ is also a projection which maps $\mathfrak{H}$ onto $\mathfrak{M}^{\perp}$. The nullspace of P is $\mathfrak{M}^{\perp}$, that of $I - P$ is $\mathfrak{M}$. Conversely, if T is Hermitian and an idempotent and if $\mathfrak{N}$ is the nullspace of T, then T is a projection of $\mathfrak{H}$ onto $\mathfrak{N}^{\perp}$. ▮

EXERCISE 2.5

1. Prove Theorem 2.5.3.
2. Prove Theorem 2.5.4.
3. Prove Theorem 2.5.5.
4. When is the series (2.5.13) convergent in norm?

5. State and prove the analogue of Theorem 1.7.1 for a Hilbert space.
6. Same question for Theorem 1.7.2.
7. Prove the existence of a transformation T^* satisfying (2.5.31). Why is T^* unique?
8. Show that the decomposition (2.5.34) is unique.
9. Verify that the mapping $\mathbf{y} \to x^*$ of Theorem 2.5.11 from $\mathfrak{H}$ into $\mathfrak{H}^*$ is skew self-adjoint.
10. If a Hermitian operator T has a characteristic value λ_0 and $T(\mathbf{x}_0) = \lambda_0 \mathbf{x}_0, \|\mathbf{x}_0\| = 1$, show that λ_0 is real and find bounds for λ_0.
11. Prove that if T is a Hermitian idempotent and its nullspace $\mathfrak{N}$ is neither $\{\mathbf{0}\}$ nor $\mathfrak{H}$, then T is a projection of $\mathfrak{H}$ onto $\mathfrak{N}^\perp$.
12. Why is an automorphism of $\mathfrak{H}$ defined by a unitary operator?
13. Let $\mathfrak{H} = \mathfrak{M}_1 \oplus \mathfrak{M}_2 \oplus \cdots \oplus \mathfrak{M}_n$ be a decomposition of $\mathfrak{H}$ into closed linear subspaces with $\mathfrak{M}_j \cap \mathfrak{M}_k = \{\mathbf{0}\}$, $j \neq k$. Let P_k be the projection of $\mathfrak{H}$ onto $\mathfrak{M}_k$. Show that $P_j P_k = O$ and $P_1 + P_2 + \cdots + P_n = I$.
14. If $\{\mathbf{0}\} \subset \mathfrak{M}_1 \subset \mathfrak{M}_2 \subset \mathfrak{H}$ and P_1, P_2 are the corresponding projections, find $P_1 P_2$ and $P_2 P_1$.
15. Show that a projection P has the characteristic values 0 and 1 ($P \neq O, I$) and no others and find the characteristic vectors.

2.6 BANACH ALGEBRAS

This is essentially a preview of things to come. Some basic properties are stated and proved but the more sophisticated parts of the theory are to be found in Chapter 9.

Definition 2.6.1. *A Banach algebra* $\mathfrak{B}$ *(B-algebra for short) is a Banach space in which a notion of element multiplication is defined subject to the following conditions:*

(M_1) *For each ordered pair of elements* $\mathbf{x}$ *and* $\mathbf{y}$ *in* $\mathfrak{B}$ *there is a product* $\mathbf{xy} \in \mathfrak{B}$.

(M_2) *Multiplication is associative:* $\mathbf{x}(\mathbf{yz}) = (\mathbf{xy})\mathbf{z}$.

(M_3) *Multiplication is distributive with respect to addition:*
$\mathbf{x}(\mathbf{y} + \mathbf{z}) = \mathbf{xy} + \mathbf{xz}, \qquad (\mathbf{x} + \mathbf{y})\mathbf{z} = \mathbf{xz} + \mathbf{yz}.$

(M_4) *Multiplication commutes with scalar multiplication:*
$(\alpha\mathbf{x})(\beta\mathbf{y}) = (\alpha\beta)(\mathbf{xy}).$

(M_5) $\|\mathbf{xy}\| \leqslant \|\mathbf{x}\| \, \|\mathbf{y}\|$.

Multiplication is not necessarily commutative. If

$$\mathbf{xy} = \mathbf{yx}, \quad \forall\, \mathbf{x}, \mathbf{y}, \tag{2.6.1}$$

we speak of a *commutative B-algebra*.

An algebra may have a unit element $\mathbf{e}$ such that

$$\mathbf{ex} = \mathbf{xe} = \mathbf{x}, \quad \forall \mathbf{x}. \tag{2.6.2}$$

Since $\mathbf{e}^2 = \mathbf{e}$, condition ($M_5$) implies that $\|\mathbf{e}\| \geqslant 1$. We shall assume

$$\|\mathbf{e}\| = 1. \tag{2.6.3}$$

Actually this assumption does not restrict the generality for it may be shown that if the original norm does not have this property, then a new norm may be found in terms of which (2.6.3) holds and, what is essential, the two norms are equivalent in the sense that they induce the same topologies in the space.

If the algebra has a unit element $\mathbf{e}$, then some elements $\mathbf{x} \in \mathfrak{B}$ may have inverses in $\mathfrak{B}$. Here $\mathbf{y}$ is said to be the *inverse* of $\mathbf{x}$ if

$$\mathbf{xy} = \mathbf{yx} = \mathbf{e}, \tag{2.6.4}$$

and we write

$$\mathbf{y} = \mathbf{x}^{-1}. \tag{2.6.5}$$

It should be noted that an element $\mathbf{x} \in \mathfrak{B}$ has at most one inverse so *the inverse is unique if it exists.* If $\mathbf{x}$ has an inverse, it is said to be *invertible* or to be *regular.* Elements without inverses are called *singular.* If $\mathfrak{B} = C$, the field of complex numbers, then all elements $\neq 0$ have inverses. On the other hand, if $\mathfrak{B} = \mathfrak{M}_n$, $n > 1$, then some elements (matrices in this case) have inverses while others have not.

Theorem 2.6.1. *The set of all regular elements of a B-algebra forms a group $\mathfrak{G}$ under multiplication.*

Proof. If $\mathbf{x}$ and $\mathbf{y} \in \mathfrak{G}$, then

$$\mathbf{xyy}^{-1}\mathbf{x}^{-1} = \mathbf{e}, \qquad \mathbf{y}^{-1}\mathbf{x}^{-1}\mathbf{xy} = \mathbf{e} \tag{2.6.6}$$

so that $\mathbf{xy} \in \mathfrak{G}$ and has the inverse $\mathbf{y}^{-1}\mathbf{x}^{-1}$ in $\mathfrak{G}$. Also $\mathbf{e} \in \mathfrak{G}$ and if $\mathbf{x} \in \mathfrak{G}$, so does $\mathbf{x}^{-1}$ since $(\mathbf{x}^{-1})^{-1} = \mathbf{x}$. ∎

Theorem 2.6.2. *If $\mathfrak{B}$ is a $\mathfrak{B}$-algebra, then all elements $\mathbf{x}$ in the sphere*

$$\|\mathbf{x} - \mathbf{e}\| < 1 \tag{2.6.7}$$

belong to $\mathfrak{G}$.

Proof. The geometric series

$$\mathbf{e} + (\mathbf{e} - \mathbf{x}) + (\mathbf{e} - \mathbf{x})^2 + \cdots + (\mathbf{e} - \mathbf{x})^n + \cdots \tag{2.6.8}$$

converges in norm to an element $\mathbf{y}$ of $\mathfrak{B}$ if (2.6.7) holds (why?). If the series is multiplied either on the left or on the right by

$$\mathbf{x} = \mathbf{e} - (\mathbf{e} - \mathbf{x}),$$

then multiplication may be carried out termwise, and if the terms are collected, then all powers cancel except $\mathbf{e}$ so that $\mathbf{y}$ is the inverse of $\mathbf{x}$ and $\mathbf{x} \in \mathfrak{G}$. ∎

Theorem 2.6.3. *In the normed metric $\mathfrak{G}$ is an open point set.*

Proof. It is to be shown that if $\mathbf{x}_0 \in \mathfrak{G}$, then there is a neighborhood of $\mathbf{x}_0$ also in $\mathfrak{G}$. This follows essentially by the argument used in proving the preceding theorem.

Note that

$$\mathbf{x} = \mathbf{x}_0 - (\mathbf{x}_0 - \mathbf{x}) = \mathbf{x}_0 [\mathbf{e} - \mathbf{x}_0^{-1}(\mathbf{x}_0 - \mathbf{x})].$$

The first factor in the last member is known to have an inverse while the second is invertible for

$$\|\mathbf{x}_0^{-1}(\mathbf{x}_0 - \mathbf{x})\| < 1.$$

This holds if, for instance,

$$\|\mathbf{x} - \mathbf{x}_0\| < [\|\mathbf{x}_0^{-1}\|]^{-1}.$$

This set is a neighborhood of $\mathbf{x}_0$. ∎

The following result is due to I. M. Gelfand. The proof will be given in Section 9.1.

Theorem 2.6.4. *If* $\mathbf{x} \in \mathfrak{B}$, *then*

$$\lim_{n \to \infty} \|\mathbf{x}^n\|^{1/n} \equiv r(\mathbf{x}) \tag{2.6.9}$$

exists and $0 \leqslant r(\mathbf{x}) \leqslant \|\mathbf{x}\|$.

This number $r(\mathbf{x})$ is known as the *spectral radius* of $\mathbf{x}$ for reasons that soon will become evident. Here we cannot exclude the possibility that $r(\mathbf{x}) = 0$ for $\mathbf{x} \neq 0$. In analogy with the matrix case we say that an element $\mathbf{x}$ of $\mathfrak{B}$ is *nilpotent of degree k* if there exists an integer k such that $\mathbf{x}^k = \mathbf{0}$, $\mathbf{x}^{k-1} \neq \mathbf{0}$. Such an element is singular and $r(\mathbf{x}) = 0$. But there are other possibilities. In the operator algebra $\mathfrak{E}(C[0, 1])$ (= linear bounded transformations from the space of functions continuous on $[0, 1]$ into itself) we may take the very simple operator

$$T[f](t) = \int_0^t f(s)\, ds. \tag{2.6.10}$$

It may be shown that $r(T) = 0$. An element $\mathbf{q}$ of $\mathfrak{B}$ such that $r(\mathbf{q}) = 0$ is said to be *quasi-nilpotent* or a *topologically nilpotent element*. Any such element is singular.

Various types of singular matrices occurred in Section 1.4 such as divisors of zero and idempotents. Such elements also occur in general B-algebras. An element $\mathbf{a} \neq \mathbf{0}$ is a *divisor of zero* if there exists an element $\mathbf{b} \neq \mathbf{0}$ such that either

$$\mathbf{ab} = \mathbf{0} \quad \text{or} \quad \mathbf{ba} = \mathbf{0}. \tag{2.6.11}$$

Both $\mathbf{a}$ and $\mathbf{b}$ are obviously singular; the contrary hypothesis would imply that the zero element is regular.

An element $\mathbf{x}$ such that

$$\mathbf{x}^2 = \mathbf{x} \tag{2.6.12}$$

is said to be *idempotent*. If $\mathbf{x} \neq \mathbf{0}, \mathbf{e}$, then $\mathbf{x}$ is a divisor of zero and hence singular. This whole discussion of regular and singular elements makes sense only if $\mathfrak{B}$ has a unit element. In Section 9.1 we shall encounter alternative definitions for B-algebras without unit element. Until then any B-algebra under consideration is assumed to have a unit element.

In such a B-algebra the regular elements are particularly important, so we shall consider the family of elements

$$\mathbf{x}_\lambda \equiv \lambda\mathbf{e} - \mathbf{x}, \ \mathbf{x} \ \text{fixed in } \mathfrak{B}, \tag{2.6.13}$$

with respect to regularity. Here λ is a complex variable. The values of λ fall into two mutually exclusive classes according as $\mathbf{x}_\lambda$ is regular or singular. The first class is called the *resolvent set* and denoted by $\rho(\mathbf{x})$, the second the *spectrum* of $\mathbf{x}$ denoted by $\sigma(\mathbf{x})$. We recognize these concepts from the matrix case as well as from operator algebras.

Theorem 2.6.5. *The resolvent set is an open unbounded set in the complex plane and contains*

$$|\lambda| > r(\mathbf{x}) \tag{2.6.14}$$

as a subset, where $r(\mathbf{x})$ is the spectral radius of $\mathbf{x}$. The spectrum is a closed bounded set confined to the disk

$$|\lambda| \leqslant r(\mathbf{x}) \tag{2.6.15}$$

and there is at least one point of $\sigma(\mathbf{x})$ on the boundary.

Proof. That $\rho(\mathbf{x})$ is open follows from Theorem 2.6.3. For large values of $|\lambda|$ we have

$$R(\lambda, \mathbf{x}) \equiv (\lambda\mathbf{e} - \mathbf{x})^{-1} = \mathbf{e}\lambda^{-1} + \mathbf{x}\lambda^{-1} + \cdots + \mathbf{x}^n\lambda^{-n-1} + \cdots. \tag{2.6.16}$$

See Lemmas 1.5.1 and 2.4.1 for matrices and operator algebras, respectively. The series converges in norm for λ satisfying (2.6.14) and diverges for $|\lambda| < r(\mathbf{x})$ since the norms of the terms are unbounded. In the convergence case we multiply the series, right or left, by $\lambda\mathbf{e} - \mathbf{x}$; the resulting product is $\mathbf{e}$, so that the series represents $R(\lambda, \mathbf{x})$ for $|\lambda| > r(\mathbf{x})$.

It is clear that the spectrum must be confined to the disk (2.6.15). It cannot be confined to a smaller disk with center at the origin of the λ-plane. To prove this rigorously we need basic properties of B-valued holomorphic functions which will be developed in Chapter 8. Such functions are representable by Cauchy integrals, Taylor's and Laurent series under essentially the same conditions as Cauchy holomorphic functions. Now $R(\lambda, \mathbf{x})$ is locally holomorphic with values in $\mathfrak{B}$. In particular, $R(\lambda, \mathbf{x})$ is holomorphic for

$$|\lambda| > r = \max\,[|\alpha|;\ \alpha \in \sigma(\mathbf{x})]$$

and as such admits of a unique power series expansion in $1/\lambda$ convergent in norm for $|\lambda| > r$. This expansion must coincide with (2.6.16) and the latter must converge for $|\lambda| > r$ and in no larger region. It follows that $r = r(\mathbf{x})$ and

$$r(\mathbf{x}) = \max\,[|\alpha|;\ \alpha \in \sigma(\mathbf{x})] \tag{2.6.17}$$

as asserted. In particular, *the spectrum is non-empty.* ∎

Further study of the resolvent is postponed until Chapter 9.

EXERCISE 2.6

1. Verify that if $\mathfrak{X}$ is a B-space and $\mathfrak{E}(\mathfrak{X})$ the set of all linear bounded transformations from $\mathfrak{X}$ into itself, then $\mathfrak{E}(\mathfrak{X})$ is a B-algebra.
2. Prove that if $\mathbf{x}$ is singular, so is any constant multiple of $\mathbf{x}$, and use this to prove that the set of singular elements is connected.
3. Is the resolvent set $\rho(\mathbf{x})$ necessarily connected?
4. Why is the spectrum a closed point set?
5. Prove that $\mathbf{x} \in \mathfrak{B}$ can have at most one inverse.
6. If $\mathbf{x}$ has an inverse, show that $\mathbf{x}^n$, n positive integer, has the same property and verify that $(\mathbf{x}^n)^{-1} = (\mathbf{x}^{-1})^n$.
7. Let $\mathbf{q}$ be a nilpotent of degree $k > 1$. Show that $\mathbf{e} + \mathbf{q}$ has an inverse and compute it. Note that $\|\mathbf{q}\|$ need not be <1.
8. Under the same assumptions, for what values of the complex number ζ does the series $\sum_{n=1}^{\infty} n(\mathbf{e} + \mathbf{q})^n \zeta^n$ converge in norm?
9. Why does (2.6.8) converge in norm in the sphere (2.6.7)?
10. If T is defined by (2.6.10) prove that $\|T^n\| \leqslant (n!)^{-1}$ and that $r(T) = 0$.
11. Suppose that $\lambda \in \rho(\mathbf{a})$, $\mathbf{a} \in \mathfrak{B}$, $\mathbf{h} \in \mathfrak{B}$, and that α is a small complex number. Use the technique of the proof of Theorem 2.6.3 to develop $R(\lambda, \mathbf{a} + \alpha\mathbf{h})$ in powers of α. Note that $\mathbf{a}$ and $\mathbf{h}$ need not commute. The expansion is of the form
$$R + \alpha R\mathbf{h}R + \cdots + \alpha^n R\mathbf{h}R\mathbf{h} \ldots \mathbf{h}R + \cdots,$$
where $R = R(\lambda, \mathbf{a})$ and the nth term contains n factors α and $\mathbf{h}$ and $n + 1$ factors R.
12. Let $\mathbf{a}$ and $\mathbf{h}$ be arbitrary non-commuting elements of $\mathfrak{B}$ and find for positive integers n
$$\lim_{\alpha \to 0} \frac{1}{\alpha} [(\mathbf{a} + \alpha\mathbf{h})^n - \mathbf{a}^n].$$

COLLATERAL READING

Supplementary information on algebra and topology is to be found, for instance, in:

HOFFMAN, K., and R. KUNZE, *Linear Algebra*, Prentice-Hall, Englewood Cliffs, N.J. (1961).

NEWMAN, M. H. A., *Elements of the Topology of Plane Sets of Points*, 2nd ed., Cambridge University Press, London (1954).

For linear vector spaces see:

BACHMAN, C., and L. NARICI, *Functional Analysis*, Academic Press, New York (1966).

KOLMOGOROV, A. N., and S. V. FOMIN, *Elements of the Theory of Functions and Functional Analysis*, Vol. 1, *Metric and Normed Spaces*, Graylock Press, Rochester, N.Y. (1957).

TAYLOR, A. E., *Introduction to Functional Analysis*, Wiley, New York (1958).

3 SOME SPECIAL LINEAR SPACES

The discussion in the preceding chapter suffered from the lack of illustrative examples. In particular, we were short on abstract spaces of infinite dimension. The spaces C^n, $\mathfrak{M}_n$, and $\mathfrak{E}(C^n)$ which served as our mainstay are finite dimensional. They were studied at some length in Chapter 1, where they served to introduce the general concepts elaborated in the Chapter 2. We shall now give some specific instances of linear vector spaces of infinite dimension. In the present chapter the examples will be of a fairly elementary nature and do not require much background on the part of the reader. The next chapter is devoted to Lebesgue spaces and will require more skill and preparation.

The present chapter has three sections: Sequence spaces; Continuous functions; and Functions of bounded variation.

3.1 SEQUENCE SPACES

These spaces are concerned with mappings of the positive integers Z^+ into one of the spaces C or R, more generally into a complete metric space.

The concept of a *sequence* involves the *ordering of the natural numbers*. These numbers are often taken for granted; the German mathematician Leopold Kronecker (1823–91) maintained that these numbers were created by God and everything else in mathematics had been achieved by man. This reverence for the natural numbers got a jolt in 1889 when the Italian logician and mathematician Guiseppe Peano (1858–1932) devised a set of postulates to characterize their properties. This event marks the beginning of a new era in mathematics, an end to the rough-and-ready pioneering period of the Calculus and a return to the axiomatic method of the Greeks. Peano dealt with an undefined set of elements Z and an undefined notion of *immediate successor of an element* x of Z subject to the following conditions:

(Z_1) *Each element x of Z has a unique immediate successor $x^+ \in Z$.*

(Z_2) *If $x^+ = y^+$, then $x = y$.*

(Z_3) $1 \in Z$ *and there is no $x \in Z$ such that $x^+ = 1$.*

(Z_4) *Any subset of Z that contains* 1 *and the successor of each of its elements coincides with Z.*

It is clear that the set Z^+ of all positive integers gives a realization of such a set Z provided $n + 1$ is taken as the immediate successor of n.

The fourth postulate implies the *principle of induction*, which we state as a theorem without a proof.

Theorem 3.1.1. *Suppose that for each natural number n there is a proposition $P(n)$. Suppose that $P(1)$ is known to be true and that $P(k^+)$ is true whenever $P(k)$ is true. Then $P(n)$ is true for all n.*

After these preliminaries we turn to sequences. Let $\mathfrak{X}$ be an abstract space and consider a mapping T of Z^+ into $\mathfrak{X}$. That is, we consider a collection of ordered pairs

$$\{(n, x_n): n \in Z^+, x_n \in \mathfrak{X}\}. \tag{3.1.1}$$

Here the domain of T is the set of natural numbers ordered by the notion of immediate successor, the range is a subset of $\mathfrak{X}$. Note that the ordering of the domain induces an order of the range. The ordered range is the associated sequence

$$\mathbf{x} = \{x_n; n = 1, 2, 3, \ldots\} = \{x_n\} \tag{3.1.2}$$

for short. Note that the sequence $\{x_n\}$ may also be regarded as a realization of the Peano postulates provided we replace 1 by x_1 and define x_{n+1} as the immediate successor of x_n.

If $\mathfrak{X}$ is a linear vector space, we can define algebraic operations on sequences with range in $\mathfrak{X}$. The sum of

$$\mathbf{a} = \{a_n\}, \qquad \mathbf{b} = \{b_n\}$$

is taken to be

$$\mathbf{a} + \mathbf{b} = \{a_n + b_n\}. \tag{3.1.3}$$

Similarly, a constant times a sequence is the sequence

$$\alpha\mathbf{a} = \{\alpha a_n\}. \tag{3.1.4}$$

If $\mathfrak{X}$ is not merely a linear vector space but actually an algebra, the product of two sequences may be defined as the sequence of the products of corresponding components

$$\mathbf{ab} = \{a_n b_n\}. \tag{3.1.5}$$

The set $S(\mathfrak{X})$ of all sequences $\{x_n\}$ with range in $\mathfrak{X}$ becomes a linear vector space provided $\mathfrak{X}$ is such a space and the algebraic operations are defined by (3.1.3) and (3.1.4). If $\mathfrak{X}$ is an algebra, so is $S(\mathfrak{X})$ if multiplication of elements is defined by (3.1.5). It should be observed, however, that other definitions of element multiplication are feasible and more appropriate for the applications. Some instances are given in Exercise 3.1.

If $\mathfrak{X}$ is an algebra with unit element e and multiplication is defined by (3.1.5), then $S(\mathfrak{X})$ also has a unit element, namely the vector $\mathbf{e}$ all components of which equal e, and now

$$\mathbf{ae} = \mathbf{ea} = \mathbf{a}. \tag{3.1.6}$$

Multiplication is commutative in $S(\mathfrak{X})$ if it is commutative in $\mathfrak{X}$. The element

$\mathbf{x} \in S(\mathfrak{X})$ has an inverse provided all components x_n are regular elements of $\mathfrak{X}$, in which case

$$\mathbf{x}^{-1} = \{x_n^{-1}\}. \tag{3.1.7}$$

We have defined an algebraic structure for $S(\mathfrak{X})$. Is it also possible to define a topological structure? Here it seems reasonable to restrict oneself to the case where $\mathfrak{X}$ is a complete metric space. We can then also define a distance $d(\mathbf{a}, \mathbf{b})$ between two sequences $\mathbf{a}$ and $\mathbf{b}$; in fact, this can be done in infinitely many ways. Following Maurice Fréchet (1906), we set

$$d(\mathbf{a}, \mathbf{b}) = \sum_{n=1}^{\infty} 2^{-n} \frac{d(a_n, b_n)}{1 + d(a_n, b_n)}. \tag{3.1.8}$$

The multipliers 2^{-n} may be replaced by any sequence of positive numbers $\{\alpha_n\}$ such that $\Sigma\alpha_n$ is convergent. It is clear that $d(\mathbf{a}, \mathbf{b}) \geqslant 0$ and $d(\mathbf{a}, \mathbf{b}) = 0$ iff $\mathbf{a} = \mathbf{b}$. Now two sequences $\mathbf{a}$ and $\mathbf{b}$ are equal by definition iff corresponding components are equal, i.e. $a_n = b_n$ for all n. Further, $d(\mathbf{a}, \mathbf{b}) = d(\mathbf{b}, \mathbf{a})$. This takes care of postulates (D_1) and (D_3) for distance. It is not clear that the triangle inequality (D_3) is satisfied. That it holds is a simple consequence of the triangle inequality in $\mathfrak{X}$ together with the elementary inequality

$$\frac{s + t}{1 + s + t} < \frac{s}{1 + s} + \frac{t}{1 + t}, \tag{3.1.9}$$

valid for all positive s and t. The verification is left to the reader.

The space $S(\mathfrak{X})$ is not of much general interest. The special case $\mathfrak{X} = C$, i.e. the algebra of complex-valued sequences, is more interesting, but to obtain concepts of greater importance we have to specialize still further. The following possibilities are noteworthy:

i) *bounded sequences, i.e.* $|a_n|$ *bounded;*

ii) $\lim a_n$ *exists;*

iii) $\sum_{n=1}^{\infty} |a_n|^p$ *converges for some fixed* $p \geqslant 1$.

In all three cases the components a_n are complex numbers.

Sequences which satisfy condition (i) form the space l_∞ of *bounded sequences*. Sequences satisfying (ii) form the space c of *convergent sequences*. The theory of infinite series is concerned primarily with the characterization of the elements of the space c. Finally, the space l_p is the set of all sequences satisfying condition (iii). Each of these spaces is linear. They are also algebras, in the case of l_p without unit element, since the vector $\mathbf{1}$ belongs to no l_p-space with a finite p. It does belong to l_∞, however.

The metric defined by (3.1.8), of course, also makes sense in these special cases, but it is not well adjusted to any one of them. In each of the three cases listed above it is possible to introduce a norm. In cases (i) and (ii) the so-called *sup-norm* is appropriate:

$$\|\mathbf{a}\| = \sup_n |a_n|. \tag{3.1.10}$$

In case (iii) we use the l_p-norm

$$\|\mathbf{a}\|_p = \left\{ \sum_{n=1}^{\infty} |a_n|^p \right\}^{1/p}. \tag{3.1.11}$$

For $p = 2$ this is the natural generalization of the Euclidean norm. The case $p \neq 2$ was also considered in Section 1.6 for finite sequences, namely the coordinates of a point in C^n.

That (3.1.10) defines a norm is clear. The reader will have no difficulty in verifying conditions (N_1) to (N_3). This statement holds also for $p = 1$ in case (iii). The case $p > 1$ is different. First it is not obvious that the space is linear as asserted, i.e. $\mathbf{a} \in l_p$, $\mathbf{b} \in l_p$ implies $\mathbf{a} + \mathbf{b} \in l_p$. Secondly, the triangle inequality must be verified. The reader should also prove completeness of l_p in the normal metric.

The first step is simple enough. If s and t are non-negative

$$s + t \leqslant 2 \max(s, t),$$

so that

$$(s + t)^p \leqslant 2^p [\max(s, t)]^p \leqslant 2^p [s^p + t^p].$$

Hence

$$\sum_{n=1}^{\infty} |a_n + b_n|^p \leqslant \sum_{n=1}^{\infty} [|a_n| + |b_n|]^p \leqslant 2^p \sum_{n=1}^{\infty} [|a_n|^p + |b_n|^p] \tag{3.1.12}$$

and $\mathbf{a} \in l_p$, $\mathbf{b} \in l_p$ implies $\mathbf{a} + \mathbf{b} \in l_p$.

For the second step we need Hölder's inequality (Otto Hölder, 1860–1937). This inequality involves the notion of *conjugate* or *adjoint* sequence spaces. *The spaces l_p and l_q are conjugate if*

$$\frac{1}{p} + \frac{1}{q} = 1, \quad 1 < p. \tag{3.1.13}$$

This means that l_2 is *self-conjugate* and we extend the definition to admit l_∞ as the conjugate of l_1.

Theorem 3.1.2. *Suppose that $\{a_n\} \in l_p$, $\{b_n\} \in l_q$, $1 < p < \infty$. Then $\{a_n b_n\} \in l_1$ and*

$$\sum_{n=1}^{\infty} |a_n b_n| \leqslant \left\{ \sum_{n=1}^{\infty} |a_n|^p \right\}^{1/p} \left\{ \sum_{n=1}^{\infty} |b_n|^q \right\}^{1/q}. \tag{3.1.14}$$

Equality holds here iff there exists a fixed number $\mu > 0$ such that

$$|a_n|^p = \mu |b_n|^q, \quad \forall n. \tag{3.1.15}$$

We shall use

Lemma 3.1.1. *If A and B are positive numbers and $0 < s < 1$, then*

$$A^s B^{1-s} \leqslant As + B(1 - s) \tag{3.1.16}$$

with equality iff $A = B$.

Proof. If $A = B$, there is nothing to prove. Suppose $A \neq B$. Then the arc

$$t = A^s B^{1-s}, \quad 0 < s < 1,$$

in the (s, t)-plane is concave upwards and hence lies below the chord joining the endpoints the equation of which is

$$t = As + B(1 - s)$$

and (3.1.16) follows. ▮

Proof of Theorem 3.1.2. In (3.1.16) we set

$$A = |a_n|^p, \qquad B = |b_n|^q, \qquad s = \frac{1}{p},$$

and obtain

$$|a_n b_n| \leqslant \frac{1}{p}|a_n|^p + \frac{1}{q}|b_n|^q.$$

Summing for n we get

$$\sum_{n=1}^{\infty} |a_n b_n| \leqslant \frac{1}{p}\sum_{n=1}^{\infty} |a_n|^p + \frac{1}{q}\sum_{n=1}^{\infty} |b_n|^q. \tag{3.1.17}$$

This inequality is of some interest in itself and it provides an alternative proof that $\mathbf{a} \in l_p$, $\mathbf{b} \in l_q$ implies $\mathbf{ab} \in l_1$. Note that equality here holds iff

$$|a_n|^p = |b_n|^q, \quad \forall\, n.$$

The desired inequality (3.1.14) is obtained from (3.1.17) by a peculiar artifice. In (3.1.17) we replace a_n by $a_n w^{-1}$ and b_n by $b_n w$ for all n. Here w is an arbitrary positive number independent of n. The left member is unchanged but the inequality now becomes

$$\sum_{n=1}^{\infty} |a_n b_n| \leqslant \frac{1}{p} w^{-p} \sum_{n=1}^{\infty} |a_n|^p + \frac{1}{q} w^q \sum_{n=1}^{\infty} |b_n|^q \tag{3.1.18}$$

with equality iff

$$|a_n|^p = w^{p+q} |b_n|^q, \quad \forall\, n. \tag{3.1.19}$$

Now w is at our disposal and we may choose it so that the right member of (3.1.18) is as small as possible. To simplify the notation let us write

$$\sum_{n=1}^{\infty} |a_n|^p = S, \qquad \sum_{n=1}^{\infty} |b_n|^q = T \neq 0.$$

Thus the problem is to minimize

$$F(w) = \frac{1}{p} S w^{-p} + \frac{1}{q} T w^q.$$

The minimum is reached for

$$w = w_0 = \left(\frac{S}{T}\right)^{1/(p+q)}$$

and is found to be

$$S^{\frac{q}{p+q}} T^{\frac{p}{p+q}} = S^{\frac{1}{p}} T^{\frac{1}{q}}$$

since $p + q = pq$. This is the right member of (3.1.14).

If equality is to hold in (3.1.14), it must hold in one of the secondary inequalities (3.1.18), i.e. there exists a w such that (3.1.19) holds. Conversely, if (3.1.19) holds, then we have equality in (3.1.18) as well as in (3.1.14). ∎

Next we have *Minkowski's inequality.*

Theorem 3.1.3. *We have*

$$\left\{\sum_{n=1}^{\infty} [|a_n| + |b_n|]^p\right\}^{1/p} \leqslant \left\{\sum_{n=1}^{\infty} |a_n|^p\right\}^{1/p} + \left\{\sum_{n=1}^{\infty} |b_n|^p\right\}^{1/p} \tag{3.1.20}$$

if either side is finite.

Proof. Formula (3.1.12) shows that both sides are finite if **a** and $\mathbf{b} \in l_p$. We have then for $p > 1$

$$\sum_{n=1}^{\infty} [|a_n| + |b_n|]^p = \sum_{n=1}^{\infty} [|a_n| + |b_n|][|a_n| + |b_n|]^{p-1}$$

$$= \sum_{n=1}^{\infty} |a_n|[|a_n| + |b_n|]^{p-1} + \sum_{n=1}^{\infty} |b_n|[|a_n| + |b_n|]^{p-1}.$$

We use Hölder's inequality on each of the two series in the last member. We take the inequality in the form

$$\sum_{n=1}^{\infty} x_n y_n \leqslant \left\{\sum_{n=1}^{\infty} x_n^{\,r}\right\}^{1/r} \left\{\sum_{n=1}^{\infty} y_n^{\,s}\right\}^{1/s}, \qquad \frac{1}{r} + \frac{1}{s} = 1, \quad x_n \geqslant 0, \quad y_n \geqslant 0.$$

We take first

$$x_n = |a_n|, \qquad y_n = [|a_n| + |b_n|]^{p-1}, \qquad r = p, \qquad s = q,$$

and obtain

$$\sum_{n=1}^{\infty} |a_n|[|a_n| + |b_n|]^{p-1} \leqslant \left\{\sum_{n=1}^{\infty} |a_n|^p\right\}^{1/p} \left\{\sum_{n=1}^{\infty} [|a_n| + |b_n|]^p\right\}^{1/q}.$$

The second series requires only to interchange a_n and b_n. Adding the two estimates gives

$$\sum_{n=1}^{\infty} [|a_n| + |b_n|]^p \leqslant \left\{\sum_{n=1}^{\infty} [|a_n| + |b_n|]^p\right\}^{1/q} \left\{\left(\sum_{n=1}^{\infty} |a_n|^p\right)^{1/p} + \left(\sum_{n=1}^{\infty} |b_n|^p\right)^{1/p}\right\}.$$

Simplification leads to (3.1.20). ∎

Corollary. $\|\mathbf{a}+\mathbf{b}\|_p \leqslant \|\mathbf{a}\|_p + \|\mathbf{b}\|_p$, $1 \leqslant p$.

We can rewrite (3.1.14) as follows:

$$\|\mathbf{ab}\|_1 \leqslant \|\mathbf{a}\|_p \, \|\mathbf{b}\|_q \tag{3.1.21}$$

assuming the right member to be meaningful. The special case $p = q = 2$ is Cauchy's inequality.

The space l_2 is an inner product space, in fact a Hilbert space, the first Hilbert space of infinite dimension that we have encountered so far. With obvious notation we define

$$(\mathbf{a}, \mathbf{b}) = \sum_{n=1}^{\infty} a_n \bar{b}_n. \tag{3.1.22}$$

The proof that this actually defines an inner product is left to the reader. The series is absolutely convergent for every **a** and **b** in l_2. By (3.1.21)

$$|(\mathbf{a}, \mathbf{b})| \leqslant \|\mathbf{a}\|_2 \, \|\mathbf{b}\|_2. \tag{3.1.23}$$

A complete orthonormal system is furnished by the unit vectors $\mathbf{u}_j$ where $\mathbf{u}_j$ has a one in the jth place and zeros elsewhere. If

$$(\mathbf{x}, \mathbf{u}_j) = \hat{x}_j, \tag{3.1.24}$$

then the partial sums

$$\sum_{j=1}^{n} \hat{x}_j \mathbf{u}_j$$

converge in the metric of the space to the limit

$$\sum_{j=1}^{\infty} \hat{x}_j \mathbf{u}_j = \{\hat{x}_k\}, \tag{3.1.25}$$

which is an element of the space since $\sum_{k=1}^{\infty} |\hat{x}_k|^2$ is convergent by Bessel's inequality.

Linear functionals were mentioned in Sections 1.1 and 1.6 and discussed for B-spaces in Section 2.3. Here we shall give a complete determination of the linear bounded functionals on l_p. The construction is essentially based on (3.1.21). We recall that a functional is a function on vectors to numbers, $F(\mathbf{x})$. It is linear if the domain is a linear vector space $\mathfrak{X}$ and if

$$F(\mathbf{x} + \mathbf{y}) = F(\mathbf{x}) + F(\mathbf{y}), \tag{3.1.26}$$

$$F(\alpha \mathbf{x}) = \alpha F(\mathbf{x}). \tag{3.1.27}$$

If $\mathfrak{X}$ is a B-space, the functional is bounded if there is a finite M such that $|F(\mathbf{x})| \leqslant M\|\mathbf{x}\|$, $\forall\, \mathbf{x}$. Its norm is then given by

$$\|F\| = \sup_{\|\mathbf{x}\|=1} |F(\mathbf{x})|. \tag{3.1.28}$$

If now $\mathfrak{X} = l_p$, $1 < p < \infty$, we proceed as follows. To start with, let $\mathbf{b}$ be a fixed vector in l_q, the conjugate sequence space, and let $\mathbf{x}$ be an arbitrary vector in l_p. If

$$\mathbf{x} = \{x_n\}, \qquad \mathbf{b} = \{b_n\},$$

we form the infinite series

$$\sum_{n=1}^{\infty} b_n x_n \equiv F(\mathbf{x}, \mathbf{b}). \tag{3.1.29}$$

By (3.1.21) the series is absolutely convergent for all $\mathbf{x} \in l_p$. $F(\mathbf{x}, \mathbf{b})$ is a function from vectors to numbers defined on l_p. It is a linear functional on l_p and by (3.1.21) it is bounded and its norm is at most equal to $\|\mathbf{b}\|_q$ since

$$|F(\mathbf{x}, \mathbf{b})| \leqslant \|\mathbf{b}\|_q \, \|\mathbf{x}\|_p. \tag{3.1.30}$$

To prove that the norm is exactly equal to $\|\mathbf{b}\|_q$, observe that it is possible to choose a vector $\mathbf{x} \in l_p$ with $\|\mathbf{x}\|_p = 1$ such that

$$F(\mathbf{x}, \mathbf{b}) = \|\mathbf{b}\|_q. \tag{3.1.31}$$

These conditions will be met if there exists a fixed $w > 0$ such that for each n

$$|x_n|^p = w|b_n|^q \quad \text{and} \quad b_n x_n \geqslant 0. \tag{3.1.32}$$

If $b_n = 0$ for some particular value of n, set $x_n = 0$. For other values of n we choose

$$|x_n| = w^{1/p} \, |b_n|^{q/p}, \qquad x_n = \frac{|b_n|}{b_n} \, |x_n| \tag{3.1.33}$$

and, finally, we adjust w so that $\|\mathbf{x}\|_p = 1$. This gives a vector $\mathbf{x}$ such that (3.1.31) holds. It follows that

$$\|F(\cdot, \mathbf{b})\| = \|\mathbf{b}\|_q \tag{3.1.34}$$

as asserted. Thus we have shown that to each fixed vector $\mathbf{b} \in l_q$ corresponds a linear bounded functional $F(\mathbf{x}, \mathbf{b})$ on l_p. It is natural to ask if all linear bounded functionals on l_p are of this form. The answer is in the affirmative.

Theorem 3.1.4. *If $F(\mathbf{x})$ is a linear bounded functional on l_p, then there exists a vector $\mathbf{b} \in l_q$ such that* (3.1.30) *holds.*

Proof. Consider the unit vectors $\mathbf{u}_k$ which obviously belong to l_p. We recall that $\mathbf{u}_k$ is the sequence with a one in the kth place and zeros elsewhere. Suppose that

$$F(\mathbf{u}_k) = b_k, \quad k = 1, 2, 3, \ldots . \tag{3.1.35}$$

This determines a vector $\mathbf{b} = \{b_k\}$ and it is to be shown that $\mathbf{b} \in l_q$ and

$$F(\mathbf{x}) = \sum_{k=1}^{\infty} b_k x_k. \tag{3.1.36}$$

Restrict $\mathbf{x}$ temporarily to the subspace $\mathfrak{X}_N$ where $x_k = 0$ for $k > N$. This is a linear subspace on which $F(\mathbf{x})$ is linear so that

$$F(\mathbf{x}) = \sum_{k=1}^{N} b_k x_k, \ \mathbf{x} \in \mathfrak{X}_N.$$

We now proceed as in the proof of (3.1.34). We choose $\mathbf{x} \in \mathfrak{X}_N$ such that

$$b_k x_k = |b_k x_k|, \ |x_k|^p = |b_k|^q, \qquad k = 1, 2, \ldots, N.$$

This can be done and by (3.1.21) applied to $\mathfrak{X}_N$

$$F(\mathbf{x}) = \sum_{k=1}^{N} b_k x_k = \left\{ \sum_{k=1}^{N} |b_k|^q \right\}^{1/q} \|\mathbf{x}\|_p.$$

By assumption $F(\mathbf{x})$ is a bounded functional, i.e. there exists a finite M such that

$$|F(\mathbf{x})| \leqslant M \|\mathbf{x}\|_p$$

for all $\mathbf{x} \in l_p$. This implies that

$$\left\{ \sum_{k=1}^{N} |b_k|^q \right\}^{1/q} \leqslant M.$$

Since this holds for all N, it follows that $\mathbf{b} \in l_q$ and

$$F(\mathbf{x}, \mathbf{b}) \equiv \sum_{n=1}^{\infty} b_n x_n$$

is a linear bounded functional on l_p. Here $F(\mathbf{x})$ and $F(\mathbf{x}, \mathbf{b})$ coincide on each subspace $\mathfrak{X}_N$. To finish the proof we need only observe that a linear bounded functional on a B-space is continuous in the normed topology. For l_p this is expressed by

Lemma 3.1.2. *If $G(\mathbf{x})$ is a linear bounded functional on l_p, then*

$$\lim_{n\to\infty} \|\mathbf{x}_n - \mathbf{x}_0\|_p = 0 \quad \textit{implies} \quad \lim_{n\to\infty} G(\mathbf{x}_n) = G(\mathbf{x}_0). \tag{3.1.37}$$

The proof is left to the reader.

We now take an arbitrary $\mathbf{x} = \{x_n\} \in l_p$ and define a sequence of vectors $\mathbf{x}_N$ where the first N coordinates of $\mathbf{x}_N$ coincide with those of $\mathbf{x}$ and the remaining ones are zero. Here

$$\lim_{N\to\infty} \|\mathbf{x}_N - \mathbf{x}\|_p = 0.$$

Hence

$$\lim_{N\to\infty} F(\mathbf{x}_N) = F(\mathbf{x}), \qquad \lim_{N\to\infty} F(\mathbf{x}_N, \mathbf{b}) = F(\mathbf{x}, \mathbf{b}).$$

Since $F(\mathbf{x}_N) = F(\mathbf{x}_N, \mathbf{b})$ for all N, their limits are also equal and $F(\mathbf{x}) = F(\mathbf{x}, \mathbf{b})$ for all $\mathbf{x}$. ∎

The cases $p = 1$ and $p = \infty$ are left to the reader.

It remains to say a few words about linear bounded transformations from l_p into itself. We can construct such transformations by an extension of the method used in Section 1.3, but here new features arise. We normally have to use infinite matrices, say

$$\mathcal{A} = (a_{jk}) \tag{3.1.38}$$

and form

$$y_j = \sum_{k=1}^{\infty} a_{jk} x_k, \quad j = 1, 2, 3, \ldots . \tag{3.1.39}$$

Here we have to satisfy two requirements. For every $\mathbf{x} \in l_p$ the infinitely many series for the y_j's must converge and the resulting vector $\{y_j\}$ must be an element of l_p. This imposes heavy restrictions on the coefficients a_{jk}. Let us just state one result as a sample:

Theorem 3.1.5. *Let* $1 < p < \infty$, *let* $1/p + 1/q = 1$, *and let*

$$\|\mathcal{A}\|_p = \left\{ \sum_{j=1}^{\infty} \left[\sum_{k=1}^{\infty} |a_{jk}|^q \right]^{p/q} \right\}^{1/p} \tag{3.1.40}$$

exist as a finite number. Then (3.1.39) *defines a linear bounded transformation from* l_p *to* l_p *and*

$$\|y\|_p \leqslant \|\mathcal{A}\|_p \, \|\mathbf{x}\|_p. \tag{3.1.41}$$

The proof is based on Hölder's inequality and the details are left to the reader.

EXERCISE 3.1

1. As an illustration of the principle of induction prove that for all n
$$1^2 + 2^2 + \cdots + n^2 = \tfrac{1}{3} n \left(n + \tfrac{1}{2}\right)(n + 1).$$
2. Verify (3.1.7).
3. An alternate way of defining multiplication of sequences $\mathbf{a} = \{a_n\}$, $\mathbf{b} = \{b_n\}$, where $n = 0, 1, 2, \ldots$, is to set $\mathbf{ab} = \{c_n\}$, where
$$c_n = a_0 b_n + a_1 b_{n-1} + \cdots + a_n b_0.$$
This may be called the *Cauchy product* in analogy with the Cauchy product for infinite series. It is understood that a_n and b_n belong to a non-commutative B-algebra with unit element e. Show that $(e, 0, \ldots, 0, \ldots)$ acts as unit element in the Cauchy sequence algebra.
4. If multiplication of sequences is defined as in the preceding problem and the sequence algebra is l_1, show that $\|\mathbf{ab}\|_1 \leqslant \|\mathbf{a}\|_1 \|\mathbf{b}\|_1$ and determine when equality holds.
5. Another definition of multiplication is based on the so-called *Dirichlet product*. Here $\mathbf{ab} = \{d_n\}$, where
$$d_n = \Sigma\, a_j b_{n|j}, \quad 0 < n,$$
and the summation is extended over the divisors of n. Is there a unit element?
6. Discuss the analogue of Problem 4 for the Dirichlet product.

7. Prove that $s/(s+1)$ is increasing for $s > 0$ and verify (3.1.9).
8. Prove the triangle inequality if $d(\mathbf{a}, \mathbf{b})$ is given by (3.1.8).
9. Fill in missing details in the proof of Theorem 3.1.2.
10. Under what conditions on the vector $\mathbf{b}$ does (3.1.29) define a linear bounded functional on l_1?
11. Same question for l_∞.
12. Construct linear bounded functionals on c. Show that $\lim x_n$ is such a functional if $\mathbf{x} = \{x_n\}$.
13. Prove Lemma 3.1.2.
14. Show that
$$y_n = \frac{1}{n}(x_1 + x_2 + \cdots + x_n), \qquad n = 1, 2, 3, \ldots$$
defines a linear bounded transformation from the space c into itself. Show that limits are preserved, i.e. $\lim x_n = s$ implies $\lim y_n = s$. This fact is known as Cauchy's First Theorem. It is the transformation that defines so-called $(C, 1)$-*summability* [after Ernesto Cesàro (1859–1906)], also known as the *arithmetic means of order one*. See Chapters 14 and 15.
15. Show that the transformation in Problem 14 is also linear and bounded from l_∞ to itself. Show that $\lim y_n$ exists for the sequence $\{x_n\}$, where $x_{2k-1} = 0$, $x_{2k} = 1$, $k = 1, 2, 3, \ldots$. What does this fact signify?
16. Show that the transformation of the preceding problems is linear but not bounded on l_1. Exhibit a sequence $\{x_n\} \in l_1$ such that the transformed sequence $\{y_n\}$ is not in l_1.
17. Prove Theorem 3.1.5.
18. When is the form $(\mathcal{A}\mathbf{x}, \mathbf{y})$, $\mathbf{x}, \mathbf{y} \in l_2$ Hermitian?
19. Find sufficient conditions on the a_{jk} so that (3.1.29) defines a linear bounded transformation from l_1 to itself.
20. Prove that l_p is separable for $1 \leqslant p < \infty$.
21. Prove that l_∞ is not separable.
22. Is the space c separable?

The remaining problems deal with certain mean values of numbers and are related to formulas for integrals in Section 4.4. See also Chapter 15, where these notions are elaborated.

23. Given n positive numbers $a_1, a_2, \ldots, a_n$. For $0 < r$ set
$$M_r(a) = \left\{\frac{1}{n}\sum_{j=1}^{n}(a_j)^r\right\}^{1/r}$$
and show that
$$\min a_j \leqslant M_r(a) \leqslant \max a_j$$
with equality iff all a_j are equal.

24. Show that $M_r(a) \leqslant M_s(a)$ for $r < s$ with equality iff all a_j are equal.

25. Show that $\lim_{r \to \infty} M_r(a) = \max a_j$.

3.2 CONTINUOUS FUNCTIONS

The next system discussed in this chapter is the set of complex-valued functions continuous in a finite closed interval $[a, b]$. This set is denoted by $C[a, b]$. The reader learnt in the Calculus that the sum of two continuous functions is a continuous function, so is a constant multiple of a continuous function as well as the product of two such functions. In other words, $C[a, b]$ is an algebra if we define

$$(f + g)(t) = f(t) + g(t), \tag{3.2.1}$$

$$(\alpha f)(t) = \alpha f(t), \tag{3.2.2}$$

$$(fg)(t) = f(t)\, g(t). \tag{3.2.3}$$

This algebra is commutative and it has a unit element, the function $1(t)$, which is identically equal to 1 in $[a, b]$. There are regular elements: we have clearly

$$f(t)[f(t)]^{-1} = [f(t)]^{-1}f(t) = 1(t) \tag{3.2.4}$$

iff $f(t) \neq 0$ in $[a, b]$. All functions $f(t)$ which assume the value 0 anywhere in $[a, b]$ are singular. There are no proper nilpotents. There is a zero element $0(t)$, the function which is identically 0 in $[a, b]$. These functions $0(t)$ and $1(t)$ are the only idempotents. On the other hand, there are divisors of zero since

$$f(t)\, g(t) = 0(t)$$

can hold for all t without either factor being $0(t)$.

In this case it is quite easy to characterize the spectrum and the resolvent set of one of the elements of the algebra. We recall that the spectrum is the set of complex numbers λ such that

$$\lambda 1(t) - f(t)$$

is singular. This is the case iff this function can take on the value zero. Hence

$$\sigma[f] = \Re[f], \tag{3.2.5}$$

the range of f. This is a bounded closed connected set in the complex plane. The resolvent set is the complement of the spectrum and hence unbounded and open and need not be connected. The resolvent of f is simply

$$R(\lambda, f)(t) = \frac{1}{\lambda - f(t)}. \tag{3.2.6}$$

This is a continuous function of t for any fixed λ in the resolvent set. For any fixed t it is a piece-wise holomorphic function of λ.

The function $f(t) = \exp(it)$ has a range which is an arc of the unit circle or the whole unit circle according as $b - a < 2\pi$ or $\geqslant 2\pi$. In the latter case the resolvent set has two disjoint components, the interior of the unit circle and the exterior. The resolvent is holomorphic in each component but nowhere on the unit circle.

We can endow $C[a, b]$ with a metric based on the sup-norm

$$\|f\| = \sup_{a \leqslant t \leqslant b} |f(t)| \tag{3.2.7}$$

if we set

$$d(f, g) = \|f - g\|. \tag{3.2.8}$$

Actually we could replace "sup" by "max" in (3.2.7) since we deal with a finite closed interval and a continuous function attains its maximum in such a set.

We recall another classical theorem concerning continuous functions: A sequence of functions in $C[a, b]$ which converges uniformly with respect to t in $[a, b]$ converges to a continuous function. Let us translate this statement into our new language. There is given a sequence $\{f_n\} \subset C[a, b]$. Uniform convergence implies and is implied by

$$\lim \|f_m - f_n\| = 0 \tag{3.2.9}$$

when m and n tend to infinity independently of each other. This says that $\{f_n\}$ is a Cauchy sequence in the space and the classical convergence theorem asserts the existence of an $f_0 \in C[a, b]$ such that

$$\lim_{n \to \infty} \|f_0 - f_n\| = 0. \tag{3.2.10}$$

It follows that the normed algebra $C[a, b]$ is complete in the metric and we are dealing with a B-algebra since clearly

$$\|fg\| \leqslant \|f\| \, \|g\|.$$

Continuous functions on a closed interval are *uniformly continuous*. This property leads to the notion of a modulus of continuity. Set

$$\mu(h; f) = \sup |f(t_1) - f(t_2)|, \qquad t_1, t_2 \in [a, b], \quad |t_1 - t_2| \leqslant h. \tag{3.2.11}$$

This is called the *modulus of continuity* of f. *As a function of its first argument* $\mu(h; f)$ *is an element of* $C[0, b - a]$. *Further,* $\mu(0; f) = 0$ *and* $\mu(h; f)$ *is non-decreasing and subadditive in* h *so that*

$$\mu(h_1 + h_2; f) \leqslant \mu(h_1; f) + \mu(h_2; f). \tag{3.2.12}$$

The verification of these properties is left to the reader.

What we have just discussed is uniform continuity with respect to t for a fixed element f of $C[a, b]$. But we can also have uniform continuity for a subset of elements $\{f_\alpha\}$. Such a set is said to be *equicontinuous* if, given any $\varepsilon > 0$, there exists an $h > 0$ such that

$$t_1, t_2 \in [a, b], \quad |t_1 - t_2| \leqslant h \quad \text{implies} \quad |f_\alpha(t_1) - f_\alpha(t_2)| \leqslant \varepsilon \tag{3.2.13}$$

for all α in the index set. If the latter is finite, equicontinuity of the set $\{f_\alpha\}$ holds trivially. If, however, the set is infinite, a restriction is imposed on the corresponding set of moduli of continuity $\mu(h; f_\alpha)$. Set

$$\mu(h) = \sup_{\alpha} \mu(h; f_\alpha). \tag{3.2.14}$$

We have $\mu(0) = 0$, but normally $\mu(h) = +\infty$ for $h > 0$. The function μ is bounded on the interval $0 \leqslant h \leqslant b - a$ iff $\mu(b - a; f_\alpha)$ is a bounded function of α. If, in addition,

$$\lim_{h \downarrow 0} \mu(h) = 0, \tag{3.2.15}$$

then $\mu(h)$ is continuous and subadditive as well as non-decreasing. In order to satisfy (3.2.13) we merely choose h so that $\mu(h) = \varepsilon$ and the equicontinuity follows.

The Bolzano-Weierstrass theorem holds in C^n: any bounded infinite set in C^n has at least one cluster-point which is an element of C^n. The theorem does not hold in $C[a, b]$. To illustrate this, take $a = 0$, $b = 1$, and $f_n = t^n$. The only point-wise limit of the sequence $\{t^n\}$ on $[0, 1]$ is the discontinuous function $f(t)$, which is 0 for $0 \leqslant t < 1$ and 1 for $t = 1$. The sequence does not converge uniformly; it is not a Cauchy sequence, nor does it contain a subsequence with the Cauchy property, and the limit is not in $C[0, 1]$. The sequence is uniformly bounded in $[0, 1]$ but it is not equicontinuous (why?) and this is decisive.

The following theorem is due to Cesare Arzelà (1847–1912). It is based on the notion of equicontinuity introduced by Guilio Ascoli (1843–96). See references at the end of the chapter for a proof.

Theorem 3.2.1. *Let* $\mathsf{F} = \{f_\alpha\}$ *be a family of functions in* $C[a, b]$ *which is* (1) *uniformly bounded* $\|f_\alpha\| \leqslant M, \forall \alpha$, *and* (2) *equicontinuous. Then each sequence in* F *contains a uniformly convergent subsequence.*

It is clear that the limit of the uniformly convergent subsequence is in $C[a, b]$, but it need not be an element of F. If F is closed in $C[a, b]$, then F is said to be (sequentially) *compact*, otherwise just *conditionally compact*.

There are any number of interesting families of functions with special properties belonging to $C[a, b]$. Among these we single out (1) the family L of piece-wise linear and continuous functions, and (2) the family P of polynomials. We say that a family F is *dense* in the space $C[a, b]$ if every $f \in C[a, b]$ is the limit in the sense of the metric of a sequence $\{f_n\} \subset \mathsf{F}$ so that

$$\lim_{n \to \infty} \|f - f_n\| = 0. \tag{3.2.16}$$

Both L and P are dense in $C[a, b]$. For L this is intuitively obvious, but we shall give a proof since this brings out an important property of the function $t \to |t|$, which will be used below in one of the proofs for P being dense.

Theorem 3.2.2. *The family* L *of piece-wise linear continuous functions, restricted to the interval* $[a, b]$, *is dense in* $C[a, b]$.

Proof. We may take $a = 0$, $b = 1$ since the mapping

$$s = a + t(b - a) \tag{3.2.17}$$

takes $f(s) \in C[a, b]$ into $f[a + t(b - a)] \in C[0, 1]$. Divide the interval $[0, 1]$ into n equal parts and set

$$L_n(t) = \sum_{j=0}^{n-1} A_j \left| t - \frac{j}{n} \right| + A_n, \tag{3.2.18}$$

where the constants A_0 to A_n will be determined. First, note that for any choice of the constants $L_n(t) \in \mathsf{L}$. Since $|t|$ is continuous, so is $L_n(t)$. Moreover, in the interval $\left[\frac{k}{n}, \frac{k+1}{n}\right]$

$$L_n(t) = -\sum_{j=0}^{k-1} A_j \left(t - \frac{j}{n}\right) + \sum_{j=k}^{n-1} A_j \left(t - \frac{j}{n}\right) + A_n, \quad k = 0, 1, \ldots, n-1,$$

and hence is a linear function of t.

Next, if $f(j/n) = y_j$ and $f(t) \in C[0, 1]$, we choose the coefficients A_j in such a manner that

$$L_n\left(\frac{j}{n}\right) = y_j, \quad j = 0, 1, 2, \ldots, n. \tag{3.2.19}$$

This gives a non-homogeneous system of $n + 1$ linear equations for the $n + 1$ unknowns $A_0, A_1, \ldots, A_n$. The determinant of this system is seen to be different from zero so the A_j's are uniquely determined and not all zero. The function L_n is that unique element of the family L which interpolates f at the division points of the partition of $[0, 1]$ into n equal parts. It consequently agrees with f at the division points, and since f is continuous in $[0, 1]$ we expect $L_n(t)$ to give a good approximation to $f(t)$ in the whole interval.

This is proved as follows. We know that f is uniformly continuous in $[0, 1]$. Consider its modulus of continuity $\mu(h; f)$ and set $\mu(1/n; f) = \eta_n$. This number goes to zero as $n \to \infty$. Hence in the interval $k/n \leqslant t \leqslant (k+1)/n$ we have

$$|f(t) - y_k| \leqslant \eta_n \quad \text{and} \quad |L_n(t) - y_k| \leqslant |y_{k+1} - y_k| \leqslant \eta_n, \tag{3.2.20}$$

so that

$$|f(t) - L_n(t)| \leqslant 2\eta_n \tag{3.2.21}$$

uniformly in t as asserted. ∎

The corresponding theorem for P is the famous theorem of Karl Weierstrass (1815–97).

Theorem 3.2.3. *The family of polynomials in t, restricted to the interval $[a, b]$, $b - a < \infty$, is dense in $C[a, b]$.*

We shall give two proofs in the text and a third one is indicated in Exercise 3.2 below. The first proof is due to Henri Lebesgue (1875–1941). It is based on the preceding theorem and on

Lemma 3.2.1. *The function* $x \to |x - c|$, $a \leqslant c \leqslant b$, *can be approximated uniformly by polynomials in the interval* $[a, b]$.

Proof. As above, we take $a = 0$, $b = 1$. We use the fact that the binomial series

$$\sum_{k=0}^{\infty} (-1)^k \binom{\frac{1}{2}}{k} z^k \equiv (1 - z)^{1/2} \tag{3.2.22}$$

converges absolutely and uniformly for $|z| \leqslant 1$. The binomial coefficients stay bounded after multiplication by $k^{3/2}$, hence the convergence. To fix the ideas, suppose that $\frac{1}{2} \leqslant c < 1$. Then

$$|t - c| = \{c^2 - [c^2 - (t - c)^2]\}^{1/2}, \tag{3.2.23}$$

where the positive square root is taken. This gives

$$\begin{aligned} |t - c| &= c\left\{1 - \left[1 - \left(\frac{t - c}{c}\right)^2\right]\right\}^{1/2} \\ &= c \sum_{k=0}^{\infty} (-1)^k \binom{\frac{1}{2}}{k} \left[1 - \left(\frac{t - c}{c}\right)^2\right]^k, \end{aligned} \tag{3.2.24}$$

convergent for $|t - c| \leqslant c$ or $0 \leqslant t \leqslant 2c$. The convergence is uniform in this interval so that the polynomial partial sums converge uniformly to $|t - c|$ in $[0, 2c]$, *a fortiori* in $[0, 1]$. If $c < \frac{1}{2}$ we replace c^2 by $(1 - c)^2$ in (3.2.23) and proceed in the same manner. ∎

First proof of Theorem 3.2.3. If $f \in C[0, 1]$, we determine L_n as in the proof of Theorem 3.2.2 in such a manner that

$$|f(t) - L_n(t)| < \varepsilon$$

for all t in $[0, 1]$ where ε is a preassigned small positive number. Here $L_n(t)$ is given by (3.2.18) for a suitable choice of the constants A_j. Lemma 3.2.1 shows that each of the components in (3.2.18) can be approximated arbitrarily closely in the interval $[0, 1]$ by polynomials in t with A_n approximated by itself. The sum of these polynomial approximations multiplied by the appropriate A_j gives a polynomial approximation of $L_n(t)$ of any desired accuracy in $[0, 1]$ and hence a polynomial approximation of $f(t)$. Hence P is dense in $C[0, 1]$. ∎

The second proof is due to Sergeĭ Natanovič Bernštein (1912).

Theorem 3.2.4. *If $f(t) \in C[0, 1]$ set*

$$B_n(t; f) = \sum_{k=0}^{n} f\left(\frac{k}{n}\right) \binom{n}{k} t^k(1 - t)^{n-k}. \tag{3.2.25}$$

Then

$$\lim_{n\to\infty} B_n(t; f) = f(t)$$

uniformly in $[0, 1]$.

The proof requires a lemma.

Lemma 3.2.2. *For $f(t) = 1, t, t^2$ the nth Bernstein polynomial is, respectively,*

$$1, t, t^2 + \frac{1}{n} t(1 - t). \tag{3.2.26}$$

Proof. By the binomial theorem

$$(p + q)^n = \sum_{k=0}^{n} \binom{n}{k} p^k q^{n-k}. \tag{3.2.27}$$

For $p = t$, $q = 1 - t$ we obtain $B_n(t; 1) \equiv 1$. Differentiation of the last identity with respect to p and multiplication of the result by p gives

$$np(p + q)^{n-1} = \sum_{k=0}^{n} \binom{n}{k} kp^k q^{n-k}.$$

Division by n and substitution give $B_n(t; t) = t$. A second differentiation gives

$$n(n - 1)(p + q)^{n-2} = \sum_{k=0}^{n} \binom{n}{k} k(k - 1) p^{k-2} q^{n-k}$$

and

$$n(n - 1) p^2(p + q)^{n-2} + np(p + q)^{n-1} = \sum_{k=0}^{n} \binom{n}{k} k^2 p^k q^{n-k}.$$

The substitution $p = t$, $q = 1 - t$ gives $B_n(t; t^2)$ as the last expression under (3.2.27). ∎

Corollary. *We have*

$$\sum_{k=0}^{n} \binom{n}{k} \left(t - \frac{k}{n}\right)^2 t^k(1 - t)^{n-k} = \frac{1}{n} t(1 - t). \tag{3.2.28}$$

Proof of Theorem 3.2.4. The value of $B_n(t; 1) \equiv 1$ gives

$$B_n(t;f) - f(t) = \sum_{k=0}^{\infty} \left[f\left(\frac{k}{n}\right) - f(t)\right] \binom{n}{k} t^k(1 - t)^{n-k}. \tag{3.2.29}$$

It is required to show that the difference is uniformly small in absolute value for large values of n and this requires each term in the right member to be small. This is true but not for the same reason for all terms. For k such that k/n is close to the particular value of t under consideration $|f(k/n) - f(t)|$ is small since f is continuous, for k/n away from t it is the weight factor $\binom{n}{k} t^k (1-t)^{n-k}$ which is small. This follows from the Corollary. For a given $\varepsilon > 0$ and a given t, separate the values of k into two disjoint subsets, N_1 and N_2. Here

$$N_1 = \left\{k; \left|t - \frac{k}{n}\right| < \varepsilon, \quad 0 \leqslant k \leqslant n\right\},$$

$$N_2 = \left\{k; \left|t - \frac{k}{n}\right| \geqslant \varepsilon, \quad 0 \leqslant k \leqslant n\right\}.$$

Let S_1 denote that part of the sum in (3.2.28) where k runs over N_1 and S_2 the rest. Suppose that δ is the value of $\mu(h; f)$ for $h = \varepsilon$. Then

$$|S_1| < \delta \sum_{N_1} \binom{n}{k} t^k(1-t)^{n-k} < \delta$$

since the summation from 0 to n equals 1.

Next we have

$$|S_2| \leqslant 2\|f\| \sum_{N_2} \binom{n}{k} t^k(1-t)^{n-k}.$$

We now resort to (3.2.28) and obtain

$$\sum_{N_2} \left(t - \frac{n}{k}\right)^2 \binom{n}{k} t^k(1-t)^{n-k} < \frac{1}{n} t(1-t).$$

In the left member $|t - k/n| \geqslant \varepsilon$ so that

$$\sum_{N_2} \binom{n}{k} t^k(1-t)^{n-k} < (4\,\varepsilon^2 n)^{-1}$$

since $t(1-t) \leqslant \frac{1}{4}$ in $[0, 1]$. Hence

$$|S_2| \leqslant \|f\|\,(2\varepsilon^2 n)^{-1}$$

and

$$|B_n(t; f) - f(t)| < \delta + \|f\|(2\varepsilon^2 n)^{-1}.$$

Here we can take $\varepsilon = n^{-1/4}$ (any exponent $<\frac{1}{2}$ will do) and obtain, finally,

$$|B_n(t;f) - f(t)| < \mu(n^{-\frac{1}{4}}; f) + \tfrac{1}{2}\|f\| n^{-1/2}. \tag{3.2.30}$$

This shows that the Bernstein polynomials are dense in $C[0, 1]$. ■

Corollary. *Theorem 3.2.3.*

We now turn to the question of the existence of linear bounded functionals on the space $C[a, b]$. That is, we want to find mappings of $C[a, b]$ into the space C of complex numbers such that the mappings are linear and bounded.

A trivial but nevertheless important choice of such a mapping is the following. Take a fixed number t_0 with $a \leqslant t_0 \leqslant b$ and set

$$F(f) = f(t_0). \tag{3.2.31}$$

This is clearly a linear functional and its norm is seen to be 1 (why?). In fact, the functional is not merely linear but also *multiplicative* since

$$F(fg) = F(f)\,F(g). \tag{3.2.32}$$

It will be seen later that every bounded linear and multiplicative functional on $C[a, b]$ is of this form, i.e. there exists a t_0 in $[a, b]$ such that (3.2.31) holds.

But even for the non-multiplicative case this approach is suggestive. Suppose that $\{t_n\}$ is a countable set of points in $[a, b]$ and that $\sum_1^\infty c_n$ is an absolutely convergent series. Then

$$\sum_1^\infty c_n f(t_n) \tag{3.2.33}$$

is an absolutely convergent series for any choice of f in $C[a, b]$. The series defines a linear functional on $C[a, b]$ and this is bounded since

$$\left|\sum_1^\infty c_n f(t_n)\right| \leqslant \sum_1^\infty |c_n|\,\|f\| \equiv C\|f\|. \tag{3.2.34}$$

The norm of the functional is $\leqslant C$ and equality may very well hold. We can generalize still further using the Riemann–Stieltjes integral instead of infinite series.

Suppose that $g \in BV[a, b]$, i.e. g is a function of bounded variation on $[a, b]$ for which see next section. This means that for any partition of $[a, b]$ by points t_k,

$$a = t_0 < t_1 < t_2 < \cdots < t_n = b,$$

the finite sum

$$\sum_{k=1}^n |g(t_k) - g(t_{k-1})| \leqslant M, \tag{3.2.35}$$

where M is independent of the partition.

Consider now an arbitrary $f \in C[a, b]$ and a fixed $g \in BV[a, b]$. We can then define the Riemann–Stieltjes integral

$$\int_a^b f(s)\,dg(s) \tag{3.2.36}$$

as the limit of Riemann–Stieltjes sums

$$\sum_{k=1}^n f(t_k)\,[g(t_k) - g(t_{k-1})].$$

The limit is clearly linear in f, and its absolute value does not exceed

$$V_a^b[g]\,\|f\|. \tag{3.2.37}$$

Here the first factor is the *total variation* of g over $[a, b]$, that is the least value of

M for which (3.2.35) can hold. Thus (3.2.36) defines a linear bounded functional on $C[a, b]$. In fact, we have the theorem of F. Riesz (1909).

Theorem 3.2.5. *The adjoint space of* $C[a, b]$ *is the space* $BV[a, b]$. *Every linear bounded functional on* $C[a, b]$ *is of the form* (3.2.36) *and its norm is at most equal to the total variation of* g *in* $[a, b]$. *Equality holds if* g *is suitably normalized. Every such integral defines a linear bounded functional on* $C[a, b]$.

The uncertainty in the value of the norm is due to the fact that g is not uniquely determined by the functional, and this goes back to the fact that in (3.2.36) we can modify g in infinitely many ways without changing the value of the integral for any choice of f in $C[a, b]$. That we can add a constant to g is trivial and does not affect the total variation. But a function of bounded variation may well have a countable set of discontinuities, though of the first kind, so that $g(t)$ has right and left hand limits at all points, commonly denoted by $g(t - 0)$ and $g(t + 0)$. If for some $t = t_0$ we have $g(t_0 - 0) \neq g(t_0 + 0)$, we have a point of discontinuity at $t = t_0$. The actual value of $g(t_0)$ does not affect the value of the integral but it does affect the total variation. If $g(b - 0) = g(b)$ and $g(t)$ is right hand continuous at all other points, we obtain the normalization of F. Riesz for which the norm of the functional equals the total variation of g. We shall not pursue this topic any further.

The next question is how to obtain linear bounded transformation on $C[a, b]$ into itself. The possibilities are legion. The following is an example of an important class of such transformations. Take a function $K(s, t)$ of two variables, defined and continuous for (s, t) in the square $[a, b] \times [a, b]$ and form the integral

$$\int_a^b K(s, t) f(s)\, ds \equiv g(t). \tag{3.2.38}$$

For any choice of $f \in C[a, b]$ this is a continuous function of t for $a \leqslant t \leqslant b$. Moreover,

$$\|g\| \leqslant \|f\| \sup_{a \leqslant t \leqslant b} \int_a^b |K(s, t)|\, ds. \tag{3.2.39}$$

The integral (3.2.38) is clearly linear in f and the last inequality shows that the integral defines a bounded linear transformation. The conditions on $K(s, t)$ may be relaxed in various ways.

Before leaving the continuous functions let us consider various related spaces. One of these is $\mathfrak{M}_n C[a, b]$, the space of n by n matrices the entries of which are elements of $C[a, b]$. Thus $\mathcal{F}(t) = [f_{jk}(t)]$ with $f_{jk} \in C[a, b]$. The algebraic operations of addition, scalar multiplication and element multiplication carry over from $\mathfrak{M}_n$ and C and become meaningful in $\mathfrak{M}_n C[a, b]$. There are several possible choices of the norm of $\mathcal{F}$. One is

$$\|\mathcal{F}(\cdot)\| = \sup_j \sum_{k=1}^n \|f_{jk}\|. \tag{3.2.40}$$

Under this norm $\mathfrak{M}_n C[a, b]$ becomes a B-algebra. It is clear that $\mathcal{F}(t_0)$ for t_0 fixed, $a \leqslant t_0 \leqslant b$, is a linear bounded functional on the space. More generally, we can replace $f(t)$ in (3.2.36) by $\mathcal{F}(t)$ and obtain a linear bounded functional. Here the integral is interpreted as the matrix having in the place (j, k) the entry

$$\int_a^b f_{jk}(s)\, dg(s). \tag{3.2.41}$$

Formula (3.2.38) can also be generalized in an obvious manner.

There is another function algebra of considerable interest, namely the restriction of $C[a, b]$ to functions having continuous derivatives of order k in $[a, b]$. We denote this set by $C^k[a, b]$. It is clearly an algebra with the operations defined as in $C[a, b]$. Let

$$\| f \|_{0,k} = \sum_{j=0}^{k} \frac{1}{j!} \| f^{(j)} \| \tag{3.2.42}$$

be taken as the norm of f in $C^k[a, b]$, where the norm used in the right member is a sup-norm. It may be shown that (3.2.42) defines a norm and that $C^k[a, b]$ is complete under this metric. Hence it is a B-space. It is even a B-algebra since the weight factors $1/j!$ ensure that

$$\| fg \|_{0,k} \leqslant \| f \|_{0,k} \| g \|_{0,k} \tag{3.2.43}$$

by virtue of Leibniz's formula for the derivatives of a product. By an obvious extension of formula (3.2.36) we can define linear bounded functionals on $C^k[a, b]$:

$$L[f] = \sum_{j=1}^{k} \int_a^b f^{(j)}(s)\, dg_j(s), \qquad g_j(s) \in BV[a, b]. \tag{3.2.44}$$

Finally, we mention the space $C^\infty[a, b]$ of functions having derivatives of all order in $[a, b]$. It is clearly a linear vector space, even an algebra. It may be made into a complete metric space by a Fréchet type of a distance

$$d(f, g) = \sum_{n=0}^{\infty} 2^{-n} \frac{\| f^{(n)} - g^{(n)} \|}{1 + \| f^{(n)} - g^{(n)} \|}\,. \tag{3.2.45}$$

EXERCISE 3.2

1. Suppose that $\lim c_n = 0$ and set $f_n = c_n \sin nt$. Is this a Cauchy sequence in $C[0, 2\pi]$?
2. Verify (3.2.6).
3. Verify the statements made concerning the modulus of continuity defined by (3.2.11).
4. Verify that $\mu(h)$ is bounded iff $\mu(b - a; f_\alpha)$ is bounded.
5. Prove that if μ is bounded and satisfies (3.2.15), then $\mu(h)$ is (i) continuous, (ii) subadditive, and (iii) non-decreasing.

6. Fill in omitted details in the proof of Lemma 3.2.1 and its corollary.
7. Verify (3.2.22) and the statements made concerning binomial coefficients and nature of convergence of the series. [*Hint:* The formula of Wallis for π would help.]
8. [Edmund Landau] An alternative proof of Theorem 3.2.3 may be based on the formula

$$\lim_{n\to\infty} \frac{(2n+1)!!}{(2n)!!} \int_0^1 f(s)\,[1-(s-t)^2]^n\,ds = f(t).$$

Here $f \in C(0, 1]$ and the symbol $(2n)!!$ means the product of the even integers $\leqslant 2n$. Similarly for $(2n+1)!!$. The limit exists uniformly with respect to t for $0 < \varepsilon \leqslant t \leqslant 1 - \varepsilon$. Start with $f \equiv 1$. The proof is analogous to that of Theorem 3.2.4 inasmuch as a neighborhood of $s = t$ gives the main contribution and the integral over the rest of the interval goes to zero, again by the formula of Wallis.
9. Suppose that h is a fixed bounded measurable function in the sense of Lebesgue (see Section 4.2). Assume the existence of the integral and show that

$$\int_a^b h(t)\,f(t)\,dt$$

is a linear bounded functional on $C[a, b]$. Get an upper bound for the norm of the functional.
10. How should g be chosen in (3.2.36) in order to give the functional of the preceding problem?
11. Write (3.2.31) as an integral of type (3.2.36).
12. Do the same for (3.2.34) if $\{t_n\}$ is a strictly increasing sequence and $\lim_{n\to\infty} t_n = b$.
13. Verify the stated properties of the transformation (3.2.38).
14. If $|K(s, b)| \leqslant B$ for all (s, t) in $[a, b] \times [a, b]$, show that $B(b - a)$ is an upper bound for the norm of the transformation. Find an example where this bound is actually the norm.
15. Verify that $\mathfrak{M}_n C[a, b]$ becomes a B-algebra under the norm (3.2.40).
16. Suggest alternative norms for this space.
17. Show that $\{\mathfrak{F}_m\}$ is a Cauchy sequence under the norm (3.2.40) iff each of the n^2 sequences $\{f_{jk}{}^m\}$, $j, k = 1, 2, \ldots, n$, is a Cauchy sequence in $C[a, b]$. Here $\mathfrak{F}_m(t) = \{f_{jk}{}^m(t)\}$.
18. Verify that (3.2.41) defines a linear bounded functional on $\mathfrak{M}_n C[a, b]$.
19. How should (3.2.38) be generalized so as to define a linear bounded transformation from $\mathfrak{M}_n C[a, b]$ into itself?
20. Show that $C^k[a, b]$ is a B-space under the norm (3.2.42).
21. Verify (3.2.43).
22. Verify that (3.2.44) defines a linear functional on $\mathfrak{M}_n C[a, b]$.

23. Verify that $C^\infty[a, b]$ becomes a complete metric space if distances are defined by (3.2.45).

24. $C[a, b]$ is a separable vector space. How could this be proved on the basis of Theorem 3.2.2?

25. Prove the same assertion on the basis of Theorem 3.2.3.

3.3 FUNCTIONS OF BOUNDED VARIATION

The class of functions $BV[a, b]$ was introduced towards the end of the preceding section. It is the set of complex-valued functions defined on the interval $[a, b]$ such that the total variation of f over $[a, b]$ is finite. Here the total variation, denoted by $V_a^b[f]$, is the greatest lower bound of the numbers M for which

$$\sum_{k=1}^{n} |f(t_k) - f(t_{k-1})| \leqslant M, \tag{3.3.1}$$

where

$$a = t_0 \leqslant t_1 < t_2 < \cdots < t_n = b$$

is an arbitrary partition of $[a, b]$ and (3.3.1) is supposed to hold with a fixed M for all possible partitions.

This restriction on the variation of f does not force f to be continuous but it does put a restriction on the nature and the number of discontinuities. A discontinuity always makes a positive contribution to the total variation. The sum of these contributions must be a convergent series. This means (i) that the set of discontinuities is countable and (ii) that each discontinuity is of the first kind, i.e.

$$\lim_{h \downarrow 0} f(t-h) = f(t-0), \qquad \lim_{h \downarrow 0} f(t+h) = f(t+0) \tag{3.3.2}$$

exist. Any discontinuity of the second kind would lead to unbounded variation in any interval containing such a point. We recall that any bounded real-valued monotone function is of bounded variation. Moreover, if $f \in BV[a, b]$, then one can find four real-valued non-decreasing bounded functions such that

$$f = f_1 + if_2 - f_3 - if_4. \tag{3.3.3}$$

If f itself is real-valued, we may take $f_2 = f_4 = 0$. After this review of the properties of the elements of the class $BV[a, b]$ we can proceed to the structural properties of the space.

The class $BV[a, b]$ is a linear vector space over C if addition and scalar multiplication are defined in the obvious manner

$$(f + g)(t) = f(t) + g(t), \tag{3.3.4}$$

$$(\alpha f)(t) = \alpha f(t). \tag{3.3.5}$$

Actually the product of two elements in $BV[a, b]$ also belongs to the class. For

$$\sum_{k=1}^{n} |f(t_k)\, g(t_k) - f(t_{k-1})\, g(t_{k-1})|$$

$$\leqslant \sum_{k=1}^{n} |f(t_k)|\, |g(t_k) - g(t_{k-1})| + \sum_{k=1}^{n} |g(t_{k-1})|\, |f(t_k) - (t_{k-1})|$$

$$\leqslant \sup_t |f(t)| \sum_{k=1}^{n} |g(t_k) - g(t_{k-1})| + \sup_t |g(t)| \sum_{k=1}^{n} |f(t_k) - f(t_{k-1})|$$

$$\leqslant \sup_t |f(t)|\, V_a^b[g] + \sup_t |g(t)|\, V_a^b[f]. \tag{3.3.6}$$

Here we have

$$\sup_t |f(t)| \leqslant |f(a)| + V_a^b[f], \qquad \sup_t |g(t)| \leqslant |g(a)| + V_a^b[g], \tag{3.3.7}$$

so that, finally

$$V_a^b[fg] \leqslant |f(a)|\, V_a^b[g] + |g(a)|\, V_a^b[f] + 2\, V_a^b[f]\, V_a^b[g]. \tag{3.3.8}$$

On the other hand, the reciprocal of a function of bounded variation f belongs to $BV[a, b]$ iff $\inf_t |f(t)|$, $a \leqslant t \leqslant b$, is >0, in which case the total variation of f^{-1} is at most

$$\left[\inf_t |f(t)|\right]^{-2} V_a^b[f]. \tag{3.3.9}$$

The proof is left to the reader.

It follows that $BV[a, b]$ is a commutative algebra. It has a unit element, the function $1(t)$ which is identically 1 in $[a, b]$. We say that f is a regular element of the algebra if $f^{-1} \in BV[a, b]$, otherwise singular.

There are obviously divisors of zero since we can have

$$f(t)\, g(t) \equiv 0$$

without either factor being identically zero.

In this algebra there is a profusion of idempotents. The equation

$$[f(t)]^2 = f(t) \tag{3.3.10}$$

shows that the range of f must be one of the three sets $\{0\}$, $\{1\}$, and $\{0, 1\}$. The first two possibilities give continuous idempotents, the functions $0(t)$ and $1(t)$. Any idempotent with the range $\{0, 1\}$ is necessarily discontinuous. If $t = t_0$, $a < t < b$, is a point of discontinuity, each of the three numbers $f(t_0 - 0)$, $f(t_0)$, $f(t_0 + 0)$ can be either 0 or 1 and there are four combinations which correspond to a discontinuity. Two of these correspond to one-sided continuity and contribute one unit to the total variation. The other two combinations involve $f(t_0)$ different from both $f(t_0 - 0)$ and $f(t_0 + 0)$ and contribute two units to the total variation. At the end-points only the first type of a jump can occur. Since the total variation of f has to be an integer, only a finite number of discontinuities can occur but the number may be arbitrarily large. Thus there is a non-denumerable set of idempotents.

On the other hand, there are no nilpotents except for zero.

We can introduce a norm in $BV[a, b]$ based on the characteristic property of the elements to be of bounded variation. The first choice that comes to mind is to take $V_a^b[f]$ as the norm of f. This does not work since all constants would then have the norm zero while only the zero element $0(t)$ can be allowed to have this property. To get a better separation we could use some fixed linear combination of $|f(a)|$ and $V_a^b[f]$, say

$$\|f\| = |f(a)| + 2\, V_a^b[f]. \tag{3.3.11}$$

Here the factor "2" is suggested by (3.3.8) and our desire to make sure that

$$\|fg\| \leqslant \|f\|\ \|g\| \tag{3.3.12}$$

as required for a B-algebra. The choice (3.3.11) ensures this.

Since an element of $BV[a, b]$ is regular iff it is bounded away from zero, we see that the *spectrum of f coincides with the closure of the range of f*. Note that the range itself need not be closed. The spectrum is a bounded closed set in the complex plane which need not be connected and may contain isolated points. The situation here is totally different from that holding in $C[a, b]$. We still have

$$R(\lambda, f)(t) = \frac{1}{\lambda - f(t)}, \tag{3.3.13}$$

however.

The problem of constructing linear bounded functionals is harder for $BV[a, b]$ than for $C[a, b]$. The theory of the Riemann–Stieltjes integral is still suggestive, however. The point of departure is the formula for *integration by parts*:

$$\int_a^b f(t)\, dg(t) = f(b)\, g(b) - f(a)\, g(a) - \int_a^b g(t)\, df(t). \tag{3.3.14}$$

Here f is an arbitrary element of $BV[a, b]$ and g is a fixed element of $C[a, b]$. The formula serves to define the integral of a function of bounded variation with respect to a continuous function. To us it gives a means of defining linear bounded functionals on $BV[a, b]$. The left member is clearly linear in f and its value is a number. Thus to each $g \in C[a, b]$ corresponds a linear functional on $BV[a, b]$ defined by (3.3.14). This functional is bounded since

$$\left|\int_a^b f(t)\, dg(t)\right| \leqslant |f(b)|\, |g(b)| + |f(a)|\, |g(a)| + \|g\|_C\, V_a^b[f]$$

$$\leqslant 2\, \|g\|_C \{|f(a)| + V_a^b[f]\} \leqslant 2\, \|g\|_C\, \|f\| \tag{3.3.15}$$

by (3.3.11). Here the subscript C indicates the sup-norm used in $C[a, b]$ while the unmarked norm is that defined for $BV[a, b]$.

In this manner we can define a class of linear bounded functionals on $BV[a, b]$. There are linear bounded functionals, however, which do not fit into this pattern, so we have given only a partial solution of the problem. The adjoint space of $BV[a, b]$ contains $C[a, b]$ as a proper subspace.

Finally, we shall define a class of linear bounded transformations from $BV[a, b]$ into itself analogous to those defined in $C[a, b]$ by formula (3.2.38). Set

$$g(t) = \int_a^b K(s, t)\, df(s), \tag{3.3.16}$$

where $f \in BV[a, b]$ and the kernel $K(s, t)$ is to be chosen so that $g \in BV[a, b]$ whenever f does. Since

$$\sum_{k=1}^{n} |g(t_k) - g(t_{k-1})| \leqslant \int_a^b \sum_{k=1}^{n} |K(s, t_k) - K(s, t_{k-1})|\, dV_a^s[f], \tag{3.3.17}$$

it is sufficient for our purpose if (i) $K(s, t)$ is continuous in s for each fixed t, and (ii) $K(s, t)$ is of bounded variation in t for fixed s, the total variation being a bounded function of s. Both conditions are satisfied if, for instance, $K(s, t)$ and $K_t(s, t)$ are continuous functions of (s, t) in $[a, b] \times [a, b]$. In formula (3.3.17) the symbol $V_a^s[f]$ stands for the total variation of $f(t)$ in the interval $[a, s]$.

EXERCISE 3.3

1. Show that $V_a^s[f]$ is a non-negative, non-decreasing, and bounded function if $f \in BV[a, b]$, and hence also belongs to this space.
2. Let f be a real-valued element of $BV[a, b]$. Then
$$f(t) = V_a^t[f] - \{V_a^t[f] - f(t)\}.$$
Show that each of the components is real-valued and non-decreasing. Use this to prove (3.3.3) for complex-valued elements of $BV[a, b]$.
3. If f is a real-valued, non-decreasing element of $BV[a, b]$, show that the number of discontinuities is denumerable, that one-sided limits exist everywhere, and that $f(t - 0) \leqslant f(t) \leqslant f(t + 0)$. Use this to prove the assertions about discontinuities made in the text. Is $BV[a, b]$ a separable space?
4. Prove the assertions made concerning f^{-1} and, in particular, the estimate (3.3.9) for the total variation of f^{-1}.
5. If f is a non-trivial idempotent, show that it is a divisor of zero by exhibiting a g in $BV[a, b]$ such that g is also a non-trivial idempotent and their product is the zero element.
6. Verify that (3.3.11) defines a norm and that (3.3.12) holds.
7. Verify (3.3.14).

8. Fill in omitted details in the proof of (3.3.15).
9. The inequality (3.3.17) is based on the estimate

$$\left| \int_a^b G(s)\, df(s) \right| \leqslant \int_a^b |G(s)|\, dV_a^s[f],$$

where $G(s)$ is continuous in $[a, b]$. Verify this inequality by considering the corresponding Riemann-Stieltjes sums.

10. Fill in other omitted details in the discussion of (3.3.16).
11. What conditions should $K(s, t)$ satisfy in order that (3.3.16) define a linear bounded transformation from $BV[a, b]$ to $C[a, b]$? Get an upper bound for the norm of such a transformation.
12. Discuss the space $\mathfrak{M}_n BV[a, b]$ along the lines of the corresponding discussion for $\mathfrak{M}_n C[a, b]$ in Section 3.2. The elements of $\mathfrak{M}_n BV[a, b]$ are n by n matrices $\{f_{jk}(t)\}$ with entries in $BV[a, b]$.

COLLATERAL READING

For a review of the elementary properties of sequences, continuous functions, and functions of bounded variation the reader may find one of the following treatises useful:

BARTLE, R. G., *The Elements of Real Analysis*, Wiley, New York (1964).

HILLE, E., *Analysis*, 2 vols., Blaisdell, New York (1965).

See also the references under Chapter 2 for the functional analytical aspects of the spaces in question.

4 LEBESGUE SPACES*

This chapter is a continuation of the study of special linear vector spaces. Here we shall be concerned with integrable functions. Integrable in what sense? As an orientation consider the space of continuous functions on $[a, b]$, where, however, we introduce a different metric than that defined by the sup-norm. Set

$$d(f, g) = \int_a^b |f(t) - g(t)| \, dt,$$

where the integral is taken in the Riemann sense. This clearly defines a normed metric, but the space is no longer complete. If we complete the space by adjoining the limit functions of all Cauchy sequences, we obtain a much larger space, namely $L(a, b)$, the space of Lebesgue integrable functions. More generally, we could consider the corresponding problem in C^n and for a different original metrization. There is a continuum of Lebesgue spaces L_p, $1 \leqslant p \leqslant \infty$, even on the line. These spaces will be the center of attention of this chapter. To make the discussion self-sufficient we start with a fairly detailed discussion of Lebesgue measure and integration. We also include the elements of the theory of Fourier series which, in the L_2 case, will give another concrete realization of a Hilbert space of infinite dimensions.

There are five sections: The m-dim Lebesgue measure; Lebesgue measurable functions; Lebesgue integration; Lebesgue spaces; and Remarks on Fourier series.

4.1 THE m-DIM LEBESGUE MEASURE

We start with the notion of a *σ-algebra* (or a *σ-field*).

Definition 4.1.1. *A non-empty collection* A *of subsets of a set X is called a σ-algebra iff*

1) *$\emptyset$, X belong to* A.

2) *$S \in$ A implies $X \ominus S \equiv S^c \in$ A.*

3) *If $\{S_n\}$ is a sequence of sets in* A, *then the union $\bigcup_{n=1}^{\infty} S_n$ belongs to* A.

---

* I am indebted to Dr. Ih-Ching Hsü for a thorough revision of this chapter.

An ordered pair (X, A) consisting of a set X and a σ-algebra A of subsets of X is called a *measurable space*. Any set in A is called a *measurable set* (more exactly, A-measurable set).

Definition 4.1.2. *A measure is an extended real-valued function μ defined on a σ-algebra A such that* (1) $\mu(\emptyset) = 0$, (2) $\mu(S) \geqslant 0$ *for all* $S \in \mathsf{A}$, *and* (3) μ *is countably additive in the sense that if* $\{S_n\}$ *is any sequence of disjoint sets in* A, *then*

$$\mu\left(\bigcup_{n=1}^{\infty} S_n\right) = \sum_{n=1}^{\infty} \mu(S_n).$$

Since we permit μ to take on $+\infty$, the series $\sum_{n=1}^{\infty} \mu(S_n)$ may be a divergent one. If a measure does not take on $+\infty$, we say that it is a *finite measure*. We shall now list, without proof, a few simple results that will be needed later.

Lemma 4.1.1. *Let μ be a measure defined on a σ-algebra A. If S_1 and S_2 belong to A and $S_1 \subseteq S_2$, then $\mu(S_1) \leqslant \mu(S_2)$. If $\mu(S_1) < +\infty$, then $\mu(S_2 \ominus S_1) = \mu(S_2) - \mu(S_1)$.*

Lemma 4.1.2. *Let μ be a measure defined on a σ-algebra A.*

1) *If $\{S_n\}$ is an increasing sequence in A, then* $\mu\left(\bigcup_{n=1}^{\infty} S_n\right) = \lim_{n\to\infty} \mu(S_n)$.

2) *If $\{T_n\}$ is a decreasing sequence in A and if $\mu(T_1) < +\infty$, then*

$$\mu\left(\bigcap_{n=1}^{\infty} T_n\right) = \lim_{n\to\infty} \mu(T_n).$$

Definition 4.1.3. *A measure space is a triple (X, A, μ) consisting of a set X, a σ-algebra A of subsets of X, and a measure μ defined on A.*

As acceptable sets of exception, sets of measure zero are of great importance in the theory of measure. Before we see the role of sets of measure of zero, we introduce the terminology *almost everywhere*. A property is said to hold almost everywhere (abbreviated a.e.) if the set of points where it fails to hold is a set of measure zero. Thus in particular we say that two functions f, g are equal μ-almost everywhere or that they are equal for μ-almost all x if f and g have the same domain and $\mu\{x; f(x) \neq g(x)\} = 0$. In this case we will often write $f = g$, μ-a.e. Similarly, we often write $f = \lim_{n\to\infty} f_n$ μ-a.e, if there is a set S of measure zero such that $f_n(x)$ converges to $f(x)$ for each x not in S.

After Definitions 4.1.1, 4.1.2, and 4.1.3 are given, examples can follow easily. As an illustration, let X be a given non-empty set, then apparently the *power set* of X, 2^X (i.e. the set of all subsets of X) is a σ-algebra. So $(X, 2^X)$ is a measurable space. Moreover, let P be a fixed element of X. A function μ on 2^X can be defined as follows:

$$\mu(S) = \begin{cases} 0 & \text{if } P \notin S, \\ 1 & \text{if } P \in S. \end{cases}$$

It is readily seen that μ is a finite measure; it is called the *unit measure* concentrated at P. By definition $(X, 2^X, \mu)$ is a measure space.

A book on measure theory may easily go further on methods of constructing σ-algebras and measures. To make our material self-sufficient, we introduce one method of defining Lebesgue measure on the real line R, then apply the same technique to get the Lebesgue m-dim measure.

The procedure that we employ is the following: we shall obtain a set function μ^* defined for all subsets of R, and then pick up a collection of sets which forms a σ-algebra and on which μ^* becomes a measure. The length $l(I)$ of an interval I is defined, as usual, to be the difference of the coordinates of the endpoints of the interval. For each set S of the real line consider the countable collections $\{I_n\}$ of open intervals which cover S, that is, collections for which $S \subset \bigcup_{n=1}^{\infty} I_n$, and for each such collection consider the sum, $\sum_{n=1}^{\infty} l(I_n)$, of the lengths of the intervals in the collection. This sum is well defined, independently of the order of the terms, since the lengths are positive numbers. If S is an arbitrary subset of R, we define $\mu^*(S) = \inf \sum_{n=1}^{\infty} l(I_n)$, where the infimum is extended over all countable collections $\{I_n\}$ of open intervals such that $S \subset \bigcup_{n=1}^{\infty} I_n$. Though μ^* is not generally a measure, μ^* does have a few properties reminiscent of a measure.

Without proofs, we list the following:

Lemma 4.1.3. *The above-defined function μ^* on the power set of R satisfies*

1) $\mu^*(\emptyset) = 0$.
2) $\mu^*(S) \geqslant 0$, *for* $S \subset R$.
3) *If* $S \subset T \subset R$, *then* $\mu^*(S) \leqslant \mu^*(T)$.
4) *If* $\{S_n\}$ *is a sequence of subsets of* R, *then* $\mu^*\left(\bigcup_{n=1}^{\infty} S_n\right) \leqslant \sum_{n=1}^{\infty} \mu^*(S_n)$.

Lemma 4.1.4. *For every interval I, half-open or open or closed, finite or infinite,* $\mu^*(I) = l(I)$.

Lemma 4.1.3 motivates the following definition of outer measures in an arbitrarily given set X.

Definition 4.1.4. *An outer measure in a set X is a non-negative function defined on the set 2^X of all subsets of X such that*

1) $\mu^*(\emptyset) = 0$.
2) μ^* *is monotonic, that is,* $\mu^*(S) \leqslant \mu^*(T)$ *whenever* $S \subset T \subset X$.
3) μ^* *is countable subadditive, that is,*

$$\mu^*\left(\bigcup_{n=1}^{\infty} S_n\right) \leqslant \sum_{n=1}^{\infty} \mu^*(S_n) \textit{ for every sequence } \{S_n\} \textit{ of subsets of } X.$$

It is very seldom that the domain of a non-trivial measure consists of all the subsets of a set X. It is just as seldom that one picks a σ-algebra *a priori* and then defines directly a measure on it. Almost always one starts with some family B of subsets of X (for example, intervals) and a non-negative function l on B (for example, length of intervals) and then one tries to generate a measure from these. We now discuss a fundamental process for doing so which is due (1918) to C. Carathéodory (1875–1950). It consists of two parts. First, it produces from l and B an outer measure μ^* defined on all the subsets of X. This μ^*, however, is not countably additive in general, but only subadditive. The second part of the process then selects a σ-algebra M on which μ^* is countably additive.

Definition 4.1.5. *Let μ^* be an outer measure in a given set X. A subset S of X is said to be μ^*-measurable iff for each subset T of X we have*

$$\mu^*(T) = \mu^*(T \cap S) + \mu^*(T \ominus S).$$

While a proof is asked in Exercise 4.1, we state the following.

Theorem 4.1.1. *Let μ^* be an outer measure in a given set X, and let* M *denote the collection of all μ^*-measurable sets. The following statements are true.*

1) *If $\mu^*(S) = 0$, then $S \in$* M.
2) M *is a σ-algebra in X.*
3) *μ^* is countably additive on* M. *(Hence if μ denotes the restriction of μ^* to* M, *then μ is a measure on* M.)

Starting with the family B of all open intervals of the real line R, and using the length of intervals as a non-negative function l on B, we have already constructed an outer measure μ^* in R. As a special case of Theorem 4.1.1, we present the following.

Theorem 4.1.2. *The family* M *of all μ^*-measurable subsets of R is a σ-algebra and the restriction of μ^* to* M, *denoted by μ, is a measure on* M. *This μ is called the Lebesgue measure on R. A subset S of R is called Lebesgue measurable iff S is a member of* M.

For the construction of Lebesgue measure in Euclidean m-space, R^m, the basic ideas are the same as in the case of R. Only the details are more complicated. A point $P \in R^m$ is an m-tuple $(x_1, x_2, \ldots, x_m)$ of real numbers. An *open block* (or *open interval, open parallelepiped*) in R^m is a set of the form $B = \{P | a_i < x_i < b_i;$ $i = 1, 2, \ldots, m\}$. By convention, b_i $(1 \leqslant i \leqslant m)$ or a_i $(1 \leqslant i \leqslant m)$ may take on $+\infty$ or $-\infty$ to form infinite open blocks. By introducing an obvious modification, a closed block, finite or infinite, in R^m can be defined similarly. We shall also be interested in half-open blocks: $\{P | a_i < x_i \leqslant b_i, i = 1, 2, \ldots, m\}$. For blocks in R^m, half-open, or open or closed, finite or infinite, the quantity

$$v(B) = (b_1 - a_1) \ldots (b_m - a_m),$$

which may be $+\infty$, will be called the volume of the block B. The reader will recognize that $v(B)$ agrees with the usual definitions of length, area, or volume, according as B is an interval of R, a rectangle of R^2, or a rectangular parallelepiped of R^3. If S is an arbitrary subset of R^m, we define $\mu^*(S) = \inf \sum_{n=1}^{\infty} v(B_n)$, where the infimum is extended over all countable collections $\{B_n\}$ of open blocks in R^m such that $S \subset \bigcup_{n=1}^{\infty} B_n$. Without giving proofs, we state the following, which are general cases of Lemmas 4.1.3 and 4.1.4.

Lemma 4.1.5. *μ^* is an outer measure in R^m.*

Lemma 4.1.6. *For every block B, half-open, or open or closed, finite or infinite, $\mu^*(B) = v(B)$.*

Theorem 4.1.1 suggests the following special case.

Theorem 4.1.3. *The family M_m of all μ^*-measurable subsets of R^m is a σ-algebra, and the restriction of μ^* to M_m, $\mu = \mu^*|\mathsf{M}_m$, is a measure on M_m. This μ is called the* m-dim *Lebesgue measure on R^m. A subset S of R^m is called Lebesgue measurable iff S is a member of M_m.*

Lemma 4.1.7. *Let c be any finite real number, and r any one of the integers $1, 2, \ldots, m$; then the sets $\{P|(x_1, x_2, \ldots, x_r, \ldots, x_m) = P \in R^m, x_r > c\}$ and $\{P|(x_1, x_2, \ldots, x_r, \ldots, x_m) = P \in R^m, x_r < c\}$, so-called special half-spaces of R^m, are Lebesgue measurable.*

Proof. Let H be a special half-space of R^m. In order to establish the equality $\mu^*(T) = \mu^*(T \cap H) + \mu^*(T \ominus H) = \mu^*(T \cap H) + \mu^*(T \cap H^c)$, for any subset T of R^m, it suffices to prove $\mu^*(T \cap H) + \mu^*(T \cap H^c) \leqslant \mu^*(T)$, since μ^* is subadditive. If $\mu^*(T) = \infty$, then there is nothing to prove. We assume $\mu^*(T) < \infty$. In case $\{B_n\}$ is a sequence of open blocks such that $T \subset \bigcup_{n=1}^{\infty} B_n$, then $T \cap H \subset \bigcup_{n=1}^{\infty} (B_n \cap H)$ and $T \cap H^c \subset \bigcup_{n=1}^{\infty} (B_n \cap H^c)$. Hence, by monotonicity and countable subadditivity of μ^*,

$$\mu^*(T \cap H) + \mu^*(T \cap H^c) \leqslant \sum_{n=1}^{\infty} \mu^*(B_n \cap H) + \sum_{n=1}^{\infty} \mu^*(B_n \cap H^c)$$

$$= \sum_{n=1}^{\infty} [\mu^*(B_n \cap H) + \mu^*(B_n \cap H^c)].$$

If we show that for every open block B in R^m

$$\mu^*(B) = \mu^*(B \cap H) + \mu^*(B \cap H^c), \tag{4.1.1}$$

then $\mu^*(T \cap H) + \mu^*(T \cap H^c) \leqslant \sum_{n=1}^{\infty} \mu^*(B_n)$, $\mu^*(T \cap H) + \mu^*(T \cap H^c) \leqslant \inf \sum_{n=1}^{\infty} \mu^*(B_n) = \mu^*(T)$, and the theorem is proved.

Suppose B is given by $\{P|P \in R^m, a_i < x_i < b_i, i = 1, 2, \ldots, m\}$. Without loss of generality, we may assume that H is given by $\{P|P \in R^m, x_1 < c\}$. Since (4.1.1) is obviously true if $c < a_1$ or $c \geqslant b_1$, let us suppose $a_1 < c < b_1$. Consequently, $B \cap H$ is an open block given by

$$\{P | P \in R^m,\ a_1 < x_1 < c \quad \text{and} \quad a_i < x_i < b_i, \quad i = 2, 3, \ldots, m\},$$

$B \cap H^c$ is the set given by

$$\{P | P \in R^m,\ c \leqslant x_1 < b_1 \quad \text{and} \quad a_i < x_i < b_i, \quad i = 2, 3, \ldots, m\}.$$

It can be verified easily that

$$\mu^*(B \cap H) = (c - a_1) \prod_{i=2}^{m} (b_i - a_i),$$

$$\mu^*(B \cap H^c) = (b_1 - c) \prod_{i=2}^{m} (b_i - a_i), \text{ and}$$

$$\mu^*(B \cap H) + \mu^*(B \cap H^c) = (b_1 - a_1) \prod_{i=2}^{m} (b_i - a_i) = \mu^*(B). \blacksquare$$

Lemma 4.1.8. *Every half-open block B in R^m is Lebesgue measurable and* $\mu^*(B) = \mu(E) = \nu(B)$.

Proof. By previous lemma, for each i, $1 \leqslant i \leqslant m$, H_i is Lebesgue measurable, where $H_i = \{P | P \in R^m,\ c_i < x_i\}$. The complement K_i of H_i, $K_i = \{P | P \in R^m,\ c_i \geqslant x_i\}$, is therefore also Lebesgue measurable. Without loss of generality, suppose a half-open block B is given by $B = \{P | b_i < x_i \leqslant c_i,\ i = 1, 2, \ldots, m\}$. Then $B = \bigcap_{i=1}^{m} (J_i \cap K_i)$, where $J_i = \{P | P \in R^m,\ b_i < x_i\}$, $i = 1, 2, \ldots, m$. As a finite intersection of Lebesgue measurable sets, B is therefore Lebesgue measurable. It follows from the definition of Lebesgue measurability and Lemma 4.1.6 that $\mu^*(B) = \mu(B) = \nu(B)$. $\blacksquare$

Lemma 4.1.9. *Every open set in R^m is the union of a countable collection of disjoint half-open blocks.*

Proof. Let G be an open set in R^m. For each positive integer k, the hyperplanes

$$x_i = n \cdot 2^{-k}, \quad n = \ldots, -1, 0, 1, 2, \ldots; \quad i = 1, 2, \ldots, m. \tag{4.1.2}$$

partition R^m into a countable collection of disjoint half-open blocks. Let $B_1{}^1, B_1{}^2, B_1{}^3, \ldots$ be a collection of such blocks generated by (4.1.2) for $k = 1$ that are contained in G. We use recursion to define suitable B_k^j's. For $k > 1$, let $B_k{}^1, B_k{}^2, B_k{}^3, \ldots$ be the collection of half-open blocks generated by (4.1.2) which are contained in G but not contained in any block $B_q{}^p$ with $1 \leqslant q < k$. If $P \in G$, then P is an interior point; so there is a partition of R^m given by (4.1.2) such that the block containing P is contained in G. Therefore

$$\bigcup_{k=1}^{\infty} \bigcup_{j} B_k^j \supset G.$$

Since $B_k^j \subset G$ for each j, k, we have

$$\bigcup_{k=1}^{\infty} \bigcup_{j} B_k^j \subset G.$$

This is clearly a countable collection of half-open blocks, and we have constructed them so they are disjoint. ∎

Theorem 4.1.4. *An open set in R^m is Lebesgue measurable and so is a closed set, which is the complement of an open set.*

Proof. The proof follows from Lemmas 4.1.9 and 4.1.8. ∎

Any Lebesgue measurable set can be approximated in measure by open sets from above and by closed sets from below in the following sense.

Theorem 4.1.5. *Let S be any Lebesgue measurable set in R^m. Given $\varepsilon > 0$, then there exists an open set G_ε in R^m and a closed set F_ε in R^m such that $F_\varepsilon \subset S \subset G_\varepsilon$ and $\mu(G_\varepsilon \ominus S) < \varepsilon$, $\mu(S \ominus F_\varepsilon) < \varepsilon$.*

Proof. First, assume that S is bounded. Then $\mu^*(S) = \mu(S)$ is finite. Given $\varepsilon > 0$, then, by definition of μ^*, there is a countable collection of open blocks $\{B_n\}$ such that $S \subset \bigcup_{n=1}^{\infty} B_n$ and $\sum_{n=1}^{\infty} v(B_n) = \sum_{n=1}^{\infty} \mu(B_n) < \mu(S) + \varepsilon$. Put $\bigcup_{n=1}^{\infty} B_n = G_\varepsilon$, then G_ε is an open set in R^m. Furthermore, $\mu(G_\varepsilon) = \mu(\bigcup_{n=1}^{\infty} B_n) \leqslant \sum_{n=1}^{\infty} \mu(B_n) < \mu(S) + \varepsilon$. Since $S \subset G_\varepsilon$ and $\mu(S) < +\infty$, it follows from Lemma 4.1.1 that $\mu(G_\varepsilon \ominus S) = \mu(G_\varepsilon) - \mu(S) < \varepsilon$. Secondly, suppose that S is an unbounded measurable set in R^m. For each positive integer n, consider the following open block ${}_nD$ in R^m

$${}_nD = \{P | P \in R^m, \ -n < x_i < n, \quad i = 1, 2, \ldots, m\}.$$

As an open set in R^m, ${}_nD$ is measurable, by Theorem 4.1.4. Clearly, for each n, ${}_nD$ is bounded. From $R^m = \bigcup_{n=1}^{\infty} {}_nD$, it follows that $S = S \cap R^m = S \cap [\bigcup_{n=1}^{\infty} {}_nD] = \bigcup_{n=1}^{\infty} (S \cap {}_nD) = \bigcup_{n=1}^{\infty} S_n$, where $S_n = S \cap {}_nD$. The previous results now apply to the bounded measurable set S_n: For $\varepsilon > 0$, and for each positive integer n, there is an open set $G_{\varepsilon 2^{-n}} \equiv G_n$ in R^m such that $S_n \subset G_n$ and $\mu(G_n \ominus S_n) < \varepsilon 2^{-n}$. Let $G = \bigcup_{n=1}^{\infty} G_n$, then G is open in R^m, $G \supset S$ and $G \ominus S = \bigcup_{n=1}^{\infty} G_n \ominus \bigcup_{n=1}^{\infty} S_n \subset \bigcup_{n=1}^{\infty} (G_n \ominus S_n)$. Thus $\mu(G \ominus S) \leqslant \mu[\bigcup_{n=1}^{\infty} (G_n \ominus S_n)] \leqslant \sum_{n=1}^{\infty} \mu(G_n \ominus S_n) \leqslant \sum_{n=1}^{\infty} \varepsilon 2^{-n} = \varepsilon$.

To obtain the rest of the assertion, let $T = R^m \ominus S$, then T is measurable. Therefore for each $\varepsilon > 0$ there is an open set G_ε in R^m such that $T \subset G_\varepsilon$ and $\mu(G_\varepsilon \ominus T) < \varepsilon$. Let $F_\varepsilon = R^m \ominus G_\varepsilon$, then F_ε is closed in R^m and $F_\varepsilon \subset S$. Moreover, since $T \cap F_\varepsilon = \emptyset$, $S \ominus F_\varepsilon = (R^m \ominus T) \ominus F_\varepsilon = (R^m \ominus F_\varepsilon) \ominus T = G_\varepsilon \ominus T$. Thus $\mu(S \ominus F_\varepsilon) = \mu(G_\varepsilon \ominus T) < \varepsilon$. ∎

EXERCISE 4.1

1. Let X be any non-empty set. Call a subset S of X *cocountable* (in X) if its complement $X \ominus S$ is countable. Prove that the collection of all those subsets of X that are either countable or cocountable is a σ-algebra of subsets of X.

2. Let B be a non-empty collection of subsets of X. Clearly, the family of all subsets of X is a σ-algebra containing B. Prove that the intersection of all the σ-algebras containing B is also a σ-algebra containing B. This is the smallest σ-algebra generated by B.

3. Prove that $$[a, b] = \bigcap_{n=1}^{\infty} \left(a - \frac{1}{n}, b + \frac{1}{n}\right), \quad (a, b) = \bigcup_{n=1}^{\infty} \left[a + \frac{1}{n}, b - \frac{1}{n}\right].$$
   Hence any σ-algebra of subsets of R which contains all open intervals also contains all closed intervals; any σ-algebra containing all closed intervals also contains open intervals.

4. If μ is a measure on a σ-algebra A and A is a fixed set in A, show that the function λ, defined for $S \in$ A by $\lambda(S) = \mu(S \cap A)$, is a measure on A.

5. Prove Lemma 4.1.2. Show that part (2) of Lemma 4.1.2 may fail if the finiteness condition $\mu(T_1) < +\infty$ is dropped.

6. Let $S_1, S_2, \ldots$ be a sequence of subsets of X. The set of all points $P \in X$ such that $P \in S_n$ for infinitely many values of n is called the *limit superior* of $\{S_n\}$ and denoted by $\limsup_{n\to\infty} S_n$. It may be verified that
   $$\limsup_{n\to\infty} S_n = \bigcap_{n=1}^{\infty} \left[\bigcup_{k=n}^{\infty} S_k\right].$$
   The set of all points $P \in X$ such that $P \in S_n$ for all sufficiently large values of n (how large may depend on P) is called the *limit inferior* of $\{S_n\}$ and denoted by $\liminf_{n\to\infty} S_n$. It may be verified that
   $$\liminf_{n\to\infty} S_n = \bigcup_{n=1}^{\infty} \left[\bigcap_{k=n}^{\infty} S_k\right].$$
   Let (X, A, μ) be a measure space and let $\{S_n\}$ be a sequence in A. Show that $\mu(\liminf_{n\to\infty} S_n) \leqslant \liminf_{n\to\infty} \mu(S_n)$. Also show that $\limsup_{n\to\infty} \mu(S_n) \leqslant \mu(\limsup_{n\to\infty} S_n)$ when
   $$\mu\left(\bigcup_{n=1}^{\infty} S_n\right) < +\infty.$$

7. Prove Lemma 4.1.4 and Theorem 4.1.1.

8. Let X be any non-empty set. Let $\mu^*(\emptyset) = 0$, $\mu^*(X) = 2$, and $\mu^*(S) = 1$ for all other sets. Show that μ^* is an outer measure, and determine the class of μ^*-measurable sets.

9. Let μ^* be an outer measure in a set X. Suppose that $\{S_n\}$ is a sequence of disjoint μ^*-measurable sets and $S = \bigcup_{n=1}^{\infty} S_n$. Show that $\mu^*(A \cap S) = \sum_{n=1}^{\infty} \mu^*(A \cap S_n)$ for any subset A of X.

4.2 LEBESGUE MEASURABLE FUNCTIONS

We shall now take up the theory of extended real-valued measurable functions with domains in R^m.

Definition 4.2.1. *An extended real-valued function f defined on a subset of R^m is called (Lebesgue) measurable if its domain is Lebesgue measurable and if for each real number α, the set $\{P|f(P) > \alpha\}$ is Lebesgue measurable.*

Lemma 4.2.1. *Let f be an extended real-valued function with a measurable domain. Then the following statements are equivalent:*

1) *For each real number α, the set $\{P|f(P) > \alpha\}$ is measurable.*
2) *For each real number α, the set $\{P|f(P) \leqslant \alpha\}$ is measurable.*
3) *For each real number α, the set $\{P|f(P) \geqslant \alpha\}$ is measurable.*
4) *For each real number α, the set $\{P|f(P) < \alpha\}$ is measurable.*

These statements imply:

5) *For each extended real number α^*, the set $\{P|f(P) = \alpha^*\}$ is measurable.*

Proof. Since $\{P|f(P) \leqslant \alpha\}$ and $\{P|f(P) > \alpha\}$ are complements of each other, statement (1) is equivalent to statement (2). Similarly, statements (3) and (4) are equivalent. If (1) holds, then

$$\{P|f(P) \geqslant \alpha\} = \bigcap_{n=1}^{\infty} \left\{P|f(P) > \alpha - \frac{1}{n}\right\}$$

is measurable, since the intersection of a sequence of measurable sets is measurable. Hence (1) implies (3). Similarly, (3) implies (1), since

$$\{P|f(P) > \alpha\} = \bigcup_{n=1}^{\infty} \left\{P|f(P) \geqslant \alpha + \frac{1}{n}\right\},$$

and the union of a sequence of measurable sets is measurable. This shows that the first four statements are equivalent. If α^* is a real number,

$$\{P|f(P) = \alpha^*\} = \{P|f(P) \leqslant \alpha^*\} \cap \{P|f(P) \geqslant \alpha^*\},$$

and so (2) and (3) imply (5) for α^* real. Since

$$\{P|f(P) = \infty\} = \bigcap_{n=1}^{\infty} \{P|f(P) \geqslant n\},$$

(3) implies (5), for $\alpha^* = \infty$. Similarly, (2) implies (5) for $\alpha^* = -\infty$, and we have (2) and (3) imply (5). ∎

Lemma 4.2.2. *Let f and g be real-valued measurable functions and let c be a real number. Then the functions $cf, f^2, f + g, f - g, fg, |f|$ are also measurable.*

Proof. Let D and E be the domain of f and g respectively. Then $D \cap E$, which is the domain of $f + g$ and fg, is measurable, since D and E are measurable.

If $c = 0$, the statement is trivial. If $c > 0$, then $\{P|cf(P) > \alpha\}$ is the same as $\{P|f(P) > \alpha/c\}$, which is measurable for each real α. The case $c < 0$ can be handled similarly.

The function f^2 is measurable, since

$$\{P|f^2(P) > \alpha\} = \{P|f(P) > \sqrt{\alpha}\} \cup \{P|f(P) < -\sqrt{\alpha}\} \quad \text{for } \alpha \geqslant 0$$

and

$$\{P|f^2(P) > \alpha\} = D \quad \text{if} \quad \alpha < 0.$$

If $f(P) + g(P) < \alpha$, then $f(P) < \alpha - g(P)$ and by the density of rational numbers there is a rational number q such that $f(P) < q < \alpha - g(P)$. Hence $\{P|f(P) + g(P) < \alpha\} = \bigcup_q (\{P|f(P) < q\} \cap \{P|g(P) < \alpha - q\})$. Since the rationals are countable, this set is measurable and so $f + g$ is measurable. Since $-g = (-1)g$ is measurable when g is, we have $f - g$ measurable.

The measurability of fg follows from $fg = \frac{1}{4}[(f+g)^2 - (f-g)^2]$ and those results we just obtained. Finally

$$\{P; |f(P)| > \alpha\} = D \quad \text{for} \quad \alpha < 0,$$

and

$$\{P; |f(P)| > \alpha\} = \{P|f(P) > \alpha\} \cup \{P|f(P) < -\alpha\} \quad \text{if} \quad \alpha \geqslant 0.$$

Thus the function $|f|$ is measurable. ∎

If f is any real-valued function, let f^+ and f^- be the non-negative functions defined by $f^+(P) = \max\{f(P), 0\}$, $f^-(P) = \max\{-f(P), 0\}$. The function f^+ is called the *positive part* of f, f^- the *negative part* of f. Clearly, $f = f^+ - f^-$ and $|f| = f^+ + f^-$. Consequently, $f^+ = \frac{1}{2}(|f| + f)$ and $f^- = \frac{1}{2}(|f| - f)$. It follows from the previous lemma that f is measurable if and only if f^+ and f^- are measurable. If f is an extended real-valued measurable function, it should be noted that $|f| + f$ is not defined at points where $|f| = \infty$ and $f = -\infty$. However, the measurability of f^+ and f^- can still be proved by appealing to the definitions:

$$f^+(P) = \max\{f(P), 0\} \quad \text{and} \quad f^-(P) = \max\{-f(P), 0\}.$$

Applying the previous lemma, we can easily check the following

Theorem 4.2.1. *Let S be a Lebesgue measurable set. Then the set of all real-valued measurable functions defined on S forms an algebra over the reals.*

In order to see that measurability is preserved in "passing to the sequential limit," we state the following lemma, whose proof can easily be supplied.

Lemma 4.2.3. *Let $\{f_n\}$ be a sequence of extended real-valued measurable functions with common domain a measurable set S. Define the functions*

$$f(P) = \inf f_n(P), \qquad F(P) = \sup f_n(P),$$
$$f^*(P) = \liminf f_n(P), \qquad F^*(P) = \limsup f_n(P).$$

Then f, F, f^*, and F^* are measurable.

Theorem 4.2.2. *If $\{f_n\}$ is a sequence of extended real-valued measurable functions with a measurable set S as common domain, if $\{f_n\}$ converges to f on S, then f is measurable.*

Proof. In this case $f(P) = \lim f_n(P) = \lim\inf f_n(P)$. ∎

Theorem 4.2.3. *Let D be a measurable subset of R^m. If an extended real-valued function f is defined and continuous on D, then f is measurable.*

Proof. If α is any real number, then $\{P|f(P) > \alpha\} = f^{-1}(\alpha, \infty]$. $(\alpha, \infty]$ is an open subset of the extended real line. $\{P|f(P) > \alpha\}$, as the preimage of $(\alpha, \infty]$ under the continuous function f, is therefore open in D (with the relative topology). There exists an open set G_α of R^m such that $\{P|f(P) > \alpha\} = G_\alpha \cap D$. The measurability of f follows from Theorem 4.1.3. ∎

So far we have restricted ourselves to extended real-valued function f. If f should be complex-valued, we say f is measurable iff its real and imaginary parts are measurable in the previously defined sense. In Exercise 4.2 it is asked to prove that sums, differences, and products of complex-valued measurable functions are measurable. Note that for a complex-valued measurable function f we have a decomposition of the form

$$f(P) = f_1(P) + if_2(P) - f_3(P) - if_4(P),$$

where f_1 to f_4 are real non-negative measurable functions. For example: Let $g(P)$ be the real part of $f(P)$, $h(P)$ the imaginary part of $f(P)$, take

$$f_1(P) = g^+(P), \qquad f_3(P) = g^-(P), \qquad f_2(P) = h^+(P), \qquad f_4(P) = h^-(P).$$

The decomposition is clearly not unique: we can add the same constant c_1 to f_1 and f_3 and the same constant c_2 to f_2 and f_4 without affecting the result.

EXERCISE 4.2

In Exercise 4.2, by measurable functions we shall mean Lebesgue measurable functions.

1. Let D be a dense set of real numbers, that is, a set of real numbers such that every interval contains an element of D. If f is an extended real-valued function on R such that $\{P|f(P) > d\}$ is measurable for each $d \in D$, prove that f is measurable.
2. Suppose $f = g$ a.e., show that f is measurable iff g is measurable.
3. (a) Let f be an extended real-valued function with measurable domain D, D a subset of R^m. Let $D_1 = \{P|f(P) = \infty\}$, $D_2 = \{P|f(P) = -\infty\}$. Then f is measurable iff D_1 and D_2 are measurable and the restriction of f to $D \ominus (D_1 \cup D_2)$ is measurable.

   (b) Prove that the product of two measurable extended real-valued functions is measurable.

   (c) If f and g are measurable extended real-valued functions and α a fixed number, then $f + g$ is measurable if we define $f + g$ to be α whenever it is of the form $\infty - \infty$ or $-\infty + \infty$.

4. Let f be an extended real-valued measurable function and let c be a fixed positive number. Prove that the *truncation* f_c defined by

$$\begin{aligned} f_c(P) &= f(P) \quad \text{if } |f(P)| \leqslant c, \\ &= c \quad \text{if } f(P) > c, \\ &= -c \quad \text{if } f(P) < -c \end{aligned}$$

is measurable.

5. If f and g are extended real-valued measurable functions with the same domain of definition D, show that the extended real-valued functions h and k defined by

$$\begin{aligned} h(P) &= \max\{f(P), g(P)\}, \quad \forall\, P \in D, \\ k(P) &= \min\{f(P), g(P)\}, \quad \forall\, P \in D, \end{aligned}$$

are measurable. In particular, f^+ and f^- are measurable if f is an extended real-valued measurable function with

$$f^+(P) = \max\{f(P), 0\} \quad \text{and} \quad f^-(P) = \max\{-f(P), 0\}.$$

6. Prove Lemma 4.2.3.

7. If $f(P)$ is measurable and non-negative, show that its positive square root has the same properties.

8. Let f be a real-valued measurable function with domain R^m, and let ϕ be a continuous function on R^m to R^m. Show that the composition $\phi \circ f$, defined by $(\phi \circ f)(P) = \phi[f(P)]$, is measurable.

9. Show that sums, products, and limits of complex-valued measurable functions are measurable.

4.3 LEBESGUE INTEGRATION

With the Lebesgue theory of measure at our disposal, we can generalize the problem of "measuring the area under a curve." First we introduce the notion of a *set of ordinates* basic in the geometric treatment of integration.

Definition 4.3.1. *Let S be any subset of R^m and let $f(P)$ be an extended real-valued non-negative function on S; let*

$$\Omega_0(f;S) = \{(x_1, x_2, \ldots, x_m, x_{m+1}) | P = (x_1, x_2, \ldots, x_m) \in S \text{ and } 0 < x_{m+1} < f(P)\},$$
$$\Omega_1(f;S) = \{(x_1, x_2, \ldots, x_m, x_{m+1}) | P = (x_1, x_2, \ldots, x_m) \in S \text{ and } 0 < x_{m+1} \leqslant f(P)\};$$

then $\Omega_0(f; S)$ is called the least, *and $\Omega_1(f; S)$ is called the* greatest *set of ordinates of $f(P)$ over S. The symbol $\Omega(f; S)$ is used to denote any set T satisfying $\Omega_0(f; S) \subset T \subset \Omega_1(f; S)$, and T is called a set of ordinates of $f(P)$ over S.*

It is readily seen that $\Omega(f; S) \subset R^{m+1}$. We must now also consider $(m + 1)$-dimensional sets. Later we shall write μ_m and μ_{m+1} for Lebesgue measures in R^m

and R^{m+1}, respectively. At this stage we restrict ourselves to extended real-valued non-negative functions. After we extend to arbitrary real-valued functions, there will be a natural way to deal with complex-valued functions.

Lemma 4.3.1. *Let Z be a cylinder in R^{m+1} with height h and base B, i.e.*

$$Z = \{(x_1, x_2, \ldots, x_m, x_{m+1}) | (x_1, x_2, \ldots, x_m) \in B \quad \textit{and} \quad 0 < x_{m+1} < h\}.$$

If B is measurable, then Z is measurable and $\mu_{m+1}(Z) = h\mu_m(B)$.

Proof. Suppose B is a half-open block. It can be verified easily (Lemmas 4.1.7 and 4.1.8 should be helpful) that in this case Z is measurable and

$$\mu_{m+1}(Z) = h\mu_m(B). \tag{4.3.1}$$

If B is an open set, B can be written as the union of a countable collection of disjoint half-open blocks by Lemma 4.1.9. Consequently, (4.3.1) holds for any open base B by the countable additivity of the measure. It is then also true for a closed base. (Why? *Hint:* Consider two cases, i.e. B is bounded or unbounded.) To deal with an arbitrary measurable base B, we apply Theorem 4.1.5, which states: *For each $\varepsilon > 0$ there exists an open set G_ε and a closed set F_ε such that $F_\varepsilon \subset B \subset G_\varepsilon$ with $\mu_m(G_\varepsilon \ominus B) < \varepsilon$ and $\mu_m(B \ominus F_\varepsilon) < \varepsilon$.* From this it follows that $h\mu_m(B) - h\varepsilon \leqslant \mu^*(Z) \leqslant h\mu_m(B) + h\varepsilon$ for each $\varepsilon > 0$. Therefore $\mu^*(Z) = h\mu_m(B)$. To prove that Z, with a bounded measurable base B, is an $(m+1)$-dim measurable set, we introduce another criterion of measurability. (Cf. Exercise 4.3.3.) Given a subset S of R^{m+1} with $\mu^*(S) < \infty$, S is measurable iff $\mu^*(S) = \mu_*(S)$, where $\mu_*(S)$ is so-called Lebesgue *inner measure* of S and is defined by

$$\mu_*(S) = \sup \{\mu_{m+1}(E) | E \text{ is measurable and } E \subset S\}.$$

By Theorem 4.1.5, for any $\varepsilon > 0$ there is a closed set F_ε such that $F_\varepsilon \subset B$ and $\mu_m(B \ominus F_\varepsilon) < \varepsilon$. From this it follows that $\mu_*(Z) = h\mu_m(B) = \mu^*(Z)$. Thus Z is measurable with $\mu(Z) = \mu^*(Z) = h\mu_m(B)$, provided B is bounded and measurable. Finally, suppose Z has an unbounded measurable base B. Then, as shown in the proof of Theorem 4.1.5, there is an increasing sequence of bounded measurable sets whose union is B. Apply Lemma 4.1.1 to infer that (4.3.1) holds for an unbounded measurable base B. ∎

Theorem 4.3.1. *Let $f(P)$ be an extended real-valued non-negative function defined on an m-dim measurable set B. Suppose that $f(P)$ is measurable on B, then $\Omega_0(f; B)$ is an $(m+1)$-dim measurable set.*

Proof. We start with the proof that $\Omega_0(f; B)$ is the union of a sequence of cylinders (except when $f(P) = 0$ for every P of B, in which case $\Omega_0(f; B) = \emptyset$). Consider the set of all positive rationals. This set is countable; hence its elements may be listed $h_1, h_2, h_3, \ldots, h_n, \ldots$. Let $B_r = \{P | P \in B, h_r < f(P)\}$ $(r = 1, 2, \ldots)$. Let Z_r denote a cylinder in R^{m+1} with height h_r and base B_r, i.e.

$$Z_r = \{(x_1, x_2, \ldots, x_m, x_{m+1}) | (x_1, x_2, \ldots, x_m) \in B_r, 0 < x_{m+1} < h_r\}.$$

Clearly $\bigcup_{r=1}^{\infty} Z_r \subset \Omega_0(f; B)$. On the other hand, if

$$(P, x_{m+1}) \equiv (x_1, x_2, x_3, \ldots, x_m, x_{m+1}) \in \Omega_0(f; S),$$

then there is a positive rational number, say h_ρ, such that $0 < x_{m+1} < h_\rho < f(P)$, which means that $(P, x_{m+1}) \in Z_\rho$; hence $\Omega_0(f; B) \subset \bigcup_{r=1}^{\infty} Z_r$. Thus $\Omega_0(f;B)$ is the union of a sequence of cylinders.

To prove the measurability of $\Omega_0(f; B)$, it suffices to prove that Z_r is measurable for each r. Since f is measurable, B_r is measurable for each r. By Lemma 4.3.1, Z_r is measurable. ∎

The measurability of $\Omega_0(f; B)$ enables us to give the following

Definition 4.3.2. *Let f be an extended real-valued non-negative function defined and measurable on an m-*dim *measurable set S. Then the Lebesgue integral of f over S, $\int_S f(P)\, dP$, is defined by*

$$\int_S f(P)\, dP = \mu_{m+1}[\Omega_0(f; S)].$$

If $\int_S f(P)\, dP < \infty$, then f is said to be integrable (or summable) over S.

By Lemma 4.3.1 it is clear that there are integrable functions. For example, the constant function $f(P) \equiv 1$ defined on a set S of finite measure is integrable over S and its integral is $\mu_m(S)$. In this case the least set of ordinates, $\Omega_0(f; S)$ is a cylinder in R^{m+1} with base S and height 1.

We have considered in some detail the case where the least set of ordinates is a cylinder with a measurable base. We can now extend to the case where the least set of ordinates is the union of a finite number of disjoint cylinder sets with measurable bases. The corresponding functions $f(P)$ are known as *simple functions.*

Let a bounded measurable set S in R^m have a partition into disjoint measurable subsets as follows.

$$S = S_1 \cup S_2 \cup \cdots \cup S_l, \qquad S_j \cap S_k = \varnothing, \qquad j \neq k. \tag{4.3.2}$$

It is assumed that f has a constant value on each S_j, say

$$f(P) = \alpha_j, \qquad P \in S_j, \qquad j = 1, 2, \ldots, l. \tag{4.3.3}$$

Theorem 4.3.2. *A non-negative simple function f as defined by* (4.3.2) *and* (4.3.3) *is integrable over S and*

$$\int_S f(P)\, dP = \sum_{j=1}^{l} \alpha_j \mu_m(S_j). \tag{4.3.4}$$

Proof. Clearly, such a simple function f is measurable. We assume $\alpha_j > 0$, $j = 1, 2, \ldots, l$. By Lemma 4.3.1, $\alpha_j \mu_m(S_j) < \infty$ is the $(m + 1)$-dim measure of the jth cylinder with height α_j and measurable base S_j. As the union of l disjoint measurable cylinders, $\Omega_0(f; S)$ therefore yields

$$\int_S f(P)\,dP = \mu_{m+1}[\Omega_0(f; S)] = \sum_{j=1}^{l} \alpha_j \mu_m(S_j) < \infty,$$

by the countable additivity of μ_{m+1}. ∎

Formula (4.3.4) may be extended in various directions. First the restriction to simple functions may be dropped. A countable partition of S, $S = \bigcup_{j=1}^{\infty} S_j$ each S_j measurable $S_j \cap S_k = \emptyset$, $j \neq k$, can be used, if a convergence condition $\sum_{j=1}^{\infty} \alpha_j \mu_m(S_j) < \infty$ is imposed. If the α's are merely assumed to be real, α_j is to be replaced by $|\alpha_j|$ for $j = 1, 2, \ldots, n \ldots$.

We have seen that a partition on S introduces an associated partition on $\Omega_0(f; S)$. This suggests other generalizations of (4.3.4), which will show that the fact that the components of $\Omega_0(f; S)$ sometimes are all cylinders is really immaterial. We present the following

Theorem 4.3.3. *Let S be a bounded measurable set in R^m, $f(P)$ a non-negative integrable function over S. Suppose that $S = \bigcup_j S_j$, $S_j \cap S_k = \emptyset$, $j \neq k$, is a finite or countable infinite partition of S into disjoint measurable subsets. Then f is integrable over each of sets S_j (more precisely, the restriction of f to S_j is integrable over S_j) and*

$$\int_S f(P)\,dP = \sum_j \int_{S_j} f(P)\,dP.$$

Proof. For each j, let f_j denote the restriction of f to S_j, i.e. $f_j = f|S_j$. It can be verified easily that f_j is measurable. This implies that for each j, $\Omega_0(f_j; S_j)$ is measurable, by Theorem 4.3.1. Since $S_j \cap S_k = \emptyset$ for $j \neq k$,

$$\Omega_0(f_j; S_j) \cap \Omega_0(f_k; S_k) = \emptyset \quad \text{for} \quad j \neq k.$$

On the other hand, $\Omega_0(f; S) = \bigcup_j \Omega_0(f_j; S_j)$. By the countable additivity of μ_{m+1}, $\mu_{m+1}[\Omega_0(f; S)] = \sum_j \mu_{m+1}[\Omega_0(f_j; S_j)]$. Since f is integrable over S, $\mu_{m+1}[\Omega_0(f; S)] < \infty$ and $\int_S f(P)\,dP = \sum_j \int_{S_j} f(P)\,dP < \infty$. Here, without causing any ambiguity, $\int_{S_j} f(P)\,dP$ denotes $\int_{S_j} f_j(P)\,dP$, which is obviously finite. ∎

Theorem 4.3.4. *Let f_1 and f_2 be two extended real-valued non-negative measurable functions defined on a measurable set S. If f_2 is integrable over S and if $f_1(P) \leqslant f_2(P)$, $\forall\, P \in S$, then f_1 is integrable over S and $\int_S f_1(P)\,dP \leqslant \int_S f_2(P)\,dP$.*

Proof. The conclusion follows from $\Omega_0(f_1; S) \subset \Omega_0(f_2; S)$ and Lemma 4.1.1. ∎

Theorem 4.3.5. (*Monotone Convergence Theorem.*) *Let $\{f_n\}$ be a monotone increasing sequence of extended real-valued non-negative measurable functions with a measurable set S as common domain. Then $\int_S f(P)\,dP = \lim_{n\to\infty} \int_S f_n(P)\,dP$, where $f(P) \equiv \lim_{n\to\infty} f_n(P)$. (This limit function always exists. Why?) Furthermore, f is integrable over S iff the sequence $\{\int_S f_n(P)\,dP\}$ is bounded.*

Proof. Since f_n is measurable for each n, the sequential limit f of $\{f_n\}$ is therefore measurable, by Theorem 4.2.2. From $f_n(P) \leqslant f_{n+1}(P)$ for each n and for each $P \in S$, and from $f(P) = \lim_{n\to\infty} f_n(P)$, it follows that for each n, $\Omega_0(f_n; S) \subset \Omega_0(f_{n+1}; S)$ and $\Omega_0(f; S) = \bigcup_{n=1} \Omega_0(f_n; S)$. Apply Lemma 4.1.2 to infer

$$\mu_{m+1}[\Omega_0(f; S)] = \lim_{n\to\infty} \mu_{m+1}[\Omega_0(f_n; S)],$$

that is, $\int_S f(P)\, dP = \lim_{n\to\infty} \int_S f_n(P)\, dP$. Here $\int_S f(P)\, dP$ may be $+\infty$. Clearly it is finite iff the sequence $\{\int_S f_n(P)\, dP\}$ is bounded. ∎

Theorem 4.3.6. *Let $f(P)$ be an extended real-valued non-negative measurable function defined on a measurable set S. Then there exists a sequence of simple functions $\{f_n(P)\}$ such that*

1) $0 \leqslant f_n(P) \leqslant f_{n+1}(P), \quad \forall P \in S, \forall n \in N.$

2) $f(P) = \lim_{n\to\infty} f_n(P)$, *for each* $P \in S$.

Proof. Let n be a fixed natural number. If $k = 1, 2, \ldots, n2^n$, let E_{nk} be the set

$$E_{nk} = \{P | P \in S,\ (k-1)2^{-n} \leqslant f(P) < k2^{-n}\},$$

and if $k = (n+1)2^n$, let

$$E_{nk} = \{P | P \in S,\ n \leqslant f(P)\}.$$

We observe that the sets $\{E_{nk} | k = 1, 2, \ldots, (n+1)2^n\}$ are disjoint with their union equal to S, and the measurability of f implies the measurability of each E_{nk}.

A sequence of simple functions $\{f_n(P)\}$ can be constructed by defining $f_n(P) = (k-1)2^{-n}$ for P in E_{nk}. Clearly, we have $f_n(P) \leqslant f(P)$ for all P in S and all n in N. To prove (1) and (2), we observe first that, if $f_n(P) = k2^{-n}$, then $k2^{-n} \leqslant f(P)$ and $0 < 1 \leqslant k \leqslant n2^n$; hence $2k2^{-n-1} \leqslant f(P)$ and

$$0 < 2k \leqslant n2^{n+1} < (n+1)2^{n+1},$$

which by the definition of $f_{n+1}(P)$, implies $f_n(P) = 2k2^{-n-1} \leqslant f_{n+1}(P)$. Now if $f(P) < \infty$, it follows from the construction of $f_n(P)$ that $n > f(P)$ implies $0 \leqslant f(P) - f_n(P) < 2^{-n}$, and so $\lim_{n\to\infty} f_n(P) = f(P)$.

Finally, by the construction of $f_n(P)$, if $f(P) = +\infty$, then $f_n(P) = n$ $(n = 1, 2, \ldots)$, and so (1) and (2) still hold. ∎

Corollary 1. *An extended real-valued non-negative measurable function f is integrable over its domain S iff the sequence $\{J_n\}$ is bounded, where*

$$J_n = \sum_{k=1}^{(n+1)2^n} (k-1)2^{-n} \mu_m(E_{n,k}). \qquad \textit{Further,} \int_S f(P)\, dP = \lim_{n\to\infty} J_n.$$

Proof. It follows from the previous theorem and from the Monotone Convergence Theorem that

$$\int_S f(P)\,dP = \lim_{n\to\infty} \int_S f_n(P)\,dP$$

$$= \lim_{n\to\infty} \sum_{k=1}^{(n+1)2^n} (k-1)2^{-n}\,\mu_m(E_{n,k})$$

$$= \lim_{n\to\infty} J_n.$$

Clearly, $\int_S f(P)\,dP$ is finite iff the sequence $\{J_n\}$ is bounded. ∎

Corollary 2. *Let α be any real number such that $\alpha \geqslant 0$. If f is an extended real-valued non-negative function integrable over the measurable set S, then αf is integrable over S and $\int_S (\alpha f)(P)\,dP = \alpha \int_S f(P)\,dP$.*

Proof. It is clear that αf is measurable. (Cf. Lemma 4.2.1.) If $\alpha = 0$, then $\alpha f = 0$, $\Omega_0(\alpha f; S) = \varnothing$ and

$$\int_S (\alpha f)(P)\,dP = \mu_{m+1}[\Omega_0(\alpha f; S)] = \mu_{m+1}(\varnothing) = 0 = \alpha \int_S f(P)\,dP.$$

Assume now $\alpha > 0$. Apply the previous theorem to infer that there exists a monotone increasing sequence of simple functions $\{f_n(P)\}$ such that $\lim_{n\to\infty} f_n(P) = f(P)$ for each $P \in S$. Then $\{\alpha f_n(P)\}$ is a monotone increasing sequence of simple functions such that $\{\alpha f_n(P)\}$ converges to $(\alpha f)(P)$. It follows from Theorems 4.3.5 and 4.3.2 that αf is integrable over S and

$$\int_S (\alpha f)(P)\,dP = \lim_{n\to\infty} \int_S (\alpha f_n)(P)\,dP$$

$$= \lim_{n\to\infty} \alpha \int_S f_n(P)\,dP$$

$$= \alpha \lim_{n\to\infty} \int_S f_n(P)\,dP = \alpha \int_S f(P)\,dP.$$ ∎

Corollary 3. *If f and g are real-valued non-negative functions both integrable over the measurable set S, then $f + g$ is integrable over S and*

$$\int_S (f+g)(P)\,dP = \int_S f(P)\,dP + \int_S g(P)\,dP.$$

Proof. By Lemma 4.2.2, $f + g$ is measurable. If $\{f_n\}$ and $\{g_n\}$ are monotone increasing sequences of simple functions converging to f and g, respectively, then $\{f_n + g_n\}$ is a monotone increasing sequence of simple functions converging to

$f+g$. It follows from Theorem 4.3.2 and the Monotone Convergence Theorem that $f+g$ is integrable over S and

$$\begin{aligned}\int_S (f+g)(P)\,dP &= \lim_{n\to\infty}\int_S (f_n+g_n)(P)\,dP\\ &= \lim_{n\to\infty}\left[\int_S f_n(P)\,dP + \int_S g_n(P)\,dP\right]\\ &= \lim_{n\to\infty}\int_S f_n(P)\,dP + \lim_{n\to\infty}\int_S g_n(P)\,dP\\ &= \int_S f(P)\,dP + \int_S g(P)\,dP. \blacksquare\end{aligned}$$

We shall now discuss the integration of extended real-valued measurable functions which may take on both positive and negative values. Let f be an extended real-valued function. In Section 4.2 we have set $f = f^+ - f^-$ with $f^+(P) = \max\{f(P), 0\}$ and $f^-(P) = \max\{-f(P), 0\}$. This decomposition of f into the difference of two extended real-valued non-negative functions suggests the following.

Definition 4.3.3. *An extended real-valued measurable function $f(P)$ defined on a measurable set S is integrable over S iff $f^+(P)$ and $f^-(P)$ are integrable over S. In this case, we define its Lebesgue integral over S to be*

$$\int_S f(P)\,dP = \int_S f^+(P)\,dP - \int_S f^-(P)\,dP.$$

Although the integral of f is defined to be the difference of the integrals of f^+, f^-, we remark that if $f = f_1 - f_2$, where f_1, f_2 are any non-negative measurable functions with finite integrals over S, then

$$\int_S f(P)\,dP = \int_S f_1(P)\,dP - \int_S f_2(P)\,dP.$$

In fact, since $f^+ - f^- = f = f_1 - f_2$, it follows that $f^+ + f_2 = f_1 + f^-$. Apply Corollary 3 to Theorem 4.3.6 to infer that

$$\int_S f^+(P)\,dP + \int_S f_2(P)\,dP = \int_S f_1(P)\,dP + \int_S f^-(P)\,dP.$$

Since all these terms are finite, the following is obtained:

$$\int_S f(P)\,dP = \int_S f^+(P)\,dP - \int_S f^-(P)\,dP = \int_S f_1(P)\,dP - \int_S f_2(P)\,dP.$$

The next result is sometimes referred to as the *property of absolute integrability* of the Lebesgue integral.

Theorem 4.3.7. *A real-valued measurable function f defined on a measurable set S is integrable over S iff $|f|$ is integrable over S. In this case*

$$\left|\int_S f(P)\,dP\right| \leqslant \int_S |f(P)|\,dP, \tag{4.3.5}$$

where the equality holds iff $f(P)$ has the same sign almost everywhere on S.

Proof. By definition f is integrable over S iff f^+ and f^- are integrable over S. Since $|f|^+ = |f| = f^+ + f^-$, $|f|^- = 0$, $0 \leqslant f^+ \leqslant |f|$ and $0 \leqslant f^- \leqslant |f|$, the first part of the assertion follows from Theorem 4.3.4 and Corollary 3 to Theorem 4.3.6. Moreover,

$$\left|\int_S f(P)\,dP\right| = \left|\int_S f^+(P)\,dP - \int_S f^-(P)\,dP\right|$$

$$\leqslant \int_S f^+(P)\,dP + \int_S f^-(P)\,dP = \int_S |f|(P)\,dP.$$

For the proof of the statement on equality we shall find the following helpful.

Theorem 4.3.8. *Let f be a real-valued non-negative function defined and integrable on a measurable set S. Then the integral of f over S is positive iff f is positive in a subset of S of positive measure.*

Proof. Suppose that the measurable set $S^+ = \{P|P \in S, f(P) > 0\}$ has positive measure. For every integer n define the set

$$E_n = \{P|P \in S, 2^{n-1} < f(P) \leqslant 2^n\}.$$

We observe that the sets E_n $(n = 0, \pm 1, \pm 2, \ldots)$ are disjoint measurable sets with their union equal to S. The countable additivity of the measure implies $\mu_m(S^+) = \sum_{-\infty}^{\infty} \mu_m(E_n)$. This is a convergent series of non-negative terms with a positive sum. Hence at least one term is positive, say $\mu_m(E_k) > 0$. Then $\int_S f(P)\,dP = \int_{S^+} f(P)\,dP \geqslant \int_{E_k} f(P)\,dP \geqslant 2^{k-1}\mu_m(E_k) > 0$ as asserted. On the other hand, if all terms in $\mu_m(S^+) = \sum_{-\infty}^{\infty} \mu_m(E_n)$ are zero, then the least ordinate set associated with the restriction of f to E_n is a subset of the cylinder set with base E_n and height 2^n. Since $\mu_m(E_n) = 0$, it follows from Lemma 4.3.1 that the $(m+1)$-dim measure of such a cylinder set is 0. Hence $\mu_{m+1}[\Omega_0(f; S)] = 0$. Thus the condition is necessary as well as sufficient. ∎

We are now ready to complete the proof of Theorem 4.3.7.

Put $\int_S f^+(P)\,dP = \alpha$, $\int_S f^-(P)\,dP = \beta$. Put $S^+ = \{P|P \in S, f(P) > 0\}$, $S^- = \{P|P \in S, f(P) < 0\}$. Then the right member of (4.3.5) is $\alpha + \beta$ while the left is either $\alpha - \beta$ or $\beta - \alpha$. Now $\alpha - \beta = \alpha + \beta$ iff $\beta = 0$; $\beta - \alpha = \alpha + \beta$ iff $\alpha = 0$. By Theorem 4.3.8, $\beta = 0$ iff $\mu_m(S^-) = 0$, i.e. $f(P)$ is $\geqslant 0$ almost everywhere on S; $\alpha = 0$ iff $\mu_m(S^+) = 0$, i.e. $f(P)$ is $\leqslant 0$ almost everywhere on S. ∎

The linearity of the Lebesgue integration should become clear through the following

Theorem 4.3.9. *Let α be any real number. If f and g are real-valued functions defined and integrable over a measurable set S, then αf and $f + g$ are integrable over S with*

$$\int_S (\alpha f)(P)\,dP = \alpha \int_S f(P)\,dP,$$

$$\int_S [f(P) + g(P)]\,dP = \int_S f(P)\,dP + \int_S g(P)\,dP.$$

Proof. The measurability of functions αf and $f + g$ is established in Lemma 4.2.2. If $\alpha = 0$, then $\alpha f = 0$ everywhere so that

$$\int_S \alpha f(P)\,dP = \mu_{m+1}[\Omega_0(\alpha f; S)] = \mu_{m+1}(\emptyset)$$

$$= 0 = \alpha \int_S f(P)\,dP.$$

If $\alpha > 0$, then $(\alpha f)^+ = \alpha f^+$ and $(\alpha f)^- = \alpha f^-$, whence

$$\int_S (\alpha f)(P)\,dP = \int_S \alpha f^+(P)\,dP - \int_S \alpha f^-(P)\,dP.$$

By Corollary 3 to Theorem 4.3.6

$$\int_S \alpha f^+(P)\,dP = \alpha \int_S f^+(P)\,dP, \qquad \int_S \alpha f^-(P)\,dP = \alpha \int_S f^-(P)\,dP.$$

Thus

$$\int_S (\alpha f)(P)\,dP = \alpha \int_S f^+(P)\,dP - \alpha \int_S f^-(P)\,dP = \alpha \int_S f(P)\,dP.$$

If $\alpha < 0$, let $\alpha = -\beta$ with $\beta > 0$. Then $(\alpha f)^+ = (\beta f)^- = \beta f^-$ and $(\alpha f)^- = (\beta f)^+ = \beta f^+$. Thus

$$\int_S (\alpha f)(P)\,dP = \int_S (\alpha f)^+(P)\,dP - \int_S (\alpha f)^-(P)\,dP$$

$$= \int_S (\beta f^-)(P)\,dP - \int_S (\beta f^+)(P)\,dP$$

$$= \beta \int_S f^-(P)\,dP - \beta \int_S f^+(P)\,dP$$

$$= -\beta \int_S f(P)\,dP = \alpha \int_S f(P)\,dP.$$

If f and g are integrable over S, so are $|f|$ and $|g|$. Since $|f+g| \leqslant |f| + |g|$, it follows from Corollary 3 to Theorem 4.3.6 and from Theorem 4.3.4 that $f+g$ is integrable. To establish the desired relation, we observe that

$$f + g = (f^+ + g^+) - (f^- + g^-).$$

Since $f^+ + g^+$ and $f^- + g^-$ are non-negative integrable functions, it follows from the remark made after Definition 4.3.3 that

$$\int_S (f+g)(P)\,dP = \int_S (f^+ + g^+)(P)\,dP - \int_S (f^- + g^-)(P)\,dP.$$

Using Corollary 3 to Theorem 4.3.6, we obtain

$$\int_S (f+g)(P)\,dP = \int_S f^+(P)\,dP - \int_S f^-(P)\,dP + \int_S g^+(P)\,dP - \int_S g^-(P)\,dP$$

$$= \int_S f(P)\,dP + \int_S g(P)\,dP. \quad \blacksquare$$

Recall that a complex-valued function $f(P)$ defined on a measurable set S is said to be measurable iff both the real part, $g(P)$, and the imaginary part, $h(P)$, of $f(P)$ are measurable. Similarly, we have

Definition 4.3.4. *A complex-valued measurable function f defined on a measurable set S is said to be integrable over S iff both g and h are integrable over S. In this case, we define the Lebesgue integral of $f(P) = g(P) + ih(P)$ to be*

$$\int_S f(P)\,dP = \int_S g(P)\,dP + i\int_S h(P)\,dP.$$

With this definition in mind, we can easily extend the entire theory of integration to complex-valued measurable functions. For example:

Theorem 4.3.10. *A complex-valued function $f(P)$ is integrable over the measurable set S iff $|f(P)|$ is integrable over S, and then*

$$\left|\int_S f(P)\,dP\right| \leqslant \int_S |f(P)|\,dP \tag{4.3.6}$$

with equality iff there exists a constant c of unit modulus such that $f = c|f|$ almost everywhere on S.

Proof. Since $f = g + ih$, $|f| = \sqrt{g^2 + h^2} \leqslant |g| + |h|$. If f is integrable over S, then $|g|$ and $|h|$ are integrable over S. Consequently, $|g| + |h|$ is integrable over S by Corollary 3 to Theorem 4.3.6. On the other hand, as the non-negative square root of a non-negative measurable function, $|f|$ is measurable. (Cf. Exercise 4.2.7.) From $|f| \leqslant |g| + |h|$ and from Theorem 4.3.4 it follows that $|f|$ is integrable over S.

Conversely, suppose that $|f|$ is integrable over S. From $|g| \leqslant \sqrt{g^2 + h^2} = |f|$,

$|h| \leqslant \sqrt{g^2 + h^2} = |f|$, and from Theorem 4.3.4 it follows that g and h are integrable over S. This implies f is integrable over S.

To prove (4.3.6), suppose $\int_S f(P)\,dP = re^{i\theta}$ with r, θ real; consider the function $e^{-i\theta}f(P)$. Since $|f(P)| = |e^{-i\theta}f(P)|$ is integrable over S, so is $e^{-i\theta}f(P)$. Let $G(P)$ and $H(P)$ be the real part and imaginary part of $e^{-i\theta}f(P)$ respectively. Then

$$\begin{aligned} e^{-i\theta}f(P) &= (\cos\theta - i\sin\theta)[g(P) + ih(P)] \\ &= g(P)\cos\theta + h(P)\sin\theta + i[h(P)\cos\theta - g(P)\sin\theta] \\ &= G(P) + iH(P). \end{aligned} \tag{4.3.7}$$

This implies

$$\int_S e^{-i\theta}f(P)\,dP = \int_S [g(P)\cos\theta + h(P)\sin\theta]\,dP + i\int_S [h(P)\cos\theta - g(P)\sin\theta]\,dP.$$

By the linearity of the integral (cf. Theorem 4.3.10),

$$\begin{aligned} \int_S e^{-i\theta}f(P)\,dP &= \cos\theta \int_S [g(P) + ih(P)]\,dP - i\sin\theta \int_S [g(P) + ih(P)]\,dP \\ &= (\cos\theta - i\sin\theta)\,re^{i\theta} = r. \end{aligned}$$

On the other hand,

$$\int_S e^{-i\theta}f(P)\,dP = \int_S G(P)\,dP + i\int_S H(P)\,dP.$$

This implies that $r = \int_S G(P)\,dP$.

Since $|f(P)| = |e^{-i\theta}f(P)| = \sqrt{G^2(P) + H^2(P)} \geqslant |G(P)|$,

$$\begin{aligned} \int_S |f(P)|dP &= \int_S |e^{-i\theta}f(P)|dP = \int_S \sqrt{G^2(P) + H^2(P)}\,dP \\ &\geqslant \int_S |G(P)|dP \\ &\geqslant \left|\int_S G(P)\,dP\right| = r = \left|\int_S f(P)\,dP\right|. \end{aligned} \tag{4.3.8}$$

We have established (4.3.6).

To prove the statement on equality, first we assume $f = c|f|$ for some complex number $c = a + ib$ with $|c| = 1$. Then

$$\begin{aligned} \int_S f(P)\,dP &= \int_S a|f|(P)\,dP + i\int_S b|f|(P)\,dP \\ &= (a + ib)\int_S |f|(P)\,dP \\ &= c\int_S |f|(P)\,dP. \end{aligned}$$

Thus

$$\left|\int_S f(P)\,dP\right| = |c|\int_S |f|(P)\,dP = \int_S |f|(P)\,dP.$$

Conversely, let us assume $|\int_S f(P)\,dP| = \int_S |f(P)|dP$. From (4.3.8) it follows that

$$\int_S |G(P)|dP = \left|\int_S G(P)\,dP\right| \tag{4.3.9}$$

and

$$\int_S \sqrt{G^2(P) + H^2(P)}\,dP - \int_S |G(P)|dP$$

$$= \int_S [\sqrt{G^2(P) + H^2(P)} - \sqrt{G^2(P)}]\,dP = 0. \tag{4.3.10}$$

By Theorem 4.3.7, (4.3.9) yields $G(P) \geqslant 0$ a.e. on S or $G(P) \leqslant 0$ a.e. on S. By Theorem 4.3.8, (4.3.10) yields $H(P) = 0$ a.e. on S and $\sqrt{G^2(P) + H^2(P)} = \sqrt{G^2(P)}$ a.e. on S.

If $G(P) \geqslant 0$ a.e. on S, then

$$|f(P)| = \sqrt{G^2(P) + H^2(P)} = G(P) \quad \text{a.e. on } S. \tag{4.3.11}$$

From (4.3.7) and from $H(P) = 0$ a.e. on S, it follows that

$$H(P) = h(P)\cos\theta - g(P)\sin\theta = 0 \quad \text{a.e. on } S$$

and

$$\begin{aligned} G(P) &= g(P)\cos\theta + h(P)\sin\theta \\ &= (\cos\theta - i\sin\theta)\,[g(P) + ih(P)] \quad \text{a.e. on } S \\ &= e^{-i\theta}f(P). \end{aligned} \tag{4.3.12}$$

Combine (4.3.11) and (4.3.12) to infer

$$e^{i\theta}|f(P)| = f(P).$$

The case of $G(P) \leqslant 0$ a.e. on S can be handled similarly and equally easily. ∎

Applying the notion of almost everywhere, which was first introduced in Section 4.1, we may define an equivalence relation on the set of all extended real-valued measurable functions defined on the measurable set S. Two such functions $f(P)$ and $g(P)$ are said to be equivalent, denoted by $f \sim g$, iff f and g are equal μ_m-almost everywhere on S, i.e. $\mu_m\{P; f(P) \neq g(P)\} = 0$. It can be verified easily that (i) $f \sim f$, (ii) $f \sim g$ iff $g \sim f$, (iii) $f \sim g$ and $g \sim h$ implies $f \sim h$. We note that $f \sim g$ implies $\int_S f(P)\,dP = \int_S g(P)\,dP$ if either side exists. This result is important in the next section.

The last range of ideas to be discussed in this section centers around the theorem of Fubini (1907), which will be stated without a proof. Suppose that $f(P)$

is a non-negative function integrable over the measurable set S. Without loss of generality, we may extend the definition of $f(P)$ by setting $f(P) = 0$ for $P \in R^m \ominus S$. The extended function is clearly measurable and its least set of ordinates has the same measure so the integral is unchanged. Now $\int_{R^m} f(P)\, dP$ has a meaning and the question arises: Does the repeated integral

$$\int_{-\infty}^{\infty} \int_{-\infty}^{\infty} \cdots \int_{-\infty}^{\infty} f(x_1, x_2, \ldots, x_m)\, dx_1\, dx_2 \cdots dx_m$$

have a meaning, preferably one which is independent of the order of the integration and which agrees with $\int_{R^m} f(P)\, dP$? An answer to this question is given by the theorem of Guido Fubini (1879–1943), which we formulate as follows.

Theorem 4.3.11. *If $f(P)$ is non-negative integrable over R^m, then there exists at least one function $g(P)$ with $0 \leqslant g(P) \leqslant f(P)$ and $g \sim f$, such that all repeated integrals*

$$\int_{-\infty}^{\infty} \cdots \int_{-\infty}^{\infty} g(x_1, x_2, \ldots, x_m)\, dx_1\, dx_2 \cdots dx_m,$$

for any order of integration, exist and are equal, the common value being that of $\int_{R^m} f(P)\, dP$. Conversely, the existence of such a function $g(P)$, for which one of the repeated integrals exists and has a finite value, is sufficient for the integrability of $f(P)$ over R^m, provided $f(P)$ is measurable.

EXERCISE 4.3

1. Let $f(P)$ be a non-negative function defined on a set B in R^m. Show that the outer measure of $\Omega_0(f; B)$ is equal to that of $\Omega_1(f; B)$, i.e. $\mu^*[\Omega_0(f; B)] = \mu^*[\Omega_1(f; B)]$.

2. Let $\{f_n(P)\}$ be a monotone increasing sequence of non-negative functions defined on a set B in R^m. Let $f(P) = \lim_{n\to\infty} f_n(P)$ for each P in B. Show that
$$\Omega_0(f; B) = \bigcup_{n=1}^{\infty} \Omega_0(f_n; B).$$

3. For an arbitrary subset S of R^{m+1}, define the Lebesgue inner measure $\mu_*(S)$ of S by taking $\mu_*(S) = \sup \{\mu_{m+1}(E) | E$ is measurable and $E \subset S\}$. Prove that
   (a) If S is measurable, then $\mu^*(S) = \mu_*(S)$.
   (b) If $\mu_*(S) = \mu^*(S) < \infty$, then S is measurable.

4. Show that the sum, scalar multiple, and product of simple functions are simple functions.

5. Show that the Monotone Convergence Theorem need not hold for a decreasing sequence of non-negative measurable functions.

6. Use the Monotone Convergence Theorem to prove Fatou's Lemma, which is very useful for it enables us to handle sequences of functions that are not monotone. Fatou's Lemma: Let $\{f_n\}$ be a sequence of extended real-valued non-negative measurable functions (with S as the same domain of definition), then

$$\int_S (\liminf f_n)\, dP \leqslant \liminf \int_S f_n\, dP.$$

7. Use Fatou's Lemma and Theorem 4.3.10 to establish the following Lebesgue Dominated Convergence Theorem: Let $\{f_n\}$ be a sequence of integrable functions which converges to a real-valued measurable function f almost everywhere on S over which each f_n is integrable. If there exists an integrable (over S) function g such that $|f_n| \leqslant g$ for all n, then f is integrable over S and $\int_S f(P)\, dP = \lim \int_S f_n(P)\, dP$. [*Hint:* $|f_n| \leqslant g$ implies $g + f_n \geqslant 0$ and $g - f_n \geqslant 0$].
8. If an extended real-valued function $f(P)$ is integrable over S, show that f is finite almost everywhere in S, that is, the subsets $f^{-1}(\infty)$ and $f^{-1}(-\infty)$ are of measure 0.
9. Suppose that for each n, f_n is an extended real-valued function integrable over S. Suppose further $\sum_{n=1}^{\infty} \int_S |f_n(P)| dP < \infty$. Prove that the series $\sum_{n=1}^{\infty} f_n(P)$ converges almost everywhere (on S) to a real-valued integrable function f. Moreover,

$$\int_S f(P)\, dP = \sum_{n=1}^{\infty} \int_S f_n(P)\, dP.$$

4.4 LEBESGUE SPACES

After this discussion of Lebesgue measure and integration we proceed to a survey of the linear vector spaces formed by equivalence classes of functions integrable over a measurable set S in R^m. Usually the set S plays only a minor role and may be taken as all of R^m.

We consider vector spaces $L_p(S)$, where $1 \leqslant p \leqslant \infty$. For p finite the elements of one of these spaces are the equivalence classes of functions measurable over S such that

$$\|f\|_p = \left[\int_S |f(P)|^p\, dP\right]^{1/p} \tag{4.4.1}$$

exists as a finite number. The pth root is extracted to obtain homogeneity since it is planned to use this expression as the norm of f in L_p. The case $p = \infty$ is deferred for the present.

We start with $p = 1$. Any function $P \to f(P)$ which is measurable and integrable over S defines a class of equivalent functions. We recall that g is equivalent to f, if $f(P) = g(P)$ for almost all P, i.e. except for a set of m-dimensional measure zero. The equivalence classes are the elements of $L_1(S)$, usually written $L(S)$. Each equivalence class is represented by any one of its elements and the representative need be defined only almost everywhere.

Operations on equivalence classes are defined by operations on representative elements. If f and g define two equivalence classes $\{f\}$ and $\{g\}$, then $f + g$ defines an equivalence class $\{f + g\}$ which, by definition, is the sum of the equivalence classes $\{f\}$ and $\{g\}$. Similarly, we set $\alpha\{f\} = \{\alpha f\}$. Since $f + g$ and αf are measurable and, by Theorem 4.3.9, integrable over S, the set $L(S)$ is a linear vector space under this definition of addition and scalar multiplication.

We introduce a norm in $L(S)$ by (4.4.1) with $p = 1$. It is to be thought of as the norm of the equivalence class, though usually we omit the braces. It is clear that this convention leads to a norm. It is non-negative and zero iff $f \sim 0$. Properties (N_2) and (N_3) are also obvious. It will be shown later that $L(S)$ is complete under this norm.

Suppose now that p is fixed, $1 < p < \infty$. The elements of $L_p(S)$ are the equivalence classes of functions f measurable over S for which the integral in (4.4.1) has a finite value. The definitions of addition and scalar multiplication are unchanged. While αf obviously is in $L_p(S)$, it is not clear that $f + g$ always has this property. It is required to prove that if f and g generate equivalence classes in $L_p(S)$, so does $f + g$. This follows from the analogue of (3.1.12),

$$|f(P) + g(P)|^p \leqslant [|f(P)| + |g(P)|]^p \leqslant 2^p[|f(P)|^p + |g(P)|^p], \qquad (4.4.2)$$

provided f and g are both defined and finite. Here the third member is the sum of two integrable functions and hence integrable. By Theorem 4.3.4 the first member is also integrable. Hence $L_p(S)$ is a linear vector space under the adopted definition of addition and multiplication. In this case the triangle inequality does not come out as a byproduct of linearity. We shall require analogues of the theorems of Hölder and Minkowski (for l_p) in order to prove that (4.4.1) actually defines a norm. This will be done later.

$L_p(S)$ and $L_q(S)$ are said to be *conjugate Lebesgue spaces* over S if

$$\frac{1}{p} + \frac{1}{q} = 1. \qquad (4.4.3)$$

This makes L_2 *self-conjugate*. Note that the relation of conjugation is an *involution*: the conjugate of the conjugate space is the original space.

We have still to define $L_\infty(S)$. Its elements shall be the equivalence classes of measurable 'essentially bounded'' functions. A function f is *essentially bounded* if it is equivalent to a bounded function. Let the *essential supremum* of f, written ess sup $|f|$, be the infimum of the suprema of the absolute values of $|g(P)|$ for all g belonging to the equivalence class determined by f. We set

$$\|f\|_\infty = \text{ess sup}\,|f(P)|. \qquad (4.4.4)$$

It is easily seen that this is an acceptable norm. The space $L_\infty(S)$ is regarded as the conjugate space of $L_1(S)$, and vice versa. Note that

$$\|f\|_\infty = \inf_{g \in \{f\}} \sup_{P \in S} |g(P)|. \qquad (4.4.5)$$

In passing we ask: Could any Lebesgue space be an algebra? Under the usual definition of multiplication of elements only L_∞ is an algebra. For finite values of p the product in the ordinary sense of two elements of $L_p(S)$ usually fails to be in $L_p(S)$. Thus $t \to t^{-c}$ is in $L_p(0, 1)$ if $cp < 1$, while its square is in L_p iff $2cp < 1$.

In $L_1(R^m)$ there is, however, an alternate mode of defining a product in terms of which the space becomes an algebra. The important operation of *convolution* leads to such a product. We shall sketch the argument which is based on the Fubini theorem. Take two elements f and g in $L(R^m)$ and form

$$(f * g)P \equiv \int_{R^m} f(Q)\, g(P - Q)\, dQ. \tag{4.4.6}$$

Here

$$f(Q)\, g(P - Q) = f(s_1, s_2, \ldots, s_m)\, g(t_1 - s_1, t_2 - s_2, \ldots, t_m - s_m)$$

and the integration is carried out with respect to the s-variables over R^m. Here it may be questioned if the integral exists at least for almost all P, and if so, the resulting function is an element of $L(R^m)$. The answer is in the affirmative on both counts and this follows from Theorem 4.3.11. This shows that the integral of $f(Q)\, g(P - Q)$ over $R^m \times R^m$ exists at least if f and g are replaced by suitable equivalent functions. Moreover,

$$\int_{R^m}\int_{R^m} |f(Q)|\, |g(P - Q)|\, dP\, dQ = \int_{R^m} |f(Q)|\, dQ \cdot \int_{R^m} |g(P)|\, dP,$$

so that the star product is well defined and is an element of $L(R^m)$. Further,

$$\|f * g\|_1 \leqslant \|f\|_1\, \|g\|_1 \tag{4.4.7}$$

so that $L(R^m)$ forms a normed algebra under convolution. It is a commutative algebra for the product does not depend upon the order of the factors. This is shown by a change of variables in (4.4.6).

To prove that the proposed norm (4.4.1) has the triangle property we proceed as in Section 3.1 and start with Hölder's inequality for integrals.

Theorem 4.4.1. *If f and g belong to conjugate Lebesgue spaces over S, say $f \in L_p(S)$, $g \in L_q(S)$, $1 < p < \infty$, then $fg \in L_1(S)$ and*

$$\int_S |f(P)|\, |g(P)|\, dP \leqslant \left\{\int_S |f(P)|^p\, dP\right\}^{1/p} \left\{\int_S |g(P)|^q\, dP\right\}^{1/q} \tag{4.4.8}$$

with equality iff there exists a fixed constant $k > 0$ such that for almost all P we have

$$|f(P)|^p = k|g(P)|^q. \tag{4.4.9}$$

Proof. We may assume that the integrals in the right member of (4.4.8) are $\neq 0$ since otherwise f or g or both are equivalent to zero and there is nothing to prove.

Formula (3.1.16) gives

$$|f(P)|\,|g(P)| \leqslant \frac{1}{p}|f(P)|^p + \frac{1}{q}|g(P)|^q$$

for almost all P and hence by Theorem 4.3.4

$$\int_S |f(P)|\,|g(P)|\,dP \leqslant \frac{1}{p}\int_S |f(P)|^p\,dP + \frac{1}{q}\int_S |g(P)|^q\,dP. \tag{4.4.10}$$

This shows that $fg \in L_1$ and the inequality is the analogue of (3.1.17). Similarly, the analogue of (3.1.18) is

$$\int_S |f(P)|\,|g(P)|\,dP \leqslant \frac{1}{p}w^{-p}\int_S |f(P)|^p\,dP + \frac{1}{q}w^q\int_S |g(P)|^q\,dP. \tag{4.4.11}$$

This function of w has a unique minimum attained for

$$w = \left[\int_S |f(P)|^p\,dP\right]^{1/(p+q)}\left[\int_S |g(P)|^q\,dP\right]^{-1/(p+q)}. \tag{4.4.12}$$

Substitution of this value of w in (4.4.10) gives (4.4.8). In order to get equality there, we must have equality in (4.4.11) for some value of w and this requires that (4.4.9) holds for some fixed k. ∎

The special case $p = 2$ is known as the Bounyakovsky–Schwarz inequality after Viktor Jakovlevič Bounyakovsky (1804–89) and Hermann Amandus Schwarz (1843–1921), the former professor at the University of St. Petersburg (now Leningrad), the latter professor at the University of Berlin (now the Humboldt Universität) where he had been a pupil of Karl Weierstrass and became his successor.

Minkowski's inequality for integrals follows, again in the same manner as in l_p.

Theorem 4.4.2. *If f and g belong to $L_p(S)$, so does $f + g$ and*

$$\left\{\int_S |f(P) + g(P)|^p\,dP\right\}^{1/p} \leqslant \left[\int_S |f(P)|^p\,dP\right]^{1/p} + \left[\int_S |g(P)|^p\,dP\right]^{1/p}. \tag{4.4.13}$$

Proof. It is only the inequality that must be proved; the integrals are already known to exist. We have

$$[|f(P)| + |g(P)|]^p = [|f(P)| + |g(P)|][|f(P)| + |g(P)|]^{p-1}.$$

Next

$$\int_S |f(P)|\,[|f(P)| + |g(P)|]^{p-1}\,dP$$

$$\leqslant \left\{\int_S |f(P)|^p\,dP\right\}^{1/p}\left\{\int_S [|f(P)| + |g(P)|]^p\,dP\right\}^{1/q}.$$

Here we interchange f and g, add the results and simplify to obtain (4.4.13). ∎

Thus it has been shown that (4.4.1) defines a norm in $L_p(S)$.

There are various consequences of (4.4.8) where until further notice S is assumed to be bounded so that $\mu_m(S)$ is finite. Then for all $p > 1$

$$L_p(S) \subset L_1(S) \tag{4.4.14}$$

and it is fairly easy to see that the inclusion is proper. To prove (4.4.14), observe that the function $1(P)$ which is identically one in S belongs to $L_q(S)$ for all q where p and q satisfy (4.4.3). Since the q-norm of $1(P)$ is $[\mu_m(S)]^{1/q}$, it is seen that

$$\|f\|_1 \leqslant [\mu_m(S)]^{1/q} \|f\|_p. \tag{4.4.15}$$

We can rewrite this and similar inequalities in a more elegant form by introducing the *pth mean of f over S*,

$$\mathbf{M}_p[f] \equiv \left[[\mu_m(S)]^{-1} \int_S |f(P)|^p \, dP \right]^{1/p}. \tag{4.4.16}$$

We then get

$$\mathbf{M}_1[f] \leqslant \mathbf{M}_p[f] \tag{4.4.17}$$

with equality iff f is a constant. More generally

$$\mathbf{M}_a[f] \leqslant \mathbf{M}_b[f], \; a < b, \tag{4.4.18}$$

again with equality iff f is (equivalent to) a constant.

Once more, let $\mu_m(S)$ be finite and let S_1 be a measurable subset of S. Consider the characteristic function of S_1. This is the function which equals one in the set S_1 and is zero elsewhere. It is clearly an element of $L_q(S)$ for any $q > 1$ and (4.4.8) gives

$$\int_{S_1} |f(P)| \, dP \leqslant [\mu_m(S_1)]^{1/q} \|f\|_p, \tag{4.4.19}$$

where the norm is with respect to $L_p(S)$. This inequality has an interesting implication. The left member of (4.4.19) is a set function defined on any measurable subset of S and for any element of $L_1(S)$. All we can say in general is that this non-negative function of sets goes to zero with the measure of the set. On the subspace $L_p(S)$ the sharper estimate (4.4.19) holds and this is the best possible in the sense that the exponent $1/q$ can be replaced by no larger number. See Problem 6, Exercise 4.4.

We have introduced a metric in each of the spaces L_p. It remains to prove that the space is complete in its metric. We no longer insist that $\mu_m(S)$ be finite.

Theorem 4.4.3. *The space $L_1(S)$ is complete with respect to the metric defined by*

$$d(f, g) = \|f - g\|_1. \tag{4.4.20}$$

Proof. Suppose that a Cauchy sequence is given. There exists then a subsequence $\{f_n\}$ such that

$$\|f_0\|_1 + \sum_0^\infty \|f_{k+1} - f_k\|_1 \equiv A < \infty. \tag{4.4.21}$$

Consider the infinite series

$$|f_0(P)| + \sum_0^\infty |f_{k+1}(P) - f_k(P)| \equiv F(P) \tag{4.4.22}$$

with the partial sums

$$F_n(P) = |f_0(P)| + \sum_{k=0}^{n} |f_{k+1}(P) - f_k(P)|.$$

The sequence $\{F_n(P)\}$ is non-decreasing and made up of non-negative elements. Further,

$$\lim_{n\to\infty} \int_S F_n(P)\, dP = A.$$

By Theorem 4.3.5 the series (4.4.22) converges for almost all P and defines an element of $L_1(S)$.

Consider the series

$$f_0(P) + \sum_{k=0}^{\infty} [f_{k+1}(P) - f_k(P)] \equiv f(P), \tag{4.4.23}$$

obtained by omitting the absolute value signs in (4.4.22). This series is absolutely convergent for almost all P and for such points

$$|f(P)| \leqslant F(P)$$

so that $f(P) \in L_1(S)$. Moreover,

$$\|f - f_n\|_1 \leqslant \|F - F_n\|_1 \to 0 \quad \text{as} \quad n \to \infty. \tag{4.4.24}$$

What we have proved so far is that the given Cauchy sequence contains a subsequence which converges almost everywhere to a limit which is in $L_1(S)$. If the original Cauchy sequence is $\{g_k\}$, it is to be proved that

$$\lim_{k\to\infty} \|f - g_k\|_1 = 0. \tag{4.4.25}$$

Suppose that

$$f_n = g_{k(n)},$$

where $k(n)$ is a strictly increasing sequence of positive integers. We can choose n so large that for a given $\varepsilon > 0$

$$\|f - f_n\|_1 = \|f - g_{k(n)}\|_1 < \tfrac{1}{2}\varepsilon.$$

If now $k(n) \leqslant k < k(n+1)$, we may assume that

$$\|g_k - g_{k(n)}\|_1 < \tfrac{1}{2}\varepsilon$$

since $\{g_k\}$ is a Cauchy sequence. This gives

$$\|f - g_k\|_1 \leqslant \|f - g_{k(n)}\|_1 + \|g_{k(n)} - g_k\|_1 < \tfrac{1}{2}\varepsilon + \tfrac{1}{2}\varepsilon = \varepsilon.$$

This proves (4.4.25) and shows that $L_1(S)$ is complete in the normed metric. ∎

A similar result holds for $1 < p < \infty$. We shall only sketch the proof.

Theorem 4.4.4. *The space $L_p(S)$ is complete under the norm defined by* (4.4.1) *for* $1 < p < \infty$.

Proof. We can restrict ourselves to the case where S is bounded. If $\{g_k\}$ is the given Cauchy sequence, we can find a subsequence $\{f_n\}$ such that

$$\|f_0\|_p + \sum_{n=1}^{\infty} \|f_{n+1} - f_n\|_p \equiv A < \infty. \tag{4.4.26}$$

As above we set

$$|f_0(P)| + \sum_{n=0}^{\infty} |f_{n+1}(P) - f_n(P)| \equiv F(P) \tag{4.4.27}$$

when the series converges. Then by (4.4.19) and (4.4.26)

$$\int_S F(P)\, dP \leqslant [\mu_m(S)]^{1/q} A,$$

whence it follows that the series (4.4.27) converges almost everywhere. Thus F is measurable and

$$\|F\|_p \leqslant A.$$

Next we set

$$f_0(P) + \sum_{n=0}^{\infty} [f_{n+1}(P) - f_n(P)] \equiv f(P). \tag{4.4.28}$$

Again this series converges absolutely almost everywhere in S and its sum is dominated by $F(P)$ which is in $L_1(S)$. It follows that $f(P)$ is also in $L_1(S)$. Furthermore, the statement that the series converges to f in the pth norm becomes

$$\lim_{n\to\infty} \|f - f_n\|_p = 0. \tag{4.4.29}$$

The proof is then completed as in the case $p = 1$. ∎

Theorem 4.4.5. *The space $L_\infty(S)$ is complete under the normed metric defined by* (4.4.4).

Proof. As above, we choose from the given Cauchy sequence $\{g_k\}$ a subsequence $\{f_n\}$ satisfying a convergence condition. In this case it is assumed that

$$\|f_{n+1} - f_n\|_\infty = a_n \quad \text{with} \quad \Sigma\, a_n < \infty. \tag{4.4.30}$$

We then consider the series

$$|f_0(P)| + \sum_{n=0}^{\infty} |f_{n+1}(P) - f_n(P)| \equiv F(P). \tag{4.4.31}$$

The terms of this series are measurable functions of P on S and each term is equivalent to a bounded function. We proceed to replace the terms of the series by equivalent bounded functions. Here $f_n(P)$ is replaced by an equivalent function $\tilde{f}_n(P)$ subject to the following conditions:

1) *$\tilde{f}_{n+1}(P) - \tilde{f}_n(P)$ is bounded in S and its supnorm is $\leqslant a_n$.*
2) *$\tilde{f}_{n+1}(P) - \tilde{f}_n(P) = f_{n+1}(P) - f_n(P)$ except in a set E_n of measure zero.*

The union of these sets $\bigcup_{n=0}^{\infty} E_n = E$ is also a set of measure zero. We now form

$$\tilde{F}(P) = |\tilde{f}_0(P)| + \sum_{n=0}^{\infty} |\tilde{f}_{n+1}(P) - \tilde{f}_n(P)| \tag{4.4.32}$$

and note that

$$\tilde{F}(P) = F(P)$$

outside the set E. The series in (4.4.32) converges uniformly in S including E. Without restricting the generality we may assume that $\tilde{f}_n(P) = 0$ for $P \in E$ and all n. Thus $\tilde{F}$ is a bounded measurable function and hence defines an equivalence class in $L_\infty(S)$. Further,

$$\lim_{n\to\infty} \tilde{f}_n(P) \equiv \tilde{f}(P)$$

exists uniformly in S and

$$\lim_{n\to\infty} f_n(P) \equiv f(P)$$

exists at least for P not in E. Moreover, $|f(P)| \leqslant |\tilde{f}(P)|$. Thus $f(P)$ is equivalent to a measurable bounded function on S, i.e. f defines an equivalence class in $L_\infty(S)$. We have

$$\text{ess sup}\, |f(P)| \leqslant \text{ess sup}\, F(P) \leqslant \Sigma\, a_n.$$

Further,

$$\lim_{n\to\infty} \|f - f_n\|_\infty = 0$$

and an adaptation of the argument used for $p = 1$ shows that this limit relation may be strengthened to

$$\lim_{k\to\infty} \|f - g_k\|_\infty = 0. \blacksquare$$

We know that simple functions are dense in $L_p(S)$ at least for $p = 1$. An examination of the proof of Theorem 4.3.6 shows that this assertion holds also for $1 < p < \infty$. What is more significant, however, is

Theorem 4.4.6. *Continuous functions are dense in $L_p(S)$ for $1 \leqslant p < \infty$.*

Proof. We start with the case $p = 1$. The assertion is that if $f \in L_1(S)$, then there exists a sequence of continuous functions $\{f_n\}$ such that

$$\lim_{n \to \infty} \|f - f_n\|_1 = 0. \tag{4.4.33}$$

We can make several preliminary reductions of this problem. We can restrict ourselves to bounded non-negative functions f vanishing outside a large but bounded m-dim block B. The norm can be taken with respect to $L_1(B)$. Next we know that f can be approximated arbitrarily closely by a simple non-negative function defined in B. Hence it is enough to prove (4.4.33) when f is a simple function. Now every simple function is of the form

$$g(P) = \Sigma\, \alpha_j\, \chi(P, S_j), \tag{4.4.34}$$

where the α's are positive numbers (since $g \geqslant 0$) and $\chi(P, S_j)$ is the characteristic function of the set S_j, the subset of B where g takes the value α_j. This means that the problem is reduced to proving (4.4.33) when f is a characteristic function, say $f(P) = \chi(P, E)$, where E is a measurable subset of B. Here we are back to the elements of measure theory. From the definition of outer measure we conclude that we can find a finite partial covering E^+ of E with the following properties: (i) E^+ is the union of a finite number of disjoint blocks (they may have boundary points in common); (ii) $\mu_m(E \ominus E^+) + \mu_m(E^+ \ominus E) < \varepsilon$, a preassigned positive number. It is clear that the L_1-norm of

$$\chi(P, E) - \chi(P, E^+)$$

is $< \varepsilon$. Now the characteristic function of E^+ is the sum of the characteristic functions of the constituent blocks since they are disjoint. This means that all we have to do is to prove (4.4.33) when f is the characteristic function of a block in B.

Suppose the block is B_0 and let $f(P) = \chi(P; B_0)$, $P \in B$. Let $\varepsilon > 0$ be given and define a continuous function f_ε as follows. Let B_ε be a block concentric with B_0 and with edges parallel to those of B_0 obtained from B_0 by a similitude with respect to the center P_0 with ratio $(1 + \varepsilon) : 1$. Draw an arbitrary ray from P_0 and mark the points Q_0 and Q_ε where the ray meets the boundaries of B_0 and B_ε, respectively. Let P be any point on this ray and define $f_\varepsilon(P)$ to have the value 1 if P lies between P_0 and Q_0, the value 0 for P outside of B_ε, and let f_ε vary linearly for P between Q_0 and Q_ε. Here $P \to f_\varepsilon(P)$ is obviously continuous on each ray from $P = P_0$ and it is not hard to see that it is not necessary to restrict oneself to a ray and f_ε is a continuous function of P in B. Further,

$$\|f - f\|_1 < \mu_m(B_\varepsilon \ominus B_0),$$

and this is $O(\varepsilon)$ uniformly for all blocks in B. Retracing the steps, we see that any element of $L_1(S)$ can be approximated arbitrarily closely by a continuous function.

If S is bounded, a continuous function f_ε which approximates f in the L_1-metric will have the same property with respect to the L_p-metric. If S is unbounded we first replace f by g_k, where $g_k(P) = f(P)$ or 0 according as the distance of P from the origin is $\leqslant k$ or $>k$. For large values of k the L_p-norm of $f - g_k$ is arbitrarily small and we can proceed as above. ∎

The elements of $L_p(S)$ have certain generalized continuity properties which lead to a definition of a *Lebesgue modulus of continuity*. The definition requires that f is defined for all P, so we assume that $S = R^m$. The reader will find another alternative in Section 4.5. If $H \in R^m$, we can consider

$$g(H) \equiv \left\{ \int_{R^m} |f(P + H) - f(P)|^p \, dP \right\}^{1/p}. \tag{4.4.35}$$

This is a bounded function of H and $0 \leqslant g(H) \leqslant 2\|f\|_p$. For a given positive number h we define

$$\mu_p(h;f) = \sup g(H), \tag{4.4.36}$$

where H ranges over the sphere $\|H\| \leqslant h$ in R^m.

Theorem 4.4.7. *The Lebesgue p-modulus of continuity of f is a bounded, non-negative, non-decreasing continuous function of h such that*

$$\lim_{h \downarrow 0} \mu_p(h;f) = 0 \tag{4.4.37}$$

and

$$\mu_p(h_1 + h_2; f) \leqslant \mu_p(h_1; f) + \mu_p(h_2; f). \tag{4.4.38}$$

Proof. It is clear that $\mu_p(h; f)$ is well defined and positive for $0 < h$ unless f is a constant on its compact support. Further,

$$\mu_p(h_1; f) \leqslant \mu_p(h_2; f), \quad \text{if} \quad 0 < h_1 < h_2, \tag{4.4.39}$$

since the set of H's which have to be taken into account in computing the left member form a proper subset of those relevant for the right member. Since

$$\begin{aligned}\|f(\cdot + H_1 + H_2) - f(\cdot)\|_p &\leqslant \|f(\cdot + H_1 + H_2) - f(\cdot + H_1)\|_p \\ &\qquad + \|f(\cdot + H_1) - f(\cdot)\|_p \\ &= \|f(\cdot + H_2) - f(\cdot)\|_p + \|f(\cdot + H_1) - f(\cdot)\|_p,\end{aligned}$$

formula (4.4.38) results by taking the appropriate suprema.

It remains to prove the continuity and (4.4.37). The latter relation is trivial if f is continuous and of compact support. But by the preceding theorem we can always approximate an $f \in L_p(S)$ in the sense of the norm by a function f_ε which is continuous and of compact support and this implies (4.4.37) for all f.

We have now by (4.4.38) for any h and t, $0 < h < t$,

$$0 \leqslant \mu_p(t; f) - \mu_p(t - h; f) \leqslant \mu_p(h; f) \downarrow 0,$$

$$0 \leqslant \mu_p(t + h; f) - \mu_p(t; f) \leqslant \mu_p(h; f) \downarrow 0,$$

as $h \downarrow 0$. Since $\mu_p(t; f)$ is non-decreasing, these inequalities imply continuity. See Section 3.2 for the ordinary modulus of continuity, formula (3.2.1). ∎

Our next topic is linear functionals on $L_p(S)$. The discussion is attached to formula (4.4.8). It says that if $f \in L_p(S)$, $g \in L_q(S)$, where $1 < p < \infty$ and L_p and L_q are conjugate spaces, then

$$\|fg\|_1 \leqslant \|f\|_p \|g\|_q. \tag{4.4.40}$$

Hence, for any fixed choice of $g \in L_q$

$$\langle f, g\rangle \equiv \int_S f(P)\, g(P)\, dP \tag{4.4.41}$$

defines a linear bounded functional on $L_p(S)$ with bound $\|g\|_q$ so that

$$|\langle f, g\rangle| \leqslant \|g\|_q \|f\|_p. \tag{4.4.42}$$

Here $\|g\|_q$ is the best possible bound, i.e. it gives the norm of the functional. This follows from the fact that it is possible to choose f in L_p so that

i) $f(P)\, g(P) = |f(P)|\, |g(P)|$, and

ii) there is equality in (4.4.42).

To satisfy (ii) it is necessary and sufficient that $|f(P)|^p$ equals a constant multiple of $|g(P)|^q$ for almost all P. These conditions determine f uniquely up to a constant multiplier.

Actually all linear bounded functionals on $L_p(S)$ are of this form, that is, if x^* is such a functional, then there exists $g \in L_q$ such that

$$x^*[f] = \int_S f(P)\, g(P)\, dP. \tag{4.4.43}$$

For $p = 2$ this was proved by M. Fréchet in 1907; the general case is due to F. Riesz in 1910. The formula still makes sense if $f \in L_1$ and $g \in L_\infty$, or vice versa. In the first case, it was proved by H. Steinhaus in 1918 that every linear bounded functional on $L_1(S)$ is given by (4.4.43) with a $g \in L_\infty(S)$ and the norm of the functional is $\|g\|_\infty$. In the second case where $f \in L_\infty(S)$ and $g \in L_1(S)$ the formula defines a linear bounded functional on L_∞ of norm $\|g\|_1$, but this is not the most general functional on $L_\infty(S)$. The latter involves a finitely additive set function $E \to h(E)$ defined on Borel sets of S and is given by a Radon–Stieltjes integral

$$x^*[f] = \int_S f(P)\, dh. \tag{4.4.44}$$

For further details see the literature.

Let us observe that the case $p = 2$ is a special case of Theorem 2.5.11, which says that any linear bounded functional on a Hilbert space is given by an inner product

$$x^*[f] = (f, h), \tag{4.4.45}$$

where h is a fixed uniquely determined element of the space. Now $L_2(S)$ is a Hilbert space with inner product

$$(f, h) = \int_S f(P)\,\overline{h(P)}\,dP, \tag{4.4.46}$$

and this is equivalent to (4.4.43) for $p = 2$.

In conclusion a remark concerning linear bounded transformations from $L_p(S)$ into itself. For $m = 1$ a frequently encountered type is

$$T[f](t) = \int_S K(s, t) f(s)\,ds, \tag{4.4.47}$$

where the kernel K satisfies suitable integrability conditions to ensure the integrability of the transform. These conditions are normally adjusted to the use of Hölder's inequality and the theorem of Fubini. Thus take $p = 2$ and suppose that $K(s, t)$ is measurable and square integrable over $S \times S$. Set

$$\left\{\int_S \int_S |K(s, t)|^2\,ds\,dt\right\}^{1/2} = \|K\|. \tag{4.4.48}$$

Then (4.4.47) exists for almost all t and defines a mapping from $L_2(S)$ into itself with norm $\|T\| \leqslant \|K\|$. The proof is left to the reader.

EXERCISE 4.4

1. A function $t \to f(t)$ is defined on [0, 1] and takes on the three values 0, 1, 2, namely 0 if t is rational, 1 if t is transcendental, and 2 if t is algebraic but not rational. What is the L_∞-norm?
2. Verify that convolution is a commutative product.
3. Verify (4.4.7).
4. Verify that the inclusion (4.4.14) is proper.
5. Prove (4.4.18) and discuss when equality holds.
6. Given the function $t \to f(t) = t^{-1/p}\,[\log(\frac{1}{2}t)]^{-2}$ with $1 < p < \infty$. Consider

$$\max \int_E f(t)\,dt,$$

where E is any interval of length h, $0 < h < 1$, located in the interval $0 < t < 1$. Show that the maximum lies between constant multiples of $h^{1/q}\,[\log(\frac{1}{2}h)]^{-2}$ for small values of h. What bearing does this result have on (4.4.19)?
7. Complete the proof of Theorem 4.4.4.
8. The series (4.4.32) converges uniformly on S. Does this imply continuity of its sum?
9. Complete the proof of Theorem 4.4.5.
10. Fill in missing details in the proof of Theorem 4.4.6.

11. Let $\mathfrak{X} = L(-\infty, \infty)$ and define for $h > 0$

$$f(t; h) = (2h)^{-1} \int_{-h}^{h} f(t + s)\, ds.$$

Show that this "moving average" is a continuous function of t which belongs to $L(-\infty, \infty)$ and its norm is $\leqslant \| f \|$. Actually $f(t; h)$ converges to $f(t)$ as $h \downarrow 0$, pointwise almost everywhere and in the L_1-norm.

12. Prove the last clause of the preceding problem under the added assumption that f itself is uniformly continuous for $-\infty < t < \infty$.

13. Fill in missing details in the proof of Theorem 4.4.7.

14. Show that (4.4.41) defines a linear bounded functional on $L_1(S)$ if $f \in L_1(S)$ and g is fixed in $L_\infty(S)$. Find its norm.

15. Show that the same formula defines a linear bounded functional on $L_\infty(S)$ if $f \in L_\infty$ and g is fixed in L_1. Find its norm.

16. Let χ_n be the characteristic function of the interval $[n, n + 1)$, $n = 0, \pm 1, \pm 2, \ldots$. Let x^* be a linear bounded functional on $L_p(-\infty, \infty)$ and suppose that $x^*[\chi_n] = a_n$. Prove that $\Sigma\, |a_n|^q$ is convergent and

$$g(t) = \sum_{-\infty}^{\infty} a_n\, \chi_n(t) \in L_q(-\infty, \infty).$$

Prove that

$$\int_{-\infty}^{\infty} f(t)\, g(t)\, dt$$

is a linear bounded functional on $L_p(-\infty, \infty)$ which coincides with x^* on the linear subspace spanned by the χ_n's. Here $1 < p < \infty$ and $1/p + 1/q = 1$.

17. With the same notation as in the preceding problem, take $p = 1$ and determine what conditions $\{a_n\}$ and $g(t)$ must satisfy. Show that a linear functional is defined on $L_1(-\infty, \infty)$ and find its norm.

18. Carry out the same investigation for $p = \infty$.

19. Take $1 < p < \infty$ and show that the null space of the functional determined by $g(t)$ contains every f such that $\int_k^{k+1} f(t)\, dt = 0$ and $f(t) = 0$ outside of $[k, k + 1]$. Do the values of x^* on the set $\{\chi_n\}$ determine x^* uniquely?

20. The functions χ_n form an orthogonal system in $L_2(-\infty, \infty)$. Can it be complete?

21. Show that the nth Fourier coefficient of f

$$f_n = \frac{1}{2\pi} \int_{-\pi}^{\pi} f(s)\, e^{-nis}\, ds,$$

n integer or zero, defines a linear bounded functional on any space $L_p(-\pi, \pi)$. Find a bound for this functional.

22. Take $p = 1$ and define products in $L_1(-\pi, \pi)$ by convolution. How is this done? Show that the Fourier coefficients are not merely linear functionals but also *multiplicative*, i.e. $x^*[f * g] = x^*[f]\,x^*[g]$. A constant multiplier should figure before the integral defining the star product and the Fubini theorem is required.
23. Verify the statements made in connection with the transformation (4.4.47).
24. Modify the conditions on K if it be required that T is a mapping from $L_1(S)$ into itself.

4.5 REMARKS ON FOURIER SERIES

With any Lebesgue space $L_p(a, b)$ on the real line are associated orthonormal systems and corresponding expansions which may represent, in one sense or another, the elements of the space in question. At this juncture we restrict ourselves to classical Fourier series

$$f(t) \sim \sum_{-\infty}^{\infty} f_n e^{nit} \tag{4.5.1}$$

with

$$f_n = \frac{1}{2\pi} \int_{-\pi}^{\pi} f(s)\, e^{-nis}\, ds. \tag{4.5.2}$$

For the moment the sign of equivalence in (4.5.1) indicates merely that the coefficients f_n are associated with a given function f in $L_1(-\pi, \pi)$ by the formulas (4.5.2). Any other connection with f remains to be proved.

Here the orthonormal system is

$$\{(2\pi)^{-1/2} e^{nit}\}. \tag{4.5.3}$$

In this theory it is customary to suppose f continued with period 2π outside the interval $(-\pi, \pi)$ and that $f(-\pi) = f(\pi)$. This implies that in (4.5.2) the integral may be taken over any interval of length 2π. The Fourier coefficients f_n are defined for f in any space $L_p(-\pi, \pi)$ and form a bounded double sequence. Much more information will become available in the subsequent discussion.

Theorem 4.4.6 applies to the present situation and shows that continuous functions are dense in $L_p(-\pi, \pi)$ for any p with $1 \leqslant p < \infty$. Such a continuous function is supposed to satisfy the condition $f(-\pi) = f(\pi)$ and be continued with period 2π. The notion of a Lebesgue p-modulus of continuity has to be modified slightly in this case since we cannot integrate a periodic function over the whole real line. Instead we set

$$\mu_p(h; f) = \sup_{|s| \leqslant h} \left\{ \int_{-\pi}^{\pi} |f(t + s) - f(t)|^p\, dt \right\}^{1/p}, \tag{4.5.4}$$

and it is an easy matter to show that this function has all the properties stated in Theorem 4.4.7 for the function, denoted by the same symbol there but defined by formulas (4.4.35) and (4.4.36).

We can utilize these moduli to get estimates for the Fourier coefficients.

Theorem 4.5.1. *If $n \neq 0$, we have for any p, $1 \leqslant p < \infty$,*

$$|f_n| \leqslant \tfrac{1}{2}(2\pi)^{-1/p} \mu_p\left(\frac{\pi}{|n|}; f\right). \tag{4.5.5}$$

Proof. We make a change of variable in (4.5.2), replacing s by $s + \pi/n$. There is no need of changing the interval of integration. This gives

$$f_n = -\frac{1}{2\pi} \int_{-\pi}^{\pi} f\left(s + \frac{\pi}{n}\right) e^{-nis}\, ds,$$

which, added to (4.5.2), gives

$$2f_n = -\frac{1}{2\pi} \int_{-\pi}^{\pi} \left[f\left(s + \frac{\pi}{n}\right) - f(s)\right] e^{-nis}\, ds.$$

Hölder's inequality together with the definition of the pth modulus gives (4.5.5). ∎

Corollary (*Theorem of Riemann–Lebesgue*). *We have*

$$\lim_{|n| \to \infty} f_n = 0. \tag{4.5.6}$$

Proof. This follows from the fact that for any p, $\lim \mu_p(h; f) = 0$. ∎

In the L_2-case the Fourier coefficients have certain extremal properties, as may be expected from Theorem 2.5.6 when applied to $\mathfrak{H} = L_2(-\pi, \pi)$ and the orthonormal system (4.5.3).

Theorem 4.5.2. *In $L_2(-\pi, \pi)$ the partial sums of the Fourier series* (4.5.1) *form a sequence of best approximations to f in the sense that for any n and any choice of $2n + 1$ numbers $c_{-n}, c_{-n+1}, \ldots, c_0, \ldots, c_{n-1}, c_n$*

$$\int_{-\pi}^{\pi} |f(s) - \sum_{-n}^{n} c_k e^{kis}|^2\, ds \geqslant \int_{-\pi}^{\pi} |f(s)|^2\, ds - 2\pi \sum_{-n}^{n} |f_k|^2 \tag{4.5.7}$$

with equality iff $c_k = f_k$ for all k.

Proof. Translate (2.5.14) into L_2-language, keeping in mind the normalization factor in (4.5.3). ∎

Corollary [*Bessel's inequality*]. *We have*

$$2\pi \sum_{-\infty}^{\infty} |f_k|^2 \leqslant \int_{-\pi}^{\pi} |f(s)|^2\, ds. \tag{4.5.8}$$

Proof. This is (2.5.15) in the L_2-case. ∎

This shows, in particular, that the series in the left member of (4.4.21) is convergent. Actually we have the equality

$$2\pi \sum_{-\infty}^{\infty} |f_k|^2 = \int_{-\pi}^{\pi} |f(s)|^2 \, ds \tag{4.5.9}$$

for all $f \in L_2(-\pi, \pi)$. This is known as the *trigonometric closure relation* or *Parseval's identity*.

The reader should not be shocked by the word "trigonometric" in this connection. In classical analysis Fourier series were actually trigonometric series

$$f(t) \sim \tfrac{1}{2} a_0 + \sum_{n=1}^{\infty} [a_n \cos nt + b_n \sin nt] \tag{4.5.10}$$

with

$$\left.\begin{matrix} a_n \\ b_n \end{matrix}\right\} = \frac{1}{\pi} \int_{-\pi}^{\pi} f(s) \left\{\begin{matrix} \sin \\ \cos \end{matrix}\right\} (ns) \, ds. \tag{4.5.11}$$

Using the formula

$$e^{nit} = \cos nt + i \sin nt,$$

we can pass from (4.5.1) to (4.5.10), where for $n \geqslant 0$

$$c_n = \tfrac{1}{2}(a_n - ib_n), \qquad c_{-n} = \tfrac{1}{2}(a_n + ib_n). \tag{4.5.12}$$

We come now to the question of proving (4.5.9). Theorem 4.5.2 asserts that the partial sums of the Fourier series give the best L_2-approximation to f by trigonometric polynomials of a given degree n. Now if we can exhibit a sequence of trigonometric polynomials

$$T_n(t) = \sum_{k=-n}^{n} c_{k,n} e^{kit}$$

for which the left member of (4.5.7) goes to zero as $n \to \infty$, then the right member must have the same property and (4.5.9) results.

There are a large number of such polynomials available in the literature. The oldest and simplest example is due to the Hungarian mathematician Leopold Fejér (1880–1959), who in 1900 showed that the arithmetic means of the partial sums of the Fourier series of a continuous periodic function converge uniformly to the function for all t. Later it was shown that they also have the property of mean convergence in L_2 and this is what is needed here.

We start with the partial sums

$$S_n(t; f) = \sum_{k=-n}^{n} f_k e^{kit} = \int_{-\pi}^{\pi} D_n(s) f(s + t) \, ds. \tag{4.5.13}$$

Here $D_n(s)$ is a trigonometric polynomial of degree n. It is known as the *Dirichlet kernel* [Peter Gustav Lejeune-Dirichlet (1805–59), German mathematician of Huguenot descent]. Formulas (4.5.13) and (4.5.2) show that

$$2\pi D_n(s) = \tfrac{1}{2} + \cos s + \cos 2s + \cdots + \cos ns. \tag{4.5.14}$$

Multiplying both sides by $\sin \frac{1}{2}s$ and using the trigonometric identity

$$2 \sin a \cos b = \sin (a + b) - \sin (a - b),$$

we obtain, finally,

$$D_n(s) = \frac{1}{2\pi} \frac{\sin (n + \frac{1}{2})s}{\sin \frac{1}{2}s}. \tag{4.5.15}$$

The Dirichlet kernel is not of constant sign and the integral of its absolute value over a period is an unbounded function of n, in fact

$$\int_{-\pi}^{\pi} |D_n(s)| \, ds = O[\log n]. \tag{4.5.16}$$

It has been shown that the various convergency defects which plague the theory of Fourier series are ultimately due to (4.5.16). The arithmetic means of the partial sums have much better properties. We form

$$T_n(t; f) = \frac{1}{n + 1} [S_0(t; f) + S_1(t; f) + \cdots + S_n(t; f)]. \tag{4.5.17}$$

This is again a trigonometric polynomial in t of degree n, namely

$$T_n(t; f) = \sum_{-n}^{n} \left(1 - \frac{|k|}{n + 1}\right) f_k \, e^{kit} \tag{4.5.18}$$

with the integral representation

$$T_n(t; f) = \int_{-\pi}^{\pi} F_n(s) f(s + t) \, ds, \tag{4.5.19}$$

where $F_n(s)$ is the *Fejér kernel*, which is the arithmetic mean of the Dirichlet kernels. Using (4.5.15) and the trigonometric identity

$$2 \sin a \sin b = \cos (a - b) - \cos (a + b),$$

we find, after simplification,

$$F_n(s) = \frac{1}{2\pi(n + 1)} \left\{ \frac{\sin \frac{1}{2}(n + 1)s}{\sin \frac{1}{2}s} \right\}^2. \tag{4.5.20}$$

The properties of this kernel are much more favorable for our purposes than those of the Dirichlet kernel. The Fejér kernel is non-negative and this is the crux of the matter. We list the properties that will be used below:

$$F_n(s) \geqslant 0, \tag{4.5.21}$$

$$\int_{-\pi}^{\pi} F_n(s) \, ds = 1, \qquad \forall \, n, \tag{4.5.22}$$

$$\lim_{n \to \infty} F_n(s) = 0, \qquad 0 < |s| \leqslant \pi. \tag{4.5.23}$$

Here (4.5.21) and (4.5.23) follow from (4.5.20), and in the last formula the limit holds uniformly in s provided $|s| \geqslant \delta$, a fixed positive number. See Problem 9, Exercise 4.5. Formula (4.5.14) shows that

$$\int_{-\pi}^{\pi} D_n(s)\,ds = 1, \qquad \forall\, n, \tag{4.5.24}$$

and this implies (4.5.22).

We shall now prove that $T_n(t; f)$ has the desired convergency properties. From this will follow (1) that the partial sums $S_n(t; f)$ also converge to $f(t)$ in L_2 and (2) that the closure relation holds.

Theorem 4.5.3. *For $f \in L_2(-\pi, \pi)$*

$$\lim_{n\to\infty} \int_{-\pi}^{\pi} |f(t) - T_n(t; f)|^2\,dt = 0. \tag{4.5.25}$$

Proof. We have

$$\int_{-\pi}^{\pi} |f(t) - T_n(t; f)|^2\,dt = \int_{-\pi}^{\pi} \left| f(t) - \int_{-\pi}^{\pi} F_n(s) f(s+t)\,ds \right|^2 dt$$

$$= \int_{-\pi}^{\pi} \left| \int_{-\pi}^{\pi} F_n(s)\,[f(s+t) - f(t)]\,ds \right|^2 dt,$$

where (4.5.22) has been used. Here we apply the Bounyakovsky–Schwarz inequality to the s-integral, using as the two factors in L_2

$$[F_n(s)]^{1/2} \quad \text{and} \quad [F_n(s)]^{1/2}\,|f(s+t) - f(t)|$$

so that

$$\left[\int_{-\pi}^{\pi} F_n(s)\,|f(s+t) - f(t)|\,ds\right]^2 \leqslant \int_{-\pi}^{\pi} F_n(s)\,ds \cdot \int_{-\pi}^{\pi} F_n(s)\,|f(s+t) - f(t)|^2\,ds$$

where the first factor is 1 by (4.5.22). Substituting and changing the order of integration (the Fubini theorem!) we obtain

$$\int_{-\pi}^{\pi} F_n(s) \left\{ \int_{-\pi}^{\pi} |f(s+t) - f(t)|^2\,dt \right\} ds \leqslant \int_{-\pi}^{\pi} F_n(s)\,[\mu_2(|s|; f)]^2\,ds \tag{4.5.26}$$

by (4.5.4) with $p = 2$.

Here $\mu_2(|s|; f)$ is a continuous function which tends to zero with s by the analogue of Theorem 4.4.7. Let

$$\mu_2(|s|; f) < M, \quad \forall\, s, \quad \text{and} \quad \mu_2(|s|; f) < \varepsilon, \quad |s| < \delta.$$

It follows that (4.5.26) is dominated by

$$2\varepsilon \int_0^{\delta} F_n(s)\,ds + 2M^2 \int_{\delta}^{\pi} F_n(s)\,ds < \varepsilon + 2M^2 \int_{\delta}^{\pi} F_n(s)\,ds,$$

again by (4.5.22). On the other hand, (4.5.23) shows that the integral goes to 0 as $n \to \infty$ since the integrand converges uniformly to 0. It follows that the superior limit of the integral in (4.5.25) does not exceed ε. Since ε is arbitrary, (4.5.25) holds. ∎

Corollary 1. *The partial sums of the Fourier series of a function $f \in L_2(-\pi, \pi)$ converge to f in the sense of the metric, i.e.*

$$\lim_{n\to\infty} \int_{-\pi}^{\pi} \left| f(t) - \sum_{k=-n}^{n} f_k e^{kit} \right|^2 dt = 0. \tag{4.5.27}$$

Proof. This follows from the inequality

$$\int_{-\pi}^{\pi} |f(t) - T_n(t;f)|^2 \, dt \geqslant \int_{-\pi}^{\pi} |f(t) - S_n(t;f)|^2 \, dt,$$

which is a special case of (4.5.7), together with Theorem 4.5.3. ∎

Corollary 2. *Under the same assumptions*

$$\int_{-\pi}^{\pi} |f(s)|^2 \, ds = 2\pi \sum_{-\infty}^{\infty} |f_k|^2. \tag{4.5.28}$$

Proof. For this is equivalent to (4.5.27). ∎

Theorem 4.5.4. *The closure relation* (4.5.28) *is equivalent to either of the following statements:*

1) *The only $f \in L_2(-\pi, \pi)$ which is orthogonal to each of the functions e^{kit} is equivalent to* 0.
2) *The linear combinations of the functions e^{kit} are dense in $L_2(-\pi, \pi)$.*

Proof. This is a special case of Theorem 2.5.8. ∎

Property (1) expresses that the orthogonal system e^{kit} is *maximal*, while property (2) is often expressed by saying that the set of elements e^{kit} is *fundamental* in L_2. In L_2 a system of elements is maximal iff it is fundamental. In other Lebesgue spaces the corresponding notions are not necessarily equivalent.

The following is an important consequence of (4.5.28).

Theorem 4.5.5. *If $f \in L_2(-\pi, \pi)$, $g \in L_2(-\pi, \pi)$, and if their Fourier coefficients are f_n and g_n, respectively, then*

$$\int_{-\pi}^{\pi} f(s)\overline{g(s)} \, ds = 2\pi \sum_{-\infty}^{\infty} f_n \bar{g}_n, \tag{4.5.29}$$

the series being absolutely convergent.

Proof. Write the analogue of (4.5.28) for $\alpha f + \beta g$ and equate the coefficients of $\alpha\bar{\beta}$ on the two sides of the equation. ∎

We apply this result to the case where $s \to g(s) = \chi(s; I)$, the characteristic function of the interval $I = (a, b) \subset [-\pi, \pi]$. Here

$$g_0 = \frac{1}{2\pi}(b - a), \qquad g_n = -\frac{1}{2\pi i n}(e^{-nib} - e^{-nia}), \qquad n \neq 0, \tag{4.5.30}$$

so that

$$\int_a^b f(s)\, ds = f_0(b - a) + \sum_{-\infty}^{\infty}{}' \frac{f_n}{ni}(e^{nib} - e^{nia}), \tag{4.5.31}$$

where the prime after the summation sign indicates that $n = 0$ is to be omitted. Here the series is absolutely convergent. Hence we have proved

Theorem 4.5.6. *The Fourier series of any function f in $L_2(-\pi, \pi)$ may be integrated term by term between any two limits a and b, $(a, b) \in [-\pi, \pi)$. The result is absolutely convergent and the sum equals $\int_a^b f(s)\, ds$.*

The same result holds in L_p, $1 < p < 2$, and trivially for $2 < p$. For $p = 1$ the integrated series still converges to the integral but not necessarily absolutely.

This result is really astounding since the Fourier series itself may diverge everywhere. This and other facts proved above show that the association between the function and its Fourier series is much closer than what is implied by the formulas (4.5.2) from which we started. The series may diverge or converge to a sum different from the local value of the function. On the average the series converges to the function if $f \in L_2$. "On the average" here means in the sense of (4.5.27). We can evaluate

$$\int_a^b f(s)\, ds \quad \text{and} \quad \int_{-\pi}^{\pi} f(s)\, g(s)\, ds$$

by termwise integration of Fourier series regardless of divergence.

Every $f \in L_2$ gives rise to a sequence of Fourier coefficients which belongs to a space l_2 of bilateral infinite sequences. It is convenient to define the norm in the latter space by

$$\|\{f_n\}\| = \left\{2\pi \sum_{-\infty}^{\infty} |f_n|^2\right\}^{1/2}. \tag{4.5.32}$$

In effect, this means replacing the ordinary Fourier coefficients by the expansion coefficients with respect to the orthonormal system (4.5.3). We have now a mapping T from L_2 into l_2

$$T: f \to \{f_n\} \tag{4.5.33}$$

which is linear. Since

$$\|f\| = \|\{f_n\}\| \tag{4.5.34}$$

the mapping preserves length, i.e. it is an *isometry*. Moreover, Theorem 4.5.5 may be interpreted as stating that inner products are preserved. Here by definition the left member of (4.5.29) is the inner product in L_2 while the right member is the inner product in l_2. This also implies that orthogonality is preserved, since if one side of the equality is zero, so is the other side.

It is clear that T is 1–1 since the zero elements of L_2 and of l_2 are in unique correspondence under T. It is natural to ask if T is "onto." In other words: Is every element $\{f_n\} \in l_2$ the image under T of an element f of L_2? The answer is in the affirmative and is known as the *Riesz–Fischer theorem*, found in 1907 independently by F. Riesz and the German mathematician Ernst Fischer (1875–1954).

Theorem 4.5.7. *If $\{f_n\}$ is an element of l_2, then there exists an element f of L_2 whose Fourier coefficients coincide with $\{f_n\}$.*

Proof. Cf. the proof of Theorem 3.5.4. Form the polynomials

$$S_n(t) = \sum_{-n}^{n} f_k\, e^{kit}. \tag{4.5.35}$$

They form a Cauchy sequence in $L_2(-\pi, \pi)$. Let f be the limit of this sequence. Then $f \in L_2$ and for $|k| < n$

$$\frac{1}{2\pi} \int_{-\pi}^{\pi} S_n(s)\, e^{-kis}\, ds = f_k.$$

As $n \to \infty$ the left member of this relation tends to

$$\frac{1}{2\pi} \int_{-\pi}^{\pi} f(s)\, e^{-kis}\, ds,$$

so the kth Fourier coefficient of f is indeed f_k. ∎

As the last topic of this section we consider *convolution* of two elements in L_2 defined by

$$(f * g)(t) = \frac{1}{2\pi} \int_{-\pi}^{\pi} f(t - s)\, g(s)\, ds. \tag{4.5.36}$$

Cf. formula (4.4.6) for similar constructs. The integral exists for almost all t and the product is symmetric in the two factors.

Theorem 4.5.8. *If f and $g \in L_2(-\pi, \pi)$, then $f * g$ is continuous and periodic with period 2π. The Fourier series*

$$(f * g)(t) = \sum_{-\infty}^{\infty} f_n\, g_n\, e^{nit} \tag{4.5.37}$$

is absolutely convergent for all t.

Proof. Here we have

$$\frac{1}{2\pi}\int_{-\pi}^{\pi} e^{-nit}\left\{\frac{1}{2\pi}\int_{-\pi}^{\pi} f(t-s)\,g(s)\,ds\right\}dt$$

$$= \frac{1}{2\pi}\int_{-\pi}^{\pi} e^{-niu} f(u)\,du \cdot \frac{1}{2\pi}\int_{-\pi}^{\pi} e^{-nis}\,g(s)\,ds = f_n\,g_n.$$

This gives the Fourier series in (4.5.37). The series is absolutely convergent by Cauchy's inequality, uniformly with respect to t. Hence the sum of the series is a continuous function of t; moreover, the series is the Fourier series of its sum (see Problem 2, Exercise 4.5). On the other hand, the series is the Fourier series of the convolution product. Thus the latter is a continuous function of t and the series converges point-wise to $(f * g)(t)$. It may be shown directly that this function is continuous. For if $t_1 < t_2$, then

$$4\pi^2\,|(f * g)(t_2) - (f * g)(t_1)|^2 \leqslant \|g\|^2 \int_{-\pi}^{\pi} |f(t_2 - s) - f(t_1 - s)|^2\,ds$$

and the integral is dominated by $[\mu_2(t_2 - t_1; f)]^2$, which goes to zero with $t_2 - t_1$.

Thus we see that the transformation T takes a convolution product in L_2 into a product of Fourier coefficients in l_2. ∎

EXERCISE 4.5

1. Verify formulas (4.5.11) and (4.5.12). What is the corresponding form of (4.5.9)?
2. Show that if a series $\sum_{-\infty}^{\infty} a_k e^{kit}$ is uniformly convergent for all t and if the sum of the series is $F(t)$, then the series is the Fourier series of F.
3. If f and f' are both in $L_2(-\pi, \pi)$, what relations hold between their Fourier coefficients? Show that $\sum k^2\,|f_k|^2$ converges.
4. We have $\frac{1}{2}(\pi - t) \sim \sum_{k=1}^{\infty} (1/k)\sin kt$ for $0 < t < 2\pi$. Find $\mu_1(h; f)$ and discuss the corresponding inequality (4.5.5). How good is it?
5. If f is a periodic function in $BV[-\pi, \pi]$ having Fourier coefficients f_n, show that
$$2|n|\,|f_n| \leqslant V_{-\pi}^{\pi}[f].$$
Show that the estimate is the best possible of its kind.
6. Let $\{n_k\}$ be a strictly increasing sequence of positive integers such that $n_k/k \to \infty$. A series of the form $\sum_1^{\infty} a_k \cos(n_k t)$ is known as a *cosine gap series* or *lacunary series*. When is this the Fourier series of a function in L_2? Does this throw some light on how fast the Fourier coefficients have to tend to zero?
7. Prove Fejér's theorem: If f is a continuous function of period 2π, then the sequence $\{T_n(t; f)\}$ converges uniformly to $f(t)$.
8. Under the same assumptions, if $m \leqslant f(t) \leqslant M$, $\forall t$, show that $m \leqslant T_n(t; f) \leqslant M$, $\forall t, \forall n$.

9. Prove (4.5.23) by showing that

$$F_n(u) < \frac{3\pi(n+1)}{[1+(n+1)|u|]^2}, \quad |u| < \pi, \quad n > 1.$$

[*Hint:* Consider the interval $(n+1)|u| < \pi$ separately.]

10. Verify the expressions for $D_n(s)$ and $F_n(s)$.

11. [Lebesgue] If $f(t; h) = (2h)^{-1}\int_{-h}^{h} f(t+s)\,ds$ where $f \in L_2$, show that $f(t; h)$ is a continuous function of t which converges to $f(t)$ in mean square as $h \downarrow 0$.

12. Show that

$$f(t; h) = f_0 + \Sigma' f_n \frac{\sin nh}{nh} e^{nit},$$

where the prime indicates that $n = 0$ is excluded. Show that the series is absolutely convergent.

13. In (4.5.31) take $a = 0$, $b = x$. Show that the closure relation for this characteristic function $\chi(t; I)$ simplifies to

$$\tfrac{1}{4}x^2 - \tfrac{1}{2}\pi\,|x| + \tfrac{1}{6}\pi^2 = \sum_{1}^{\infty} n^{-2} \cos nx, \qquad -\pi < x < \pi.$$

Verify directly this identity.

14. In the preceding problem, form the integral of the square of the left member and sum the series $\Sigma\, n^{-4}$. Verify the closure relation.

15. Under what conditions does the equation $f * f = g$ have a solution $f \in L_2(-\pi, \pi)$?

COLLATERAL READING

CARATHÉODORY, C., *Vorlesungen über reelle Funktionen*, 2nd ed., B. G. Teubner, Leipzig (1927).

EPSTEIN, B., *Linear Functional Analysis; Introduction to Lebesgue Integration and Infinite-Dimensional Problems*, Saunders, Philadelphia (1970).

HARTMAN, S. and J. MIKUSIŃSKI, *The Theory of Lebesgue Measure and Integration*, Pergamon Press, London (1961).

5 METRIC SPACES AND FIXED POINT THEOREMS

The beginning of this chapter is an elaboration and review of notions introduced in Chapter 2. This is followed by partial ordering and fixed point theorems. The latter form the hard core of the chapter.

After the early topologists got through with what might be called the taxonomy of the discipline, attention was devoted to the nature of topological mapping. A basic question is whether or not such a mapping of a space into itself shifts all points, or, possibly, leaves one or more points invariant. Both possibilities occur. See Exercise 5.4 for examples. The earliest fixed point theorem is that of L. E. J. Brouwer who proved in 1912 that a continuous map T of the closed unit ball in R^m has at least one fixed point, i.e. a point $\mathbf{x}_0$ such that $T\mathbf{x}_0 = \mathbf{x}_0$. Brouwer's fixed point theorem was used by G. D. Birkhoff and O. D. Kellogg in 1922 to prove existence theorems in the theory of differential equations. At the same time, S. Banach found a contraction fixed point theorem that rapidly manifested its importance. Thus in 1930 R. Caccioppoli proved that the Birkhoff–Kellogg theorem could be derived from the Banach theorem. In the present chapter we shall prove some fixed point theorems of importance in analysis.

There are seven sections in the chapter: Metric spaces; Partial ordering; Contraction mappings; Contraction fixed point theorems; Contractive mappings; The Volterra fixed point theorem; and Some applications.

5.1 METRIC SPACES

We refer back to Section 2.1 for the basic definitions. A metric space is an abstract space in which a notion of distance $d(x, y)$ between any two elements x and y is defined subject to postulates (D_1) to (D_3). See Definition 2.1.2. In terms of distance we could define the concepts of open set, ε-neighborhood, cluster point, closure, closed set, and being dense in the space. Definition 2.1.3 introduced convergence, convergence to a point, Cauchy sequence, and complete metric space. Metric defined by a norm in a linear vector space came in Definition 2.1.4.

In Chapters 3 and 4 a number of metric spaces were introduced mostly with a normed metric, and stress was laid on the completeness of the space. The completeness proofs varied with the space, though certain common features will have been apparent to the attentive reader.

We do not have far to go to find a metric space which is not complete. We can take Q, the space of all rational numbers, with the usual metric

$$d(x, y) = |x - y|.$$

A Cauchy sequence of rational numbers does not necessarily converge to a rational number, so Q is not complete. On the other hand, we can embed Q as a dense subspace, in the space R^1, where all Cauchy sequences, in particular the sequences which are in Q, converge to limits. In fact, one way of introducing the real number system, when the rationals are taken as known, is to consider all Cauchy sequences of rational numbers. Two such sequences are equivalent, say $\{x_n\} \sim \{y_n\}$, iff

$$\lim_{n \to \infty} |x_n - y_n| = 0.$$

Equivalent Cauchy sequences form *equivalence classes.* The set of all equivalence classes defines an abstract space in which a metric may be introduced by setting

$$d(\{x_n\}, \{y_n\}) = \lim_{n \to \infty} |x_n - y_n|.$$

This is a complete metric space isomorphic to R^1.

If a metric space is not complete, it may be completed by embedding it densely in the set of equivalence classes formed by the Cauchy sequences in the manner exhibited above for the space Q. It will be assumed in the following that the metric spaces under consideration are complete.

We like to add some further concepts taken over from real analysis. In Section 2.1 the concept of *nowhere dense* was defined for a metric space $\mathfrak{X}$. A set S in $\mathfrak{X}$ had this property if every open ball in $\mathfrak{X}$ contains a ball having no points in common with S. *A set is of the first category in the sense of Baire if it is the union of a finite or countably infinite set of subsets each of which is nowhere dense in* $\mathfrak{X}$. *A set is of the second category if it is not of the first.* In R^1 the set Q is of the first category and so is the set of real algebraic numbers, while that of the real transcendental numbers is of the second category.

The classical *Bolzano–Weierstrass theorem* asserts that any bounded infinite set in R^1 has at least one cluster point. The latter is an element of R^1, but need not be an element of the set. The same theorem holds for sets in C^n. It need not hold in a general metric space. An instance of this was exhibited in Section 3.2, where it was shown that the powers of t form an infinite bounded set in $C[0, 1]$ without cluster point; a still simpler instance is furnished by the space l_1. Here the unit vectors $\mathbf{u}_k = \{\delta_{jk}\}$ form a bounded infinite sequence without cluster points since $\|\mathbf{u}_k - \mathbf{u}_l\| = 2$ for all distinct k and l.

The Heine–Borel theorem is also apt to go by the board in the transit from finite to infinite dimensional space. Let $\mathfrak{X}$ be a complete metric space and S a set in $\mathfrak{X}$. A system of open sets G_α is a *covering* of S if each point of S belongs to at least one set G_α. *S is said to have the Borel property if every system of open sets $\{G_\alpha\}$ which covers S contains a finite subsystem which also covers S.* In R^1 every closed bounded set has the Borel property and this fact is known as the *Heine–Borel theorem.* It is also true in C^n. Such a set S is said to be *compact.* It is *conditionally* *compact* if its closure is compact. It is *sequentially compact* if every sequence of points in S contains a subsequence which converges to a point in S. It may be shown that *in a complete metric space a set S is sequentially compact if it is compact.*

The theorem of Arzelà, Theorem 3.2.1, is an example of a compactness theorem holding in $C[a, b]$.

EXERCISE 5.1

1. The metric in l_p is defined in terms of the norm (3.1.11). Show that l_p is a complete metric space under the normed topology.
2. Consider the set S in l_1 consisting of all vectors such that all but a finite number of the coordinates are zero. Show that S is dense in the space.
3. Find an infinite subset of l_1 which is nowhere dense in the space.
4. Find two vectors $\mathbf{x}$ and $\mathbf{y}$ in l_1 of norm 1 such that all points $t\mathbf{x} + (1 - t)\mathbf{y}$, $0 \leqslant t \leqslant 1$, are also of norm 1. In other words, show that the unit sphere in l_1 contains (a continuum of) line segments.
5. Let S be a family of vectors in l_1, say $\mathbf{v}_\alpha = (v_{\alpha,n})$, and suppose that there is a fixed vector $\mathbf{x} = (x_n)$ in l_1 such that $|v_{\alpha,n}| \leqslant |x_n|$ for all α and n. Show that S is conditionally compact.
6. Let $\mathsf{F} = \{f_\alpha\}$ be a family of elements in $L_1(-\pi, \pi)$. Show that if F is conditionally compact, then its elements are uniformly bounded, i.e. there is a constant M such that $\| f_\alpha\| \leqslant M, \forall \alpha$.
7. The family $\{e^{nit}; n = 0, \pm 1, \pm 2, \ldots\}$ is uniformly bounded in $L_1(-\pi, \pi)$. Show that it contains no convergent subsequences.
8. Is the result of Problem 5 true in l_p?
9. Show that the unit sphere in $C[a, b]$ also contains a continuum of line segments.
10. If $E \subset F \subset \mathfrak{X}$ and E is of the second category in the complete metric space $\mathfrak{X}$, show that F is also of the second category.
11. The boundary of a set E in a metric space $\mathfrak{X}$ is the intersection of its closure with the closure of its complement in $\mathfrak{X}$. If E is open, show that its boundary is a set of the first category.
12. If a set E is nowhere dense in a metric space $\mathfrak{X}$, show that the interior of its closure is void, $\text{Int}(\bar{E}) = \emptyset$. Is this true if the set is merely of the first category?

5.2 PARTIAL ORDERING

The real number system is not merely the prototype of all linear vector spaces; it is also the prototype of an ordered system. There exists in this system a *binary order relation* expressed by the symbol "$a \leqslant b$" (less than or equal to) or, equivalently,

"$b \geqslant a$" with the following properties:

(O_1) *For all* a, $a \leqslant a$ *(reflexive).*

(O_2) *If* $a \leqslant b$ *and* $b \leqslant a$, *then* $a = b$ *(antisymmetric).*

(O_3) *If* $a \leqslant b$ *and* $b \leqslant c$, *then* $a \leqslant c$ *(transitive).*

The real number system is *linearly ordered* or *totally ordered* in the sense that for every pair of elements a and b either $a \leqslant b$ or $b < a$.

A notion of order can be introduced in much more general systems than the real numbers.

Definition 5.2.1. *A set S of elements $a, b, \ldots$ is said to be partially ordered if there exists a binary relation $a \leqslant b$ defined for certain pairs (a, b) in S for which (O_1) to (O_3) hold.*

Just as in the theory of distance significant results are obtainable even if the symmetry condition (D_2) of page 40 is dropped, the anti-symmetry condition (O_2) may be dropped, as shown in a recent investigation by Ih-Ching Hsü (University of New Mexico dissertation, 1969). We shall, however, find it convenient to invoke all three conditions.

Some examples will clarify these notions.

Example 1. Take $S = C^r[0, 1]$, the space of real-valued functions f defined and continuous in $[0, 1]$, and define

$$f \leqslant g \quad \text{to mean} \quad f(t) \leqslant g(t) \quad \text{for all } t \text{ in } [0, 1]. \tag{5.2.1}$$

This is a meaningful partial ordering. It cannot be extended to a total ordering for there are elements of the space for which neither $f \leqslant g$ nor $g \leqslant f$ is true. On the other hand, for any f in S, the set $[g; f \leqslant g]$ is never void and the same is true for the set $[g; g \leqslant f]$. Condition (5.2.1) is equivalent to

$$g - f \geqslant 0, \tag{5.2.2}$$

that is, $g(t) - f(t) \geqslant 0$ for all t. Such "positive" elements play a basic role in the theory of partially ordered function spaces. See further below.

Example 2. S is the set of subsets of a given set E. Here $E_1 \leqslant E_2$ is defined to mean $E_1 \subseteq E_2$, i.e. *ordering under set inclusion.* Here for any pair of sets E_1 and E_2 the intersection $E_1 \cap E_2$ may very well be void and an inclusion $E_1 \subseteq E_2$ is accidental. On the other hand, given any pair (E_1, E_2) in S, there exists at least one set E_3 containing them both, so that $E_1 \leqslant E_3$, $E_2 \leqslant E_3$. We can take $E_3 = E_1 \cup E_2$, for instance.

Example 3. A partial ordering may be established in R^m as follows. Let

$$\mathbf{x} = (x_1, x_2, \ldots, x_m), \qquad \mathbf{y} = (y_1, y_2, \ldots, y_m)$$

and define $\mathbf{x} \leqslant \mathbf{y}$ to mean that

$$x_j \leqslant y_j, \quad j = 1, 2, \ldots, m. \tag{5.2.3}$$

This is equivalent to saying that $\mathbf{y} - \mathbf{x}$ shall be a positive vector, i.e. one with non-negative coordinates.

Definition 5.2.2. *Let S be a partially ordered set. If $a \leqslant c$ and $b \leqslant c$, then c is called an upper bound for a and b. Moreover, if $c \leqslant d$ whenever d is an upper bound of a and b, we call c the least upper bound or the supremum of a and b, in symbols*

$$c = \sup(a, b) \quad \text{or} \quad c = a \vee b. \tag{5.2.4}$$

This element of S is unique if its exists. This fact depends essentially upon (O_2). Similarly, we define the *greatest lower bound* or *infimum* of a and b

$$p = \inf(a, b) \quad \text{or} \quad p = a \wedge b. \tag{5.2.5}$$

In all three cases listed above infima and suprema exist. Thus in Example 1 we define

$$c(t) = \max\,[a(t), b(t)], \qquad p(t) = \min\,[a(t), b(t)]. \tag{5.2.6}$$

These are elements of $C^r[0, 1]$. Here $c(t)$ is an upper bound of $a(t)$ and $b(t)$ and clearly the least upper bound. Similarly, $p(t)$ is seen to be the greatest lower bound. The other examples are left to the reader.

In a partially ordered set S the subset T is said to have b for an upper bound if $t \leqslant b$ for all $t \in T$. A subset T may have a least upper bound, i.e. an element b of S which is an upper bound of T and such that if t is any upper bound of T, then $b \leqslant t$. Similarly for greatest lower bounds. Both are unique if they exist.

Definition 5.2.3. *An element m of a partially ordered set S is said to be maximal if $x \in S$ together with $m \leqslant x$ implies $x = m$.*

For the existence of maximal elements we have the *maximal principle* of Max Zorn, one form of which is

Zorn's Lemma. *Let S be a non-empty partially ordered set such that every totally ordered subset of S has an upper bound in S, then S contains at least one maximal element.*

It is known that Zorn's Lemma is equivalent to Zermelo's Axiom of Choice in set theory.

Definition 5.2.4. *A lattice is a partially ordered set S any two elements of which have a least upper bound and a greatest lower bound.*

The examples listed above involve lattices.

To the analyst a more interesting situation arises if the set S is a linear vector space as well as partially ordered. Here the partial ordering should satisfy the requirements that the two transformations

$$\mathbf{x} \to \mathbf{x} + \mathbf{a} \quad \text{and} \quad \mathbf{x} \to \alpha\mathbf{x}, \quad \forall\, \mathbf{x}, \quad \alpha > 0, \tag{5.2.7}$$

are *order preserving*, i.e.

(O_4) $\mathbf{x} \leqslant \mathbf{y}$ *implies* $\mathbf{x} + \mathbf{a} \leqslant \mathbf{y} + \mathbf{a}$ *for all* $\mathbf{a}$,

(O_5) $\mathbf{x} \leqslant \mathbf{y}$ *implies* $\alpha\mathbf{x} \leqslant \alpha\mathbf{y}$ *for all* $\alpha > 0$.

In partially ordered vector spaces where (O_1) to (O_5) are valid we can define positive elements: $\mathbf{x}$ is *positive* if $\mathbf{0} \leqslant \mathbf{x}$. The set of positive elements forms the *positive cone* $\mathfrak{X}^+$ of $\mathfrak{X}$. This is a proper cone in the sense of Problem 15, Exercise 1.2.

The positive cone $\mathfrak{X}^+$ *contains the zero element and is invariant under addition and multiplication by positive scalars.*

If, in addition, $\mathfrak{X}$ is a complete metric space, it is customary to require

(O_6) *A convergent sequence of positive elements converges to a positive element.*

We shall see later that transformations of a linear partially ordered vector space into itself which map the positive cone into itself are particularly important for analysis.

EXERCISE 5.2

1. Prove the existence of infima and suprema in Example 2.
2. Do the same for Example 3.
3. Why are infima and suprema unique when they exist?
4. Verify that the sets figuring in Examples 1 to 3 are lattices.
5. Verify that the functions $c(t)$ and $p(t)$ of (5.2.6) have the stated properties.
6. Let $K(s, t)$ be a positive continuous function on the unit square $[0, 1] \times [0, 1]$ and define

$$T[f](t) = \int_0^1 K(s, t)\, f(s)\, ds.$$

   This is a linear bounded transformation from $C[0, 1]$ to itself. See (3.2.38) and following. Show that T is order preserving in the sense that $f \leqslant g$ implies $T[f] \leqslant T[g]$ and that the positive cone $\mathfrak{C}^+[0, 1]$ is mapped into itself.
7. Formula (3.1.39) together with (3.1.40) defines a linear bounded transformation from l_p into itself. When is this transformation order preserving? Here $\mathbf{x} \leqslant \mathbf{y}$ means $x_k \leqslant y_k, \forall\, k$.

8. A linear bounded functional on l_p is defined by (3.1.29). Under what conditions on the vector $\mathbf{b} \in l_q$ is the functional order preserving?
9. When is the functional defined by (3.2.36) order preserving?

5.3 CONTRACTION MAPPINGS

Let $\mathfrak{X}$ and $\mathfrak{Y}$ be complete metric spaces and let T be a mapping from $\mathfrak{X}$ into $\mathfrak{Y}$. Then to every $\mathbf{x} \in \mathfrak{X}$ corresponds a unique $\mathbf{y} \in \mathfrak{Y}$, the *image* of $\mathbf{x}$ under the mapping T, and we write

$$\mathbf{y} = T(\mathbf{x}). \tag{5.3.1}$$

The mapping is said to be *bounded* if there exists a fixed finite M such that for $\mathbf{x}_1$ and $\mathbf{x}_2$ belonging to $\mathfrak{X}$

$$d[T(\mathbf{x}_1), T(\mathbf{x}_2)] \leqslant M d(\mathbf{x}_1, \mathbf{x}_2). \tag{5.3.2}$$

Note that in this formula d stands for distance and may mean one thing in the left member and another in the right if $\mathfrak{X} \neq \mathfrak{Y}$.

A bounded transformation is necessarily continuous for (5.3.2) is a type of Lipschitz condition and hence implies continuity.

A mapping T is *onto* if every point $\mathbf{y}$ of $\mathfrak{Y}$ is the image of at least one point of $\mathfrak{X}$. It is 1–1 if

$$\mathbf{x}_1 \neq \mathbf{x}_2 \quad \text{implies} \quad T(\mathbf{x}_1) \neq T(\mathbf{x}_2). \tag{5.3.3}$$

Suppose now that $\mathfrak{Y} = \mathfrak{X}$ so that T is a mapping from $\mathfrak{X}$ into itself. The condition for boundedness is still given by (5.2.2), but now distance means the same thing on both sides of the inequality. Suppose that T_1 and T_2 are two bounded transformations from $\mathfrak{X}$ into itself. Just as in the linear case, we can then define products of transformations as follows. Set

$$(T_1 T_2)(\mathbf{x}) = T_1[T_2(\mathbf{x})], \qquad (T_2 T_1)(\mathbf{x}) = T_2[T_1(\mathbf{x})]. \tag{5.3.4}$$

The products are well defined and normally $T_1 T_2 \neq T_2 T_1$. The bounded operators from $\mathfrak{X}$ into itself form a *semi-group* $\mathsf{S}(\mathfrak{X})$ *under multiplication.* Since multiplication may possibly be non-associative, we are using the term "semi-group" simply as a designation for a system where one operation may be performed on any ordered pair of elements, the result being again an element of the system. A semi-group may have a *neutral element* U such that

$$UT = TU = T, \quad \forall\, T. \tag{5.3.5}$$

In our case this is the identity mapping I,

$$I(\mathbf{x}) = \mathbf{x}, \quad \forall\, \mathbf{x}. \tag{5.3.6}$$

Mappings may have *inverses*. We say that a mapping S of $\mathfrak{X}$ into itself is the inverse of the mapping T of $\mathfrak{X}$ into itself if

$$ST = TS = I. \tag{5.3.7}$$

We note that if $T \in \mathsf{S}(\mathfrak{X})$, then all the powers of T are well defined and also belong to $\mathsf{S}(\mathfrak{X})$. We set

$$T^n(\mathbf{x}) = T[T^{n-1}(\mathbf{x})], \quad n = 2, 3, \ldots . \tag{5.3.8}$$

The Law of Exponents is valid, so that

$$T^m T^n = T^{m+n}. \tag{5.3.9}$$

Note that

$$T^m T^n = T^n T^m. \tag{5.3.10}$$

If T^{-1} (the S of (5.3.7)) should exist, then (5.3.8) to (5.3.10) hold for all integers, not merely the positive ones, with the convention that $T^0 = I$, $T^1 = T$.

We are now going to specialize still further and consider bounded transformations where the bound M is $\leqslant 1$.

Definition 5.3.1. *Consider a bounded mapping T of a complete metric space $\mathfrak{X}$ into itself so that $T \in \mathsf{S}(\mathfrak{X})$. Then T is a contraction mapping in the wide sense if*

$$d[T(\mathbf{x}_1), T(\mathbf{x}_2)] \leqslant d(\mathbf{x}_1, \mathbf{x}_2). \tag{5.3.11}$$

T is a strict contraction if

$$d[T(\mathbf{x}_1), T(\mathbf{x}_2)] \leqslant k d(\mathbf{x}_1, \mathbf{x}_2) \tag{5.3.12}$$

where k is a fixed number, $0 < k < 1$. Finally, T is contractive if

$$d[T(\mathbf{x}_1), T(\mathbf{x}_2)] < d(\mathbf{x}_1, \mathbf{x}_2). \tag{5.3.13}$$

In each case the inequality is to hold for all $\mathbf{x}_1$ and $\mathbf{x}_2$ in $\mathfrak{X}$ and in (5.3.13) $\mathbf{x}_1 \neq \mathbf{x}_2$.

We illustrate by some examples.

Example 1. Take $\mathfrak{X} = C^+[0, 1]$ with the sup-norm and set

$$T[f](t) = k \int_0^t f(s)\, ds, \quad 0 \leqslant t \leqslant 1, \tag{5.3.14}$$

with a fixed positive k. If $k = 1$, this is a contraction in the wide sense since obviously

$$\|T[f] - T[g]\| \leqslant k\|f - g\|. \tag{5.3.15}$$

If $0 < k < 1$, this is a contraction in the strict sense.

Example 2. We take $\mathfrak{X} = \{x; 1 \leqslant x\}$ with the Euclidean metric and define

$$T(x) = x + \frac{1}{x}. \tag{5.3.16}$$

Here $\mathfrak{X}$ is a complete metric space and

$$d[T(x_1), T(x_2)] = \frac{x_1 x_2 - 1}{x_1 x_2} |x_1 - x_2| < |x_1 - x_2| = d(x_1, x_2)$$

since at least one of the numbers x_1 and x_2 may be taken to be >1. Thus this is a contractive mapping and it is easy to see that it cannot be a strict contraction.

Example 3. We take $\mathfrak{X} = C[0, \infty]$, the space of functions continuous for all t, $0 \leqslant t \leqslant \infty$. Continuity at infinity means that $f(t)$ tends to a finite limit as $t \to \infty$. The mapping is to be defined by the *shift operator* $T(s)$, where

$$T(s)[f](t) = f(t+s), \quad 0 \leqslant s. \tag{5.3.17}$$

This is a contraction in the wide sense since

$$\|T(s)\| \equiv 1 \tag{5.3.18}$$

for all s. Note that $f \equiv 1$ is left invariant by $T(s)$. Here $\{T(s)\}$ is a *one-parameter family of contraction operators forming a semi-group* in the sense that

$$T(s+t) = T(s)\,T(t) = T(t)\,T(s). \tag{5.3.19}$$

In connection with Example 3 we shall discuss an inequality which goes back to Edmund Landau (1877–1938) who in 1913 showed that if f, f', and f'' belong to $C[a, b]$, then

$$\|f'\|^2 \leqslant 4\|f\|\,\|f''\|. \tag{5.3.20}$$

In particular, the inequality holds for $C[0, \infty]$ and the same inequality was later found to be valid in $L_p(0, \infty)$ for any $p \geqslant 1$. The intrinsic reason for these inequalities was discovered in 1967 by R. R. Kallman and G.–C. Rota, who showed that Landau's inequality is a special case of an inequality that holds for any contraction semi-group of linear operators on a B-space $\mathfrak{X}$ into itself. If $T(s)$ is strongly continuous in a sense to be defined in Chapter 7 and if $T(0) = I$, then a linear, in general unbounded, operator A from $\mathfrak{X}$ to $\mathfrak{X}$ is defined by the condition

$$A[\mathbf{x}] = \lim_{h \downarrow 0} \frac{1}{h}[T(h) - I](\mathbf{x}). \tag{5.3.21}$$

The operator A is known as the *infinitesimal generator* of the semi-group. Its domain of definition $\mathfrak{D}[A]$ is dense in $\mathfrak{X}$ and so are the domains of the iterates A^n. For $\mathbf{x} \in \mathfrak{D}[A^2]$ Kallman and Rota proved that

$$\|A(\mathbf{x})\|^2 \leqslant 4\|\mathbf{x}\|\,\|A^2(\mathbf{x})\|. \tag{5.3.22}$$

This holds for any contraction semi-group, in particular for the semi-group of shift operators (5.3.17). In the latter case A is the operation of differentiation with respect to t and (5.3.20) results. The proof of (5.3.22) is based on Taylor's theorem with remainder, a form of which is valid for semi-group operators as shown by E. Hille and K. Yosida independently of each other in the 1940's. As applied to the shift operator this argument reduces to the classical Taylor's theorem.

Theorem 5.3.1. *Let $\mathfrak{X}$ be one of the spaces $C[0, \infty]$ or $L_p(0, \infty)$ and let f be an element of $\mathfrak{X}$ for which f' and f'' also belong to $\mathfrak{X}$. Then* (5.3.20) *holds where the norm is that of the space in question.*

Proof. By Taylor's theorem, for $s > 0$, $t > 0$,

$$f(t+s) = f(t) + sf'(t) + \int_0^s (s-u)f''(t+u)\,du,$$

whence

$$f'(t) = \frac{1}{s}[f(t+s) - f(t)] - \frac{1}{s}\int_0^s (s-u)f''(t+u)\,du.$$

Here all functions involved belong to $\mathfrak{X}$ and so does the integral for all s, since

$$\left\| \int_0^s (s-u)f''(t+u)\,du \right\| \leqslant \|f''\| \int_0^s (s-u)\,du = \tfrac{1}{2}s^2\|f''\|.$$

This is trivial for $C[0, \infty]$ and for L_p it follows from the approximation of the integral by a Riemann sum, the summands of which are L_p, plus Minkowski's inequality. This argument gives

$$\|f'\| \leqslant \frac{2}{s}\|f\| + \tfrac{1}{2}s\|f''\| \tag{5.3.23}$$

since we deal with contraction operators. Here s is arbitrary, $s > 0$. If $\|f''\| = 0$, we let $s \to \infty$ and obtain $\|f'\| = 0$ so that (5.3.20) is trivially true. If $\|f''\| > 0$, we minimize the right member of (5.3.23) by taking

$$s = 2(\|f\|)^{1/2}(\|f''\|)^{-1/2}.$$

Substitution of this value in (5.3.23) and simplification give (5.3.20). ∎

EXERCISE 5.3

1. If $\mathfrak{X} = C^+[0, a]$, when is

$$T[f](t) = \int_0^t (t-s)f(s)\,ds, \quad 0 \leqslant t \leqslant a,$$

a contraction? Find by inspection an element left invariant by T.

2. In $L_2(0, 2\pi)$ a linear transformation from L_2 to L_2 is determined by its action on the Fourier coefficients; if

$$f(t) \sim \sum_{-\infty}^{\infty} f_n e^{nit}, \qquad T[f](t) \sim \sum_{-\infty}^{\infty} F_n e^{nit},$$

take

$$F_n = \lambda_n f_n, \quad \forall n,$$

where $\{\lambda_n\}$ is a given sequence of complex numbers. What condition on the λ_n will make T (i) bounded, (ii) a contraction, (iii) contractive?

3. With the same notation, suppose that $\{\lambda_n\}$ is a bounded sequence and determine the nullspace of T. When does $\mathfrak{N}[T]$ reduce to $\{\mathbf{0}\}$?

4. What condition on the λ's will ensure the existence of T^{-1} as an element of $\mathfrak{E}(L_2)$? Find T^{-1} when it exists.

5. Is there a fixed point in Example 2 above?

6. [J. B. Miller] With $\mathfrak{X}$ as in Theorem 5.3.1, assume that f, f', f'', and f''' belong to $\mathfrak{X}$ and prove that (with associated non-optimal numerical constants)

$$\|f''\| \leqslant \frac{2}{3}\frac{\|f\|\ \|f'''\|}{\|f'\|} + \{6\|f'\|\ \|f'''\|\}^{1/2}.$$

7. [S. Kurepa] With $\mathfrak{X}$ equal to one of the spaces $C[-\infty, \infty]$ or $L_p(-\infty, \infty)$, assume that f and its derivatives of order $\leqslant 4$ belong to $\mathfrak{X}$. Show that

$$\tfrac{1}{2}[f(t+s) + f(t-s)] \equiv T(s)[f](t)$$

is a family of contraction operators and use the Kallman–Rota device to prove the inequality

$$\|f''\|^2 \leqslant \tfrac{4}{3}\|f\|\ \|f^{(iv)}\|.$$

8. The following transformation is associated with the name of Émile Picard. Take $\mathfrak{X}$ as one of the spaces $C[-\infty, \infty]$ or $L_p(-\infty, \infty)$ and form

$$T(s)[f](x) = \tfrac{1}{2}s \int_{-\infty}^{\infty} \exp[-s|x-u|]\, f(u)\, du, \quad 0 < s.$$

Show that $T(s)$ is a linear contraction mapping from $\mathfrak{X}$ into itself.

9. If $\mathfrak{X} = C[-\infty, \infty]$ show that

$$\lim_{s\to\infty} T(s)[f](x) = f(x)$$

uniformly on compact sets.

10. With $\mathfrak{X}$ as in the preceding problem show that $T(s)[f](x)$ satisfies the differential equation

$$F''(x) - s^2F(x) = -s^2f(x)$$

and is the only solution of this equation in $\mathfrak{X}$.

11. Prove that

$$(s^2 - t^2)\, T(s)[T(t)f] = s^2T(t)[f] - t^2T(s)[f].$$

12. Suppose that f is a measurable function of x such that $T(s)[f]$ exists for all $s > 0$. Suppose that $T(s)[f] = f$ for all $s > 0$. Show that $f(x) = ax + b$ for some choice of the constants a and b. When is such a function in (1) $C[-\infty, \infty]$, (2) $L_p(-\infty, \infty)$?

5.4 CONTRACTION FIXED POINT THEOREMS

We refer to Definition 5.3.1 for the terminology. In the present section "contraction" shall mean "strict contraction." We shall prove the contraction fixed point theorem of S. Banach.

Thus we are concerned with a complete metric space $\mathfrak{X}$ which is mapped into itself by a transformation T such that

$$d[T(\mathbf{x}_1), T(\mathbf{x}_2)] \leqslant kd(\mathbf{x}_1, \mathbf{x}_2) \tag{5.4.1}$$

for all $\mathbf{x}_1, \mathbf{x}_2 \in X$, where k is a fixed number, $0 < k < 1$.

Theorem 5.4.1. *If T is a contraction of $\mathfrak{X}$ into itself in the sense of* (5.4.1), *then T has a unique fixed point, i.e. there exists one and only one point $\mathbf{x}_0 \in \mathfrak{X}$ such that*

$$T(\mathbf{x}_0) = \mathbf{x}_0. \tag{5.4.2}$$

Proof. We start with an arbitrary point $\mathbf{x}_1$ in $\mathfrak{X}$ and form the successive transforms of $\mathbf{x}_1$ by the powers of T,

$$\mathbf{x}_{n+1} = T(\mathbf{x}_n), \quad n = 1, 2, 3, \ldots . \tag{5.4.3}$$

It will be proved first that $\{\mathbf{x}_n\}$ is a Cauchy sequence in $\mathfrak{X}$. For $m < n$

$$d(\mathbf{x}_m, \mathbf{x}_n) \leqslant d(\mathbf{x}_m, \mathbf{x}_{m+1}) + d(\mathbf{x}_{m+1}, \mathbf{x}_{m+2}) + \cdots + d(\mathbf{x}_{n-1}, \mathbf{x}_n) \tag{5.4.4}$$

by the triangle inequality. Formula (5.4.1) shows that for any positive integer

$$d(\mathbf{x}_p, \mathbf{x}_{p+1}) = d[T(\mathbf{x}_{p-1}), T(\mathbf{x}_p)] \leqslant kd(\mathbf{x}_{p-1}, \mathbf{x}_p),$$

and by repeated use of this device,

$$d(\mathbf{x}_p, \mathbf{x}_{p+1}) \leqslant k^{p-1}d(\mathbf{x}_1, \mathbf{x}_2).$$

Summing these inequalities from $p = m$ to $p = n - 1$ we obtain

$$d(\mathbf{x}_m, \mathbf{x}_n) \leqslant [k^{m-1} + k^m + \cdots + k^{n-2}]\, d(\mathbf{x}_1, \mathbf{x}_2).$$

Thus

$$d(\mathbf{x}_m, \mathbf{x}_n) \leqslant \frac{k^{m-1}}{1-k} d(\mathbf{x}_1, \mathbf{x}_2). \tag{5.4.5}$$

Since this goes to zero as $m \to \infty$, it is seen that $\{\mathbf{x}_n\}$ is a Cauchy sequence in $\mathfrak{X}$. Now $\mathfrak{X}$ is a complete metric space, so we can aver the existence of

$$\lim_{n\to\infty} \mathbf{x}_n \equiv \mathbf{x}_0.$$

From

$$\mathbf{x}_{n+1} = T(\mathbf{x}_n)$$

and the continuity of T which is a consequence of the Lipschitz condition (5.4.1) it follows that

$$\mathbf{x}_0 = \lim_{n\to\infty} \mathbf{x}_{n+1} = \lim_{n\to\infty} T(\mathbf{x}_n) = T\left(\lim_{n\to\infty} \mathbf{x}_n\right) = T(\mathbf{x}_0), \tag{5.4.6}$$

so that (5.4.2) holds.

Furthermore, $\mathbf{x}_0$ is the only fixed point. For suppose that

$$\mathbf{y}_0 = T(\mathbf{y}_0), \quad \mathbf{y}_0 \in \mathfrak{X}.$$

Then

$$d(\mathbf{x}_0, \mathbf{y}_0) = d[T(\mathbf{x}_0), T(\mathbf{y}_0)] \leqslant kd(\mathbf{x}_0, \mathbf{y}_0)$$

and this is possible iff $\mathbf{y}_0 = \mathbf{x}_0$. ▌

As an illustration we may take Example 1 of the preceding section. Thus $\mathfrak{X} = C^+[0, 1]$ and

$$T[f](t) = k\int_0^t f(s)\,ds, \quad 0 \leqslant t \leqslant 1,$$

where $0 < k < 1$. This is obviously a contraction, so there is a unique fixed point. On the other hand, by inspection the zero element is left invariant, so it must be the unique fixed point. This example is instructive inasmuch as the zero element is a fixed point for any value of k, not necessarily restricted to the interval $(0, 1)$.

Since the powers of T are used in the proof of Theorem 5.4.1, it would seem plausible that the contractive properties of the powers of T may play some role for the existence and uniqueness of fixed points. This is, indeed, the case.

Theorem 5.4.2. *T has a unique fixed point if one of the powers of T is a contraction.*

Proof. Suppose that T^m is a contraction. By the preceding theorem there is a unique point $\mathbf{x}_0$ such that

$$T^m(\mathbf{x}_0) = \mathbf{x}_0. \tag{5.4.7}$$

In the proof of Theorem 5.4.1 the fixed point was obtained as the limit of a Cauchy sequence $\{\mathbf{x}_n\}$ where the first element $\mathbf{x}_1$ is arbitrary. We apply this observation to the present case, taking $\mathbf{x}_1 = T(\mathbf{x}_0)$ and replacing T by T^m in the iterative process. This gives

$$\lim_{n\to\infty} (T^m)^n[T(\mathbf{x}_0)] = \mathbf{x}_0. \tag{5.4.8}$$

Since the powers of T commute, we can rewrite this as

$$T\left[\lim_{n\to\infty} (T^m)^n\,\mathbf{x}_0\right] = \mathbf{x}_0.$$

Here $(T^m)^n\,\mathbf{x}_0 = \mathbf{x}_0$ for all n, so we are left with

$$T(\mathbf{x}_0) = \mathbf{x}_0.$$

Thus $\mathbf{x}_0$ is also a fixed element under T itself. Moreover, it must be the only fixed point under T, for any fixed point of T is also a fixed point of T^m, which by assumption is a contraction and has a unique fixed point. It follows that T has one and only one fixed point, namely $\mathbf{x}_0$. ∎

This theorem is a powerful extension of the Banach contraction fixed point theorem since the contractive power of a bounded transformation often increases upon iteration. One of our standard examples illustrates this. We take $\mathfrak{X} = C[a, b]$, $0 < b - a < \infty$ and

$$T[f](t) = \int_a^t f(s)\,ds, \quad a \leqslant t \leqslant b. \tag{5.4.9}$$

Here $\|T\| = b - a$ so that T is not a contraction if $b - a > 1$. On the other hand,

$$T^m[f](t) = \frac{1}{(m-1)!} \int_a^t (t - s)^{m-1} f(s)\, ds \tag{5.4.10}$$

with norm

$$\|T^m\| = \frac{1}{m!}(b - a)^m, \tag{5.4.11}$$

so that T^m is a contraction for all large values of m. Since the zero element is obviously invariant, this is the only fixed point.

EXERCISE 5.4

1. Let $\mathfrak{X} = R^1$ and define $T(x) = \frac{1}{2}(x + \sin x)$. Show that T is contractive. Are there any fixed points?
2. Consider the transformation on R^2 defined by
$$T(x, y) = (x, ay), \quad a \neq 0, 1.$$
Show that there are infinitely many fixed points. Is T contractive for any value of a? Use the Euclidean metric.
3. [G. Szekeres] Given the transformation on R^2 defined by
$$T(x, y) = [x + y, y - (x + y)^3].$$
T is odd in the sense that $T(-x, -y) = -T(x, y)$. Show that $(0, 0)$ is the only fixed point of T. Show that T^2 has in addition the two fixed point $(2, -4)$ and $(-2, 4)$ which form a fixed point cycle of order two, i.e. the two fixed points of T^2 are permuted by T.
4. The surface S of a torus (anchor ring) is given by the parametric equations
$$x = \cos\alpha\,[R + r\cos\beta], \quad y = \sin\alpha\,[R + r\cos\beta], \quad z = r\sin\beta,$$
where $0 < r < R$. Show that there are two families of continuous transformations of S into itself which are without fixed points. [*Hint*: What is the geometric meaning of the parameters?]
5. Take $\mathfrak{X} = C[-\infty, \infty]$, let $s > 0$ and consider the Poisson transformation
$$P_s[f](t) = \frac{s}{\pi} \int_{-\infty}^{\infty} \frac{f(t + u)\, du}{s^2 + u^2}.$$
Show that P_s maps $\mathfrak{X}$ into itself and, in particular, maps the positive cone of $\mathfrak{X}$ into itself. Show that P_s is a contraction in the wide sense. There are infinitely many fixed points. Find some of them.
6. Show that $P_s[P_t] = P_t[P_s] = P_{s+t}$, the semi-group property.
7. Does the Picard transformation of Problem 8, Exercise 5.3, have any fixed points in $C[-\infty, \infty]$?

The next four problems give examples of transformations T from $C^+[0, b]$ into itself. Determine the nature of the transformation (contraction in the wide sense or in the strict sense or contractive mapping or neither one nor the other) and determine the fixed points. Let $T[f](t)$ be

8. $f(t) - \log[1 + f(t)]$.
9. $1 - \exp[-f(t)]$.
10. $\int_0^t [f(s)]^{1/2}\, ds$.
11. $\int_0^t [f(s)]^2\, ds$.
12. Is the mapping defined by Problem 10 bounded in the sense of (5.3.1)? Show that there are infinitely many fixed points and that the corresponding curves fill out a region in the plane with one and only one curve through each interior point of the region.

5.5 CONTRACTIVE MAPPINGS

We have seen that a contractive mapping need not have a fixed point. Thus the transformation

$$T(x) = x + x^{-1}$$

on $\mathfrak{X} = \{x;\ 1 \leqslant x\}$ with the Euclidean metric exhibits this phenomenon. It is necessary to impose additional conditions on the mapping. Such a condition was found in 1964 by M. Edelstein, who proved

Theorem 5.5.1. *Let T be a contractive mapping of a complete metric space $\mathfrak{X}$ into itself and suppose that there is a point $\mathbf{x}_1 \in \mathfrak{X}$ such that the sequence $\{T^n(\mathbf{x}_1)\}$ contains a subsequence converging to a point $\mathbf{x}_0 \in \mathfrak{X}$, then $\mathbf{x}_0$ is a fixed point of T and it is unique.*

Proof. The proof requires three steps. We start by observing that if $\{n_j\}$, $n_j < n_{j+1}$, $j = 1, 2, 3, \ldots$, is a sequence of integers such that

$$\mathbf{x}_j = T^{n_j}(\mathbf{x}_1) \to \mathbf{x}_0 \tag{5.5.1}$$

is a convergent subsequence, then $d(\mathbf{x}_j, \mathbf{x}_0) < \varepsilon$ for $j > N_\varepsilon$. Take $i > N_\varepsilon$ and set $m = n_{i+1} - n_i$. Then

$$d[T^m(\mathbf{x}_0), \mathbf{x}_{i+1}] = d[T^m(\mathbf{x}_0), T^m(\mathbf{x}_i)] \leqslant d(\mathbf{x}_0, \mathbf{x}_i) < \varepsilon,$$

since T^m is also contractive. Hence

$$d[T^m(\mathbf{x}_0), \mathbf{x}_0] \leqslant d[T^m(\mathbf{x}_0), \mathbf{x}_{i+1}] + d[\mathbf{x}_{i+1}, \mathbf{x}_i] < 2\varepsilon$$

by the triangle inequality. It is true that here ε is arbitrary, but the choice of ε will affect the choice of i and hence that of m. As a result of the first step we see, however, that there exist integers m such that $T^m(\mathbf{x}_0)$ is arbitrarily close to $\mathbf{x}_0$.

For the second step we fix ε, choose i and m as above, and consider

$$T^m(\mathbf{x}_0) \equiv \mathbf{y}_0. \tag{5.5.2}$$

If it should turn out that $\mathbf{y}_0 = \mathbf{x}_0$, then we can pass directly to the final step. We now assume that $\mathbf{y}_0 \neq \mathbf{x}_0$ and will show that this assumption leads to a contradiction. Consider the mapping

$$(\mathbf{x}, \mathbf{y}) \to \frac{d[T(\mathbf{x}), T(\mathbf{y})]}{d(\mathbf{x}, \mathbf{y})} \equiv f(\mathbf{x}, \mathbf{y}). \tag{5.5.3}$$

The quotient is defined and continuous as long as $\mathbf{y} \neq \mathbf{x}$, hence, in particular, for $\mathbf{x} = \mathbf{x}_0$, $\mathbf{y} = \mathbf{y}_0$. Since T is contractive, $f(\mathbf{x}, \mathbf{y}) < 1$ wherever it is defined. If now $f(\mathbf{x}_0, \mathbf{y}_0) = q < 1$, then we can find a $\delta > 0$ and a number k, $q \leqslant k < 1$, such that

$$d[T(\mathbf{x}), T(\mathbf{y})] \leqslant kd(\mathbf{x}, \mathbf{y}) \quad \text{for} \quad d(\mathbf{x}, \mathbf{x}_0) < \delta, \quad d(\mathbf{y}, \mathbf{y}_0) < \delta. \tag{5.5.4}$$

In order to achieve this it is enough to choose δ so small that

$$|f(\mathbf{x}, \mathbf{y}) - f(\mathbf{x}_0, \mathbf{y}_0)| < \tfrac{1}{2}(1 - q)$$

in the neighborhood of $(\mathbf{x}_0, \mathbf{y}_0)$ in question. This is possible since f is continuous. We have then also

$$f(\mathbf{x}, \mathbf{y}) < q + \tfrac{1}{2}(1 - q) = \tfrac{1}{2}(1 + q) < 1.$$

We take $k = \frac{1}{2}(1 + q)$. Since

$$\lim_{j \to \infty} T^m(\mathbf{x}_j) = T^m(\lim \mathbf{x}_j) = T^m(\mathbf{x}_0) = \mathbf{y}_0,$$

we see that

$$d(\mathbf{x}_j, \mathbf{x}_0) < \delta, \quad d[T^m(\mathbf{x}_j), \mathbf{y}_0] < \delta, \quad \text{for} \quad j > N,$$

some large positive integer. Hence

$$d[T(\mathbf{x}_j), TT^m(\mathbf{x}_j)] \leqslant kd[\mathbf{x}_j, T^m(\mathbf{x}_j)].$$

T being contractive, we have for any positive integer p

$$d[T^p(\mathbf{x}_j), T^{m+p}(\mathbf{x}_j)] \leqslant d[T(\mathbf{x}_j), T^{m+1}(\mathbf{x}_j)] \leqslant kd[\mathbf{x}_j, T^m(\mathbf{x}_j)].$$

In particular, for $p = n_{j+1} - n_j$,

$$d[\mathbf{x}_{j+1}, T^m(\mathbf{x}_{j+1})] \leqslant kd[\mathbf{x}_j, T^m(\mathbf{x}_j)]$$

for any $j > N$. Thus for $r > j$,

$$d[\mathbf{x}_r, T^m(\mathbf{x}_r)] \leqslant k^{r-j}d[\mathbf{x}_j, T_m(\mathbf{x}_j)]. \tag{5.5.5}$$

Here as $r \to \infty$, the left member goes to $d(\mathbf{x}_0, \mathbf{y}_0)$ while the right member goes to zero. This contradiction shows that $\mathbf{y}_0$ cannot be distinct from $\mathbf{x}_0$. Hence we have $\mathbf{x}_0 = \mathbf{y}_0 = T^m(\mathbf{x}_0)$, and this completes the second step.

The third step is much simpler. We have now

$$T^m(\mathbf{x}_0) = \mathbf{x}_0 \tag{5.5.6}$$

for some positive integer m. Suppose that

$$T(\mathbf{x}_0) = \mathbf{v}_0 \neq \mathbf{x}_0.$$

Then

$$T^m(\mathbf{v}_0) = T^m T(\mathbf{x}_0) = T T^m(\mathbf{x}_0) = T(\mathbf{x}_0) = \mathbf{v}_0$$

and

$$d(\mathbf{x}_0, \mathbf{v}_0) = d[T^m(\mathbf{x}_0), T^m(\mathbf{v}_0)] < d(\mathbf{x}_0, \mathbf{v}_0)$$

since T is contractive. Thus the assumption that $\mathbf{v}_0 \neq \mathbf{x}_0$ also leads to a contradiction and we must have

$$\mathbf{v}_0 = \mathbf{x}_0, \qquad T(\mathbf{x}_0) = \mathbf{x}_0$$

and $\mathbf{x}_0$ is a fixed point of T.

To prove uniqueness, we suppose that $\mathbf{z}_0$ is a fixed point of T. Then

$$d(\mathbf{x}_0, \mathbf{z}_0) = d[T(\mathbf{x}_0), T(\mathbf{z}_0)] < d(\mathbf{x}_0, \mathbf{z}_0),$$

again because T is contractive. This contradiction shows that $\mathbf{z}_0 = \mathbf{x}_0$ and there is one and only one fixed point. ∎

There is an important corollary which we state as

Theorem 5.5.2. *If T is a contractive mapping of a complete metric space $\mathfrak{X}$ into a compact subset of itself, then T has a unique fixed point $\mathbf{x}_0$ and for every $\mathbf{x} \in \mathfrak{X}$*

$$\mathbf{x}_0 = \lim_{n\to\infty} T^n(\mathbf{x}). \tag{5.5.7}$$

Proof. To see this, note that the range $T(\mathfrak{X})$ is sequentially compact. This means that for a fixed $\mathbf{x}$ in $\mathfrak{X}$ the sequence $\{T^n(\mathbf{x})\}$ has at least one cluster point in $\mathfrak{X}$. Hence there is a convergent subsequence and by the preceding theorem T has a unique fixed point given by the limit of the convergent subsequence. Thus the original sequence can have only one cluster point. Hence (5.5.7) holds and the limit is independent of the choice of $\mathbf{x}$ in $\mathfrak{X}$. ∎

EXERCISE 5.5

1. If $\mathfrak{X} = \{x; 1 \leqslant x\}$ with the Euclidean metric and if $T(x) = x + x^{-1}$, then for fixed x the sequence $\{T^n(x)\}$ is obviously increasing. Is it bounded?
2. Take $\mathfrak{X} = R^1$ and $T(x) = \sin x$. Show that T maps R^1 onto a compact subset. Show that T is contractive and find the fixed point.
3. Take $\mathfrak{X} = R^2$ with the maximum coordinate metric and, further, let $T(x, y) = [\frac{1}{2} \arctan (x + y), \frac{1}{2} \sin (x - y)]$ where the arc tangent has its principal value. Is $T(R^2)$ compact? Show that T is contractive and find the fixed point.
4. The optimal value of k in formula (5.5.4) depends upon $\mathbf{x}_0$, $\mathbf{y}_0$ and δ. To illustrate, use Problem 2 above and set $\mathbf{x}_0 = \frac{1}{6}\pi$, $\mathbf{y}_0 = \frac{1}{3}\pi$, $\delta = \frac{1}{12}\pi$. Show that k will not exceed $\cos(\frac{1}{12}\pi)$. What happens to k when $\mathbf{x}_0$ and $\mathbf{y}_0$ approach the fixed point?
5. The Poisson transformation defined in Problem 5, Exercise 5.4, maps $L_1(-\infty, \infty)$ into itself. Prove this and show that P_s is contractive and has a unique fixed point.

5.6 THE VOLTERRA FIXED POINT THEOREM

We use this designation for a theorem which underlies the work of Vito Volterra (1860–1940) on linear integral equations in the 1890's.

Theorem 5.6.1. *Let $\mathfrak{X}$ be a B-space. Let $\mathbf{z}_0$ be a given element of $\mathfrak{X}$ and let $S \in \mathfrak{E}(\mathfrak{X})$ be such that*

$$1 + \sum_{n=1}^{\infty} \|S^n\| \equiv A < \infty. \tag{5.6.1}$$

Then the transformation T defined by

$$T(\mathbf{x}) = \mathbf{z}_0 + S(\mathbf{x}) \tag{5.6.2}$$

has a unique fixed point given by

$$\mathbf{x}_0 = \mathbf{z}_0 + \sum_{n=1}^{\infty} S^n(\mathbf{z}_0). \tag{5.6.3}$$

Proof. Here again we start with an arbitrary element $\mathbf{x}_1$ of $\mathfrak{X}$ and form the sequence $\{\mathbf{x}_n\}$ where

$$\mathbf{x}_{n+1} = T(\mathbf{x}_n), \quad n = 1, 2, 3, \ldots .$$

Thus

$$\mathbf{x}_n = \mathbf{z}_0 + \sum_{k=1}^{n-2} S^k(\mathbf{z}_0) + S^{n-1}(\mathbf{x}_1)$$

and

$$\|\mathbf{x}_{n+p} - \mathbf{x}_n\| = \| - S^{n-1}(\mathbf{x}_1) + S^{n-1}(\mathbf{z}_0) + \cdots + S^{n+p-2}(\mathbf{z}_0) + S^{n+p-1}(\mathbf{x}_1)\|.$$

This does not exceed

$$\sum_{n-1}^{n+p-1} \|S^k\| \, [\|\mathbf{z}_0\| + \|\mathbf{x}_1\|],$$

and by condition (5.6.1) this goes to zero as $n \to \infty$. Hence $\{\mathbf{x}_n\}$ is a Cauchy sequence in the complete metric space $\mathfrak{X}$ and its limit is

$$\mathbf{x}_0 = \lim_{n\to\infty} \mathbf{x}_n = \mathbf{z}_0 + \sum_{n=1}^{\infty} S^n(\mathbf{z}_0)$$

as asserted above. This is a fixed point of T since

$$\mathbf{x}_0 = \lim_{n\to\infty} \mathbf{x}_n = \lim_{n\to\infty} T(\mathbf{x}_{n-1}) = T\left(\lim_{n\to\infty} \mathbf{x}_{n-1}\right) = T(\mathbf{x}_0).$$

If $\mathbf{y}_0$ is a fixed point of T, then

$$\mathbf{x}_0 - \mathbf{y}_0 = T(\mathbf{x}_0) - T(\mathbf{y}_0) = S(\mathbf{x}_0 - \mathbf{y}_0) = \cdots = S^n(\mathbf{x}_0 - \mathbf{y}_0)$$

and this goes to $\mathbf{0}$ as $n \to \infty$. Hence $\mathbf{y}_0 = \mathbf{x}_0$ and the fixed point is unique. ∎

The solution (5.6.3) is known as the *Neumann series*. It is named after the Carl Neumann (1832–1925) who did important work in potential theory and in algebraic functions and their integrals. The fixed point of (5.6.2) is the solution of the functional equation

$$\mathbf{x} = \mathbf{z}_0 + S(\mathbf{x}) \tag{5.6.4}$$

or

$$[I - S](\mathbf{x}) = \mathbf{z}_0, \tag{5.6.5}$$

whence, formally,

$$\mathbf{x} = [I - S]^{-1}(\mathbf{z}_0). \tag{5.6.6}$$

Here $[I - S]^{-1}$ is the inverse transformation of $I - S$ or, in the notation of (2.4.24), $R(1, S)$, the resolvent of S evaluated at $\lambda = 1$. The proof of Theorem 5.6.1 shows that these formal considerations make sense: the resolvent does exist for $\lambda = 1$ and is given by the Neumann series. First, the series (5.6.3) converges in norm since

$$\|\mathbf{z}_0\| + \sum_{n=1}^{\infty} \|S^n(\mathbf{z}_0)\| \leqslant A\|\mathbf{z}_0\| \tag{5.6.7}$$

by condition (5.6.1). Secondly, we have

$$(I - S)\left[\mathbf{z}_0 + \sum_{k=1}^{n} S^k(\mathbf{z}_0)\right] = \left[I + \sum_{k=1}^{n} S^k\right][I - S](\mathbf{z}_0) = \mathbf{z}_0 - S^{n+1}(\mathbf{z}_0) \to \mathbf{z}_0$$

as $n \to \infty$. This shows that the series (5.6.3) represents $R(\lambda, S)(z_0)$ for $\lambda = 1$ and we see that the spectrum of S is confined to the open disk $|\lambda| < 1$. The spectral radius of S is <1 and no stronger assertion can be made without imposing further conditions on S. Actually the Neumann series is no novelty to us; its first occurrence in this treatise was in formula (1.5.38) for the resolvent of a matrix.

Example 1. We take our standard transformation with $\mathfrak{X} = C[a, b]$ and S defined by integration from a to t. Further, $\mathbf{z}_0 = g(t)$ is in $C[a, b]$. Equation (5.6.4) now becomes

$$f(t) = g(t) + \int_a^t f(s)\,ds, \quad a \leqslant t \leqslant b, \tag{5.6.8}$$

with the solution

$$f(t) = g(t) + \sum_{m=1}^{\infty} \frac{1}{(m-1)!} \int_a^t (t-s)^{m-1} g(s)\,ds$$

or

$$f(t) = g(t) + \int_a^t \exp(t-s)\, g(s)\,ds. \tag{5.6.9}$$

In this case the spectral radius is 0 (why?) and $\lambda = 0$ is the only spectral value.

EXERCISE 5.6

1. Find the solution of the matrix vector equation

$$\mathbf{y} = \mathbf{z}_0 + \mathfrak{Q} \cdot \mathbf{y}, \quad \mathfrak{Q} \in \mathfrak{M}_n, \quad \mathfrak{Q}^p = \mathfrak{O}, \quad \mathfrak{Q}^{p-1} \neq \mathfrak{O}.$$

Here $\mathbf{z}_0 \in C^n$ and the unique solution $\mathbf{y} \in C^n$ is requested.

2. An apparently trivial application of Theorem 5.6.1 is to find the solution of

$$f(t) = g(t) + \mu F(t) f(t),$$

where g and F are given elements of $C[0, 1]$ and μ is a complex parameter. Get the formal solution and decide what values of μ must be excluded in order that $f \in C[0, 1]$.

3. Verify the assertion that $\sigma(S) = \{0\}$ in the case of (5.6.8).

The next three problems deal with a special Fredholm integral equation [Ivar Fredholm (1866–1927)]:

$$f(t) = g(t) + \frac{\mu}{2\pi} \int_{-\infty}^{\infty} K(t - s) f(s) \, ds.$$

Here μ is a complex parameter; the function g and the kernel K are given elements of $L_2(-\pi, \pi)$,

$$g(t) \sim \sum_{-\infty}^{\infty} g_n e^{nit}, \qquad K(u) \sim \sum_{-\infty}^{\infty} k_n e^{nit}.$$

This is a functional equation to which Theorem 5.6.1 applies for sufficiently small values of $|\mu|$. Here $\mathfrak{X} = L_2(-\pi, \pi)$, $\mathbf{z}_0 = g(t)$, and

$$S[f](t) = \frac{\mu}{2\pi} \int_{\pi}^{-\pi} K(t - s) f(s) \, ds.$$

4. Show that

$$S^p[f](t) = \left(\frac{\mu}{2\pi}\right)^p \int_{-\pi}^{\pi} K_p(t - s) f(s) \, ds,$$

where

$$K_p(u) \sim \sum_{-\infty}^{\infty} (k_n)^p e^{niu}.$$

Show that the series converges absolutely and uniformly when $p > 1$.

5. If $f(t; \mu)$ denotes the solution of the Fredholm equation, show that

$$f(t; \mu) - g(t) = \mu \sum_{-\infty}^{\infty} \frac{k_n g_n}{1 - \mu k_n} e^{nit}$$

where the series is absolutely and uniformly convergent with respect to t. What values of μ must be excluded in making this statement?

6. Show that the excluded values of μ are the reciprocals of the characteristic values of the kernel as an operator in L_2, i.e. those values of λ for which

$$\lambda h(t) - \frac{1}{2\pi}\int_{-\pi}^{\pi} K(t-s)\,h(s)\,ds = 0$$

has a non-trivial solution in $L_2(-\pi, \pi)$. [*Hint:* Postulate a solution in terms of a Fourier series and determine the conditions on λ.]

5.7 SOME APPLICATIONS

As a matter of fact, all of Chapter 6 and most of Chapter 12 are given to applications of fixed point theorems to basic questions of analysis. In this section we consider some questions of a somewhat special nature.

I. Volterra's Equation

It is appropriate to start with a Volterra equation,

$$f(t) = g(t) + \int_0^t K(s, t) f(s)\,ds. \tag{5.7.1}$$

To simplify matters, we restrict ourselves to a finite interval and continuous functions. Let $g \in C[0, b]$ and let $K(s, t)$ be continuous for (s, t) in $[0, b] \times [0, b]$, where $\|K\| = B$. Then for $f \in C[0, b]$

$$S[f](t) = \int_0^t K(s, t) f(s)\,ds \tag{5.7.2}$$

belongs to $C[0, b]$ and the norm of the transformation S is at most bB. To be able to apply Theorem 5.6.1 we must show that

$$\Sigma\,\|S^n\| \tag{5.7.3}$$

converges. Now

$$S^n[f](t) = \int_0^t K_n(s, t) f(s)\,ds, \tag{5.7.4}$$

where K_n is the nth iterated kernel. Compare Problem 4, Exercise 5.6, where, however, the interval of integration is fixed and the kernel is periodic. Here

$$S^2[f](t) = \int_0^t K(s, t)\left\{\int_0^s K(u, s) f(u)\,du\right\} ds$$

$$= \int_0^t f(u)\left\{\int_u^t K(s, t)\,K(u, s)\,ds\right\} du,$$

where the interchange of the order of integration is permitted since the functions are continuous. This gives

$$K_2(s, t) = \int_s^t K(s, u)\, K(u, t)\, du, \tag{5.7.5}$$

as is seen by permuting the variables in the preceding integral. In general,

$$K_n(s, t) = \int_s^t K(s, u)\, K_{n-1}(u, t)\, du. \tag{5.7.6}$$

From (5.7.5) we obtain

$$\|S^2\| \leqslant \tfrac{1}{2}(bB)^2$$

and by complete induction

$$\|S^n\| \leqslant \frac{1}{n!}(bB)^n. \tag{5.7.7}$$

Thus condition (5.7.3) is amply satisfied. Since the other conditions of Theorem 5.6.1 obviously hold, it follows that the transformation $T[f] = g + S[f]$ has a unique fixed point or, equivalently, the Volterra equation (5.7.1) has a unique solution. The solution is given by the Neumann series, which now becomes

$$f(t) = g(t) + \int_0^t \mathsf{K}(s, t)\, g(s)\, ds, \tag{5.7.8}$$

where the so-called *resolvent kernel* is

$$\mathsf{K}(s, t) = \sum_{n=1}^{\infty} K_n(s, t) \tag{5.7.9}$$

with $K_1(s, t) = K(s, t)$.

The assumptions on f, g, and K may be relaxed. Thus we may replace $C[0, b]$ by $L_2(0, b)$ and appeal to the Fubini theorem.

In the special case

$$K(s, t) = K(s) \in L(0, b) \cap C(0, b] \tag{5.7.10}$$

the solution can be given explicitly. Here

$$K_2(s, t) = \int_s^t K(s)\, K(u)\, du = K(s) \int_s^t K(u)\, du,$$

$$K_3(s, t) = K(s) \int_s^t K(u) \left\{ \int_u^t K(v)\, dv \right\} du = K(s)\, \tfrac{1}{2} \left\{ \int_s^t K(u)\, du \right\}^2$$

. . . .

$$K_n(s, t) = \frac{1}{(n-1)!} K(s) \left\{ \int_s^t K(u)\, du \right\}^{n-1}.$$

It follows that the resolvent kernel is given by

$$\mathsf{K}(s, t) = K(s) \exp\left[\int_s^t K(u)\, du\right] \tag{5.7.11}$$

and

$$f(t) = g(t) + \int_0^t K(s) \exp\left[\int_s^t K(u)\, du\right] g(s)\, ds. \tag{5.7.12}$$

The special case $K(s) \equiv 1$ figured in (5.6.9): we note that $g(t) \equiv 0$ implies $f(t) \equiv 0$. It should be observed that the spectrum of this operator as well as the general Volterra operator reduces to $\{0\}$. Volterra developed an extensive theory of functions of composition in connection with his integral equation.

II. Fredholm's Equation

A particular case of this equation was studied in Exercise 5.6 where, thanks to the special nature of the kernel, the solution as a function of the parameter μ was obtained for all non-singular values of μ. The resulting series in Problem 5 is of a type familiar in analytic function theory as the Mittag–Leffler expansion of the meromorphic function $f(t; \mu) - g(t)$. In the case to be studied here, the solution is still a meromorphic function of μ as shown by Fredholm, but all we obtain in general is a power series for the function for small values of μ.

We shall consider

$$f(t) = g(t) + \mu \int_a^b K(s, t) f(s)\, ds. \tag{5.7.13}$$

Here $g \in C[a, b]$ and K is continuous for (s, t) in the square $[a, b] \times [a, b]$. We set

$$S[f](t) = \mu \int_a^b K(s, t) f(s)\, ds. \tag{5.7.14}$$

This is a linear bounded transformation from $C[a, b]$ into itself and

$$\|S\| \leqslant |\mu|(b - a)\, B, \quad \text{if} \quad B = \max |K(s, t)|. \tag{5.7.15}$$

Thus the series $\Sigma \, \|S^n\|$ is certainly convergent for

$$|\mu| < [(b - a)B]^{-1}. \tag{5.7.16}$$

Hence the solution is of the form

$$f(t) = g(t) + \int_a^b \mathsf{K}(s, t; \mu)\, g(s)\, ds \tag{5.7.17}$$

with the resolvent kernel

$$\mathsf{K}(s, t; \mu) = \sum_{n=1}^{\infty} K_n(s, t)\, \mu^n. \tag{5.7.18}$$

The series converges at least for μ satisfying (5.7.16) and again $K_n(s, t)$ is an iterated kernel

$$K_n(s,t) = \int_a^b K(s,u)\,K_{n-1}(u,t)\,du, \quad n = 2, 3, \ldots . \tag{5.7.19}$$

For fixed (s, t) the resolvent kernel is an analytic function of μ. It turns out that, just as in the special case considered in Exercise 5.6, the singularities of this function are fixed, independent of (s, t), and those in the finite plane, if any, are poles. The singularities are the reciprocals of the values of λ for which the homogeneous equation

$$\lambda h(t) = \int_a^b K(s,t)\,h(s)\,ds \tag{5.7.20}$$

has a non-trivial solution in $C[a, b]$. For such a value, λ_0 say, the non-homogeneous equation

$$\lambda_0 f(t) = g(t) + \int_a^b K(s,t) f(s)\,ds \tag{5.7.21}$$

does not possess a solution for an arbitrary choice of $g \in C[a, b]$. The situation is analogous to that holding for systems of linear equations in C^n. Fredholm suspected this analogy to be valid and it served him as a guiding line in his work.

The restriction to continuous functions is by no means necessary. If it is dropped, then we can consider Volterra's equation as a special case of Fredholm's, where the kernel is zero above the diagonal $t = s$ and the upper limit of integration is taken as t.

To illustrate the theory, let us consider two special examples where explicit solutions are available. Let $\{\omega_n(t)\}$ be a complete orthonormal system for $L_2(a, b)$. To simplify matters, we assume $\omega_n(t) \in C[a, b]$ for all n and to be real and of uniformly bounded sup norm. Such systems exist. Form the kernel

$$K(s,t) = \sum_{m=1}^{\infty} k_m \omega_m(s)\,\omega_m(t), \tag{5.7.22}$$

where $k_m \geqslant 0$, $\forall\, m$, and the series

$$\sum_{m=1}^{\infty} k_m \tag{5.7.23}$$

is convergent. These assumptions imply that (5.7.22) is absolutely convergent, uniformly in (s, t), and its sum is a continuous function of (s, t). We can now compute the iterated kernels and find that

$$K_n(s,t) = \sum_{m=1}^{\infty} (k_m)^n \omega_m(s)\,\omega_m(t) \tag{5.7.24}$$

and the resolvent kernel becomes

$$\mathsf{K}(s,t;\mu) = \sum_{n=1}^{\infty} \mu^n \sum_{m=1}^{\infty} (k_m)^n \omega_m(s)\,\omega_m(t). \tag{5.7.25}$$

If now

$$|\mu| < \inf(|k_m|^{-1}), \tag{5.7.26}$$

then the double series is absolutely convergent, uniformly in (s, t), and the order of summation may be interchanged. This gives, after simplification,

$$\mathsf{K}(s, t; \mu) = \sum_{m=1}^{\infty} \frac{\mu k_m}{1 - \mu k_m} \omega_m(s)\,\omega_m(t). \tag{5.7.27}$$

Denote the set

$$\{k_m^{-1};\ k_m \neq 0,\ 1 \leqslant m < \infty\}$$

by Σ and suppose that μ has a positive distance from Σ. Then the series (5.7.27) is absolutely convergent, uniformly in (s, t). For fixed (s, t) the resolvent kernel is a meromorphic function of μ with simple poles at the points of Σ. Note that Σ has no finite cluster points. We have obviously

$$k_m \cdot \omega_m(t) = \int_a^b K(s, t)\,\omega_m(s)\,ds, \tag{5.7.28}$$

that is, the k_m are all in the point spectrum (= set of characteristic values) of the corresponding operator. Cf. (5.7.20).

Note that the partial fraction series (5.7.27) is valid for all μ not in Σ, while the power series (5.7.27) presupposes that (5.7.26) holds, and in this circular disk the expansions represent the same analytic function of μ.

We can now form the solution (5.7.17). Here g shall belong to $L_2(a, b)$ and have Fourier coefficients $\{g_n\}$ with respect to the system $\{\omega_n\}$. Substitution of (5.7.27) and termwise integration finally yield

$$f(t; \mu) = g(t) + \mu \sum_{m=1}^{\infty} \frac{g_m k_m}{1 - \mu k_m} \omega_m(t), \tag{5.7.29}$$

where the series is absolutely convergent for all μ not in Σ and the convergence is uniform in t. Thus the non-homogeneous equation is solvable for every μ not in Σ and the solution is given by (5.7.29). On the other hand, for $\mu = (k_m)^{-1}$ the non-homogeneous equation has a solution in L_2 iff the corresponding Fourier coefficient g_m of $g(t)$ is zero so that (5.7.29) remains valid after the indeterminate form has been suppressed.

The kernel (5.7.22) is symmetric, i.e. $K(s, t) = K(t, s)$. Such kernels were studied in great detail by Hilbert and his school during the first decade of this century after Fredholm's results became known. Erhardt Schmidt, one of Hilbert's pupils, proved that such a kernel necessarily has at least one characteristic value. The minimum number one is reached by the kernel

$$k_m \omega_m(s)\,\omega_m(t).$$

Let us consider briefly the non-symmetric kernel

$$K(s, t) = \sum_{m=1}^{\infty} k_m \omega_m(s)\,\omega_{m+1}(t) \tag{5.7.30}$$

under the same assumptions on k_m and ω_m. Here the situation is radically different. *There are no characteristic values.* The reader should note that the term "characteristic value" means one thing in the theory of integral equations and something different in linear operator theory. The relation is simply $\lambda = 1/\mu$. Cf. comments to equations (5.7.20) and (5.7.21).

We now obtain

$$K_n(s, t) = \sum_{m=1}^{\infty} k_m k_{m+1} \cdots k_{m+n-1} \omega_m(s)\, \omega_{m+n}(t) \tag{5.7.31}$$

and

$$\mathsf{K}(s, t; \mu) = \sum_{n=1}^{\infty} \mu^n \sum_{m=1}^{\infty} k_m k_{m+1} \cdots k_{m+n-1}\, \omega_m(s)\, \omega_{m+n}(t). \tag{5.7.32}$$

Here the double series is absolutely convergent, uniformly in (s, t), for any finite μ so that $\mathsf{K}(s, t; \mu)$ is an entire function of μ. We shall not prove this and note merely the solution of the non-homogeneous equation, valid for all μ,

$$f(t; \mu) = g(t) + \sum_{n=1}^{\infty} \mu^n \sum_{m=1}^{\infty} k_m k_{m+1} \cdots k_{m+n-1} g_m \omega_{m+n}(t), \tag{5.7.33}$$

where, as above, the g_m's are the Fourier coefficients of g in the system $\{\omega_m\}$.

Integral equations are of great importance in applied mathematics (potential theory, boundary value problems for differential equations which usually lead to kernels of type (5.7.22), etc.) and this theory was one of the forerunners of functional analysis.

III. *Infinitely Many Equations in Infinitely Many Unknowns*

The problem of solving an infinite system of linear equations presented itself quite early in classical analysis and in many different connections. An early instance occurs in the work of J. Fourier on heat conduction. The American astronomer G. W. Hill (1838–1914) encountered such systems in lunar theory in 1877 and solved these systems formally by infinite determinants. Henri Poincaré (1854–1912) in his revision of celestial mechanics, 1892–9, put this work on a firm basis and Helge von Koch (1870–1924) developed an elaborate theory of infinite determinants with applications to linear differential equations and other fields, starting in 1892. His work was basic for Fredholm's investigations. Quadratic forms in infinitely many unknowns and the related linear equations formed a fundamental part in Hilbert's work on integral equations. Here we shall take up a few cases of linear systems which may be studied with the aid of the tools developed in this chapter,

We consider first the class $\mathfrak{M}_\infty$ of infinite matrices $\mathcal{A} = (a_{jk})$ with the norm

$$\|\mathcal{A}\|_\infty = \sup_j \sum_{k=1}^{\infty} |a_{jk}| < \infty, \tag{5.7.34}$$

which is a natural generalization of (1.4.43). We define the algebraic matrix operations in the natural manner. The product is the only operation which needs justification. We have $\mathcal{A}\mathcal{B} = (c_{jk})$ with

$$c_{jk} = \sum_{m=1}^{\infty} a_{jm} b_{mk}.$$

Hence

$$\sum_{k=1}^{\infty} |c_{jk}| \leqslant \sum_{k=1}^{\infty} \sum_{m=1}^{\infty} |a_{jm}|\,|b_{mk}| = \sum_{m=1}^{\infty} |a_{jm}| \sum_{k=1}^{\infty} |b_{mk}| \leqslant \|\mathcal{A}\|_{\infty}\,\|\mathcal{B}\|_{\infty}.$$

Thus $\mathcal{A}\mathcal{B}$ exists and is an element of $\mathfrak{M}_{\infty}$: moreover,

$$\|\mathcal{A}\mathcal{B}\|_{\infty} \leqslant \|\mathcal{A}\|_{\infty}\,\|\mathcal{B}\|_{\infty}. \tag{5.7.35}$$

The reader should verify that $\mathfrak{M}_{\infty}$ is complete in the metric so that it is a B-algebra. The unit element is $\mathcal{E} = (\delta_{jk})$.

We now consider the equation

$$\lambda \mathbf{y} - \mathcal{A}\mathbf{y} = \mathbf{x}, \tag{5.7.36}$$

where $\mathbf{x} = (x_j) \in l_{\infty}$; that is,

$$\|\mathbf{x}\|_{\infty} = \sup_j |x_j| < \infty. \tag{5.7.37}$$

A solution is desired in the same space. Here λ is a complex parameter and formally the solution is given by the resolvent of $\mathcal{A}$ operating on $\mathbf{x}$ or

$$\mathbf{y} = \mathcal{R}(\lambda, \mathcal{A})\,\mathbf{x}. \tag{5.7.38}$$

This is true whenever the resolvent makes sense, in particular for

$$|\lambda| > \|\mathcal{A}\|_{\infty}. \tag{5.7.39}$$

For such values the usual Neumann series will give the solution.

The special case

$$a_{jk} = \begin{cases} j^{-1}, & 1 \leqslant k \leqslant j, \\ 0, & j < k, \end{cases} \tag{5.7.40}$$

is of some interest. It is the matrix $\mathcal{C}$ corresponding to Cesàro summability of order one, usually denoted by $(C, 1)$. Cf. Problem 14, Exercise 3.1. Here $\|\mathcal{C}\|_{\infty} = 1$ so the resolvent converges for $|\lambda| > 1$. It is possible to find the characteristic values and corresponding vectors explicitly. They are

$$\lambda_n = \frac{1}{n}, \quad \mathbf{y}_n = (\delta_{mn}), \quad n = 1, 2, 3, \ldots, \tag{5.7.41}$$

respectively. The proof is left to the reader

Another case in which the Neumann series applies is given by the matrix (a_{jk}) with

$$a_{jk} = c_{j+k-1}. \tag{5.7.42}$$

Here $c_n > 0$ and strictly decreasing to 0 as $n \to \infty$. Further,

$$\sum_{n=1}^{\infty} c_n = 1. \tag{5.7.43}$$

Here $\|\mathcal{A}\|_\infty = 1$. Nevertheless, the solution of

$$\mathbf{y} - \mathcal{A}\mathbf{y} = \mathbf{x} \tag{5.7.44}$$

is given by the absolutely convergent series

$$\mathbf{y} = \mathbf{x} + \mathcal{A}\mathbf{x} + \mathcal{A}^2\mathbf{x} + \cdots + \mathcal{A}^n\mathbf{x} + \cdots. \tag{5.7.45}$$

This follows from the fact that $\|\mathcal{A}^n\|$ actually goes exponentially to zero. We have

$$\|\mathcal{A}^2\| = \sup_j \sum_k \sum_m c_{j+m-1}c_{m+k-1} < 1 - c_1 + c_1{}^2 \equiv r < 1.$$

Complete induction gives

$$\|\mathcal{A}^n\| \leqslant r^{[\frac{1}{2}n]}, \tag{5.7.46}$$

where $[u]$ is the largest integer $\leqslant u$. This proves the assertion on the norm of $\mathcal{A}^n$ and also the validity of (5.7.45) as the solution of (5.7.44).

EXERCISE 5.7

1. Find the resolvent kernel $\mathsf{K}(s, t)$ of the Volterra equation with kernel $(t - s)^{-\alpha}$, $\alpha > -1$. [*Hint:* Remember the β-function of Euler!]
2. Find the resolvent kernel $\mathsf{K}(s, t; \mu)$ of the Fredholm equation with $a = 0$, $b = 1$, $K(s, t) = t - s$. The power series can be summed in closed form. Show that it is a meromorphic function of μ and find the poles.
3. Verify (5.7.24) to (5.7.27).
4. Verify (5.7.28) and justify (5.7.29).
5. Are there complete orthonormal systems in $L_2(a, b)$ the elements of which are in $C[a, b]$ and of uniformly bounded sup norms?
6. Verify (5.7.31) and prove the convergence of the series.
7. Prove the absolute and uniform convergence of the series (5.7.32), assuming (5.7.23) absolutely convergent.
8. If $K(s, t)$ is defined by (5.7.30) with (5.7.23) holding and if

$$S[f](t) = \int_a^b K(s, t) f(s)\, ds, \; f \in C[a, b],$$

what is the spectrum of the operator S? [*Hint:* Prove quasi-nilpotency.]

9. Justify (5.7.33).

10. If $\mu_m = m^{-2}$ in (5.7.32), show that $\sup_{(s,t)} |\mathsf{K}(s, t; \mu)| < A \exp(B|\mu|^{1/2})$ for all μ and some constants A and B.

11. Suppose, instead, that $\mu_m = m^{-\alpha}$, $\alpha > 1$; show that the upper bound should be of the form $A \exp(B|\mu|^{1/\alpha})$. Thus the resolvent kernel is an entire function of order at most $1/\alpha$. For $\frac{1}{2} < \alpha \leqslant 1$, the kernel $K(s, t)$ is still L_2 and all iterated kernels are in $C[a, b]$. The estimate now applies to $\mathsf{K}(s, t; \mu) - \mu K(s, t)$. Prove this.

12. Prove that $\mathfrak{M}_\infty$ is complete in the normed metric.

13. The matrix $\mathfrak{C}$ of (5.7.40) has an inverse not in $\mathfrak{M}_\infty$. Find it!

14. Show that the spectrum and characteristic values of $\mathfrak{C}$ are as stated in (5.7.41). Describe the matrix $\mathfrak{R}(\lambda, \mathfrak{C})$.

15. Verify (5.7.46).

16. In 1832 the German astronomer and mathematician August Ferdinand Möbius (1790–1868) discovered an inversion formula of importance in analysis and number theory. It can be formulated in terms of two infinite matrices which are inverses of each other. Here $\mathfrak{A} = (a_{jk})$, where $a_{jk} = 1$ or 0 according as j divides k or not. The elements b_{jk} of $\mathfrak{B}$ assume the values $0, +1, -1$. They are expressible in terms of the Möbius constants $\mu(n)$. The latter is 0 if n is divisible by a square, $\mu(1) = 1$, and $\mu(n) = (-1)^k$ if n is the product of k distinct primes. Show that $b_{jk} = 0$ unless $k = jn$, in which case $b_{jk} = \mu(n)$. [Note that this is merely a convenient way of formulating the facts. Möbius wrote before the days of matrices.]

COLLATERAL READING

For the first three sections consult:

SCHAEFFER, H. H., *Topological Vector Spaces*, Macmillan, New York (1966), Partial ordering in Chapter V.

The following is a monograph on fixed point theorems:

BONSALL, F. F., *Lectures on Some Fixed Point Theorems in Functional Analysis*, Tata Institute of Fundamental Research, Bombay (1962).

For Banach's dissertation and extensions see:

BANACH, S., Sur les opérations dans les ensembles abstraits et leurs applications aux équations intégrales, *Fundamenta Math.* **3** (1922) 133–81.

BANACH, S., *Théorie des opérations linéaires*, Warsaw (1932).

For the modern theory of integral equations consult:

YOSIDA, K., *Lectures on Differential and Integral Equations*, Interscience, New York (1960).

For the extensions of Landau's inequality and contraction operators, see:

KALLMAN, R. R. and G.-C. ROTA, *The Inequality* $\| f' \|^2 \leqslant 4 \| f \| \| f'' \|$, *Inequalities. II*, Academic Press, New York (1970), pages 187–192.

HILLE, E., Notes on linear transformations, I, *Trans. Am. Math. Soc.* **39** (1936) 131–153.

HILLE, E., On the Landau–Kallman–Rota inequality, *J. Approximation Theory*, to appear in 1972.

Interesting generalizations of formulas (5.6.8) and (5.6.9) are found in:

MILLER, J. B., Some properties of Baxter operators, *Acta Math. Acad. Sci. Hung.* **17** (1966) 387–400.

6 EXISTENCE AND UNIQUENESS THEOREMS

Analysts are sceptical by instinct, training, and habitude. One way in which this attitude shows itself is through the emphasis placed on existence and uniqueness theorems in analysis. Where the physicist places his trust in intuition and in the intrinsic simplicity of natural phenomena to help him over the mathematical difficulties adherent to his problems, the mathematician holds back until he has confirmed his guess by a proof. There is something to be said for both attitudes. Mathematicians are seldom prepared at the outset to render effective help to the physicist, who is thus left to do the pioneering. On the other hand, once the mathematician has become interested in the ideas involved, he can often go further than the physicist and investigate aspects of the problem where the physicist's intuition is of little help. What will be developed in this chapter should be viewed against the background of alternating advances in mathematics and physics during the last 300 years. Most of what we are going to do are applications either of fixed point theorems or of the older method of successive approximations.

There are four sections: The implicit function theorem; The method of successive approximations; Majorants; and Applications to ordinary differential equations.

6.1 THE IMPLICIT FUNCTION THEOREM

In analysis functions are often defined implicitly, and this leads to two basic problems: (1) Is there a function that satisfies the conditions of the problem and, if so, is it unique? (2) Find a process that leads to a construction of the solution. Nowadays the second problem will presumably include a request for effective programming of the computer.

We shall start with the simplest case. There is given a real-valued function of two real variables $F(x, y)$ and a point (x_0, y_0), where

$$F(x_0, y_0) = 0. \tag{6.1.1}$$

Is it possible to solve the equation

$$F(x, y) = 0 \tag{6.1.2}$$

for y in terms of x, say $y = f(x)$, in some neighborhood of $x = x_0$, where $f(x_0) = y_0$? If it is possible, is the solution unique? The geometrical interpretation of the problem is simple. We are given a curve C in the (x, y)-plane of equation (6.1.2) and a point (x_0, y_0) on the curve. The example

$$(x^2 + y^2)^2 - x^2 - ay^2 = 0, \qquad (x_0, y_0) = (0, 0),$$

shows that such a point may be an isolated point (for $a > 0$) or a double point with two real branches ($a \leqslant 0$). What we need is a condition that excludes multiple points as well as branches with a vertical tangent where no local representation of the form $y = f(x)$ is possible. Such conditions can be found and we shall show how.

We place (x_0, y_0) at the origin and assume that $F(x, y)$ is continuous and has continuous first order partials in some neighborhood of the origin. We express this simply by saying that F is *continuously differentiable* in the neighborhood in question.

Theorem 6.1.1. *Let $F(x, y)$ be defined and continuously differentiable in some domain D of R^2 containing the origin. Suppose that* (i) $F(0, 0) = 0$ *and* (ii) $F_y(0, 0) \neq 0$. *Then there exists an interval $(-r, r)$ and a function $f(x)$ such that* (1) $f(x)$ *is continuous and differentiable in $(-r, r)$,* (2) $f(0) = 0$, (3)

$$F[x, f(x)] = 0, \qquad -r < x < r, \tag{6.1.3}$$

and (4)

$$f'(x) = -\frac{F_x[x, f(x)]}{F_y[x, f(x)]}. \tag{6.1.4}$$

Proof. We can give an elementary proof for the stated facts. Its only blemish is that it does not readily extend to more complicated situations. But we are concerned here with methods and one more will not be fatal. The idea of the proof is that in some square centered at the origin with sides parallel to the axes the partial $F_y(x, y)$ keeps a constant sign so that $F(x, y)$ is a monotone function of y for fixed x. If the side of the square is small, $F(x, y)$ will be of constant sign on each of the horizontal boundaries, the sign on one being the opposite of that on the other. This means that on each vertical line segment of the square $F(x, y)$ changes its sign once and only once. The value of y where this change takes place is the desired solution. The details follow.

To fix the ideas, suppose that $F_y(0, 0) > 0$. Then by the continuity of the partials $F_y(x, y) > 0$ in some neighborhood of the origin contained in D. In this neighborhood we can find a square, $|x| \leqslant a$, $|y| \leqslant a$, where $F_y(x, y)$ has a positive lower bound, say

$$F_y(x, y) \geqslant B > 0. \tag{6.1.5}$$

In the square $F_x(x, y)$ exists and is bounded, say,

$$|F_x(x, y)| \leqslant A. \tag{6.1.6}$$

We now define

$$r = \min\left(a, \frac{B}{A}a\right) \tag{6.1.7}$$

and consider the rectangle R: $|x| \leqslant r$, $|y| \leqslant a$. Since $F(x, y)$ is monotone on that part of the y-axis which lies in R, we have $F(0, a) > 0$, $F(0, -a) < 0$. More precisely, by the mean value theorem

$$F(0, a) = a\, F_y(0, t) \geqslant aB, \qquad 0 < t < a, \qquad F(0, -a) \leqslant -aB.$$

Now consider, for $0 < x < r$,

$$F(x, a) = F(0, a) + xF_x(s, a) > aB - rA \geqslant 0$$

by (6.1.7), where $0 < s < r$. Similarly, for $0 < x < r$,

$$F(x, -a) = F(0, -a) + xF_x(s_1, -a) < -aB + rA \leqslant 0,$$

where $0 < s_1 < x$. These inequalities are also true for $-r < x < 0$.

Thus $F(x, y)$ is positive on the upper edge of the rectangle R, negative on the lower edge. Further, for fixed s, $-r < s < r$, $F(s, y)$ is an increasing function of y for $-a \leqslant y \leqslant a$. Thus there is one and only one value $y = t$ where

$$F(s, t) = 0.$$

This defines t uniquely as a function of s, $t = f(s)$. We have

$$f(0) = 0, \qquad F[s, f(s)] = 0, \qquad -r < s < r.$$

It is fairly obvious that $f(x)$ is continuous. The proof goes as follows and will give differentiability as a byproduct. Suppose that $-r < x_1 < x_2 < r$ and set $y_1 = f(x_1)$, $y_2 = f(x_2)$. Then by the mean value theorem

$$\begin{aligned} 0 &= F(x_1, y_1) - F(x_2, y_2) \\ &= F_x(x_3, y_3)\,(x_1 - x_2) + F_y(x_3, y_3)\,(y_1 - y_2), \end{aligned}$$

where (x_3, y_3) is a point on the line segment joining (x_1, y_1) with (x_2, y_2). Hence

$$y_2 - y_1 = -\frac{F_x(x_3, y_3)}{F_y(x_3, y_3)}(x_2 - x_1) \tag{6.1.8}$$

and

$$|f(x_2) - f(x_1)| \leqslant \frac{A}{B}|x_2 - x_1|. \tag{6.1.9}$$

This Lipschitz condition implies continuity. But (6.1.8) also gives

$$\lim_{x_2 \to x_1} \frac{f(x_2) - f(x_1)}{x_2 - x_1} = -\lim_{x_2 \to x_1} \frac{F_x(x_3, y_3)}{F_y(x_3, y_3)}.$$

Here $x_2 \to x_1$ implies $x_3 \to x_1$, $y_3 \to y_1 = f(x_1)$ and (6.1.4) results. ∎

Various other methods are available to prove such facts. We shall consider the contraction fixed point theorem but apply it to a much more general situation, namely the existence of implicit functions in a Banach algebra $\mathfrak{B}$ with unit element e. Commutativity of multiplication is not assumed. Elements are denoted by italics to avoid uglification.

Theorem 6.1.2. *Let $G(x, y)$ be a function from $\mathfrak{B} \times \mathfrak{B}$ to $\mathfrak{B}$ defined on a domain $\mathfrak{D}$. Here $\mathfrak{D}$ is taken as the "di-sphere": $\|x\| \leqslant \alpha$, $\|y\| \leqslant \beta$, $G(0, 0) = 0$, and G shall satisfy a Lipschitz condition*

$$\|G(x_1, y_1) - G(x_2, y_2)\| \leqslant k[\|x_1 - x_2\| + \|y_1 - y_2\|], \tag{6.1.10}$$

where k is fixed, $0 < k < 1$. Then the equation

$$y = ax + G(x, y), \qquad a \in \mathfrak{B}, \tag{6.1.11}$$

has a unique solution, $y = f(x)$, in $\mathfrak{B}$, with $f(0) = 0$, defined for

$$\|x\| \leqslant \rho < \min\left\{\alpha, \frac{1-k}{\|a\| + k}\beta\right\}. \tag{6.1.12}$$

The solution satisfies a Lipschitz condition

$$\|f(x_2) - f(x_1)\| \leqslant \frac{\|a\| + k}{1 - k}\|x_2 - x_1\|. \tag{6.1.13}$$

Proof. We shall use the fixed point method. The reader will find another proof based on the method of successive approximations in the next section. For the first method we have to produce a complete metric space $\mathfrak{X}$ and a contraction mapping from $\mathfrak{X}$ into $\mathfrak{X}$ such that the expected solution of (6.1.11) comes out as the fixed point of the mapping. Let $\mathfrak{X}$ be the family of functions $g(x)$ satisfying the following conditions:

1) *$g(x)$ is defined and continuous for $\|x\| \leqslant \rho$, defined by* (6.1.12), *and its range is in $\mathfrak{B}$.*
2) *$g(0) = 0$ and $\|g(x)\| \leqslant \beta$ for $\|x\| \leqslant \rho$.*

This is the $\mathfrak{B}$-norm but we have also to introduce a metric in $\mathfrak{X}$. For $g, h \in \mathfrak{X}$ we set $d(g, h) = \sup \|g(x) - h(x)\|$ where the supremum refers to $\|x\| \leqslant \rho$. It is clear that $\mathfrak{X}$ is a complete metric space. For if $\{g_n\}$ is a Cauchy sequence in $\mathfrak{X}$, then $\lim_{n\to\infty} g_n(x) \equiv g_0(x)$ exists, is continuous, vanishes at $x = 0$ and $\|g_0(x)\| \leqslant \beta$ for all x with $\|x\| \leqslant \rho$. Hence $g_0 \in \mathfrak{X}$ and $\mathfrak{X}$ is complete under the normed metric.

We now define T as the operator that takes g into

$$T[g](x) = ax + G[x, g(x)], \qquad g \in \mathfrak{X}. \tag{6.1.14}$$

We have to show that $T[g]$ exists and belongs to $\mathfrak{X}$. First $T[g](x)$ is defined and continuous for $\|x\| \leqslant \rho$ by virtue of the conditions imposed on g. Using (6.1.10) with $x_2 = y_2 = 0$, $x_1 = x$, $y_1 = g(x)$ we get

$$\|G[x, g(x)]\| \leqslant k[\|x\| + \|g(x)\|] \leqslant k(\rho + \beta)$$

so that

$$\|T[g](x)\| \leqslant \|a\| \, \|x\| + k(\rho + \beta) \leqslant \|a\|\rho + k(\rho + \beta) \equiv \beta$$

by the choice of ρ. Hence $T[g] \in \mathfrak{X}$. Further,

$$\begin{aligned}\|T[g_1](x) - T[g_2](x)\| &= \|G[x, g_1(x)] - G[x, g_2(x)]\| \\ &\leqslant k\|g_1(x) - g_2(x)\|.\end{aligned} \tag{6.1.15}$$

Thus, in the metric of $\mathfrak{X}$ defined above,

$$d[T(g_1), T(g_2)] \leqslant kd[g_1, g_2],$$

so that T is indeed a contraction. It follows that T has a unique fixed point in $\mathfrak{X}$. Thus there exists one and only one function $f(x) \in \mathfrak{X}$ such that

$$f(x) = ax + G[x, f(x)], \qquad \|x\| \leqslant \rho, \tag{6.1.16}$$

and this is the desired solution. The estimate (6.1.13) is obtained from (6.1.10) and

$$\|f(x_2) - f(x_1)\| \leqslant \|a\| \, \|x_2 - x_1\| + k[\|x_2 - x_1\| + \|f(x_2) - f(x_1)\|]$$

is valid for all x_1, x_2 under consideration. ∎

The same technique applies to other problems involving implicit functions. We shall give two samples, omitting proofs.

Theorem 6.1.3. *Let* $\mathbf{a}$ *and* $\mathbf{x}$ *be vectors in* C^n, $\mathbf{a}$ *fixed, and let* $(\mathbf{a}, \mathbf{x})$ *denote their inner product. Let* $G(\mathbf{x}, y)$ *be a function from* $C^n \times C$ *to* C, *defined for* $\|\mathbf{x}\| \leqslant \alpha$, $|y| \leqslant \beta$ *where it satisfies a Lipschitz condition*

$$\|G(\mathbf{x}_2, y_2) - G(\mathbf{x}_1, y_1)\| \leqslant k[\|\mathbf{x}_2 - \mathbf{x}_1\| + |y_2 - y_1|] \tag{6.1.17}$$

with k *fixed,* $0 < k < 1$. *Further,* $G(\mathbf{0}, 0) = 0$. *Then the equation*

$$y = (\mathbf{a}, \mathbf{x}) + G(\mathbf{x}, y) \tag{6.1.18}$$

has a unique solution, $y = f(\mathbf{x})$, *for*

$$\|\mathbf{x}\| \leqslant \rho < \min \left\{ \alpha, \frac{1-k}{\|a\| + k} \beta \right\} \tag{6.1.19}$$

such that $f(\mathbf{0}) = 0$ *and*

$$f(\mathbf{x}) = (\mathbf{a}, \mathbf{x}) + G[\mathbf{x}, f(\mathbf{x})]. \tag{6.1.20}$$

Next we take the case where both $\mathbf{x}$ and $\mathbf{y}$ are vectors belonging to the same space C^n. The norms for C^n and the matrix space $\mathfrak{M}_n$ are arbitrary so long as $\|\mathcal{A}\mathbf{x}\| \leqslant \|\mathcal{A}\| \, \|\mathbf{x}\|$.

Theorem 6.1.4. *Let* $\mathcal{A} \in \mathfrak{M}_n$ *be a given constant matrix. Let* $\mathbf{G}(\mathbf{x}, \mathbf{y})$ *be a mapping from* $C^n \times C^n$ *to* C^n *defined for* $\|\mathbf{x}\| \leqslant \alpha$, $\|\mathbf{y}\| \leqslant \beta$, *where it satisfies the Lipschitz condition*

$$\|\mathbf{G}(\mathbf{x}_2, \mathbf{y}_2) - \mathbf{G}(\mathbf{x}_1, \mathbf{y}_1)\| \leqslant k[\|\mathbf{x}_2 - \mathbf{x}_1\| + \|\mathbf{y}_2 - \mathbf{y}_1\|] \tag{6.1.21}$$

with $0 < k < 1$ *and* $\mathbf{G}(\mathbf{0}, \mathbf{0}) = \mathbf{0}$. *Then the equation*

$$\mathbf{y} = \mathcal{A}\mathbf{x} + \mathbf{G}(\mathbf{x}, \mathbf{y}) \tag{6.1.22}$$

has a unique solution $\mathbf{y} = \mathbf{f}(\mathbf{x})$ *defined for*

$$\|\mathbf{x}\| \leqslant \rho < \min \left\{ \alpha, \frac{1-k}{\|\mathcal{A}\| + k} \beta \right\} \tag{6.1.23}$$

such that $\mathbf{f}(\mathbf{0}) = \mathbf{0}$ *and*

$$\mathbf{f}(\mathbf{x}) = \mathcal{A}\mathbf{x} + \mathbf{G}[\mathbf{x}, \mathbf{f}(\mathbf{x})]. \tag{6.1.24}$$

The general problem of solving $\mathbf{F}(\mathbf{x}, \mathbf{y}) = \mathbf{0}$ for $\mathbf{y}$, where $\mathbf{F}$ is a function from $C^n \times C^n$ to C^n, can often be reduced to the preceding case. Suppose that in a neighborhood of $(\mathbf{0}, \mathbf{0})$

$$\mathbf{F}(\mathbf{x}, \mathbf{y}) = \mathcal{B}\mathbf{y} - \mathcal{C}\mathbf{x} - \mathbf{H}(\mathbf{x}, \mathbf{y}), \tag{6.1.25}$$

where $\mathcal{B}$ and $\mathcal{C}$ are constant matrices in $\mathfrak{M}_n$ and $\mathbf{H}(\mathbf{x}, \mathbf{y})$ is continuously differentiable and

$$\|\mathbf{H}(\mathbf{x}, \mathbf{y})\| = o[\|\mathbf{x}\| + \|\mathbf{y}\|] \tag{6.1.26}$$

as $\mathbf{x} \to \mathbf{0}$, $\mathbf{y} \to \mathbf{0}$. If now $\mathcal{B}$ is a regular matrix, we can multiply $\mathbf{F}$ on the left by $\mathcal{B}^{-1}$ to obtain (6.1.22) with

$$\mathcal{A} = \mathcal{B}^{-1}\,\mathcal{C}, \qquad \mathbf{G} = \mathcal{B}^{-1}\,\mathbf{H}.$$

In the classical theory $\mathcal{B}$ would be the Jacobian matrix evaluated at the origin and we would get the familiar condition of solvability at the origin in terms of the non-vanishing of the Jacobian determinant. The extension to the case where $\mathbf{F}$ is a function on $C^n \times C^m$ to C^m follows similar lines.

EXERCISE 6.1

1. Find a solution $y = f(x)$ of
$$x^3 + y^3 + x - y = 0$$
in R such that $f(0) = 0$. Find a suitable space $\mathfrak{X}$ of real-valued continuous functions $g(x)$ and an interval $(-r, r)$ such that
$$T\colon g(x) \to x + x^3 + [g(x)]^3$$
is a contraction in $\mathfrak{X}$ if x is restricted to $(-r, r)$.
2. Solve the preceding problem when x and y are matrices in $\mathfrak{M}_n$.
3. Prove Theorem 6.1.3.
4. Prove Theorem 6.1.4.
5. Carry through the discussion for (6.1.25) and prove that the solution has continuous first order partials at $(\mathbf{0}, \mathbf{0})$.
6. Show that (6.1.26) holds for
$$\mathbf{H}(\mathbf{x}, \mathbf{y}) = (\mathbf{a}, \mathbf{x})\,\mathbf{y} + (\mathbf{a}, \mathbf{y})\,\mathbf{x},$$
where $\mathbf{a} \in C^n$ is fixed and $(\mathbf{a}, \mathbf{x})$ is the inner product.
7. Discuss the existence of partial derivatives of $\mathbf{H}(\mathbf{x}, \mathbf{y})$ at $(\mathbf{0}, \mathbf{0})$ when (6.1.26) holds.
8. Does $|(\mathbf{x}, \mathbf{y})|^{1/2}\,(\mathbf{x} + \mathbf{y})$ satisfy (6.1.26)?

6.2 THE METHOD OF SUCCESSIVE APPROXIMATIONS

Before fixed point theorems became available the method of *successive approximations* was the standard tool used for the study of functional equations. This method was launched in 1890 by the eminent French mathematician Émile Picard

(1856–1941) who applied it to the theory of differential equations. Édouard Goursat (1858–1936), also a prominent French mathematician, adapted the method to the needs of the implicit function theorem in 1903. We shall use this method in a

Second proof of Theorem 6.1.2. There is no difference in the assumptions made. Typical for the method is the successive definition of the members of a sequence of approximations $\{f_n(x)\}$ where, in the present case,

$$f_0(x) = ax, \qquad f_n(x) = ax + G[x, f_{n-1}(x)] \tag{6.2.1}$$

for $n = 1, 2, 3, \ldots$. It is to be shown that these functions are well defined, continuous in the normed metric for x restricted by the condition (6.1.13). There is no question concerning $f_0(x)$. Suppose that all functions $f_m(x)$ with $m \leqslant n$ are in order. Thus $f_n(x)$ exists, is continuous, and satisfies $\|f_n(x)\| \leqslant \beta$ for the admitted values of x. We have then

$$f_{n+1}(x) = ax + G[x, f_n(x)].$$

Here the second term on the right exists and is a continuous function of x for $\|x\| \leqslant \rho$. Since $G(0, 0) = 0$, the Lipschitz condition gives

$$\|G(x, y)\| \leqslant k[\|x\| + \|y\|] \tag{6.2.2}$$

in the domain of definition of G. Hence

$$\begin{aligned} \|f_{n+1}(x)\| &\leqslant \|a\|\,\|x\| + k[\|x\| + \|f_n(x)\|] \\ &\leqslant [\|a\| + k]\rho + k\beta \leqslant (1 - k)\beta + k\beta = \beta \end{aligned}$$

by the choice of ρ. Hence $f_{n+1}(x)$ is also in order.

To prove convergence of the sequence $\{f_n(x)\}$ we again use the Lipschitz condition. We have now

$$\begin{aligned} \|f_{n+1}(x) - f_n(x)\| &= \|G[x, f_n(x)] - G[x, f_{n-1}(x)]\| \\ &\leqslant k\|f_n(x) - f_{n-1}(x)\| \\ &\cdots\cdots\cdots\cdots \\ &\leqslant k^n\|f_1(x) - f_0(x)\|. \end{aligned} \tag{6.2.3}$$

It follows that

$$\sum_{n=1}^{\infty} \|f_{n+1}(x) - f_n(x)\|$$

converges uniformly for $\|x\| \leqslant \rho$. Since

$$\|f_n(x) - f_m(x)\| \leqslant \sum_{k=m}^{n-1} \|f_{k+1}(x) - f_k(x)\| \to 0 \quad \text{as} \quad m \to \infty,$$

it follows that $\{f_k(x)\}$ is a Cauchy sequence. Hence

$$\lim_{n \to \infty} f_n(x) \equiv f(x) \tag{6.2.4}$$

exists uniformly for $\|x\| \leqslant \rho$. Since $f_n(0) = 0, \forall n$, we have also $f(0) = 0$. Further,

$$f(x) = \lim_{n\to\infty} f_{n+1}(x) = ax + \lim_{n\to\infty} G[x, f_n(x)]$$

$$= ax + G[x, \lim f_n(x)] = ax + G[x, f(x)]$$

by the continuity of G. Thus $f(x)$ satisfies (6.1.16) for $\|x\| \leqslant \rho$.

To prove uniqueness we again appeal to the Lipschitz condition. Let $\{f_n(x)\}$ be the sequence defined above and suppose that $g(x)$ is such that $g(0) = 0$ and

$$g(x) = ax + G[x, g(x)]$$

for $\|x\| \leqslant \rho_0$. Then for $\|x\| \leqslant \rho_1 = \min(\rho, \rho_0)$ the Lipschitz condition gives

$$\|g(x) - f_{n+1}(x)\| = \|G[x, g(x)] - G[x, f_n(x)]\| \leqslant k\|g(x) - f_n(x)\|.$$

Hence, by induction,

$$\|g(x) - f_{n+1}(x)\| \leqslant k^{n+1}\|g(x) - ax\|, \tag{6.2.5}$$

and this goes to zero as $n \to \infty$. Hence

$$g(x) = \lim_{n\to\infty} f_n(x) = f(x),$$

and the solution is unique for $\|x\| < \rho_1$. ∎

We can use Problem 1 of the preceding exercise as an illustration.

Example 1. Find the solution $y = f(x)$ of

$$y = x + x^3 + y^3, \qquad f(0) = 0, \tag{6.2.6}$$

in an interval containing $x = 0$. From the point of view of analytic geometry this is a cubic passing through the origin and we want the ordinate expressed in terms of the abscissa on the largest possible arc containing the point (0, 0). A sketch of the graph will show that the curve tangent is vertical at the endpoints of the maximal arc. The analytical discussion will confirm this.

We have here

$$f_0(x) = x, \quad f_1(x) = x + 2x^3, \quad f_2(x) = x + x^3 + (x + 2x^3)^3,$$

in general

$$f_{n+1}(x) = x + x^3 + [f_n(x)]^3. \tag{6.2.7}$$

This is a sequence of polynomials with positive integral coefficients involving only odd powers of x and $f_n(x)$ is of degree 3^n. Further,

$$f_{n+1}(x) - f_n(x) = [f_n(x) - f_{n-1}(x)]$$
$$\{[f_n(x)]^2 + f_{n-1}(x)f_n(x) + [f_{n-1}(x)]^2\}. \tag{6.2.8}$$

Since $f_1(x) - f_0(x) = 2x^3 > 0$ for $x > 0$, complete induction shows that the sequence $\{f_n(x)\}$ is strictly increasing for $x > 0$. It will then converge to a finite limit iff it is bounded. In (6.2.8) the quantity between the braces exceeds

$$3[f_{n-1}(x)]^2,$$

and this quantity must be <1 if the sequence $\{f_n(x)\}$ is to be bounded. Thus a necessary condition for convergence is that there is an interval $[0, r)$ where the condition

$$f_n(x) < \tfrac{1}{3}\sqrt{3}, \quad \forall\, n, \tag{6.2.9}$$

holds. Suppose that we have found an x and an n for which (6.2.9) holds. Then by (6.2.7)

$$f_{n+1}(x) < x + x^3 + \tfrac{1}{9}\sqrt{3} < \tfrac{1}{3}\sqrt{3}$$

provided

$$x^3 + x - \tfrac{2}{9}\sqrt{3} < 0.$$

This says that x must be less than r, the positive real root of the equation

$$x^3 + x - \tfrac{2}{9}\sqrt{3} = 0, \tag{6.2.10}$$

which is approximately 0.34412. For $0 \leqslant x < r$ the inequality (6.2.9) holds for all n and

$$\lim f_n(x) \equiv f(x) \tag{6.2.11}$$

exists. By (6.2.7) we have

$$f(x) = x + x^3 + [f(x)]^3, \tag{6.2.12}$$

and we have, of course, also $f(0) = 0$.

Each function $f_n(x)$ is strictly increasing, and they are uniformly bounded in $[0, r)$. Hence the same is true for $f(x)$ and

$$\lim_{x \uparrow r} f(x) \equiv y_0 \tag{6.2.13}$$

exists as a finite number. Further,

$$y_0 = r + r^3 + y_0{}^3 = \tfrac{2}{9}\sqrt{3} + y_0{}^3,$$

which is satisfied by $y_0 = \tfrac{1}{3}\sqrt{3}$. This number, then, is the least upper bound of $f(x)$ in $[0, r]$. If we set

$$F(x, y) = x - y + x^3 + y^3,$$

then

$$F_y(x, y) = -1 + 3y^2,$$

and this is zero for $y = \pm y_0$. Thus the cubic $F = 0$ has vertical tangents at the points $(-r, -y_0)$ and (r, y_0). It follows that we have found the largest interval where a unique solution exists with $f(0) = 0$. The extension to the interval $(-r, 0)$ is trivial since $F(-x, -y) = -F(x, y)$.

We have found the implicit function defined by (6.2.6) and the condition $f(0) = 0$ and we have used the method of successive approximations. The reader may object, and with some justification, that Theorem 6.1.2 has not really been used.

There was no mention of a Lipschitz condition and no use was made of (6.1.12). It is easy to produce a Lipschitz condition. If $G(x, y) = x^3 + y^3$ we have

$$|G(x_2, y_2) - G(x_1, y_1)| \leqslant k[|x_2 - x_1| + |y_2 - y_1|] \tag{6.2.14}$$

provided we take $\alpha = \frac{1}{3}(3\,k)^{1/2}$, $\beta = \frac{1}{3}(3\,k)^{1/2}$. We can now use (6.1.12), but the results obtainable in this manner are much less satisfactory than what we obtained using positivity and the monotonic properties of the approximations.

Example 2. In Problem 2, Exercise 6.1, the question was to solve equation (6.2.6) in the matrix case where $x, y \in \mathfrak{M}_n$. It is natural to ask to what extent the discussion of Example 1 carries over to the matrix case. We replace the italic letters x and y by the corresponding roman script letters $\mathcal{X}$ and $\mathcal{Y}$. We have now a sequence of matrix polynomials $\{\mathcal{F}_m(\mathcal{X})\}$ with

$$\mathcal{F}_0(\mathcal{X}) = \mathcal{X}, \qquad \mathcal{F}_{m+1}(\mathcal{X}) = \mathcal{X} + \mathcal{X}^3 + [\mathcal{F}_m(\mathcal{X})]^3. \tag{6.2.15}$$

The numerical coefficients are positive integers and the degree of $\mathcal{F}_m(\mathcal{X})$ is still 3^m. Again, we determine r as the positive root of the equation (6.2.10) and find that $\|\mathcal{X}\| < r$ implies

$$\|\mathcal{F}_m(\mathcal{X})\| \leqslant \tfrac{1}{3}\sqrt{3}, \quad \forall\, m. \tag{6.2.16}$$

There is one case where the previously used argument carries over without further ado. We can introduce partial ordering in $\mathfrak{M}_n$. defining a matrix $\mathcal{P} = (p_{jk})$ as positive if each $p_{jk} \geqslant 0$. If we now take $\mathcal{X} = \mathcal{P}$ with $\|\mathcal{P}\| < r$, then every $\mathcal{F}_m(\mathcal{P})$ is a positive matrix and

$$\mathcal{F}_m(\mathcal{P}) \leqslant \mathcal{F}_{m+1}(\mathcal{P}), \qquad \|\mathcal{F}_m(\mathcal{P})\| \leqslant \tfrac{1}{3}\sqrt{3}.$$

If f_{mjk} denotes the entry at the place (j, k) in the matrix $\mathcal{F}_m(\mathcal{P})$, then for fixed j, k the numbers $\{f_{mjk}\}$ form an increasing bounded sequence which thus tends to a limit f_{jk} and $(f_{jk}) = \mathcal{F}(\mathcal{P})$ is the desired limit of $\mathcal{F}_m(\mathcal{P})$.

Suppose now that $\mathcal{X} = (x_{jk})$ is an arbitrary matrix with $\|\mathcal{X}\| < r$. We set $\mathcal{P} = (|x_{jk}|)$ and take, as usual,

$$\|\mathcal{A}\| = \sup_j \sum_{k=1}^{n} |a_{jk}|,$$

then

$$\|\mathcal{X}\| = \|\mathcal{P}\|, \qquad \|\mathcal{X}^k\| \leqslant \|\mathcal{P}^k\|, \qquad \|\mathcal{F}_m(\mathcal{X})\| \leqslant \|\mathcal{F}_m(\mathcal{P})\| \tag{6.2.17}$$

and also for $k < l$

$$\|\mathcal{F}_l(\mathcal{X}) - \mathcal{F}_k(\mathcal{X})\| \leqslant \|\mathcal{F}_l(\mathcal{P}) - \mathcal{F}_k(\mathcal{P})\|. \tag{6.2.18}$$

As we have just seen, the right member of this inequality goes to zero as $k \to \infty$. Hence so does the left member, so that $\{\mathcal{F}_m(\mathcal{X})\}$ is a Cauchy sequence in the complete metric space $\mathfrak{M}_n$. Then its limit is the desired solution.

This is in perfect analogy with the result for the scalar case. There is one basic difference, however, since $\lim \mathcal{F}_m(\mathfrak{X})$ may exist for some matrix $\mathfrak{X}$ with $\|\mathfrak{X}\| > r$, due to the fact that $\|\mathfrak{X}^m\|$ may go to zero much faster than $\|\mathfrak{X}\|^m$. An extreme case is that in which $\mathfrak{X}$ is a nilpotent matrix where now no limitation on $\|\mathfrak{X}\|$ is required.

EXERCISE 6.2

1. Give a proof of Theorem 6.1.3 using the method of successive approximations.
2. Do the same for Theorem 6.1.4.
3. Verify (6.2.14).
4. Show that 3^m is the degree of $f_m(x)$ in (6.2.7).
5. If $f_m(x)$ is rearranged after increasing powers of x, it is observed that for a given k a certain number of terms at the beginning of the expansion of $f_m(x)$ will be the same for all $m \geqslant k$. Thus for $k > 0$ every $f_k(x)$ starts out with $x + 2x^3$. How far is it necessary to go with k for all the terms of degree <10 to remain unchanged and what are these terms?
6. Suppose that $\mathfrak{X}$ is a nilpotent matrix in $\mathfrak{M}_n$ with $n > 8$, $\mathfrak{X}^8 = \mathcal{O}$, $\mathfrak{X}^7 \neq \mathcal{O}$. Find $\mathcal{F}(\mathfrak{X})$.
7. Let $\mathfrak{J}$ be an idempotent matrix. For what values of the complex number α would $\mathcal{F}(\mathfrak{X})$ exist if $\mathfrak{X} = \alpha\mathfrak{J}$?
8. Verify (6.2.17) and (6.2.18).

6.3 MAJORANTS

There are several valuable lessons to be drawn from the discussion of Examples 1 and 2 in the preceding section, all concerned with the importance of positivity. The operation

$$T[g](x) = x + x^3 + [g(x)]^3 \tag{6.3.1}$$

applied to $C[0, 1]$ is *positive*, i.e. it maps the positive cone into itself, and T is *order preserving*. These properties enabled us to reduce a discussion of convergence to a discussion of boundedness. In the matrix case the existence of $\lim \mathcal{F}_m(\mathfrak{X})$ for positive matrices $\mathfrak{P}$ with $\|\mathfrak{P}\| < r$ enabled us to prove convergence for general matrices of the same norm.

In this section we shall give further applications of positivity, now for the case where $F(x, y)$ is a power series in the two variables x and y. One more look at Example 1 will be instructive. It was observed in Problem 5, Exercise 6.2, that there exists a formal power series

$$P(x) = \sum_{k=0}^{\infty} c_{2k+1}\, x^{2k+1} \tag{6.3.2}$$

with the following properties. Let $P_n(x)$ be the nth partial sum of the series. Then there exists an integer $m > n$ such that for $p \geqslant m$ the terms of $f_p(x)$ of order $\leqslant 2n + 1$ coincide with $P_n(x)$. Since all coefficients are positive, we have for $0 < x < r$

$$P_n(x) < f_m(x) < f(x). \tag{6.3.3}$$

This implies the convergence of the formal power series and the representation

$$f(x) = \sum_{k=0}^{\infty} c_{2k+1} x^{2k+1}. \tag{6.3.4}$$

The coefficients of this power series may be obtained from the identity

$$\sum_{k=0}^{\infty} c_{2k+1} x^{2k+1} = x + x^3 + \left\{ \sum_{k=0}^{\infty} c_{2k+1} x^{2k+1} \right\}^3 \tag{6.3.5}$$

by expanding the third power, collecting terms and equating coefficients of like powers of x. We are now ready to generalize.

In the following the variables as well as the coefficients are complex-valued at the outset. We very soon reduce to positive variables and coefficients, however. We are given a function $F(z, w)$ of two complex variables in a neighborhood of $(0, 0)$ and

$$F(0, 0) = 0, \qquad F_w(0, 0) \neq 0. \tag{6.3.6}$$

We shall assume the relation $F(z, w) = 0$ written in the form

$$w = a_{10} z + \sum_j \sum_k a_{jk} z^j w^k, \tag{6.3.7}$$

where in the double series $j \geqslant 0$, $k \geqslant 0$, $j + k > 1$. Assuming that the double series converges for some non-trivial values of z and w, we are led to the following result.

Theorem 6.3.1. *Suppose that there exist three positive numbers M, s, t such that*

$$|a_{jk}|\, s^j t^k \leqslant M \tag{6.3.8}$$

for all j and k. Then there exists a unique power series

$$w = \sum_{n=1}^{\infty} c_n z^n, \tag{6.3.9}$$

convergent for

$$|z| \leqslant r = \left(\frac{t}{t + 2M} \right)^2 s \tag{6.3.10}$$

such that

$$\sum_{n=1}^{\infty} c_n z^n = a_{10} z + \sum_{j=1}^{\infty} \sum_{k=1}^{\infty} a_{jk} z^j (\Sigma\, c_n z^n)^k \tag{6.3.11}$$

is an identity between absolutely convergent power series for $|z| \leqslant r$.

Proof. Assuming convergence of the series involved, we pose the identity (6.3.11) with a view of determining how the coefficients c_n have to be chosen for the identity to hold. If the series (6.3.9) has a positive radius of convergence, R say, then we can form the kth power of the series using Cauchy's product theorem according to which the kth power can be written as a power series in z starting with a term in z^k and absolutely convergent for $|z| < R$. If now $|z| < s$ and s is sufficiently small, then the sum of the series (6.3.9) is less than t in absolute value. For such values of z the triple series is absolutely convergent and may be rearranged in any way we please—for instance, as a power series in z. We have now an identity between two power series which can hold iff corresponding coefficients are equal. On the left the coefficient of z^n is c_n. On the right we get a number of terms involving z^n and the sum of their coefficients must then also equal c_n.

In the right member a term in z^n can arise only in a finite number of ways. For $n = 1$ there is only one term, namely $a_{10}\, z$, so $c_1 = a_{10}$. For $n > 1$ we must have

$$0 \leqslant j \leqslant n, \quad 0 \leqslant k \leqslant n, \quad 1 \leqslant j + k \leqslant n. \tag{6.3.12}$$

There is no term in the right member with $j = 0$, $k = 1$. Hence the coefficient of z^n in the right member involves no c_p with $p \geqslant n$. This means that

$$c_n = M_n(a_{jk};\, c_1, c_2, \ldots, c_{n-1}). \tag{6.3.13}$$

Here M_n is a multinomial in $c_1, c_2, \ldots, c_{n-1}$ and is linear in the a_{jk} where j and k are subject to the condition (6.3.12). Any numerical coefficient which occurs in M_n is a positive integer. It follows that the coefficients c_n may be computed successively in terms of the preceding coefficients. The first three relations under (6.3.13) read

$$\begin{aligned} c_1 &= a_{10}, \\ c_2 &= a_{20} + a_{11}c_1 + a_{02}{c_1}^2, \\ c_3 &= a_{30} + a_{21}c_1 + a_{12}{c_1}^2 + a_{03}{c_1}^3 + a_{11}c_2 + 2\, a_{22}c_1c_2. \end{aligned} \tag{6.3.14}$$

The next problem is that of proving convergence of the formal series (6.3.9) for sufficiently small values of $|z|$. This will be achieved if we can prove the existence of a power series

$$\sum_{n=1}^{\infty} C_n z^n \tag{6.3.15}$$

with

$$|c_n| \leqslant C_n, \quad \forall\, n, \tag{6.3.16}$$

having a positive radius of convergence. Such a series is known as a *majorant* of the given series and the relation between them is expressed by the symbol $\ll$, thus

$$\sum_{n=1}^{\infty} c_n z^n \ll \sum_{n=1}^{\infty} C_n z^n. \tag{6.3.17}$$

This simple observation goes back to Cauchy, who developed what he called a *Calcul de limites* to prove convergence theorems for implicit functions and for solutions of differential equations. In this connection "*limites*" means "bounds" rather than "limits."

For the problem of finding a suitable majorant for the series (6.3.9) we take another look at the recurrence relations (6.3.13) for the determination of the c_n. Suppose that in these relations we replace the a_{jk} by positive numbers A_{jk} such that

$$|a_{jk}| \leqslant A_{jk}, \quad \forall j, k. \tag{6.3.18}$$

Let us then determine C_n from the relations

$$C_n = M_n(A_{jk}; C_1, C_2, \ldots, C_{n-1}) \tag{6.3.19}$$

for $n = 1, 2, 3, \ldots$. Then

$$|c_1| = |a_{10}| \leqslant A_{10} = C_1$$

and the formulas (6.3.14) show that

$$|c_2| \leqslant C_2, \qquad |c_3| \leqslant C_3.$$

The general inequality

$$|c_n| \leqslant C_n$$

now follows by complete induction. Here we have used the structure of the forms (6.3.13). All numerical coefficients are positive. This implies that the forms are positive for positive entries and are increasing functions of each of the arguments when the latter are positive. This type of positivity gives the inequality.

$$\begin{aligned} |M_n(a_{jk}; c_1, c_2, \ldots, c_{n-1})| &\leqslant M_n(|a_{jk}|; |c_1|, |c_2|, \ldots, |c_{n-1}|) \\ &\leqslant M_n(A_{jk}; C_1, C_2, \ldots, C_{n-1}), \end{aligned} \tag{6.3.20}$$

if (6.3.18) holds.

What we have done is to introduce an auxiliary equation

$$w = A_{10}z + \sum_j \sum_k A_{jk} z^j w^k, \tag{6.3.21}$$

where the right member is a majorant of the right member of (6.3.7). Equation (6.3.21) has the formal solution (6.3.15), where the coefficients are determined by (6.3.19). If it can be shown that the series (6.3.15) has a positive radius of convergence, R say, then it becomes an actual solution of (6.3.21). Moreover, since (6.3.15) is a majorant of (6.3.9), the latter series will also be convergent at least for $|z| < R$, and this implies that the formal series (6.3.9) is an actual solution of (6.3.7) in its domain of convergence. This observation may be stated roughly as: *A majorant of the equation gives a majorant of the solution.*

We still have the problem of finding a suitable set of coefficients A_{jk}. Such a set is furnished by condition (6.3.8). We can take

$$A_{jk} = Ms^{-j}t^{-k}, \quad \forall j, k. \tag{6.3.22}$$

Now the corresponding series

$$G(z, w) \equiv A_{10}z + \Sigma \; \Sigma A_{jk} z^j w^k \tag{6.3.23}$$

is seen to be absolutely convergent for $|z| < s$, $|w| < t$. It is essentially the product of two geometric series, one in z/s, the other in w/t. More precisely we have

$$G(z, w) = M\left\{\left(1 - \frac{z}{s}\right)^{-1}\left(1 - \frac{w}{t}\right)^{-1} - 1 - \frac{w}{t}\right\}. \tag{6.3.24}$$

Now the majorant equation

$$w = G(z, w) \tag{6.3.25}$$

is a quadratic equation in w with coefficients which are rational functions of z, s, t. We want the root of this equation which is zero for $z = 0$. It is given by

$$w = \tfrac{1}{2}\frac{t^2}{t + M} - \tfrac{1}{2}\left\{\left(\frac{t^2}{t + M}\right)^2 - 4\frac{t^2 M}{t + M}\frac{z}{s - z}\right\}^{1/2}, \tag{6.3.26}$$

where the positive determination of the root is taken for $z = 0$. This expression is of the form

$$w = A - B\,(r - z)^{1/2}\,(s - z)^{-1/2}, \tag{6.3.27}$$

where r is defined by (6.3.10) and A and B are positive rational functions of M and t of no importance for the following. Now w can be expanded in powers of z by expanding $(r - z)^{1/2}$ and $(s - z)^{-1/2}$ in binomial series in z/r and z/s, respectively. The first of these series converges absolutely for $|z| \leqslant r$, the second for $|z| < s$. Here $0 < r < s$ and the Cauchy product series converges absolutely for $|z| \leqslant r$. This says that the majorant series defined by (6.3.27) is absolutely convergent for $|z| \leqslant r$. This series must coincide with the formal series

$$\sum_{n=1}^{\infty} C_n z^n,$$

where the C_n's are determined by (6.3.19) and the A_{jk} by (6.3.22). From the convergence of the majorant series (6.3.15) for $|z| \leqslant r$ we conclude that the formal series (6.3.9) for the solution of the original problem also converges for $|z| \leqslant r$. But this means that the various operations which were performed in order to find this series are legitimate and the formal solution is an actual one and clearly unique. ∎

What we have obtained is essentially an existence proof. There exists a unique power series (6.3.9) which satisfies (6.3.11) in its domain of convergence, a circle of radius at least equal to r. The method does not determine the radius of convergence: all it gives is lower bounds for the radius in terms of the quantities M, s, and t. Normally these are not given in advance; we are given a power series (6.3.7) and from our knowledge of the coefficients numbers M, s, t have to be deduced. This can be done in infinitely many ways in general and there is a problem of optimization: What admissible choice of M, s, t will give the largest possible lower bound for r? Normally this problem cannot be solved and we are left with what is obtainable by cursory inspection.

Let us look at our standard example

$$w = z + z^3 + w^3.$$

Here a possible choice is obviously $M = s = t = 1$ since we deal with a polynomial and all coefficients are 0 or 1. This gives $r = \frac{1}{9}$, less than one-third of the actual radius of convergence.

The majorant $G(z, w)$ defined by (6.3.24) is that given by Cauchy. The choice

$$\dot{A}_{jk} = |a_{jk}|, \tag{6.3.28}$$

which may claim to be the best choice, was used by Ernst Lindelöf (1870–1946) to prove existence theorems for implicit functions and for differential equations in 1896–9.

EXERCISE 6.3

1. Verify (6.3.3).
2. What recurrence relations are implied by (6.3.5)? Show that all coefficients are divisible by 6 except c_1 and c_3.
3. Verify (6.3.14).
4. The equation $w = z^3 + w^5$ has a unique power series solution which is 0 for $z = 0$. Show that all coefficients c_n are zero except those for which $n = 3 + 12k$, $k = 0, 1, 2, \ldots$. Find a lower bound for the radius of convergence.
5. Find the solution of $w = \sin z \sin w$ with $w = 0$ for $z = 0$.
6. Show that the relation $\ll$ is transitive and preserved under addition and multiplication by a positive number.
7. Verify (6.3.26) and (6.3.27).
8. Supply details omitted in the discussion of the series expansion for (6.3.27).
9. Extend Theorem 6.3.1 to the matrix case. The a_{jk} are to be scalars but $z = \mathfrak{Z}$ and $w = \mathfrak{W}$ are matrices in $\mathfrak{M}_n$. On the whole the proof can be imitated, but there are some difficult points in connection with inverses and square roots. Assume $\mathfrak{Z} = \mathfrak{T}$ to be a positive matrix; carry out the argument for this case and then generalize.

6.4 APPLICATIONS TO ORDINARY DIFFERENTIAL EQUATIONS

In the preceding sections we have discussed several aspects of the implicit function theorem at some length. The main object was not to arrive at results by the shortest route but to describe alternate routes, to bring out their distinctive features, and to elucidate advantages and limitations.

We have now various powerful techniques at our disposal and shall apply them to a brief study of ordinary differential equations. Here again Cauchy was the pioneer both in the real and in the complex domain. Let us first describe the problems in general terms.

A first order differential equation is a relation of the form

$$y'(x) = F[x, y(x)] \tag{6.4.1}$$

with an *initial condition*

$$y(x_0) = y_0. \tag{6.4.2}$$

As to the meaning to be attached to the symbols, we have a wide choice. In the simplest case x and y are real variables: $F(x, y)$ is a function from $R^1 \times R^1$ to R^1, defined in some domain containing the point (x_0, y_0). It is customary to require that $F(x, y)$ be a continuous function of (x, y) in D. With such assumptions there is at least one solution $y(x)$ defined in some interval $(x_0 - r, x_0 + r)$ such that (6.4.1) and (6.4.2) are satisfied. To ensure uniqueness of the solution, a Lipschitz condition

$$|F(x, y_1) - F(x, y_2)| \leqslant K|y_1 - y_2| \tag{6.4.3}$$

is usually added. Here K is any positive number, not necessarily between 0 and 1.

A more general situation is that where $\mathbf{F}(x, \mathbf{y})$ is a function from $R^1 \times R^m$ to R^m or from $C^1 \times C^m$ to C^m defined and continuous in some domain D, say $|x - x_0| \leqslant a$, $\|\mathbf{y} - \mathbf{y}_0\| \leqslant b$. Again a Lipschitz condition

$$\|\mathbf{F}(x, \mathbf{y}_1) - \mathbf{F}(x, \mathbf{y}_2)\| \leqslant K\|\mathbf{y}_1 - \mathbf{y}_2\| \tag{6.4.4}$$

will ensure uniqueness of the solution of

$$\mathbf{y}'(x) = \mathbf{F}[x, \mathbf{y}(x)], \qquad \mathbf{y}(x_0) = \mathbf{y}_0. \tag{6.4.5}$$

There are other possibilities. We may take $\mathcal{F}(x, \mathfrak{Y})$ as a function from $R^1 \times \mathfrak{M}_n$ to $\mathfrak{M}_n$ defined and continuous for $|x - x_0| \leqslant a$, $\|\mathfrak{Y} - \mathfrak{Y}_0\| \leqslant b$ with a Lipschitz condition

$$\|\mathcal{F}(x, \mathfrak{Y}_1) - \mathcal{F}(x, \mathfrak{Y}_2)\| \leqslant K\|\mathfrak{Y}_1 - \mathfrak{Y}_2\| \tag{6.4.6}$$

to ensure uniqueness of the solution of

$$\mathfrak{Y}'(x) = \mathcal{F}[x, \mathfrak{Y}(x)], \qquad \mathfrak{Y}(x_0) = \mathfrak{Y}_0. \tag{6.4.7}$$

The case where the matrix algebra is replaced by a general B-algebra is not essentially different.

We could also assume $\mathbf{F}(x, \mathbf{y})$ to be a function from $R^1 \times \mathfrak{X}$ to $\mathfrak{X}$ where $\mathfrak{X}$ is a B-space. Here we run into difficulties with the meaning to be assigned to $\mathbf{y}'(x)$, the derivative of $\mathbf{y}(x)$ with respect to x. As we shall see in the next chapter, in a B-space there are weak derivatives as well as strong derivatives and other interpretations are also occasionally available. We shall not elaborate further in this direction.

On the other hand, attention should be paid to the case

$$w'(z) = \sum_0^\infty \sum_0^\infty a_{jk} z^j w^k \tag{6.4.8}$$

where, in analogy with the implicit function case, we may expect the solution which is 0 for $z = 0$ to be representable by a power series in z convergent for small values of $|z|$.

This ends the survey. We shall complete it by proving some of the results indicated, taking care to exhibit the various methods at our disposal. We start with problem (6.4.5), where to simplify we take $(x_0, \mathbf{y}_0) = (0, \mathbf{0})$. We shall use the contraction fixed point theorem.

Theorem 6.4.1. *Let* $\mathbf{F}(x, \mathbf{y})$ *be a function from* $C^1 \times C^m$ *to* C^m *defined and continuous in the domain* D: $|x| \leqslant a$, $\|\mathbf{y}\| \leqslant b$. *Further, the Lipschitz condition* (6.4.4) *shall hold and*

$$\|\mathbf{F}(x, \mathbf{y})\| \leqslant M \tag{6.4.9}$$

in D. *Then problem* (6.4.5) *has a unique solution defined for*

$$|x| \leqslant r < \min\left(a, \frac{b}{M}, \frac{1}{K}\right). \tag{6.4.10}$$

Proof. We have to construct a complete metric space $\mathfrak{X}$ and a contraction mapping from $\mathfrak{X}$ into itself such that the existing unique fixed point gives the solution of (6.4.5). To this end, let $\mathfrak{X}$ be the set of all functions $\mathbf{g}(x)$ defined and continuous in the interval $[-r, r]$, having values in C^m, and such that $\mathbf{g}(0) = \mathbf{0}$ and $\|\mathbf{g}(x)\| \leqslant b$ for all x in the interval. As distance in $\mathfrak{X}$ we take

$$d[\mathbf{g}, \mathbf{h}] = \sup_x \|\mathbf{g}(x) - \mathbf{h}(x)\|. \tag{6.4.11}$$

It may be shown that $\mathfrak{X}$ is complete in this metric. Define

$$T[\mathbf{g}](x) = \int_0^x \mathbf{F}[s, \mathbf{g}(s)]\, ds, \tag{6.4.12}$$

where, as usual, the integral of a vector is taken as the vector whose components are the integrals of the components of the integrand. Since the range of $\mathbf{g}(s)$ belongs to the domain of definition of the integrand, the latter exists and is a continuous function of s. Hence the integral exists and is a continuous function of x in $[-r, r]$. Further,

$$\|T[\mathbf{g}](x)\| \leqslant M|x| \leqslant Mr < b$$

by the choice of r. We have also $T[\mathbf{g}](0) = \mathbf{0}$ so that T maps $\mathfrak{X}$ into itself. It is to be shown that T is a contraction. Now

$$\begin{aligned}
\|T[\mathbf{g}_1](x) - T[\mathbf{g}_2](x)\| &= \left\| \int_0^x \{\mathbf{F}[s, \mathbf{g}_1(s)] - \mathbf{F}[s, \mathbf{g}_2(s)]\}\, ds \right\| \\
&\leqslant \left| \int_0^x \|\mathbf{F}[s, \mathbf{g}_1(s)] - \mathbf{F}[s, \mathbf{g}_2(s)]\|\, ds \right| \\
&\leqslant K \left| \int_0^x \|\mathbf{g}_1(s) - \mathbf{g}_2(s)\|\, ds \right| \leqslant Kr\, d[\mathbf{g}_1, \mathbf{g}_2]
\end{aligned}$$

so that

$$d\{T[\mathbf{g}_1], T[\mathbf{g}_2]\} \leqslant Kr\, d[\mathbf{g}_1, \mathbf{g}_2]. \tag{6.4.13}$$

Since Kr is a fixed positive number < 1, the mapping is a contraction. Hence there is a unique fixed point of T, i.e. there exists an $\mathbf{f}(x) \in \mathfrak{X}$ such that

$$\mathbf{f}(x) = \int_0^x \mathbf{F}[s, \mathbf{f}(s)]\, ds. \tag{6.4.14}$$

We have clearly $\mathbf{f}(0) = \mathbf{0}$. The right member is differentiable with respect to the upper limit x of the integral since the integrand is continuous. Hence $\mathbf{f}'(x)$ exists and

$$\mathbf{f}'(x) = \mathbf{F}[x, \mathbf{f}(x)].$$

This is the desired solution and it is unique in $\mathfrak{X}$. It should be observed that a solution of problem (6.4.5) must satisfy the integral equation (6.4.14) for small values of $|x|$. Under the stated assumptions on $\mathbf{F}(x, \mathbf{y})$, if a solution exists, it must belong to the space $\mathfrak{X}$. The proof shows that $\mathfrak{X}$ contains a solution, one and only one solution. ∎

The value of r is subjected to the condition $Kr < 1$. This undesirable restriction could be eliminated by showing that some power of T is a contraction regardless of the value of K. This condition does not appear when the method of successive approximations is used.

We shall now apply the latter to the matrix case in order to vary the object of investigation and the mode of approach.

Theorem 6.4.2. *Let $\mathcal{F}(x, \mathcal{Y})$ be a function from $C^1 \times \mathfrak{M}_n$ to $\mathfrak{M}_n$, defined and continuous in D: $|x| \leqslant a$, $\|\mathcal{Y}\| \leqslant b$. Further, $\mathcal{F}$ satisfies the Lipschitz condition* (6.4.6) *and is bounded in D*

$$\|\mathcal{F}(x, \mathcal{Y})\| \leqslant M. \tag{6.4.15}$$

Then the equation (6.4.7) *with $x_0 = 0$, $\mathcal{Y}_0 = \mathcal{O}$ has a unique solution in $\mathfrak{M}_n$ for*

$$|x| \leqslant r < \min\left(a, \frac{b}{M}\right). \tag{6.4.16}$$

Proof. We define a sequence of approximating matrices $\{\mathcal{Y}_m(x)\}$:

$$\mathcal{Y}_0(x) = \mathcal{O}, \quad \mathcal{Y}_m(x) = \int_0^x \mathcal{F}[s, \mathcal{Y}_{m-1}(s)]\, ds. \tag{6.4.17}$$

By definition, the derivative of a matrix is the matrix of the derivatives of the entries and the integral of a matrix is the matrix of the integrals of the entries. It should be shown that the approximations exist, are continuous, and satisfy $\|\mathcal{Y}_m(x)\| \leqslant b$ for $|x| \leqslant r$. Suppose that the first k approximations have been found to be in order. Then $\mathcal{F}[s, \mathcal{Y}_k(s)]$ is defined as a continuous matrix function for $|s| \leqslant r$; $\mathcal{Y}_{k+1}(x)$ exists and is continuous and for $|x| \leqslant r$

$$\|\mathcal{Y}_{k+1}(x)\| \leqslant \left| \int_0^x \|\mathcal{F}[s, \mathcal{Y}_k(s)]\|\, ds \right| \leqslant Mr < b$$

by the choice of r. Hence all approximations are in order.

To prove convergence of the sequence $\{\mathcal{Y}_m(x)\}$ we resort to the Lipschitz condition. We have, for $0 \leqslant |x| \leqslant r$,

$$\|\mathcal{Y}_m(x) - \mathcal{Y}_{m-1}(x)\| \leqslant \left| \int_0^x \|\mathcal{F}[s, \mathcal{Y}_{m-1}(s)] - \mathcal{F}[s, \mathcal{Y}_{m-2}(s)]\| \, ds \right|$$

$$\leqslant K \left| \int_0^x \|\mathcal{Y}_{m-1}(s) - \mathcal{Y}_{m-2}(s)\| \, ds \right|.$$

Suppose that it be known that for some integer k

$$\|\mathcal{Y}_k(x) - \mathcal{Y}_{k-1}(x)\| < MK^{k-1} \frac{|x|^k}{k!}. \tag{6.4.18}$$

This is certainly true for $k = 1$. Then

$$\|\mathcal{Y}_{k+1}(x) - \mathcal{Y}_k(x)\| < M \frac{K^k}{k!} \int_0^{|x|} s^k \, ds = MK^k \frac{|x|^{k+1}}{(k+1)!}.$$

This shows that (6.4.18) is true for all values of k. Hence

$$\sum_{m=1}^{\infty} \|\mathcal{Y}_m(x) - \mathcal{Y}_{m-1}(x)\|$$

converges uniformly in the interval $[-r, r]$ and

$$\lim_{m \to \infty} \mathcal{Y}_m(x) \equiv \mathcal{Y}(x)$$

exists. Since

$$\lim_{m \to \infty} \mathcal{Y}_m(x) = \lim_{m \to \infty} \int_0^x \mathcal{F}[s, \mathcal{Y}_{m-1}(s)] \, ds = \int_0^x \mathcal{F}\left[s, \lim_{m \to \infty} \mathcal{Y}_{m-1}(s)\right] ds$$

it follows that

$$\mathcal{Y}(x) = \int_0^x \mathcal{F}[s, \mathcal{Y}(s)] \, ds, \qquad \mathcal{Y}(0) = \mathcal{O}.$$

Again we can differentiate with respect to the upper limit and obtain

$$\mathcal{Y}'(x) = \mathcal{F}[x, \mathcal{Y}(x)]$$

as desired. A uniqueness proof may be based on the Lipschitz condition following the lines of the uniqueness proof in Section 6.2. The details are left to the reader. Actually a uniqueness proof based on a Lipschitz condition goes back to the fact that the functional inequality

$$g(x) \leqslant K \int_0^x g(s) \, ds \tag{6.4.19}$$

has in $C^+[0, r]$ the unique solution $g(x) \equiv 0$. Such questions will be explored in Chapter 12. ∎

Our last item in this section is the discussion of equation (6.4.8), which calls for the use of majorants.

Theorem 6.4.3. *Let the coefficients in the right member of* (6.4.8) *satisfy the condition*

$$|a_{jk}| \leqslant Ms^{-j}t^{-k} \tag{6.4.20}$$

for some choice of positive numbers M, s, t. Then the differential equation (6.4.8) *has a unique solution*

$$w(z) = \sum_{m=1}^{\infty} c_n z^n \tag{6.4.21}$$

where the series is absolutely convergent for

$$|z| < r = s\left[1 - \exp\left(-\frac{t}{2Ms}\right)\right]. \tag{6.4.22}$$

Proof. We know the general pattern to follow from the discussion in Section 6.3. The power series (6.4.21) is substituted in (6.4.8) to obtain

$$\sum_{n=1}^{\infty} nc_n z^{n-1} = \sum_{j=0}^{\infty}\sum_{k=0}^{\infty} a_{jk} z^j \left\{\sum_{n=1}^{\infty} c_n z^n\right\}^k. \tag{6.4.23}$$

Here the kth power is expanded in a power series starting with a term in z^k. If (6.4.21) has a positive radius of convergence R, then the kth powers give series with the same radius of convergence. For $|z| < s$ and sufficiently small, the sum of the solution series is less than t in absolute value; the triple series is absolutely convergent and may be rearranged *ad lib.* If it be rewritten as a power series in z, the two power series on the left and on the right must be identical. This implies that the coefficients of z^n on the two sides must be equal.

Now on the left the constant term is c_1, on the right a_{00}, so $c_1 = a_{00}$. For $n > 0$ the admissible values of j and k satisfy

$$0 \leqslant j \leqslant n, \quad 0 \leqslant k \leqslant n, \quad 1 \leqslant j + k \leqslant n. \tag{6.4.24}$$

Since there may be a term $a_{01}w$ in the right member of (6.4.8), the coefficient of z^n may involve c_n but no c_p with $p > n$ can occur. Hence

$$(n+1)\,c_{n+1} = P_n(a_{jk};\, c_1, c_2, \ldots, c_n), \tag{6.4.25}$$

where P_n is a multinomial in c_1 to c_n, linear in the a_{jk} where j and k are subject to (6.4.24). All numerical coefficients are positive integers. This means that the c_n's may be computed consecutively from the preceding coefficients, i.e. ultimately in terms of c_1. The first four relations are

$$\begin{aligned}
1c_1 &= a_{00},\\
2c_2 &= a_{10} + a_{01}c_1,\\
3c_3 &= a_{20} + a_{11}c_1 + a_{02}c_1{}^2 + a_{01}c_2,\\
4c_4 &= a_{30} + a_{21}c_1 + a_{12}c_1{}^2 + a_{03}c_1{}^3 + 2\,a_{02}c_1c_2 + a_{11}c_2 + a_{01}c_3.
\end{aligned} \tag{6.4.26}$$

The series (6.4.21) will converge if a convergent majorant series

$$\sum_{n=1}^{\infty} C_n z^n, \quad |c_n| \leqslant C_n, \quad \forall\, n, \tag{6.4.27}$$

can be found. We construct such a series by finding a formal solution of a majorant equation

$$w' = \sum_{j=0}^{\infty} \sum_{k=0}^{\infty} A_{jk} z^j, \quad |a_{jk}| \leqslant A_{jk}, \quad \forall\, j, k. \tag{6.4.28}$$

That such a series is a majorant of the solution is shown by an analysis of the multinomials P_n. Again we are concerned with the implications of positivity.

The conditions (6.4.20) suggest taking

$$A_{jk} = M s^{-j} t^{-k}, \quad \forall\, j, k. \tag{6.4.29}$$

Then the series

$$H(z, w) = \sum_{j=0}^{\infty} \sum_{k=0}^{\infty} A_{jk} z^j w^k \tag{6.4.30}$$

is absolutely convergent for $|z| < s$, $|w| < t$. Here $H(z, w)$ is M times the product of two geometric series or in closed form

$$H(z, w) = M \left(1 - \frac{z}{s}\right)^{-1} \left(1 - \frac{w}{t}\right)^{-1} \tag{6.4.31}$$

and the differential equation

$$w' = H(z, w)$$

is simply

$$\left(1 - \frac{w}{t}\right) w' = M \left(1 - \frac{z}{s}\right)^{-1}, \tag{6.4.32}$$

which is integrable by elementary methods. With $w(0) = 0$ we get

$$w - \frac{1}{2t} w^2 = -Ms \log\left(1 - \frac{z}{s}\right). \tag{6.4.33}$$

The logarithm is uniquely determined by its initial value 0 for $z = 0$. This is a quadratic equation in w; there are two roots, only one of which is 0 for $z = 0$, namely

$$w(t) = t - t\left\{1 + \frac{2Ms}{t} \log\left(1 - \frac{z}{s}\right)\right\}^{1/2}. \tag{6.4.34}$$

This solution may be expanded in powers of z and the resulting series is the desired majorant series for (6.4.21). The expansion may be made in two steps. We first expand the square root in terms of powers of

$$Z = \frac{2Ms}{t} \log\left(1 - \frac{z}{s}\right). \tag{6.4.35}$$

The series converges for $|Z| < 1$. The largest value of $|Z|$ is for $z = x$ where $0 < x < s$. A simple computation shows that the least upper bound for x is r, the quantity defined by (6.4.22). For $|z| < r < s$ we can now expand the logarithm in powers of z and then collect terms. The result is the majorant series which is absolutely convergent for $|z| < r$. It follows that (6.4.21) is also absolutely convergent for $|z| < r$ and in this circular disk it is a solution of (6.4.8), the only solution of this equation which is 0 for $z = 0$. $\blacksquare$

The radius of convergence of the majorant series is uniquely determined by the discussion, but that of the actual solution is not and we have the task of finding values of M, s, t which maximize the lower bound. We shall not devote any time to this task.

EXERCISE 6.4

1. Show that a power of the transformation T defined by (6.4.12) is a contraction.
2. Show that Theorem 6.4.2 remains true if matrices are replaced by the elements of a B-algebra with unit element.
3. Prove Theorem 6.4.1 by the method of successive approximations.
4. Solve the equation $\mathfrak{Y}'(t) = \mathfrak{E} + [\mathfrak{Y}(t)]^2$, $\mathfrak{Y}(0) = \mathfrak{O}$, explicitly in $\mathfrak{M}_2$ and find the largest interval $[-a, a]$ in the interior of which $\mathfrak{Y}(t)$ is continuous.
5. Write out a detailed proof of convergence of the series expansion for (6.4.35) following the lines indicated in the text.
6. If the reader has had a course in analytic function theory, he would argue instead that the majorant series converges in the largest disk in which the function (6.4.34) is holomorphic. Give details!
7. The equation $w' = z^2 + w^2$ has a power series solution which is 0 for $x = 0$. Show that the coefficients c_n are zero unless $n = 3 + 4k$, $k = 0, 1, 2, \ldots$. Find a lower bound for the radius of convergence.
8. Try to work out a convergence proof replacing the Cauchy majorant $H(z, w)$ by the Lindelöf one where $A_{jk} = |a_{jk}|$, $\forall\ j, k$.
9. In the proof of Theorem 6.4.1 show that $\mathfrak{X}$ is complete under the metric defined by (6.4.11).

COLLATERAL READING

The author's choice would be:

HILLE, E., *Analysis*, Vol. 2, Blaisdell, Waltham, Mass. (1966), Chapters 17 and 20.

HILLE, E., *Lectures on Ordinary Differential Equations*, Addison-Wesley, Reading, Mass. (1969), Chapter 2.

7 REAL ANALYSIS IN LINEAR SPACES*

There are several good reasons why an analyst should take an interest in linear spaces. The functions that he considers often form a linear space which may facilitate the discussion.

On a different plane, it is striking to what extent the classical properties of functions of a real or a complex variable remain meaningful in a more general setting. We may mention such properties as continuity, differentiability, measurability, integrability, and analyticity. This is the type of analysis that we have in mind here. There are several levels of study. We have to start by modifying the classical definition of the basic properties to adjust to the new situation. The new concepts should be studied at some length. The classical theory can serve as a pattern, but we must be prepared for deviations.

As usual, there are three types of mapping. Let $\mathfrak{X}$ and $\mathfrak{Y}$ be B-spaces, say over the complex field C. We have then to consider (1) mappings from R or C into $\mathfrak{X}$, (2) mappings from $\mathfrak{X}$ into R or C, and (3) mappings of $\mathfrak{X}$ into $\mathfrak{Y}$. All three types will occur in this chapter, but the emphasis is on functions of a real variable. Complex variables are treated at length in Chapter 8. The mappings involve a concept of continuity, in fact, there are several such concepts depending upon the underlying topology. We have a similar variety for the other concepts mentioned above. Thus, *inter alia*, we speak of weak, strong, and uniform continuity for operator functions $T(s) \in \mathfrak{E}(\mathfrak{X}, \mathfrak{Y})$, where s is a real or complex variable. In all these cases a stronger property implies a weaker one, but the converse may hold and, in fact, does hold for analyticity. In all such situations the *principle of uniform boundedness* plays a basic role and will be discussed in this chapter. We also have to pay some attention to alternate topologies. The discussion of the Bochner integral is tied up with abstract measure spaces (S, A, μ) and requires a considerable expansion of the notions presented in Chapter 4, a review of which is recommended before the reader tackles Section 7.5.

There are five sections: The principle of uniform boundedness; Topologies; Vector-valued and operator-valued functions; Abstract Riemann–Stieltjes integrals; and Bochner integrals.

7.1 THE PRINCIPLE OF UNIFORM BOUNDEDNESS

We start with a result for real functionals on a complex B-space $\mathfrak{X}$. The functionals need not be linear, nor do they have to be defined everywhere in $\mathfrak{X}$. They should be

* The author is indebted to Dr. Ih-Ching Hsü for a revision of this chapter.

defined and locally continuous on a subset $\mathfrak{X}_0$ which is not too thinly dispersed in $\mathfrak{X}$. Baire second category is a suitable assumption for $\mathfrak{X}_0$. In this subset some finiteness properties are also required.

Theorem 7.1.1. *Let $\mathfrak{X}_0$ be a set of the second category in a complex B-space $\mathfrak{X}$. Suppose that $\{f_\alpha(\mathbf{x}); \alpha \in \mathrm{A}\}$ is a family of continuous real functionals on $\mathfrak{X}$ such that for each $\mathbf{x} \in \mathfrak{X}_0$*

$$\sup_\alpha f_\alpha(\mathbf{x}) < \infty. \tag{7.1.1}$$

Then in $\mathfrak{X}$ there is a closed sphere $\mathfrak{S}_1 = \{\mathbf{x}; \mathbf{x} \in \mathfrak{X}, \|\mathbf{x} - \mathbf{x}_0\| \leqslant \rho\}$ and a finite number B such that

$$f_\alpha(\mathbf{x}) \leqslant B, \quad \forall\, \mathbf{x} \in \mathfrak{S}_1, \quad \forall\, \alpha \in \mathrm{A}. \tag{7.1.2}$$

Proof. Let n be a positive integer and consider the set

$$\mathfrak{X}_{0n} = \{\mathbf{x}; \mathbf{x} \in \overline{\mathfrak{X}}_0, \quad f_\alpha(\mathbf{x}) \leqslant n, \quad \forall\, \alpha \in \mathrm{A}\}. \tag{7.1.3}$$

For each n, $\mathfrak{X}_{0n} = \{\mathbf{x}; \mathbf{x} \in \mathfrak{X}, f_\alpha(\mathbf{x}) \leqslant n, \forall\, \alpha \in \mathrm{A}\} \cap \overline{\mathfrak{X}}_0$, and is therefore closed in $\mathfrak{X}$, by the continuity of the f_α's. For sufficiently large n, $\mathfrak{X}_{0n}$ is not empty. Further, $\mathfrak{X}_0 \subset \bigcup_{n=1}^\infty \mathfrak{X}_{0n} \subset \overline{\mathfrak{X}}_0$. If for each n the set $\mathfrak{X}_{0n}$ were a set of the first category in $\mathfrak{X}$, then $\bigcup_{n=1}^\infty \mathfrak{X}_{0n}$, as a countable union of sets of the first category, would also be of the first category in $\mathfrak{X}$. By assumption, $\mathfrak{X}_0$ is of the second category in $\mathfrak{X}$, hence there exists a positive integer m such that $\mathfrak{X}_{0m}$ is of the second category in $\mathfrak{X}$ and

$$\mathrm{Int}\,(\mathfrak{X}_{0m}) = \mathrm{Int}\,(\overline{\mathfrak{X}}_{0m}) \neq \varnothing.$$

This says that $\mathfrak{X}_{0m}$ has interior points and hence contains a sphere. Thus under the norm concerned, $\mathfrak{X}_{0m}$ contains a closed sphere, say $\mathfrak{S}_1 = \{\mathbf{x}; \mathbf{x} \in \mathfrak{X}, \|\mathbf{x} - \mathbf{x}_0\| \leqslant \rho\}$ in which (7.1.2) holds with $B = m$. ∎

For the properties of Baire categories used in this proof, see Section 5.1 and especially Problems 10–12 of Exercise 5.1.

Corollary 1. *If the family $\{f_\alpha(\mathbf{x}); \alpha \in \mathrm{A}\}$ satisfies*

$$\inf_\alpha f_\alpha(\mathbf{x}) > -\infty \tag{7.1.4}$$

for each $\mathbf{x}$ on a set $\mathfrak{X}_0$ of the second category in $\mathfrak{X}$, then there is a finite number B and a closed sphere $\mathfrak{S}_2$ in $\mathfrak{X}$ such that

$$f_\alpha(\mathbf{x}) > B, \quad \forall\, \mathbf{x} \in \mathfrak{S}_2, \quad \forall\, \alpha \in \mathrm{A}. \tag{7.1.5}$$

For $\{-f_\alpha(\mathbf{x}); \alpha \in \mathrm{A}\}$ satisfies the assumptions of Theorem 7.1.1.

Corollary 2. *If the family $\{f_\alpha(\mathbf{x}); \alpha \in \mathrm{A}\}$ satisfies*

$$\sup_\alpha |f_\alpha(\mathbf{x})| < \infty \tag{7.1.6}$$

for each $\mathbf{x}$ in a set $\mathfrak{X}_0$ of the second category in $\mathfrak{X}$, then there is a finite number C and a closed sphere $\mathfrak{S}$ in $\mathfrak{X}$ such that

$$|f_\alpha(\mathbf{x})| < C, \quad \forall\, \mathbf{x} \in \mathfrak{S}, \quad \forall\, \alpha \in \mathrm{A}.$$

For $\{|f_\alpha(\mathbf{x})|; \alpha \in \mathrm{A}\}$ satisfies the assumptions of Theorem 7.1.1.

It is desired to extend the conclusions from a sphere to all of $\mathfrak{X}$. This requires further restrictions on the functionals. Here subadditivity of the functionals together with a bound on $f_\alpha(-\mathbf{x})$ in terms of $f_\alpha(\mathbf{x})$ will provide what is needed.

Theorem 7.1.2. *Let $\mathfrak{X}_0$ be a set of the second category in a complex B-space $\mathfrak{X}$. Suppose that a family* $\mathsf{F} = \{f_\alpha(\mathbf{x}); \alpha \in \mathrm{A}\}$ *of continuous real functionals satisfies* (7.1.6) *for each* $\mathbf{x}$ *in* $\mathfrak{X}_0$. *Suppose further that* F *satisfies*

$$f_\alpha(\mathbf{x} + \mathbf{y}) \leqslant f_\alpha(\mathbf{x}) + f_\alpha(\mathbf{y}), \quad \forall\, \alpha \in \mathrm{A}, \quad \forall\, \mathbf{x} \in \mathfrak{X}, \quad \forall\, \mathbf{y} \in \mathfrak{X}, \tag{7.1.7}$$

and that there exists a constant C such that

$$|f_\alpha(-\mathbf{x})| < C|f_\alpha(\mathbf{x})|, \quad \forall\, \alpha \in \mathrm{A}, \quad \forall\, \mathbf{x} \in \mathfrak{X}. \tag{7.1.8}$$

Then there are two numbers r and M such that

$$|f_\alpha(\mathbf{x})| \leqslant \max\left(M, \frac{M}{r}\|\mathbf{x}\|\right) \tag{7.1.9}$$

for all α and all $\mathbf{x}$.

Proof. By the preceding theorem and its corollaries there exist a sphere $\mathfrak{S} = \{\mathbf{x};\ \|\mathbf{x} - \mathbf{x}_0\| \leqslant \rho\}$ and a constant B such that

$$|f_\alpha(\mathbf{x})| \leqslant B, \quad \forall\, \mathbf{x} \in \mathfrak{S}, \quad \forall\, \alpha \in \mathrm{A}.$$

By condition (7.1.8) $|f_\alpha(\mathbf{x})|$ is then also bounded in the sphere

$$-\mathfrak{S} = \{\mathbf{x};\ \|\mathbf{x} + \mathbf{x}_0\| \leqslant \rho\}.$$

If $\mathbf{x}_0 = \mathbf{0}$, the two spheres coincide and we omit the next step in the proof. If $\mathbf{x}_0 \neq \mathbf{0}$ form the difference set

$$\Sigma_0 = \mathfrak{S} - \mathfrak{S} = \{\mathbf{z};\ \mathbf{z} = \mathbf{x} - \mathbf{y},\ \mathbf{x} \in \mathfrak{S},\ \mathbf{y} \in \mathfrak{S}\}. \tag{7.1.10}$$

This set contains the origin; it is closed, connected and convex. From the inequalities

$$f_\alpha(\mathbf{x} - \mathbf{y}) \leqslant f_\alpha(\mathbf{x}) + f_\alpha(-\mathbf{y}), \qquad f_\alpha(\mathbf{x}) \leqslant f_\alpha(\mathbf{x} - \mathbf{y}) + f_\alpha(\mathbf{y}),$$

which are implied by (7.1.7), we infer that there is a finite M such that

$$|f_\alpha(\mathbf{x})| \leqslant M, \quad \forall\, \alpha \in \mathrm{A}, \quad \forall\, \mathbf{x} \in \Sigma_0. \tag{7.1.11}$$

From this basis we proceed to exhaust $\mathfrak{X}$, leaning heavily on the subadditivity.

Consider the set $\frac{1}{2}\Sigma_0 = \{\mathbf{u};\ 2\mathbf{u} \in \Sigma_0\}$ and the "shell" $\Sigma_1 = \Sigma_0 \ominus \frac{1}{2}\Sigma_0$. Form now the homothetic images of Σ_1, say $\Sigma_1, \Sigma_2, \ldots, \Sigma_n, \ldots$, where $\Sigma_n = \{\mathbf{v}; \mathbf{v}/n \in \Sigma_1\}$. The union of $\frac{1}{2}\Sigma_0$ with $\bigcup_{n=1}^{\infty} \Sigma_n$ exhausts $\mathfrak{X}$. Since Σ_0 contains the sphere $\|\mathbf{x}\| \leqslant \rho$, it follows that Σ_1 is bounded away from the origin and there exists an $r \geqslant \frac{1}{2}\rho$ such that

$$\inf \|\mathbf{z}\| = r \tag{7.1.12}$$

for $\mathbf{z} \in \Sigma_1$. Suppose that $\mathbf{x} \in \Sigma_p$ and $\mathbf{x} = p\mathbf{y}$ where $\mathbf{y} \in \Sigma_1$. Then

$$f_\alpha(\mathbf{x}) = f_\alpha(p\mathbf{y}) = pf_\alpha(\mathbf{y}) \leqslant pM \tag{7.1.13}$$

by (7.1.7) and (7.1.11). Further,

$$\begin{aligned} f_\alpha(\mathbf{y}) &= f_\alpha[\mathbf{x} - (p-1)\,\mathbf{y}] \\ &\leqslant f_\alpha(\mathbf{x}) + f_\alpha[-(p-1)\,\mathbf{y}] \\ &\leqslant f_\alpha(\mathbf{x}) + (p-1) f_\alpha(-\mathbf{y}) \end{aligned}$$

so that

$$f_\alpha(\mathbf{x}) \geqslant f_\alpha(\mathbf{y}) - (p-1) f_\alpha(-\mathbf{y})$$

and

$$f_\alpha(\mathbf{x}) \geqslant -pM.$$

Hence

$$|f_\alpha(\mathbf{x})| \leqslant pM, \quad \forall\, \alpha \in \mathrm{A}, \quad \forall\, \mathbf{x} \in \Sigma_p.$$

Here

$$p = \frac{\|\mathbf{x}\|}{\|\mathbf{y}\|} \leqslant \frac{1}{r}\|\mathbf{x}\|$$

so that

$$|f_\alpha(\mathbf{x})| \leqslant \frac{M}{r}\|\mathbf{x}\|$$

for all $\mathbf{x}$ outside of $\frac{1}{2}\Sigma_0$ where we have instead

$$|f_\alpha(\mathbf{x})| \leqslant M.$$

These two inequalities give (7.1.9). If $\mathbf{x}_0 = \mathbf{0}$, we simply replace Σ_0 by $\mathfrak{S}$ and set $r = \rho$. ∎

In the preceding theorem the functionals $f_\alpha(\mathbf{x})$ are supposed to be finite in a set $\mathfrak{X}_0$ of the second category. Properties (7.1.7) and (7.1.8) then force each $f_\alpha(\mathbf{x})$ to be finite valued everywhere and, in fact, uniformly bounded on compact sets. For this reason the theorem and its variants are referred to as the *uniform boundedness principle*. The particular case where the f_α's are linear bounded functionals leads to the following assertion.

Theorem 7.1.3. *If each f_α is a real linear bounded functional on $\mathfrak{X}$ and if in some set $\mathfrak{X}_0$ of the second category in $\mathfrak{X}$*

$$\sup_\alpha |f_\alpha(\mathbf{x})| < \infty, \quad \forall\, \alpha \in \mathrm{A}, \quad \forall\, \mathbf{x} \in \mathfrak{X}_0, \tag{7.1.14}$$

then the functionals are uniformly bounded, i.e. there exists a B such that for all α

$$\|f_\alpha\| \leqslant B. \tag{7.1.15}$$

Proof. The assumptions of Theorem 7.1.2 are clearly satisfied and give

$$\|f_\alpha\| = \sup_{\|\mathbf{x}\|=1} |f_\alpha(\mathbf{x})| \leqslant \max\{M, M/r\} \equiv B,$$

which is the assertion. ∎

At this stage it is appropriate to ask if extensions to complex-valued functionals are possible. The answer is in the affirmative. For if

$$f_\alpha(\mathbf{x}) = g_\alpha(\mathbf{x}) + ih_\alpha(\mathbf{x}) \tag{7.1.16}$$

with obvious notation, then g_α and h_α are real-valued functionals and a condition of the form

$$\sup_\alpha |f_\alpha(\mathbf{x})| < \infty \tag{7.1.17}$$

implies and is implied by

$$\sup_\alpha |g_\alpha(\mathbf{x})| < \infty, \qquad \sup_\alpha |h_\alpha(\mathbf{x})| < \infty. \tag{7.1.18}$$

This gives

Theorem 7.1.4. *If* $\mathsf{F} = \{f_\alpha(\mathbf{x})\}$ *is a family of continuous complex-valued functionals on a complex B-space* $\mathfrak{X}$ *and if there exists a subset* $\mathfrak{X}_0$ *of the second category in* $\mathfrak{X}$ *where*

$$\sup_\alpha |f_\alpha(\mathbf{x})| < \infty \tag{7.1.19}$$

for each $\mathbf{x} \in \mathfrak{X}_0$*, then there exist a sphere* $\mathfrak{S}$ *and a number B with*

$$|f_\alpha(\mathbf{x})| \leqslant B, \quad \forall\, \alpha \in A, \quad \forall\, \mathbf{x} \in \mathfrak{S}. \tag{7.1.20}$$

In order to extend Theorem 7.1.2 to complex functionals we must assume that condition (7.1.7) holds for the real and for the imaginary part of $f_\alpha(\mathbf{x})$. Then the proof goes through with trivial modification. In Theorem 7.1.3 we can replace "real" by "complex" without affecting the validity of the theorem.

The following result concerning linear bounded transformations is essentially a corollary to Theorem 7.1.2.

Theorem 7.1.5. *Given a family* F *of linear bounded operators* $\{T_\alpha\}$ *from one B-space* $\mathfrak{X}$ *to another* $\mathfrak{Y}$ *such that for all* α *and all* $\mathbf{x} \in \mathfrak{X}_0$*, a subset of the second category in* $\mathfrak{X}$*, we have*

$$\sup_\alpha \|T_\alpha(\mathbf{x})\| < \infty, \tag{7.1.21}$$

then there is a finite K such that

$$\|T_\alpha\| \leqslant K, \quad \forall\, \alpha. \tag{7.1.22}$$

Proof. We consider the corresponding family G of real functionals $\{\|T_\alpha(\mathbf{x})\|\}$. Here G satisfies the conditions of Theorem 7.1.2 since

$$\|T_\alpha(\mathbf{x}+\mathbf{y})\| \leqslant \|T_\alpha(\mathbf{x})\| + \|T_\alpha(\mathbf{y})\|, \qquad \|T_\alpha(-\mathbf{x})\| = \|T_\alpha(\mathbf{x})\|.$$

Hence

$$\| T_\alpha(\mathbf{x})\| \leqslant \max\left\{M, \frac{M}{r}\|\mathbf{x}\|\right\}$$

and for $\|\mathbf{x}\| = 1$ this does not exceed $\max(M, M/r) \equiv K$. ∎

This is known as the *Uniform Boundedness Theorem.*

EXERCISE 7.1

1. Verify that the set Σ_0 of (7.1.10) contains the origin and is closed, connected and convex.
2. Why is Σ_1 bounded away from the origin?
3. Show by an example that the conclusion in Theorem 7.1.2 cannot be replaced by $|f_\alpha(\mathbf{x})| \leqslant B\|\mathbf{x}\|$ in general.
4. Let $C_0[-\pi, \pi]$ denote the set of functions $t \to f(t)$, continuous in $[-\pi, \pi]$ with $f(-\pi) = f(\pi)$ and continued outside of $[-\pi, \pi]$ with period 2π. The metric is defined by the sup-norm. Define
$$T_n[f](t) = n\left[f\left(t + \frac{1}{n}\right) - f(t)\right].$$
Here $\lim_{n\to\infty} T_n[f](t) = f'(t)$ exists for f in a dense subset of C_0. Show that (i) $\|T_n\| = 2n$, (ii) $\lim_{n\to\infty} T_n[f]$ cannot exist in a subset of C_0 which is of the second category, and (iii) the set of non-differentiable functions is of the second category in C_0.
5. The notation used in Section 7.1 suggests that the family of functionals considered has infinitely many members. Are Theorems 7.1.1 and 7.1.2 valid and significant for finite families or for a single functional? How about the other theorems?
6. Continuous functions are dense in any Lebesgue space. What is the category of such a subset?
7. In the space $L_1(S)$, where S is bounded, a family of real-valued functionals is defined by $\|f\|_p$, $1 \leqslant p \leqslant \infty$. What can be said about the category of the subset L_0 where $\sup_p \|f\|_p$ is finite?

7.2 TOPOLOGIES

Let $\mathfrak{X}$ be a B-space over the complex field C. This implies the existence of a normed metric in terms of which $\mathfrak{X}$ is complete. This metric is often referred to as defining the *strong topology* of $\mathfrak{X}$ and if $\lim_{n\to\infty}\|\mathbf{x}_n - \mathbf{x}_0\| = 0$ we say that the sequence $\{\mathbf{x}_n\}$

converges strongly to $\mathbf{x}_0$, which is called the *strong limit of* $\{\mathbf{x}_n\}$. The relation is sometimes written $\mathbf{x}_n \xrightarrow{s} \mathbf{x}_0$ or $\mathbf{x}_0 \overset{s}{=} \lim \mathbf{x}_n$.

Besides the strong topology we shall also have occasion to use the so-called *weak topology*. We recall the existence of linear bounded functionals x^* on $\mathfrak{X}$ the totality of which form the adjoint or dual space $\mathfrak{X}^*$. This is also a B-space. We can use elements of $\mathfrak{X}^*$ to define neighborhoods in $\mathfrak{X}$ which are distinct from those defined by the strong topology. Some definitions are needed.

Definition 7.2.1. *A non-void set* $\mathfrak{H}$ *is called a Hausdorff topological space if there exists a family* $\{N_\alpha\}$ *of sets* $N_\alpha \subset \mathfrak{H}$ *called neighborhoods satisfying the following four postulates:*

(H_1) *For each* $x \in \mathfrak{H}$ *there is at least one neighborhood* $N(x)$ *containing* x.

(H_2) *If* $N_1(x)$ *and* $N_2(x)$ *are neighborhoods of* x *there is at least one neighborhood* $N_3(x)$ *of* x *such that* $N_3(x) \subset N_1(x) \cap N_2(x)$.

(H_3) *If* $N(x)$ *is a neighborhood of* x *and if* $y \in N(x)$, *then there is at least one neighborhood* $N(y)$ *of* y *such that* $N(y) \subset N(x)$.

(H_4) *If* $x \neq y$ *there are neighborhoods* $N(x)$ *of* x *and* $N(y)$ *of* y *such that* $N(x) \cap N(y) = \emptyset$.

This concept was introduced by Felix Hausdorff (1868–1942) in 1914. His work on point-set topology was fundamental and he also wrote important papers on dimension theory, Fourier series, moment problems, and summability.

It is clear that any B-space is a Hausdorff space with the spheres having centers at $\mathbf{x} = \mathbf{x}_0$ as neighborhoods of $\mathbf{x}_0$. Moreover, they are linear Hausdorff spaces in the sense of the following

Definition 7.2.2. *A space* $\mathfrak{X}$ *is a linear Hausdorff space if* (1) *it is a Hausdorff space as well as a linear space,* (2) *for each choice of* $\mathbf{x}$ *and* $\mathbf{y}$ *in* $\mathfrak{X}$ *and neighborhood* $N(\mathbf{x} - \mathbf{y})$ *of* $\mathbf{x} - \mathbf{y}$ *there are neighborhoods* $N(\mathbf{x})$ *of* $\mathbf{x}$ *and* $N(\mathbf{y})$ *of* $\mathbf{y}$ *such that the difference set* $N(\mathbf{x}) - N(\mathbf{y})$ *is contained in* $N(\mathbf{x} - \mathbf{y})$, (3) *for every* $\alpha \in C, \mathbf{x} \in \mathfrak{X}$ *and neighborhood* $N(\alpha\mathbf{x})$ *of* $\alpha\mathbf{x}$ *there are neighborhoods* $N(\alpha)$ *of* α *and* $N(\mathbf{x})$ *of* $\mathbf{x}$ *such that the scalar product set* $N(\alpha)\,N(\mathbf{x}) \subset N(\alpha\mathbf{x})$.

We recall that the *difference set* of two subsets S_1 and S_2 of a linear space is the set of differences

$$S_1 - S_2 = \{\mathbf{z};\ \mathbf{z} = \mathbf{x} - \mathbf{y},\ \mathbf{x} \in S_1,\ \mathbf{y} \in S_2\}. \tag{7.2.1}$$

The definition of the *scalar product set* is obvious:

$$N(\alpha)\,N(\mathbf{x}) = \{\mathbf{z};\ \mathbf{z} = \beta\mathbf{y},\ \beta \in N(\alpha),\ \mathbf{y} \in N(\mathbf{x})\}. \tag{7.2.2}$$

For α in C we can use Euclidean neighborhoods, $N(\alpha)$.

On a given B-space $\mathfrak{X}$ we are interested in constructing a weakest topology in which each member x^* of $\mathfrak{X}^*$ is continuous. (Cf. Problem 4, Exercise 7.2.) We proceed as follows. Let $\mathbf{x}_0 \in \mathfrak{X}$, $\varepsilon > 0$, n any positive integer, and choose any n elements of X^*, say $x_1^*, x_2^*, \ldots, x_n^*$. Denote the set

$$\{\mathbf{x}; \mathbf{x} \in \mathfrak{X}, |x_k^*(\mathbf{x} - \mathbf{x}_0)| < \varepsilon, k = 1, 2, \ldots, n\} \tag{7.2.3}$$

by $N(\mathbf{x}_0; x_1^*, x_2^*, \ldots, x_n^*; \varepsilon)$.

Theorem 7.2.1. *The family $\{N_\alpha\}$ of all sets of the type (7.2.3) obtained by varying all elements (including $\mathbf{x}_0$ and n) satisfies postulates (H_1) to (H_4). $\mathfrak{X}$ together with the family of neighborhoods $\{N_\alpha\}$ is a linear Hausdorff space.*

Proof. It is clear that $\mathfrak{X}$ is a linear space since we started out with a B-space. Further, the Hausdorff axiom (H_1) is satisfied. To prove (H_2) consider any two neighborhoods N_1 and N_2 of $\mathbf{x}_0$, say

$$\begin{aligned} N_1 &= N(\mathbf{x}_0; x_1^*, x_2^*, \ldots, x_m^*; \varepsilon_1), \\ N_2 &= N(\mathbf{x}_0; x_{m+1}^*, x_{m+2}^*, \ldots, x_{m+n}^*; \varepsilon_2). \end{aligned} \tag{7.2.4}$$

It is, of course, permitted for some functionals in N_1 to coincide with functionals in N_2. If now $0 < \varepsilon < \min(\varepsilon_1, \varepsilon_2)$ we form

$$N_3 = N(\mathbf{x}_0; x_1^*, \ldots, x_m^*, x_{m+1}^*, \ldots, x_{m+n}^*; \varepsilon). \tag{7.2.5}$$

Here

$$|x_k^*(\mathbf{x} - \mathbf{x}_0)| < \varepsilon,$$

for $k = 1, \ldots, m, m + 1, \ldots, m + n$. This implies

$$|x_i^*(\mathbf{x} - \mathbf{x}_0)| < \varepsilon_1, \qquad |x_j^*(\mathbf{x} - \mathbf{x}_0)| < \varepsilon_2$$

for $i = 1, \ldots, m, j = m + 1, \ldots, m + n$, so that $N_3 \subset N_1 \cap N_2$ and (H_2) is satisfied. If now $\mathbf{y}_0 \in N(\mathbf{x}_0)$, we take

$$N(\mathbf{y}_0) = N(\mathbf{y}_0; x_1^*, \ldots, x_m^*; \delta) \tag{7.2.6}$$

with the same functionals as in $N(\mathbf{x}_0)$ and δ sufficiently small. It suffices to take

$$\delta < \varepsilon - \max_k |x_k^*(\mathbf{x}_0 - \mathbf{y}_0)| \tag{7.2.7}$$

to ensure that $N(\mathbf{y}_0) \subset N(\mathbf{x}_0)$ and postulate (H_3) is satisfied. Verification of (H_4) calls for some knowledge of the existence of functionals meeting specific requirements. If $\mathbf{x}_0$ and $\mathbf{y}_0$ are two distinct elements of $\mathfrak{X}$, then we can find a linear bounded functional x^* such that $x^*(\mathbf{x}_0) \neq x^*(\mathbf{y}_0)$. See part (3) of Theorem 2.3.2 and the discussion in Section 10.3. Set

$$|x^*(\mathbf{x}_0) - x^*(\mathbf{y}_0)| = |x^*(\mathbf{x}_0 - \mathbf{y}_0)| \equiv \delta$$

and define the neighborhoods

$$N(\mathbf{x}_0) = N(\mathbf{x}_0; x^*; \varepsilon), \qquad N(\mathbf{y}_0) = N(\mathbf{y}_0; x^*; \varepsilon), \tag{7.2.8}$$

where $0 < \varepsilon < \frac{1}{2}\delta$. This choice implies $N(\mathbf{x}_0) \cap N(\mathbf{y}_0) = \emptyset$.

It remains to prove continuity of addition and scalar multiplication in the sense of Definition 7.2.2. Suppose that $\mathbf{x}_0$ and $\mathbf{y}_0$ are given together with a neighborhood $N(\mathbf{z}_0)$ of $\mathbf{z}_0 = \mathbf{x}_0 - \mathbf{y}_0$ defined so that $\mathbf{z} \in N(\mathbf{z}_0)$ iff

$$|x_k^*(\mathbf{z} - \mathbf{z}_0)| < \varepsilon, \qquad k = 1, 2, \ldots, n,$$

for a given set of functionals x_k^*. We then set

$$\begin{aligned} N(\mathbf{x}_0) &= [\mathbf{x};\, \mathbf{x} = \mathbf{x}_0 + \tfrac{1}{2}(\mathbf{u} - \mathbf{z}_0),\ \mathbf{u} \in N(\mathbf{z}_0)], \\ N(\mathbf{y}_0) &= [\mathbf{y};\, \mathbf{y} = \mathbf{y}_0 + \tfrac{1}{2}(\mathbf{v} - \mathbf{z}_0),\ \mathbf{v} \in N(\mathbf{z}_0)]. \end{aligned} \tag{7.2.9}$$

Then

$$N(\mathbf{x}_0) - N(\mathbf{y}_0) = [\mathbf{z};\, \mathbf{z} = \mathbf{x} - \mathbf{y} = \mathbf{z}_0 + \tfrac{1}{2}(\mathbf{u} - \mathbf{v}),\ \mathbf{u} \in N(\mathbf{z}_0),\ \mathbf{v} \in N(\mathbf{z}_0)]$$

and

$$x_k^*(\mathbf{x} - \mathbf{y} - \mathbf{z}_0) = \tfrac{1}{2}x_k^*(\mathbf{u}) - \tfrac{1}{2}x_k^*(\mathbf{v}).$$

This difference is $<\varepsilon$ in absolute value so that

$$N(\mathbf{x}_0) - N(\mathbf{y}_0) \subset N(\mathbf{z}_0) = N(\mathbf{x}_0 - \mathbf{y}_0).$$

Without restricting the generality we may assume $\alpha \neq 0$ and $\mathbf{x}_0 \neq \mathbf{0}$ in the scalar product. Let $N(\mathbf{u}_0)$ be a neighborhood of $\mathbf{u}_0 = \alpha\mathbf{x}_0$ where for certain functionals x_k^* we have

$$|x_k^*(\mathbf{u} - \mathbf{u}_0)| < \varepsilon, \qquad k = 1, 2, \ldots, n.$$

Let $m = \max_k |x_k^*(\mathbf{x}_0)|$ and define

$$\begin{aligned} N(\alpha) &= [\beta;\, |\beta - \alpha| < \varepsilon_1], \\ N(\mathbf{x}_0) &= [\mathbf{x};\, |x_k^*(\mathbf{x} - \mathbf{x}_0)| < \varepsilon_2,\ k = 1, \ldots, n] \end{aligned} \tag{7.2.10}$$

where

$$\varepsilon_1 = \frac{\varepsilon}{2m}, \qquad \varepsilon_2 = \frac{m\varepsilon}{2m|\alpha| + \varepsilon}.$$

If $\beta\mathbf{x} \in N(\alpha)\,N(\mathbf{x}_0)$, then

$$\beta\mathbf{x} - \alpha\mathbf{x}_0 = (\beta - \alpha)\mathbf{x}_0 + \beta(\mathbf{x} - \mathbf{x}_0).$$

Hence

$$x_k^*(\beta\mathbf{x} - \alpha\mathbf{x}_0) = (\beta - \alpha)\, x_k^*(\mathbf{x}_0) + \beta x_k^*(\mathbf{x} - \mathbf{x}_0),$$

the absolute value of which does not reach ε by the choice of $N(\alpha)$ and $N(\mathbf{x}_0)$, whence it follows that $N(\alpha)\,N(\mathbf{x}_0) \subset N(\alpha\mathbf{x}_0)$ and that $\mathfrak{X}$ is a linear Hausdorff space in the chosen neighborhood topology. ∎

Some comments are in order. The family $\{N_\alpha\}$ of neighborhoods generate a topology on $\mathfrak{X}$ in which the open sets are arbitrary unions of the sets N_α. Note that by virtue of (H_1), (H_2), and (H_3), the finite intersection of open sets is open; the space $\mathfrak{X}$ is open and so is the empty set. Call this topology the *weak topology*. Note, furthermore, that all the sets $N(\mathbf{x}_0; x_1^*, \ldots, x_n^*; \varepsilon)$ are open, in fact, the class of all these sets with $\mathbf{x}_0$ fixed is a local basis at $\mathbf{x}_0$ in the weak topology. It is asked in Problem 4, Exercise 7.2, to prove the weak topology in $\mathfrak{X}$ is the weakest topology in which each member x^* of $\mathfrak{X}^*$ is continuous.

While functionals, addition, and scalar multiplication are continuous in the weak topology, such a simple function as the norm of $\mathbf{x}$ is not continuous in $\mathbf{x}$ in the weak topology. This follows from the peculiar structure of the neighborhoods. For any $\mathbf{y} \in \mathfrak{X}$ we can find a linear bounded functional x^* such that $x^*(\mathbf{y}) = 0$. We exclude the trivial case $\mathbf{y} = \mathbf{0}$. It follows that

$$\mathbf{x}_0 + \alpha \mathbf{y} \in N(\mathbf{x}_0; x^*; \varepsilon) \tag{7.2.11}$$

for any choice of α, ε, $\mathbf{x}_0$. Thus this neighborhood of $\mathbf{x}_0$ contains elements of $\mathfrak{X}$ of arbitrarily large norm and this precludes continuity of the norm in the weak topology.

We have now various notions of convergence and completeness associated with the weak topology.

Definition 7.2.3. *The sequence $\{\mathbf{x}_n\} \subset \mathfrak{X}$ is said to be weakly convergent if $\{x^*(\mathbf{x}_n)\}$ is a convergent sequence for each choice of $x^* \in \mathfrak{X}^*$; it is weakly convergent to $\mathbf{x}_0 \in \mathfrak{X}$ if $\lim_{n\to\infty} x^*(\mathbf{x}_n) = x^*(\mathbf{x}_0)$ for all x^*.*

It is clear that a strongly convergent sequence is also weakly convergent, but the converse is normally not true. Thus in $L_2(-\pi, \pi)$ the sequence $\{e^{nit}\}$ converges weakly to zero but not strongly. We have

$$x^*[e^{nit}] = \int_{\pi}^{\pi} e^{nit} g(t)\, dt \tag{7.2.12}$$

for some $g \in L_2$ depending upon x^* and the Fourier coefficients of any function in L_2 go to zero. On the other hand, the norm of e^{nit} is $(2\pi)^{1/2}$ for all n.

Definition 7.2.4. *A B-space is said to be weakly complete if every weakly convergent sequence in $\mathfrak{X}$ converges weakly to an element of $\mathfrak{X}$.*

There is a class of B-spaces which have the property of weak completeness, namely the *reflexive spaces*, so named by E. R. Lorch in 1939. The earlier name was the much-overworked term *regular*, due to Hans Hahn (1879–1934) in 1927. To explain this notion we must digress on (conjugate) dual spaces.

We note that $\mathfrak{X}^*$ being a B-space also has a dual space denoted by $\mathfrak{X}^{**}$ and called the *second dual* (second conjugate) of $\mathfrak{X}$. Its elements are denoted by symbols like x^{**}. Actually we can specify some of the elements of $\mathfrak{X}^{**}$ by taking a sharper look at the symbol $x^*(\mathbf{x})$. This expression is linear in each of its two arguments $\mathbf{x}$ and x^*. For a fixed $\mathbf{x}$ we can vary x^* over $\mathfrak{X}^*$. This is now a linear functional on $\mathfrak{X}^*$. Thus we can define

$$x^{**}(x^*) = x^*(\mathbf{x}). \tag{7.2.13}$$

This functional on X^* is linear. It is also bounded for

$$|x^{**}(x^*)| = |x^*(\mathbf{x})| \leqslant \|x^*\| \; \|\mathbf{x}\|,$$

so that $\|x^{**}\| \leqslant \|\mathbf{x}\|$. Actually equality must hold here, for by Theorem 2.3.2 there is an $x^* \in \mathfrak{X}^*$ such that $\|x^*\| = 1$ and $x^*(\mathbf{x}) = \|\mathbf{x}\|$. Thus x^{**} is a linear bounded functional on $\mathfrak{X}^*$ and an element of $\mathfrak{X}^{**}$.

Formula (7.2.13) defines a mapping of $\mathfrak{X}$ onto a subset $\mathfrak{X}_0$ of $\mathfrak{X}^{**}$. This mapping is an isometry, $\|\mathbf{x}\| = \|x^{**}\|$ and thus 1–1 and onto. Further, $\mathbf{x} \leftrightarrow x^{**}$, $\mathbf{y} \leftrightarrow y^{**}$ implies $\alpha\mathbf{x} + \beta\mathbf{y} \leftrightarrow \alpha x^{**} + \beta y^{**}$. Thus the mapping is an isomorphic and isometric homeomorphism. This gives

***Theorem** 7.2.2.* $\mathfrak{X} \subseteq \mathfrak{X}^{**}$.

***Definition** 7.2.5.* *$\mathfrak{X}$ is said to be reflexive if $\mathfrak{X}^{**} = \mathfrak{X}$.*

Let us consider in passing an application of the uniform boundedness principle based on the second dual.

***Theorem** 7.2.3.* *If for every $\mathbf{x}$ in some set $S \subset \mathfrak{X}$ and for every $x^* \in \mathfrak{X}^*$ we have*

$$\sup_{\mathbf{x} \in S} |x^*(\mathbf{x})| < \infty, \tag{7.2.14}$$

then S is a bounded set.

Proof. We use the standard embedding of $\mathfrak{X}$ into $\mathfrak{X}^{**}$. Then for $\mathbf{x}$ fixed in S

$$x^*(\mathbf{x}) = x^{**}(x^*) = \mathbf{y}(x^*),$$

where $\mathbf{y}$ is an element of $\mathfrak{X}$ uniquely determined by $\mathbf{x}$. Thus for each $\mathbf{x} \in S$ there is a unique linear bounded transformation $\mathbf{y}$ on $\mathfrak{X}^*$ to C and, furthermore,

$$\sup_{\mathbf{y}} |\mathbf{y}(x^*)| < \infty$$

for each $x^* \in \mathfrak{X}^*$. By Theorem 7.1.5 this implies that the norms of this family of linear transformations are uniformly bounded. Hence there is an M such that

$$\|\mathbf{y}\| = \|x^{**}\| = \|\mathbf{x}\| \leqslant M$$

for all $\mathbf{x} \in S$. This shows that S is bounded. ∎

***Corollary** 1.* *A weakly convergent sequence in $\mathfrak{X}$ is bounded.*

For $\sup_n |x^*(\mathbf{x}_n)| < \infty$ for each x^*.

We can go one step further using

***Definition** 7.2.6.* *A subset S of a B-space is said to be sequentially weakly compact if every sequence of elements in S contains a subsequence which converges weakly to an element of S.*

***Corollary** 2.* *A sequentially weakly compact subset S of a B-space $\mathfrak{X}$ is bounded.*

Proof. The assumption implies that for each fixed $x^* \in \mathfrak{X}^*$ the set of complex numbers $\{x^*(\mathbf{x});\ \mathbf{x} \in S\}$ is sequentially compact. Now a set of complex numbers has this property iff it is bounded and closed. The boundedness condition is simply (7.2.14) so the conclusion follows. ∎

Theorem 7.2.4. *A reflexive space is weakly complete.*

Proof. Let $\{\mathbf{x}_n\}$ be a weakly convergent sequence in the reflexive space $\mathfrak{X}$. For any given $\varepsilon > 0$ there is then an integer N depending upon ε and the functional $x^* \in X^*$ such that

$$|x^*(\mathbf{x}_m) - x^*(\mathbf{x}_n)| < \varepsilon;\ m, n > N. \tag{7.2.15}$$

Since $\mathfrak{X}$ is reflexive, the embedding mapping $\mathfrak{X} \leftrightarrow \mathfrak{X}^{**} = \mathfrak{X}$ leads to a sequence $\{\mathbf{y}_n\} \subset \mathfrak{X}$ such that $\mathbf{x}_n \leftrightarrow \mathbf{y}_n$ and we have

$$x^*(\mathbf{x}_n) = \mathbf{y}_n(x^*). \tag{7.2.16}$$

Since for each x^*

$$|\mathbf{y}_m(x^*) - \mathbf{y}_n(x^*)| < \varepsilon,\ m, n > N,$$

it is seen that the sequence of complex numbers $\{\mathbf{y}_n(x^*)\}$ is convergent so that

$$\lim_{n \to \infty} \mathbf{y}_n(x^*) \tag{7.2.17}$$

exists for every x^*. By Theorem 7.2.3 this implies the existence of a finite positive M such that

$$\|\mathbf{y}_n\| \leqslant M. \tag{7.2.18}$$

Note that each $\mathbf{y}_n$ defines a linear bounded transformation from $\mathfrak{X}^*$ to C and the norms of these transformations are uniformly bounded by Theorem 7.2.3. Set

$$\lim_{n \to \infty} \mathbf{y}_n(x^*) = \mathbf{y}_0(x^*). \tag{7.2.19}$$

Here $\mathbf{y}_0$ defines a linear transformation on $\mathfrak{X}^*$ to C which is bounded by (7.2.18) so that $\mathbf{y}_0 \in \mathfrak{X}^{**} = \mathfrak{X}$. Now in the mapping $\mathbf{x} \leftrightarrow x^{**}$ there is a unique element $\mathbf{x}_0 \in \mathfrak{X}$ which corresponds to $\mathbf{y}_0$ and

$$\mathbf{y}_0(x^*) = x^*(\mathbf{x}_0). \tag{7.2.20}$$

Finally, for fixed x^*

$$\varepsilon \geqslant \lim_{m \to \infty} |\mathbf{y}_m(x^*) - \mathbf{y}_n(x^*)| = |\mathbf{y}_0(x^*) - \mathbf{y}_n(x^*)| = |x^*(\mathbf{x}_0) - x^*(\mathbf{x}_n)|.$$

Since ε is arbitrary, this shows that $\mathbf{x}_n$ converges weakly to $\mathbf{x}_0$, which is often written

$$\mathbf{x}_n \xrightarrow{\text{w}} \mathbf{x}_0. \tag{7.2.21}$$

Thus it has been shown that in a reflexive space a weakly convergent sequence has a weak limit. ∎

Theorem 7.2.5. *A B-space $\mathfrak{X}$ is reflexive iff its dual $\mathfrak{X}^*$ is reflexive.*

Proof. If $\mathfrak{X}$ is reflexive, then $\mathfrak{X} = \mathfrak{X}^{**}$ in the natural embedding. On the other hand,

$$\mathfrak{X}^* \subseteq (\mathfrak{X}^*)^{**} = (\mathfrak{X}^{**})^* = \mathfrak{X}^*. \tag{7.2.22}$$

Since the second dual of $\mathfrak{X}^*$ cannot be a proper subspace of $\mathfrak{X}^*$ under the natural embedding, the inclusion in (7.2.22) must reduce to equality so that $\mathfrak{X}^*$ is also reflexive.

Suppose now that $\mathfrak{X}^*$ is reflexive. Then so is $\mathfrak{X}^{**}$. Now we have $\mathfrak{X} \subseteq \mathfrak{X}^{**}$ and since $\mathfrak{X}^* = (\mathfrak{X}^{**})^*$, the dual spaces of $\mathfrak{X}^*$ and $\mathfrak{X}^{**}$ are identical. If $\mathfrak{X}$ were a proper linear subspace of $\mathfrak{X}^{**}$, then by Theorem 2.3.2 there would exist a linear bounded functional on $\mathfrak{X}^{**}$ of norm one which is zero on $\mathfrak{X}$. The statement that $\mathfrak{X}$ and $\mathfrak{X}^{**}$ have the same linear functionals must imply that there is one and only one linear bounded functional which takes on given values on $\mathfrak{X}$. Such a functional can be extended to all of $\mathfrak{X}^{**}$ in one and only one way. Now the zero functional on $\mathfrak{X}$ admits as extension to $\mathfrak{X}^{**}$ the zero functional on $\mathfrak{X}^{**}$. Since there is a unique extension our functional must be the zero functional on $\mathfrak{X}^{**}$. This shows that $\mathfrak{X}$ cannot be a proper linear subspace of $\mathfrak{X}^{**}$. Thus $\mathfrak{X} = \mathfrak{X}^{**}$ and $\mathfrak{X}$ is reflexive. ∎

We state without proof one more important property of reflexive spaces.

Theorem 7.2.6. *The closed unit sphere of a reflexive space is weakly sequentially compact.*

We note that a Hilbert space is reflexive by virtue of Theorem 2.5.11.

The rest of this section will be devoted to a brief discussion of operator topologies. If $\mathfrak{X}$ is a complex B-space we denote by $\mathfrak{E}(\mathfrak{X})$ as usual the B-space of all linear bounded transformations T from $\mathfrak{X}$ into $\mathfrak{X}$. We have available at the outset the normed operator topology on $\mathfrak{E}(\mathfrak{X})$. In this context this is usually referred to as the *uniform operator topology* based on the distance

$$d(T_1, T_2) = \|T_1 - T_2\|. \tag{7.2.23}$$

But in addition we can introduce various neighborhoods in analogy with Definition 7.2.3 above.

Definition 7.2.7. *A strong operator neighborhood of $T_0 \in \mathfrak{E}(\mathfrak{X})$ is any set of the form*

$$N(T_0; \mathbf{x}_1, \ldots, \mathbf{x}_n; \varepsilon) = (T; \|T(\mathbf{x}_k) - T_0(\mathbf{x}_k)\| < \varepsilon; k = 1, \ldots, n),$$

where $\mathbf{x}_1, \ldots, \mathbf{x}_n$ are n arbitrary elements of $\mathfrak{X}$.

A weak operator neighborhood of T_0 is any set of the form

$$N(T_0; \mathbf{x}_1, \ldots, \mathbf{x}_m; x_1^*, \ldots, x_n^*; \varepsilon) = [T; |x_j^*[T(\mathbf{x}_k)] - x_j^*[T_0(\mathbf{x}_k)] < \varepsilon],$$

where $\mathbf{x}_1, \ldots, \mathbf{x}_m$ and $x_1^*, \ldots, x_n^*$ are arbitrary elements of $\mathfrak{X}$ and $\mathfrak{X}^*$ respectively.

The set of all strong operator neighborhoods constitutes a neighborhood covering of $\mathfrak{E}(\mathfrak{X})$ in terms of which this space becomes a linear Hausdorff space in the sense of Definitions 7.2.1 and 7.2.2. The same holds for the set of weak operator neighborhoods. The argument, which largely parallels the proof of Theorem 7.2.1, is left to the reader.

Now $\mathfrak{E}(\mathfrak{X})$ is not merely a B-space, it is a B-algebra in terms of the uniform topology. This raises the question of corresponding structural properties of $\mathfrak{E}(\mathfrak{X})$ in terms of the strong and the weak operator topologies, respectively. Here the basic concept is that of a *topological algebra.* Since we restrict ourselves to Hausdorff topologies, we may limit ourselves to *Hausdorff algebras.*

***Definition** 7.2.8. An algebra $\mathfrak{A}$ is a Hausdorff algebra if* (1) *$\mathfrak{A}$ is an algebra as well as a linear Hausdorff space and if* (2) *to every pair $\mathbf{x}, \mathbf{y} \in \mathfrak{A}$ and each neighborhood $N(\mathbf{xy})$ of $\mathbf{xy}$ there are neighborhoods $N(\mathbf{x})$ of $\mathbf{x}$ and $N(\mathbf{y})$ of $\mathbf{y}$ such that $\mathbf{x}N(\mathbf{y}) \subset N(\mathbf{xy})$ and $N(\mathbf{x})\mathbf{y} \subset N(\mathbf{xy})$.*

It should be noted that $\mathfrak{A}$ is normally non-commutative as in the case $\mathfrak{E}(\mathfrak{X})$ under consideration. Hence we have to write $\mathbf{x}N(\mathbf{y})$ and $N(\mathbf{x})\mathbf{y}$. The first is the set of products $\mathbf{xv}$ with $\mathbf{v} \in N(\mathbf{y})$, the second the set of products $\mathbf{uy}$ with $\mathbf{u} \in N(\mathbf{x})$. Note also that we do not demand that $N(\mathbf{x})\,N(\mathbf{y}) \subset N(\mathbf{xy})$. We have now

***Theorem** 7.2.7. $\mathfrak{E}(\mathfrak{X})$ is a Hausdorff algebra in terms of the strong operator topology as well as in terms of the weak one.*

The proof is left to the reader.

EXERCISE 7.2

1. Verify that condition (7.2.7) suffices for the stated purpose, $N(\mathbf{y}_0) \subset N(\mathbf{x}_0)$.
2. Same question for (7.2.8).
3. Fill in missing details in the proof of $N(\mathbf{x}) - N(\mathbf{y}) \subset N(\mathbf{x} - \mathbf{y})$ and of $N(\alpha)\,N(\mathbf{x}) \subset N(\alpha\mathbf{x})$.
4. Let $\mathfrak{T}$ be the weak topology on a given B-space $\mathfrak{X}$. Prove that (a) With respect to T, any $x^* \in \mathfrak{X}^*$ is continuous, and (*b*) On $\mathfrak{X}$ let U be a topology in which each $x^* \in \mathfrak{X}^*$ is continuous, then $\mathsf{T} \subseteq \mathsf{U}$. Here the inclusion asserts that every set open with respect to U is open with respect to T.
5. Construct the weak topology for the space $\mathfrak{M}_n$ of n by n matrices of complex numbers.
6. Show that $\mathfrak{M}_n$ is a Hausdorff algebra in the sense of Definition 7.2.8 in the weak topology.
7. Show that the sequence space l_p is reflexive for $1 < p < \infty$.
8. Prove that the Lebesgue space $L_p(a, b)$ is reflexive for $1 < p < \infty$.
9. What is wrong with $p = 1$ in the preceding problems?

10. Show that the sequence $\{t^n; n = 1, 2, \dots\}$ in $C[0, 1]$ is weakly convergent but has no weak limit. [*Hint*: Recall the form of the linear bounded functionals on this space.]
11. Prove Theorem 7.2.6.
12. From Theorem 7.2.6 and the result in Problem 10 conclude that $C[0, 1]$, more generally $C[a, b]$, is not reflexive.
13. Is C^n reflexive?
14. Taking $\mathfrak{M}_n$ as $\mathfrak{E}(C^n)$, discuss the strong and the weak operator algebras on C^n.
15. Prove for a general $\mathfrak{E}(\mathfrak{X})$ that it is a linear Hausdorff space in terms of the strong operator topology.
16. Same question for the weak operator topology.
17. Prove Theorem 7.2.7 for the strong operator topology.
18. Give a definition of the strong operator topology for the space $\mathfrak{E}(\mathfrak{X}, \mathfrak{Y})$.
19. Same question for the weak operator topology.

7.3 VECTOR-VALUED AND OPERATOR-VALUED FUNCTIONS

We shall be concerned with mappings of R^1 into either a B-space $\mathfrak{X}$ over the complex field or into a space $\mathfrak{E}(\mathfrak{X}, \mathfrak{Y})$ of linear bounded operators from $\mathfrak{X}$ into $\mathfrak{Y}$, two B-spaces over the complex field. A function of the first kind will be called a *vector function* and denoted by $\mathbf{x}(s)$ or similar symbol. Functions of the second kind are called *operator functions* and denoted by $T(s)$, $U(s)$ or the like. We can look upon such functions as a family of elements of $\mathfrak{X}$ or of $\mathfrak{E}(\mathfrak{X}, \mathfrak{Y})$, respectively, where the individual elements are indexed by a real parameter s.

Such objects have occurred repeatedly in previous chapters of this treatise. Thus an n by n matrix $\mathcal{A}(s) = (a_{jk}(s))$, where the elements are complex-valued functions of a real variable s, occurs in a number of places, as solutions of implicit matrix equations, as solutions of matrix differential equations, or as the resolvent of some constant matrix. Here $\mathcal{A}(s)$ figures as a vector function in the B-space $\mathfrak{M}_n$. But the same matrix-valued function may be regarded as an element of $\mathfrak{E}(C^n)$ defining a vector function $\mathbf{y}(s) \in C^n$ by

$$\mathbf{y}(s) = \mathcal{A}(s)\,\mathbf{x},$$

where $\mathbf{x} \in C^n$. In this setting, $\mathcal{A}(s)$ is an operator function.

Now vector and operator functions may have continuity properties with respect to s. Since we admit two types of topology in $\mathfrak{X}$ and three in $\mathfrak{E}(\mathfrak{X})$ (the same types apply in $\mathfrak{E}(\mathfrak{X}, \mathfrak{Y})$), we have a corresponding variety of notions of continuity.

Definition 7.3.1. *A vector function* $\mathbf{x}(s)$ *which is defined on a subset* S *of* R^1 *with values in the B-space* $\mathfrak{X}$ *is* (1) *weakly continuous at* $s = s_0 \in S$ *if* $\lim_{s \to s_0} |x^*[\mathbf{x}(s)] - x^*[\mathbf{x}(s_0)]| = 0$ *for each* $x^* \in \mathfrak{X}^*$, (2) *strongly continuous at* s_0 *if* $\lim_{s \to s_0} \|\mathbf{x}(s) - \mathbf{x}(s_0)\| = 0$.

Definition 7.3.2. *An operator function $T(s)$ from $S \subset R^1$ to $\mathfrak{E}(\mathfrak{X}, \mathfrak{Y})$ is continuous at $s = s_0$ in the sense of* (1) *the weak operator topology if* $\lim_{s \to s_0} |y^*[T(s)\mathbf{x}] - y^*[T(s_0)\mathbf{x}]| = 0$ *for each* $\mathbf{x} \in \mathfrak{X}$, $y^* \in \mathfrak{Y}^*$, (2) *the strong operator topology if* $\lim_{s \to s_0} \|T(s)\mathbf{x} - T(s_0)\mathbf{x}\| = 0$ *for each* $\mathbf{x} \in \mathfrak{X}$, *and* (3) *the uniform operator topology if* $\lim_{s \to s_0} \|T(s) - T(s_0)\| = 0$. *If no confusion is likely to arise, we speak of weak, strong or uniform continuity in these cases.*

In general these notions are distinct. In this respect the special case mentioned above is misleading for the weak continuity of the vector function $\mathcal{A}(s)$ implies strong continuity and the weak continuity of the operator function $\mathcal{A}(s)$ implies strong continuity, which in its turn implies uniform continuity.

For differentiability we have a similar wealth of possibilities. We state the definition for vector functions and let the reader formulate corresponding notions for operators.

Definition 7.3.3. *A vector function $\mathbf{x}(s)$ from the interval (a, b) to the B-space $\mathfrak{X}$ is weakly (strongly) differentiable at $s = s_0$ if there is an element $\mathbf{x}'(s_0) \in \mathfrak{X}$ such that the difference quotient $h^{-1}[\mathbf{x}(s_0 + h) - \mathbf{x}(s_0)]$ tends weakly (strongly) to $\mathbf{x}'(s_0)$ as $h \to 0$. We call $\mathbf{x}'(s_0)$ the weak (strong) derivative of $\mathbf{x}(s)$ at $s = s_0$.*

Here there is still another possibility. We could demand that all the functionals $x^*[\mathbf{x}(s)]$ are differentiable at $s = s_0$. This will be the case if $\mathbf{x}(s)$ is weakly differentiable at $s = s_0$, but the stated condition does not imply the existence of a weak derivative.

Theorem 7.3.1. *If the weak derivative of $\mathbf{x}(s)$ exists and is zero everywhere in (a, b), then $\mathbf{x}(s)$ is a constant.*

Proof. The assumption is that

$$\frac{d}{ds} x^*[\mathbf{x}(s)] = 0, \qquad \forall\, x^* \in \mathfrak{X}^*.$$

Hence

$$x^*[\mathbf{x}(s)] = x^*[\mathbf{x}(s_0)]$$

for a fixed $s_0 \in (a, b)$. This says that

$$x^*[\mathbf{x}(s) - \mathbf{x}(s_0)] = 0, \qquad \forall\, x^* \in \mathfrak{X}^*.$$

But the zero element of $\mathfrak{X}$ is the only one that can be annihilated by all functionals $x^* \in \mathfrak{X}^*$. This gives $\mathbf{x}(s) = \mathbf{x}(s_0)$, a constant. ∎

We shall have to consider Riemann–Stieltjes integrals with a vector- or operator-valued function as integrator. This requires the notion of abstract-valued functions of bounded variation. We give the definitions only for vector-valued functions.

Definition 7.3.4. *A vector function* $\mathbf{x}(s)$ *from the interval* $[a, b]$ *to the B-space* $\mathfrak{X}$ *is of* (1) *weak bounded variation in* $[a, b]$ *if* $x^*[\mathbf{x}(s)] \in BV[a, b]$ *for every* $x^* \in \mathfrak{X}^*$, (2) *bounded variation if* $\sup \| \Sigma_j [\mathbf{x}(t_j) - \mathbf{x}(s_j)] \| < \infty$ *for every choice of a finite number of non-overlapping intervals* (s_j, t_j) *in* $[a, b]$ *and* (3) *strong bounded variation if* $\sup \Sigma_j \|\mathbf{x}(s_j) - \mathbf{x}(s_{j-1})\| < \infty$ *where all possible partitions of* $[a, b]$ *are considered. The two suprema are known as the total and the strong total variation of* $\mathbf{x}(s)$ *respectively.*

We leave it to the reader to verify that strong bounded variation implies bounded variation and that the latter implies weak bounded variation.

Theorem 7.3.2. *If* $\mathbf{x}(s)$ *is of strong bounded variation in* $[a, b]$, *then* $\mathbf{x}(s)$ *can have only a countable number of discontinuities and has one-sided limits everywhere in* $[a, b]$.

Proof. Let $V(t)$ denote the strong total variation of $\mathbf{x}(s)$ in the interval $[a, t]$, $a < t \leqslant b$. This is a real non-negative, non-decreasing bounded function of t. As such it has left and right hand limits everywhere in (a, b), a right hand limit at $t = a$, and a left hand limit at $t = b$. Further, $V(t)$ can have only a countable number of discontinuities. Since

$$0 \leqslant \|\mathbf{x}(s_2) - \mathbf{x}(s_1)\| \leqslant V(s_2) - V(s_1), \qquad s_1 < s_2, \tag{7.3.1}$$

we see that $\mathbf{x}(s)$ is left hand (right hand) continuous wherever $V(s)$ has this property. In particular, $\mathbf{x}(s)$ has at most a countable number of discontinuities. We have to show that one-sided limits exist.

Let $s = s_0$ be a discontinuity of $V(s)$. To fix the ideas, suppose that $V(s_0 + 0) - V(s_0) = \Delta > 0$. Suppose that $0 < \alpha < \beta$ where β is small. $V(s)$ need not be continuous in $[s_0 + \alpha, s_0 + \beta]$, but its increase in the interval is small as soon as β is small regardless of the value of $\alpha > 0$. Suppose that

$$V(s_0 + \beta) - V(s_0 + \alpha) < \delta \tag{7.3.2}$$

for all $\alpha > 0$. We now take two sequences $\{s_n\}$ and $\{t_n\}$ in $(s_0, s_0 + \beta)$ which both descend to s_0 and consider the corresponding sequences $\{\mathbf{x}(s_n)\}$ and $\{\mathbf{x}(t_n)\}$. It is desired to prove that they are convergent and, in fact, converge to the same limit in $\mathfrak{X}$.

Convergence follows from the inequality

$$\sum_m^n \|\mathbf{x}(s_{k-1}) - \mathbf{x}(s_k)\| \leqslant V(s_{m-1}) - V(s_n) < \delta,$$

which holds for all $n > m$. Since δ is arbitrary, this implies that $\lim_{k\to\infty} \mathbf{x}(s_k) \equiv \mathbf{x}_1$ exists and the existence of $\lim_{k\to\infty} \mathbf{x}(t_k) \equiv \mathbf{x}_2$ is proved in the same manner.

If, now, $\mathbf{x}_1 \neq \mathbf{x}_2$, we can use the fact that $\mathbf{x}(s_n)$ is close to $\mathbf{x}_1$ while $\mathbf{x}(t_n)$ is close to $\mathbf{x}_2$ to infer that an integer k, arbitrarily large, may be found such that

$$\|\mathbf{x}(s_k) - \mathbf{x}_1\| < \tfrac{1}{3}\|\mathbf{x}_2 - \mathbf{x}_1\|, \qquad \|\mathbf{x}(t_k) - \mathbf{x}_2\| < \tfrac{1}{3}\|\mathbf{x}_2 - \mathbf{x}_1\|.$$

Then

$$\begin{aligned}\|\mathbf{x}(t_k) - \mathbf{x}(s_k)\| &= \|\mathbf{x}(t_k) - \mathbf{x}_2 + \mathbf{x}_2 - \mathbf{x}_1 + \mathbf{x}_1 - \mathbf{x}(s_k)\| \\ &\geqslant \|\mathbf{x}_2 - \mathbf{x}_1\| - \|\mathbf{x}(t_k) - \mathbf{x}_2\| - \|\mathbf{x}(s_k) - \mathbf{x}_1\| \\ &> \tfrac{1}{3}\|\mathbf{x}_2 - \mathbf{x}_1\|.\end{aligned}$$

On the other hand,

$$\|\mathbf{x}(t_k) - \mathbf{x}(s_k)\| < \delta$$

by the choice of the sequences. Here δ is arbitrarily small so that the assumption $\mathbf{x}_1 \neq \mathbf{x}_2$ leads to a contradiction. It follows that $\mathbf{x}(s)$ has a unique right hand limit at $s = s_0$ and the existence and uniqueness of left hand limits is proved in the same manner. ∎

Weak and strong properties are never independent of each other. In the present case we know that bounded variation implies weak bounded variation. As shown by N. Dunford and I. Gelfand, independently of each other, in 1938, the converse is also true.

Theorem 7.3.3. *A function of weak bounded variation is of bounded variation.*

Proof. The argument is an application of the uniform boundedness principle. Let $V[f]$ denote the total variation in $[a, b]$ of a numerically valued function $f \in BV[a, b]$. We have then for any choice of a finite number of non-overlapping intervals $[a_j, b_j]$ in $[a, b]$

$$\left| x^* \left\{ \sum_j [\mathbf{x}(b_j) - \mathbf{x}(a_j)] \right\} \right| \leqslant \sum_j |x^*[\mathbf{x}(b_j)] - x^*[\mathbf{x}(a_j)]| \leqslant V\{x^*[\mathbf{x}(s)]\}.$$

The last member is finite for every $x^* \in \mathfrak{X}^*$. We can then use Theorem 7.2.3 to conclude that there is an $M > 0$ such that

$$\left\| \sum_j [\mathbf{x}(b_j) - \mathbf{x}(a_j)] \right\| \leqslant M$$

for each choice of intervals $[a_j, b_j]$ and this shows that $\mathbf{x}(s)$ is actually of bounded variation. ∎

On the other hand, it may be shown by examples that a function of bounded variation need not be of strong bounded variation. See Problem 7, Exercise 7.3.

EXERCISE 7.3

1. If $\mathcal{A}(s) = (a_{jk}(s))$ is an n by n matrix where $a_{jk}(s)$ is a complex-valued function defined for $a \leqslant s \leqslant b$, consider $\mathcal{A}(s)$ as a vector function with values in $\mathfrak{M}_n$. When is $\mathcal{A}(s)$ (1) weakly continuous, (2) strongly continuous?

2. Consider $\mathcal{A}(s)$ as an element of $\mathfrak{E}(C^n)$ and find necessary and sufficient conditions for weak, strong, and uniform continuity.
3. Prove that a weakly (strongly) differentiable function $\mathbf{x}(s)$ is weakly (strongly) continuous.
4. Prove that if $\mathbf{x}(s)$ is weakly continuous in $[a, b]$, then $\|\mathbf{x}(s)\|$ is bounded in $[a, b]$.
5. Prove that strong bounded variation implies bounded variation which implies weak bounded variation.
6. Consider a partition of the interval [0, 1] by $n + 1$ points $s_0 = 0 < s_1 < s_2 < \dots < s_n = 1$ and form the sum

$$\sum_{k=1}^{n} (s_k - s_{k-1})^{\frac{1}{2}}.$$

For n fixed, find the supremum of all such sums and show that this is an unbounded function of n.

7. Define $\mathbf{x}(s)$ for $0 \leqslant s \leqslant 1$ as the characteristic function of the interval $[0, s]$ so that $\mathbf{x}(s) = f(s, t)$ equals 1 for $0 \leqslant t \leqslant s$ and 0 elsewhere. Let $\mathfrak{X}$ be the space $L_2(0, 1)$. Show that $\mathbf{x}(s)$ is of weak bounded variation in [0, 1] and hence of bounded variation. Use the preceding problem to show that $\mathbf{x}(s)$ is not of strong bounded variation.
8. Is the situation different in $L(0, 1)$?

Let $\mathbf{x}_n \in \mathfrak{X}, \forall\, n$, and set $\mathbf{S}_n = \Sigma_{k=1}^{n} \mathbf{x}_k$. The infinite series $\Sigma\, \mathbf{x}_k$ is said to *converge weakly* (*strongly*) [*to* $\mathbf{s}$] *if the sequence* $\{\mathbf{S}_n\}$ *converges weakly* (*strongly*) [*to* $\mathbf{s}$]. The series is *absolutely convergent* or *convergent in norm if* $\Sigma\, \|\mathbf{x}_k\|$ *converges*, *unconditionally weakly* (*strongly*) *convergent if every subseries is weakly* (*strongly*) *convergent*.

9. If $\Sigma\, \mathbf{x}_k$ converges weakly, show that $\|\mathbf{S}_n\|$ is uniformly bounded.
10. Show that a numerical series is unconditionally convergent iff it is absolutely convergent.
11. Show that a weakly unconditionally convergent series is necessarily strongly unconditionally convergent.
12. Show that in the space l_1 weak convergence implies strong convergence.
13. A series $\Sigma\, \mathbf{x}_k(s)$, $a \leqslant s \leqslant b$, with elements in $\mathfrak{X}$ may have convergence properties holding uniformly with respect to s. Formulate a definition of uniformly weak convergence, etc.
14. Do the same for operator series $\Sigma\, T_k(s)$.

7.4 ABSTRACT RIEMANN–STIELTJES INTEGRALS

Such an integral involves two functions: an *integrand* and an *integrator*. Suppose that one of these functions has values in a B-space $\mathfrak{X}$ which is not a B-algebra. Then the other factor is normally restricted to be numerically valued. We have two types of integral according as the integrand or the integrator takes on values in $\mathfrak{X}$. The integral, just as in the classical case, is defined as the limit of a class of Riemann–Stieltjes sums. Here are various possibilities corresponding to the several topologies available in $\mathfrak{X}$ that may be used to define the limit. Actually the two types are coexisting for the same choice of limiting process, as may be shown by an obvious generalization of the classical formula for integration by parts.

Basically we start with a vector function $\mathbf{f}(s)$ with values in $\mathfrak{X}$ and a numerically valued function $g(s)$, both defined in an interval $[a, b]$, where they are to satisfy conditions to be specified later. We consider a partition of $[a, b]$

$$s_0 = a \leqslant s_1 \leqslant s_2 \leqslant \cdots \leqslant s_n = b \tag{7.4.1}$$

and a choice of intermediary points $\{t_j\}$ with

$$s_{j-1} \leqslant t_j \leqslant s_j, \qquad j = 1, 2, \ldots, n. \tag{7.4.2}$$

The collection of these $2n + 1$ real numbers is denoted by π and we set $|\pi| = \max_j (s_j - s_{j-1})$. We then form the two Riemann–Stieltjes sums

$$S_\pi(\mathbf{f}, g) = \sum_{j=1}^{n} \mathbf{f}(t_j)\,[g(s_j) - g(s_{j-1})], \tag{7.4.3}$$

$$S_\pi(g, \mathbf{f}) = \sum_{j=1}^{n} g(t_j)\,[\mathbf{f}(s_j) - \mathbf{f}(s_{j-1})], \tag{7.4.4}$$

and consider either the strong topology in $\mathfrak{X}$ or the weak one.

Definition 7.4.1. *If in the chosen topology $S_\pi(\mathbf{f}, g)$ should have a limit as $|\pi| \to 0$, then this limit is the Riemann–Stieltjes integral*

$$\int_a^b \mathbf{f}(s)\,dg(s) \tag{7.4.5}$$

with respect to this topology.

Definition 7.4.2. *If in the chosen topology $S_\pi(g, \mathbf{f})$ should have a limit as $|\pi| \to 0$, then this limit is the Riemann–Stieltjes integral*

$$\int_a^b g(s)\,d\mathbf{f}(s) \tag{7.4.6}$$

with respect to this topology.

As indicated above, we have

Theorem 7.4.1. *If either integral exists with respect to the chosen topology, then both exist in this topology and*

$$\int_a^b \mathbf{f}(s)\,dg(s) = \mathbf{f}(b)\,g(b) - \mathbf{f}(a)\,g(a) - \int_a^b g(s)\,d\mathbf{f}(s). \tag{7.4.7}$$

Proof. As in the classical case, we have

$$\sum_{j=1}^{n} \mathbf{f}(t_j)\,[g(s_j) - g(s_{j-1})] = \mathbf{f}(b)\,g(b) - \mathbf{f}(a)\,g(a) - \sum_{j=0}^{n} g(s_j)\,[\mathbf{f}(t_{j+1}) - \mathbf{f}(t_j)], \tag{7.4.8}$$

where $t_0 = a$, $t_{n+1} = b$. The second sum is of type (7.4.4) for

$$t_0 = a \leqslant t_1 \leqslant t_2 \leqslant \cdots \leqslant t_{n+1} = b$$

and $t_j \leqslant s_j \leqslant t_{j+1}$. Further, $\max |t_{j+1} - t_j| \leqslant 2|\pi|$. Hence, if one of the sums has a limit, so does the other. ∎

This, of course, raises the question of the existence of one of the limits. In the classical theory the integral (7.4.5) exists if f is continuous and g is of bounded variation, both in $[a, b]$. The following theorem is a natural extension of this prototype.

Theorem 7.4.2. *Suppose that either* (1) **f** *is a strongly continuous vector function from* $[a, b]$ *to* $\mathfrak{X}$ *and* g *is a numerically valued function of bounded variation on* $[a, b]$ *or, alternatively,* (2) **f** *is a vector-valued function from* $[a, b]$ *to* $\mathfrak{X}$ *of bounded variation in the sense of Definition* 7.3.4 *while* g *is a continuous numerically valued function on* $[a, b]$. *Then the integrals* (7.4.5) *and* (7.4.6) *exist in the normed topology of* $\mathfrak{X}$.

Proof. The argument in case (1) is patterned on the classical case. Since **f** is strongly continuous in $[a, b]$, we have uniform strong continuity. That is, given an $\varepsilon > 0$, there is a $\delta > 0$ such that

$$\|\mathbf{f}(s_1) - \mathbf{f}(s_2)\| < \varepsilon \quad \text{if} \quad |s_1 - s_2| < \delta. \tag{7.4.9}$$

If we now take two sums of type (7.4.3), one with set π_1 and the other with set π_2 with $|\pi_1|$ and $|\pi_2| < \frac{1}{2}\delta$, then the usual argument, with absolute values replaced by norms, gives

$$\|\mathbf{S}_{\pi_1} - \mathbf{S}_{\pi_2}\| \leqslant 2\varepsilon V_a^{\,b}[g]. \tag{7.4.10}$$

This proves the existence of the strong integral (7.4.5) in case (1).

In case (2) g is uniformly continuous in $[a, b]$, and choosing ε and δ as above we see that for any $x^* \in \mathfrak{X}^*$ we have

$$|x^*[\mathbf{S}_{\pi_1} - \mathbf{S}_{\pi_2}]| \leqslant 2\varepsilon V_a^{\,b}\{x^*[\mathbf{f}(s)]\}, \tag{7.4.11}$$

where, as above, $|\pi_1|$ and $|\pi_2|$ are $<\frac{1}{2}\delta$. The right hand member involves the total variation of the numerically valued function $x^*[\mathbf{f}(s)]$, and this should be compared with the total variation of $\mathbf{f}(s)$, which is known to be finite.

Some preliminary considerations will be helpful. It is known that a complex-valued function $h(s) \in BV[a, b]$ admits of a representation of the form

$$h = h_1 - h_2 + i(h_3 - h_4) \equiv h_5 + ih_6, \tag{7.4.12}$$

where h_1 to h_4 are non-negative and non-decreasing. Moreover,

$$V[h_5] = V[h_1] + V[h_2], \qquad V[h_5] = V[h_3] + V[h_4]$$

for a suitably chosen decomposition. Further, h_1 and h_2 (h_3 and h_4) are never increasing for the same values of s; if one member of a pair is increasing in an interval, the other keeps a constant value.

Consider now a finite set of non-overlapping intervals $[a_j, b_j]$ in $[a, b]$. If a particular interval $[a_j, b_j]$ is an interval of constancy of h_2, then $h_1(b_j) \geqslant h_1(a_j)$. If in

$$\left| \sum_{j=1}^{n} [h_5(b_j) - h_5(a_j)] \right|$$

we extend the summation over intervals of constancy of h_2 the sum reduces to

$$\sum_{j=1}^{n} [h_1(b_j) - h_1(a_j)].$$

By taking sufficiently many such intervals we can get arbitrarily close to $V[h_1]$. Similarly, we can approximate $V[h_2]$ by summing over intervals of constancy of h_1. This argument also applies to h_6 and hence to h itself. It follows that

$$\sup \left| \sum_{j=1}^{n} [h(b_j) - h(a_j)] \right| \geqslant \max_k V[h_k]. \tag{7.4.13}$$

But

$$V[h] \leqslant V[h_1] + V[h_2] + V[h_3] + V[h_4] \leqslant 4 \max_k V[h_k]. \tag{7.4.14}$$

This gives, finally,

$$V[h] \leqslant 4 \sup |\Sigma [h(b_j) - h(a_j)]|. \tag{7.4.15}$$

We now apply this inequality to $h(s) = x^*[\mathbf{f}(s)]$ and obtain

$$V\{x^*[\mathbf{f}(s)]\} \leqslant 4 \sup_{\Delta} \left| \sum_{j=1}^{n} x^*[\mathbf{f}(b_j)] - x^*[\mathbf{f}(a_j)] \right|$$

$$= 4 \sup_{\Delta} \left| x^* \left\{ \sum_{j=1}^{n} [\mathbf{f}(b_j) - \mathbf{f}(a_j)] \right\} \right|.$$

Here the subscript Δ indicates that the supremum is taken for all n and all choices of n non-overlapping intervals in $[a, b]$. Now

$$\|\mathbf{S}_{\pi_1} - \mathbf{S}_{\pi_2}\| = \sup_{\|x^*\|=1} |x^*[\mathbf{S}_{\pi_1} - \mathbf{S}_{\pi_2}]| \leqslant 2\varepsilon \sup_{\|x^*\|=1} V\{x^*[\mathbf{f}(s)]\}$$

$$\leqslant 8\varepsilon \sup_{\|x^*\|=1} \sup_{\Delta} \left| x^* \left\{ \sum_{j=1}^{n} [\mathbf{f}(b_j) - \mathbf{f}(a_j)] \right\} \right|$$

$$\leqslant 8\varepsilon \sup \left\| \sum_{j=1}^{n} [\mathbf{f}(b_j) - \mathbf{f}(a_j)] \right\|$$

$$= 8\varepsilon\, V_a^b[\mathbf{f}]$$

in terms of the total variation of the vector function **f**. This shows that the integral in case (2) also exists in the strong topology. ∎

The proof is much simpler in case (2) if **f** is of strong bounded variation rather than just bounded variation.

The integrals (7.4.5) and (7.4.6) are elements of the B-space $\mathfrak{X}$, and as such they may be operated on by any element T of an operator algebra $\mathfrak{E}(\mathfrak{X}, \mathfrak{Y})$. There are two operators involved here, T and $\int$, and the question arises whether or not they commute. Can T be taken under the sign of integration and applied to **f**? In integrals of the second type we want to go a step further: $T[d\mathbf{f}(s)]$ looks mysterious, while $d[T\mathbf{f}(s)]$ may possibly make sense. The following theorem gives an answer to this question.

Theorem 7.4.3. *If* **f** *is a strongly continuous mapping of* $[a, b]$ *into* $\mathfrak{X}$, *if* $g \in BV[a, b]$ *and if* $T \in \mathfrak{E}(\mathfrak{X}, \mathfrak{Y})$, *then* $T[\mathbf{f}](s)$ *exists as an element of* $\mathfrak{Y}$ *and is strongly continuous in* $[a, b]$. *Further,*

$$\int_a^b T[\mathbf{f}](s)\, dg(s) = T\left\{\int_a^b \mathbf{f}(s)\, dg(s)\right\}. \tag{7.4.16}$$

If, instead, $\mathbf{f}(s) \in \mathfrak{X}$ *and is of bounded variation in* $[a, b]$, *if* $g \in C[a, b]$, $T \in \mathfrak{E}(\mathfrak{X}, \mathfrak{Y})$, *and if* $T[\mathbf{f}](s)$ *is of bounded variation in* $[a, b]$, *then*

$$\int_a^b g(s)\, d\{T[\mathbf{f}](s)\} = T\left\{\int_a^b g(s)\, d\mathbf{f}(s)\right\}. \tag{7.4.17}$$

Proof. Here

$$\|T[\mathbf{f}](s) - T[\mathbf{f}](t)\| \leqslant \|T\|\ \|\mathbf{f}(s) - \mathbf{f}(t)\|$$

since T is linear and bounded, and **f** is strongly continuous in $[a, b]$. It follows that $T[\mathbf{f}](s)$ is also strongly continuous. Hence both sides of (7.4.16) exist. Equality then follows from the linearity of T which gives the corresponding identity for the Riemann–Stieltjes sums

$$\sum_{j=1}^{n} T[\mathbf{f}](t_j)\,[g(s_j) - g(s_{j-1})] = T\left\{\sum_{j=1}^{n} \mathbf{f}(t_j)\,[g(s_j) - g(s_{j-1})]\right\}.$$

Passing to the limit with $|\pi|$ we get (7.4.16). In the second case both sides exist by assumption and Theorem 7.4.2. Again we have equality for the Riemann–Stieltjes sums and in the limit (7.4.17) results. ∎

Corollary. *If* $\mathbf{f}(s)$ *is of strong bounded variation, so is* $T[\mathbf{f}](s)$, *and* (7.4.17) *holds.*

Actually the theorem holds under more general assumptions, but we shall not examine this possibility here.

An important special case is that in which T is a linear bounded functional $x^* \in X^*$. Here $x^*[\mathbf{f}](s)$ is automatically of bounded variation if $\mathbf{f}$ is of bounded variation, so both identities hold with T replaced by x^*.

It remains to say a few words about operator-valued integrals. Suppose $U(s) \in \mathfrak{E}(\mathfrak{X}, \mathfrak{Y})$ for each s in $[a, b]$. According as $U(s)$ is continuous in the uniform, the strong, or the weak operator topology, we can form the integrals

$$\int_a^b U(s)\,dg(s), \quad \int_a^b U(s)[\mathbf{x}]\,dg(s), \quad \int_a^b y^*\{U(s)[\mathbf{x}]\}\,dg(s). \tag{7.4.18}$$

Here $g \in BV[a, b]$ while $\mathbf{x}$ is any element of $\mathfrak{X}$, y^* any element of $\mathfrak{Y}^*$. We leave to the reader to produce integrals of the second kind where $g \in C[a, b]$ and $U(s)$ is of the appropriate type of bounded variation.

EXERCISE 7.4

1. Verify (7.4.14).

2. Why is $\|\mathbf{x}\| = \sup_{\|x^*\|=1} |x^*(\mathbf{x})|$?

3. Give a proof of Theorem 7.4.2, case (2), assuming $\mathbf{f}$ to be of strong bounded variation.

4. Prove the Corollary to Theorem 7.4.3.

5. In the proof of (7.4.13) what is the passage from h_5 and h_6 which justifies replacing them by h?

6. Prove the existence of the integrals in (7.4.18).

7. If $U(s)$ is continuous in the uniform operator topology of $\mathfrak{E}(\mathfrak{X}, \mathfrak{Y})$, prove that

$$\left\{\int_a^b U(s)\,dg(s)\right\}[\mathbf{x}] = \int_a^b U(s)[\mathbf{x}]\,dg(s).$$

8. When is it true that

$$y^*\left\{\int_a^b U(s)[\mathbf{x}]\,dg(s)\right\} = \int_a^b y^*\{U(s)[\mathbf{x}]\}\,dg(s)?$$

9. What do operator integrals of the second kind look like? Give sufficient conditions for existence.

10. Abstract Riemann–Stieltjes integrals are bilinear operations. They are linear in the integrand as well as in the integrator. Write out the formulas and give a verification.

11. Under the assumptions of Theorem 7.4.2 show that

$$\left\| \int_a^b \mathbf{f}(s)\, dg(s) \right\| \leqslant \max_{a \leqslant s \leqslant b} \|\mathbf{f}(s)\| \; V_a{}^b[g].$$

12. Is there a corresponding estimate for integrals of the second kind?

7.5 BOCHNER INTEGRALS

Before we can introduce the Bochner integral of a vector-valued function, we briefly discuss how to abstract the most important properties of Lebesgue measure and Lebesgue integration treated, in geometric language, in Chapter 4.

The notion of a measure space, here denoted by (S, A, μ), is given in Section 4.1. While the Lebesgue measurable function serves as a prototype, the concept of an A-measurable function on an abstract measure space (S, A, μ) can be formulated similarly. Consequently the lemmas and theorems that are proved in essentially the same manner as those of Chapter 4 are stated here without proof.

Definition 7.5.1. *Let (S, A, μ) be an abstract measure space. An extended real-valued function defined on S is called* A*-measurable iff for each real number α, the set $\{x; f(x) > \alpha\}$ is* A*-measurable, i.e. $\{x; f(x) > \alpha\} \in \mathsf{A}$.*

(See Definitions 4.1.1 and 4.2.1.)

Lemma 7.5.1. *Let f be an extended real-valued function defined on S. The following statements are equivalent:*

1) $\{x; f(x) > \alpha\} \in \mathsf{A}, \quad \forall\, \alpha \in R.$
2) $\{x; f(x) \geqslant \alpha\} \in \mathsf{A}, \quad \forall\, \alpha \in R.$
3) $\{x; f(x) < \alpha\} \in \mathsf{A}, \quad \forall\, \alpha \in R.$
4) $\{x; f(x) \leqslant \alpha\} \in \mathsf{A}, \quad \forall\, \alpha \in R.$

(See Lemma 4.2.1.)

Lemma 7.5.2. *Let c be a real number. Suppose that the real-valued functions f and g are* A*-measurable, then so are the functions $cf, f^2, f + g, f - g, fg, |f|$. Moreover, if $\{f_n\}$ is a sequence of* A*-measurable functions, then* $\sup f_n$, $\inf f_n$, $\liminf f_n$, *and* $\limsup f_n$ *are all* A*-measurable.*

(See Lemmas 4.2.2 and 4.2.3.)

A real-valued A*-measurable function is simple if it has only a finite number of values* (see (4.3.3)). It is clear that a simple A-measurable function g can be represented in the form

$$g = \sum_{j=1}^{m} \alpha_j \chi_{S_j}, \tag{7.5.1}$$

where $\alpha_j \in R$ and χ_{S_j} is the characteristic function of a set S_j in A. Among these representations for g there is a unique standard representation characterized by the fact that the α_j are distinct and the S_j disjoint. Indeed, if $\alpha_1, \alpha_2, \ldots, \alpha_m$ are the distinct values of g, and if $S_j = \{x; x \in S, g(x) = \alpha_j\}$, then the S_j are disjoint and $S = \bigcup_{j=1}^{m} S_j$.

It happens quite often in the process of abstraction that the formulation of a generalized definition is suggested by properties of the concrete cases. Here we present the following

Definition 7.5.2. *If g is a non-negative simple function on S with the standard representation* (7.5.1), *the integral of g with respect to μ is defined to be the extended real number*

$$\int g \, d\mu = \sum_{j=1}^{m} \alpha_j \, \mu(S_j). \tag{7.5.2}$$

(See (4.3.4).)

As a direct consequence, the following can be verified easily.

Lemma 7.5.3. (1) *If g and h are non-negative simple functions on S, and if $\alpha \geqslant 0$, then*

$$\int \alpha g \, d\mu = \alpha \int g \, d\mu, \tag{7.5.3}$$

$$\int (g + h) \, d\mu = \int g \, d\mu + \int h \, d\mu. \tag{7.5.4}$$

(2) *If ν is defined for E in A by*

$$\nu(E) = \int g \chi_E \, d\mu, \tag{7.5.5}$$

then ν is a measure on A.

Definition 7.5.3. *If f is a non-negative extended real-valued A-measurable function on S, the integral of f with respect to μ is defined to be the extended real number*

$$\int f \, d\mu = \sup \int g \, d\mu, \tag{7.5.6}$$

where the supremum is extended over all simple functions g on S satisfying $0 \leqslant g(x) \leqslant f(x)$ for all $x \in S$. If f is a non-negative extended real valued A-measurable function on S, if E belongs to A, then $f\chi_E$ is also a non-negative A-measurable function on S and the integral of f over E with respect to μ is defined to be the extended real number

$$\int_E f \, d\mu = \int f \chi_E \, d\mu.$$

(See Definition 4.3.2, Theorems 4.3.3 to 4.3.6.)

At this moment of processing abstraction, we must check if Definition 7.5.3 really extends Definition 4.3.2. Suppose that f is a non-negative extended real-valued Lebesgue measurable function defined on R^m. According to Definition 4.3.2, the Lebesgue integral of f with respect to the m-dim Lebesgue measure, μ_m is equal to $\int f(P)\, dP = \int f\, d\mu_m = \mu_{m+1}[\Omega_0(f; S)]$. By virtue of Theorems 4.3.4–4.3.6 it can be verified easily that

$$\int f\, d\mu_m = \mu_{m+1}[\Omega_0(f; S)] = \sup \int g\, d\mu_m, \tag{7.5.7}$$

where g ranges over all simple functions with $0 \leqslant g \leqslant f$.

The next lemma follows directly from Definition 7.5.3.

Lemma 7.5.4. (1) *If f and g are non-negative extended real-valued* A*-measurable functions on S with $f \leqslant g$, then*

$$\int f\, d\mu \leqslant \int g\, d\mu. \tag{7.5.8}$$

(2) *If f is a non-negative extended real-valued* A*-measurable function on S, and if E, F belong to* A *with $E \subseteq F$, then*

$$\int_E f\, d\mu \leqslant \int_F f\, d\mu. \tag{7.5.9}$$

We are now prepared to establish an important theorem which provides the key to the fundamental convergence properties of integration.

Theorem 7.5.1 (*Monotone Convergence Theorem*). *Let $\{f_n\}$ be a monotone increasing sequence of non-negative extended real-valued* A*-measurable functions defined on S. Let f be the limit function of $\{f_n\}$. Then*

$$\int f\, d\mu = \lim_{n\to\infty} \int f_n\, d\mu.$$

(See Theorem 4.3.5.)

Proof. According to Lemma 7.5.2 the limit function f is A-measurable. Since $f_n \leqslant f_{n+1} \leqslant f$, it follows from (7.5.8) of Lemma 7.5.3 that

$$\int f_n\, d\mu \leqslant \int f_{n+1}\, d\mu \leqslant \int f\, d\mu$$

for all positive integers n. Therefore

$$\lim_{n\to\infty} \int f_n\, d\mu \leqslant \int f\, d\mu.$$

To establish the opposite inequality, let α be a real number, $0 < \alpha < 1$, and let g be a simple function with $0 \leqslant g \leqslant f$. Let $S_n = \{x;\ x \in S, f_n(x) \geqslant \alpha g(x)\}$ so that $S_n \in$ A, $S_n \subseteq S_{n+1}$, and $S = \bigcup S_n$. According to (7.5.9) of Lemma 7.5.4,

$$\int_{S_n} \alpha g\, d\mu \leqslant \int_{S_n} f_n\, d\mu \leqslant \int f_n\, d\mu. \tag{7.5.10}$$

Since the sequence $\{S_n\}$ is monotone increasing and has union S, it follows from Lemmas 7.5.3 and 4.1.2 that

$$\int g\,d\mu = \lim_{n\to\infty} \int_{S_n} g\,d\mu.$$

Therefore, letting $n \to \infty$ in (7.5.10) we obtain

$$\alpha \int g\,d\mu \leqslant \lim_{n\to\infty} \int f_n\,d\mu.$$

Since this holds for arbitrary α with $0 < \alpha < 1$, it follows that

$$\int g\,d\mu \leqslant \lim_{n\to\infty} \int f_n\,d\mu,$$

and since g is an arbitrary simple function on S such that $0 \leqslant g \leqslant f$, we have

$$\int f\,d\mu = \sup_g \int g\,d\mu \leqslant \lim_{n\to\infty} \int f_n\,d\mu.$$

Combining this with the opposite inequality, we obtain the desired result. ∎

A careful study of the proof of Theorem 4.3.6 reveals that the use of Lebesgue measure is immaterial. The same process of construction as used in Theorem 4.3.6 gives

Theorem 7.5.2. *Let f be a non-negative extended real-valued* A-*measurable function on S. Then there exists a monotone increasing sequence $\{f_n\}$ of non-negative simple functions on S such that $\{f_n\}$ converges to f.*

As direct consequences of Theorem 7.5.2 and/or the Monotone Convergence Theorem, the following corollaries can be verified easily.

Corollary 1 (*Fatou's Lemma*). *Let $\{f_n\}$ be a sequence of non-negative extended real-valued* A-*measurable functions on S, then*

$$\int (\liminf_{n\to\infty} f_n)\,d\mu \leqslant \liminf_{n\to\infty} \int f_n\,d\mu. \tag{7.5.11}$$

(See Problem 6 in Exercise 4.3.)

Corollary 2. *Given a real number $\alpha \geqslant 0$. If f is a non-negative extended real-valued* A-*measurable function on S, then so is αf and*

$$\int \alpha f\,d\mu = \alpha \int f\,d\mu. \tag{7.5.12}$$

(See Corollary 2 to Theorem 4.3.6.)

Corollary 3. *If f and g are non-negative extended real-valued* A-*measurable functions on S, then so is $f + g$ and*

$$\int (f + g)\,d\mu = \int f\,d\mu + \int g\,d\mu. \tag{7.5.13}$$

(See Corollary 3 to Theorem 4.3.6.)

With respect to μ, we shall now discuss the integration of extended real-valued A-measurable functions which are not of one sign. Many definitions and proofs of Chapter 4 depend only on properties of Lebesgue measure true for an arbitrary measure in an abstract measure space and carry over to this case.

Definition 7.5.4. *An extended real-valued* A*-measurable function f on S is integrable over S iff the positive and negative parts f^+, f^- of f have finite integrals with respect to μ. In this case, the integral of f with respect to μ is defined to be*

$$\int f\,d\mu = \int f^+\,d\mu - \int f^-\,d\mu. \tag{7.5.14}$$

(See Definition 4.3.3.)

We notice that the remark made after Definition 4.3.3 also applies to this extended definition.

The *Principle of Absolute Integrability* takes the following form.

Theorem 7.5.3. *A real-valued* A*-measurable function f is integrable with respect to μ over S iff $|f|$ is integrable over S. In this case*

$$\left|\int f\,d\mu\right| \leqslant \int |f|\,d\mu. \tag{7.5.15}$$

(See Theorem 4.3.7.)

The linearity of integration with respect to μ should become clear through the following theorem, whose proof is almost identical with that of Theorem 4.3.9.

Theorem 7.5.4. *If α is a real number and the real-valued functions f and g are integrable over S, then so are αf and $f + g$. Furthermore,*

$$\int (\alpha f)\,d\mu = \alpha \int f\,d\mu, \tag{7.5.16}$$

$$\int (f + g)\,d\mu = \int f\,d\mu + \int g\,d\mu. \tag{7.5.17}$$

The following is probably the most important convergence theorem for integrable functions.

Theorem 7.5.5 (*Dominated Convergence Theorem for integrals with respect to μ*). *Let g be integrable over S, and suppose that $\{f_n\}$ is a sequence of* A*-measurable functions such that on S we have $|f_n(x)| \leqslant g(x)$ for all n, and such that $\{f_n\}$ converges μ-almost everywhere to an* A*-measurable function f. Then f is integrable and*

$$\int f\,d\mu = \lim_{n\to\infty} \int f_n\,d\mu. \tag{7.5.18}$$

Proof. Apply Fatou's Lemma to the sequences $\{g + f_n\}$ and $\{g - f_n\}$. ∎

The following will be needed in introducing the Bochner Integral.

Lemma 7.5.5. *Suppose that for each n, f_n is an extended real-valued function integrable with respect to μ over S. Suppose, further, that*

$$\sum_{n=1}^{\infty} \int |f_n|\, d\mu < \infty. \tag{7.5.19}$$

Then the series $\sum_{n=1}^{\infty} f_n$ converges μ-a.e. on S to a real-valued integrable function f. Moreover,

$$\int f\, d\mu = \sum_{n=1}^{\infty} \int f_n\, d\mu. \tag{7.5.20}$$

(See Problem 9, Exercise 4.3.)

Let (S, A, μ) be an abstract measure space. Let $\mathfrak{X}$ be a B-space. One type of integral assigned to certain vector-valued functions $\mathbf{x}(s)$ on S into $\mathfrak{X}$ was developed by S. Bochner (1933) and may be described as follows. One begins with the simple functions $\mathbf{x}(t)$ identifying any pair which differ only on a set of measure zero. The classes of such functions then form a normed linear space with norm $\|\mathbf{x}(\cdot)\|_{\mathrm{B}} = \int \|\mathbf{x}(t)\|\, d\mu$. Defining the vector-valued integral

$$\int \mathbf{x}(t)\, d\mu$$

in the obvious way, it is clear that

$$\left\|\int \mathbf{x}(t)\, d\mu\right\| \leqslant \|\mathbf{x}(\cdot)\|_{\mathrm{B}}. \tag{7.5.21}$$

If one now completes the space, one can extend the integral to all Cauchy sequences and obtain the Bochner integral. More precisely, let us start with

Definition 7.5.5. *Each vector-valued function $\mathbf{x}(s)$ on S into $\mathfrak{X}$ which assumes only a finite number of distinct values, each value $\neq \mathbf{0}$ on a μ-measurable set of finite measure, is called a simple function.*

Obviously there is a uniquely determined standard representation for each simple function. See (7.5.1).

Observe that in the previous discussion a function of the form (7.5.1) is called a simple function even though the condition $\mu(S_j) < \infty$ for $j = 1, 2, \ldots, m$ is not necessarily satisfied. Therefore we stress the fact that in the present discussion we adhere to the narrower interpretation where this condition is satisfied.

Definition 7.5.6. *If $\mathbf{g}(s)$ is a simple function on S with the standard representation*

$$\mathbf{g}(s) = \sum_{j=1}^{m} \mathbf{g}_j\, \chi_{S_j}, \tag{7.5.22}$$

the vector-valued integral of $\mathbf{g}(s)$ with respect to μ is defined to be the vector

$$\int \mathbf{g}(s)\, d\mu = \sum_{j=1}^{m} \mathbf{g}_j\, \mu(S_j). \tag{7.5.23}$$

(See Definition 7.5.2.)

It is clear that this vector-valued integral is linear on the vector space of all simple functions. Finally, (7.5.21) holds. For the proof note that if $\mathbf{g}(s)$ is given by (7.5.22) with disjoint sets S_j, then

$$\|\mathbf{g}(s)\| = \sum_{j=1}^{m} \|\mathbf{g}_j\| \chi_{S_j}, \quad \forall\, s \in S.$$

Hence

$$\|\int \mathbf{g}(s)\, d\mu\| = \left\| \sum_{j=1}^{m} \mathbf{g}_j\, \mu(S_j) \right\| \leqslant \sum_{j=1}^{m} \|\mathbf{g}_j\|\, \mu(S_j) = \int \|\mathbf{g}(s)\| d\mu = \|\mathbf{g}(\cdot)\|_{\mathrm{B}}.$$

Theorem 7.5.6. *If a sequence $\{\mathbf{x}_n(t)\}$ of vector-valued simple functions on S satisfies*

$$\lim \int \|\mathbf{x}_n(t) - \mathbf{x}_m(t)\|\, d\mu = 0 \tag{7.5.24}$$

as $m, n \to \infty$, then there exists a unique vector-valued function $\mathbf{x}(t)$ such that $\|\mathbf{x}(t)\|$ and all functions $\|\mathbf{x}(t) - \mathbf{x}_n(t)\|$ are A*-measurable and*

$$\lim_{n\to\infty} \int \|\mathbf{x}(t) - \mathbf{x}_n(t)\|\, d\mu = 0. \tag{7.5.25}$$

Uniqueness here is understood in the sense that functions differing only on a set of μ-measure zero are identified.

Proof. The argument is similar to that used to prove that the space L^1 of all numerical μ-integrable functions is a complete metric space. Given now the sequence $\{\mathbf{x}_n(t)\}$ of simple functions, satisfying (7.5.24). This implies the existence of a subsequence $\{\mathbf{x}_{n_k}(t)\}$ satisfying

$$\sum_{k=1}^{\infty} \int \|\mathbf{x}_{n_{k+1}}(t) - \mathbf{x}_{n_k}(t)\|\, d\mu < \infty. \tag{7.5.26}$$

See (4.4.21). Then, by Lemma 7.5.5,

$$\|\mathbf{x}_{n_1}(t)\| + \sum_{k=1}^{\infty} \|\mathbf{x}_{n_{k+1}}(t) - \mathbf{x}_{n_k}(t)\|$$

is μ-integrable and hence finite for almost every $t \in S$. Since in a B-space an absolutely convergent series is convergent (why?), the vector-valued series

$$\mathbf{x}_{n_1}(t) + \sum_{k=1}^{\infty} [\mathbf{x}_{n_{k+1}}(t) - \mathbf{x}_{n_k}(t)]$$

converges for the same values of t; let $\mathbf{x}(t)$ be its sum where it converges, i.e. μ-a.e., and let $\mathbf{x}(t) = 0$ elsewhere. Since $\mathbf{x}(t) = \lim_{k\to\infty} \mathbf{x}_{n_k}(t)$ μ-a.e., we have $\|\mathbf{x}(t)\| = \lim_{k\to\infty} \|\mathbf{x}_{n_k}(t)\|$ and $\|\mathbf{x}(t) - \mathbf{x}_n(t)\| = \lim_{k\to\infty} \|\mathbf{x}_{n_k}(t) - \mathbf{x}_n(t)\|$ μ-a.e. Thus $\|\mathbf{x}(t)\|$ and $\|\mathbf{x}(t) - \mathbf{x}_n(t)\|$ are A-measurable. It follows from

$$\mathbf{x}(t) - \mathbf{x}_{n_p}(t) = \sum_{k=p}^{\infty} [\mathbf{x}_{n_{k+1}}(t) - \mathbf{x}_{n_k}(t)]$$

that

$$\lim \int \|\mathbf{x}(t) - \mathbf{x}_{n_p}(t)\| \, d\mu = 0$$

as $p \to \infty$ and, hence,

$$\int \|\mathbf{x}(t) - \mathbf{x}_n(t)\| \, d\mu \leqslant \int \|\mathbf{x}_n(t) - \mathbf{x}_{n_p}(t)\| \, d\mu + \int \|\mathbf{x}(t) - \mathbf{x}_{n_p}(t)\| \, d\mu.$$

As n and $p \to \infty$, both terms on the right go to 0, so (7.5.25) holds.

Assume that

$$\int \|\mathbf{y}(t) - \mathbf{x}_n(t)\| \, d\mu \to 0$$

for another vector-valued $\mathbf{y}(t)$. Then, in particular, this holds for $n = n_k$, where $\{\mathbf{x}_{n_k}(t)\}$ is the subsequence used above. But the L^1 convergence of $\|\mathbf{y}(t) - \mathbf{x}_{n_k}(t)\|$ to zero implies the pointwise convergence almost everywhere. The sequence $\{\mathbf{x}_{n_k}(t)\}$ converges then pointwise to $\mathbf{x}(t)$ and to $\mathbf{y}(t)$ simultaneously, so $\mathbf{y}(t) = \mathbf{x}(t)$ μ-a.e. This shows that $\mathbf{x}(t)$ is uniquely determined.

Furthermore,

$$\begin{aligned} \left\|\int \mathbf{x}_n(t) \, d\mu - \int \mathbf{x}_m(t) \, d\mu\right\| &= \left\|\int [\mathbf{x}_n(t) - \mathbf{x}_m(t)] \, d\mu\right\| \\ &\leqslant \int \|\mathbf{x}_n(t) - \mathbf{x}_m(t)\| \, d\mu \to 0 \quad \text{as} \quad m, n \to \infty. \end{aligned}$$

Thus the Cauchy sequence $\{\int \mathbf{x}_n(t) \, d\mu\}$ converges in $\mathfrak{X}$. Suppose that $\{\mathbf{x}_n(t)\}$ and $\{\mathbf{y}_n(t)\}$ are two sequences of simple functions such that

$$\lim_{n \to \infty} \int \|\mathbf{x}(t) - \mathbf{x}_n(t)\| \, d\mu = 0, \qquad \lim_{n \to \infty} \int \|\mathbf{x}(t) - \mathbf{y}_n(t)\| \, d\mu = 0.$$

Then

$$\begin{aligned} \left\|\int \mathbf{x}_n(t) \, d\mu - \int \mathbf{y}_n(t) \, d\mu\right\| &\leqslant \int \|\mathbf{x}_n(t) - \mathbf{y}_n(t)\| \, d\mu \\ &\leqslant \int \|\mathbf{x}(t) - \mathbf{x}_n(t)\| \, d\mu + \int \|\mathbf{x}(t) - \mathbf{y}_n(t)\| \, d\mu. \end{aligned}$$

This shows that the limit element of $\{\int \mathbf{x}_n(t) \, d\mu\}$ is uniquely determined by $\mathbf{x}(t)$. We use $\int \mathbf{x}(t) \, d\mu$ to denote $\lim_{n \to \infty} \int \mathbf{x}_n(t) \, d\mu$ and

$$\int \mathbf{x}(t) \, d\mu \tag{7.5.27}$$

is called the *Bochner integral* of $\mathbf{x}(t)$ with respect to μ. ∎

We shall denote the collection of all $\mathbf{x}(t)$ for which the Bochner integral is thus defined by $\mathfrak{B}^1 = \mathfrak{B}(S, \mathsf{A}, \mu; \mathfrak{X})$. It is evident that $\mathfrak{B}^1$ is a complex vector space, and Bochner integration is linear on $\mathfrak{B}^1$. Any $\mathbf{x}(t) \in \mathfrak{B}^1$ is called a *Bochner integrable function*. In particular, if $\mathfrak{X} = C$, then $\mathfrak{B}^1 = L^1(S, \mathsf{A}, \mu)$ is the space of all complex-valued μ-integrable functions.

Theorem 7.5.7. *If* $\mathbf{x}(t)$ *is Bochner integrable, then*

$$\left\|\int \mathbf{x}(t) \, d\mu\right\| \leqslant \int \|\mathbf{x}(t)\| \, d\mu. \tag{7.5.28}$$

Proof. Let $\{\mathbf{x}_n(t)\}$ be such a sequence of simple functions that

$$\lim_{n \to \infty} \int \|\mathbf{x}(t) - \mathbf{x}_n(t)\| \, d\mu = 0. \tag{7.5.29}$$

Then

$$\int \|\mathbf{x}(t)\|\, d\mu = \lim_{n\to\infty} \int \|\mathbf{x}_n(t)\|\, d\mu. \tag{7.5.30}$$

Since we also have

$$\left\|\int \mathbf{x}_n(t)\, d\mu\right\| \leqslant \int \|\mathbf{x}_n(t)\|\, d\mu \tag{7.5.31}$$

the last three inequalities imply (7.5.28). ∎

Theorem 7.5.8. *The complex vector space $\mathfrak{B}^1$ of all Bochner integrable functions $\mathbf{x}(t)$ is a B-space with respect to the norm $\int \|\mathbf{x}(t)\|\, d\mu$.*

Proof. It is clear that this expression is a norm on $\mathfrak{B}^1$. Let us prove completeness under the norm. Let $\{\mathbf{y}_n(t)\}$ be a Cauchy sequence in $\mathfrak{B}^1$ such that

$$\int \|\mathbf{y}_n(t) - \mathbf{y}_m(t)\|\, d\mu \to 0 \quad \text{as} \quad m, n \to \infty.$$

By the definition of Bochner integrability there exists for each $\mathbf{y}_n(t)$ a simple function $\mathbf{x}_n(t)$ such that

$$\int \|\mathbf{y}_n(t) - \mathbf{x}_n(t)\|\, d\mu < 2^{-n}.$$

By the triangle inequality

$$\int \|\mathbf{x}_n(t) - \mathbf{x}_m(t)\|\, d\mu \to 0 \quad \text{as} \quad m, n \to \infty.$$

Hence, by Theorem 7.5.6, there exists a function $\mathbf{x}(t) \in \mathfrak{B}^1$ with

$$\lim_{n\to\infty} \int \|\mathbf{x}(t) - \mathbf{x}_n(t)\|\, d\mu = 0.$$

Once more, by the triangle inequality, we have

$$\lim_{n\to\infty} \int \|\mathbf{x}(t) - \mathbf{y}_n(t)\|\, d\mu = 0,$$

and completeness follows. ∎

Theorem 7.5.9 *(Dominated Convergence Theorem for Bochner Integrals). If $\mathbf{y}_n(t) \in \mathfrak{B}^1$ and $\|\mathbf{y}_n(t)\| \leqslant g(t)$ for $n = 1, 2, \ldots$, where $g(t)$ is μ-integrable over S, and if $\mathbf{y}_n(t)$ converges to $\mathbf{x}(t)$ μ-a.e., then $\mathbf{x}(t) \in \mathfrak{B}^1$ and*

$$\lim_{n\to\infty} \int \|\mathbf{x}(t) - \mathbf{y}_n(t)\|\, d\mu = 0. \tag{7.5.32}$$

In particular,

$$\int \mathbf{x}(t)\, d\mu = \lim_{n\to\infty} \int \mathbf{y}_n(t)\, d\mu. \tag{7.5.33}$$

Proof. From

$$\mathbf{y}_n(t) - \mathbf{x}(t) = \lim_{m\to\infty} [\mathbf{y}_n(t) - \mathbf{y}_m(t)] \qquad \mu\text{-a.e.}$$

it follows that

$$\|\mathbf{y}_n(t) - \mathbf{x}(t)\| = \lim_{m\to\infty} \|\mathbf{y}_n(t) - \mathbf{y}_m(t)\| \qquad \mu\text{-a.e.},$$

so that the left member is A-measurable. Moreover,

$$\|\mathbf{y}_n(t) - \mathbf{x}(t)\| \leqslant 2\,g(t) \quad \text{and} \quad \lim_{n\to\infty} \|\mathbf{y}_n(t) - \mathbf{x}(t)\| = 0 \qquad \mu\text{-a.e. on } S.$$

Hence, by the dominated convergence theorem for numerical functions,

$$\lim_{n\to\infty} \int \|\mathbf{y}_n(t) - \mathbf{x}(t)\|\, d\mu = 0.$$

Thus by the preceding theorem, $\mathbf{x}(t) \in \mathfrak{B}_1$. Furthermore,

$$\int \mathbf{x}(t)\, d\mu = \lim_{n\to\infty} \int \mathbf{y}_n(t)\, d\mu$$

since

$$\left\|\int \mathbf{x}(t)\, d\mu - \int \mathbf{y}_n(t)\, d\mu\right\| \leqslant \int \|\mathbf{x}(t) - \mathbf{y}_n(t)\|\, d\mu. \blacksquare$$

In Chapter 4 we have seen that the Lebesgue measurable functions and the Lebesgue integrable functions are closely related. This relation extends to A-measurable functions and μ-integrable functions when integration on an abstract measure space (S, A, μ) is treated. In the process of abstraction, Bochner's integral yields $\mathfrak{X}$-valued integration on an abstract measure space (S, A, μ). It is natural to ask: How is Bochner integrability characterized in terms of some measurability defined for functions from S to $\mathfrak{X}$ a B-space? The first step in this direction is given by

Definition 7.5.7. *Let $\mathfrak{X}$ be a B-space and (S, A, μ) a measure space. A function $\mathbf{x}(t)$ from S to $\mathfrak{X}$ is called weakly A-measurable iff for any $x^* \in \mathfrak{X}^*$, the numerical function $x^*[\mathbf{x}(t)]$ of t is A-measurable. $\mathbf{x}(t)$ is strongly A-measurable if there exists a sequence of simple functions strongly convergent to $\mathbf{x}(t)$ μ-a.e. on S.*

Definition 7.5.8. *$\mathbf{x}(t)$ is said to be separably valued if its range $\mathbf{x}(S) = \{\mathbf{x}(t);\ t \in S\}$ is separable. $\mathbf{x}(t)$ is called μ-almost separably valued if there exists an A-measurable set S_0 of μ-measure zero such that $\mathbf{x}(S \ominus S_0)$ is separable.*

The two notions of measurability for $\mathfrak{X}$-valued functions are connected by the following theorem due to B. J. Pettis (1938).

Theorem 7.5.10. *$\mathbf{x}(t)$ is strongly A-measurable iff it is weakly A-measurable and μ-almost separably valued.*

Proof. Suppose that $\mathbf{x}(t)$ is strongly A-measurable, then there exists a sequence of simple functions $\{\mathbf{x}_n(t)\}$ such that

$$\lim \mathbf{x}_n(t) \stackrel{s}{=} \mathbf{x}(t)$$

except on a set $S_0 \in \mathsf{A}$ of μ-measure zero. Clearly, each vector-valued simple function $\mathbf{x}_n(t)$ is weakly A-measurable. This implies that, for each $x^* \in \mathfrak{X}^*$, the numerical function

$$x^*[\mathbf{x}(t)] = \lim_{n\to\infty} x^*[\mathbf{x}_n(t)]$$

is A-measurable. Therefore, $\mathbf{x}(t)$ is weakly A-measurable. Furthermore, the union of the ranges of $\mathbf{x}_n(t)$ $(n = 1, 2, \ldots)$ is a countable set, and the closure of this set is separable. Being a subset of this separable set, $\mathbf{x}(S \ominus S_0)$ is therefore separable.

In proving the converse proposition, without losing the generality, the range $\mathbf{x}(S)$ may be assumed separable. Hence the space $\mathfrak{X}$ itself may be assumed separable; otherwise simply replace $\mathfrak{X}$ by the smallest closed subspace generated by $\mathbf{x}(S)$.

For any real number α, consider sets

$$A = \{t;\ \|\mathbf{x}(t)\| \leqslant \alpha\} \quad \text{and} \quad A(x^*) = \{t;\ |x^*[x(t)]| \leqslant \alpha\}$$

where $x^* \in \mathfrak{X}^*$. Clearly,

$$A \subseteq \bigcap \{A(x^*);\ \|x^*\| \leqslant 1\}. \tag{7.5.34}$$

By a direct consequence of the Hahn–Banach theorem (see Section 10.3 and Theorem 2.3.2 (1)), for a fixed t, there exists an $x_0^* \in \mathfrak{X}^*$ with $\|x_0^*\| = 1$ and $x_0^*[\mathbf{x}(t)] = \|\mathbf{x}(t)\|$. This implies that we can invert the order of inclusion in (7.5.34). Thus the inclusion should be replaced by equality.

In order to show that A can be reduced to the intersection of countably many sets $A(x^*)$, we apply the following lemma to be proved by the reader (see Problem 5, Exercise 7.5).

Lemma 7.5.6. *Let $\mathfrak{X}$ be a separable B-space, then there exists a sequence $\{x_n^*\} \subset \mathfrak{X}^*$ with $\|x_n^*\| \leqslant 1$ such that, for any $x_0^* \in \mathfrak{X}^*$ with $\|x_0^*| \leqslant 1$, a subsequence $\{x_{n_k}^*\}$ can be so chosen that*

$$\lim_{k\to\infty} x_{n_k}^*(\mathbf{x}) = x_0(\mathbf{x}), \qquad \forall\, \mathbf{x} \in \mathfrak{X}. \tag{7.5.35}$$

By the lemma we have now

$$A = \bigcap\, [A(x^*);\ \|x^*\| \leqslant 1] = \bigcap_{j=1}^{\infty} A(x_j^*).$$

From this and the weak A-measurability of $\mathbf{x}(t)$ it follows that $\|\mathbf{x}(t)\|$ is A-measurable.

For any positive integer n, the range $\{\mathbf{x}(t);\ t \in S\}$ may be covered by a countable number of open spheres $S_{j,n}$ $(j = 1, 2, \ldots)$ of radius $\leqslant 1/n$, since the range is separable by assumption. Let $\mathbf{x}_{j,n}$ denote the center of the sphere $S_{j,n}$. As proved above, $\|\mathbf{x}(t) - \mathbf{x}_{j,n}\|$ is A-measurable in t. Hence the set

$$T_{j,n} = \{t;\ t \in S,\ \mathbf{x}(t) \in S_{j,n}\}$$

is A-measurable and $S = \bigcup_{j=1}^{\infty} T_{j,n}$. Construct $\mathbf{x}_n(t)$ by setting

$$x_n(t) = \mathbf{x}_{i,n} \quad \text{if} \quad t \in \left\{T_{i,n} \ominus \bigcup_{j=1}^{i-1} T_{j,n}\right\}.$$

Since $S = \bigcup_{i=1}^{\infty} (T_{i,n} \ominus \bigcup_{j=1}^{i-1} T_{j,n})$, it follows that $\|\mathbf{x}(t) - \mathbf{x}_n(t)\| \leqslant 1/n$ for every $t \in S$. Furthermore, each $\mathbf{x}_n(t)$ is strongly A-measurable because $T_{i,n} \ominus \bigcup_{j=1}^{i-1} T_{j,n}$

is A-measurable. $\mathbf{x}(t)$ is the strong limit of the sequence $\{\mathbf{x}_n(t)\}$ and is thus A-measurable.

Theorem 7.5.11. *A strongly A-measurable function $\mathbf{x}(t)$ is Bochner μ-integrable iff $\|\mathbf{x}(t)\|$ is μ-integrable.*

Proof. The necessity has been shown in Theorem 7.5.7. To prove the sufficiency, let $\{\mathbf{x}_n(t)\}$ be a sequence of simple functions strongly convergent to $\mathbf{x}(t)$ μ-a.e. on S. Let

$$\mathbf{y}_n(t) = \mathbf{x}_n(t) \quad \text{if} \quad \|\mathbf{x}_n(t)\| \leqslant \|\mathbf{x}(t)\|\left(1 + \frac{1}{n}\right),$$

$$= 0 \quad \text{if} \quad \|\mathbf{x}_n(t)\| > \|\mathbf{x}(t)\|\left(1 + \frac{1}{n}\right).$$

Then the sequence of simple functions $\{\mathbf{y}_n(t)\}$ satisfies

$$\|\mathbf{y}_n(t)\| \leqslant \|\mathbf{x}(t)\|\left(1 + \frac{1}{n}\right) \leqslant 2\|\mathbf{x}(t)\|$$

and

$$\lim_{n\to\infty} \|\mathbf{x}(t) - \mathbf{y}_n(t)\| = 0 \qquad \mu\text{-a.e.}$$

Thus, by the μ-integrability of $\|\mathbf{x}(t)\|$ and by the Dominated Convergence Theorem for the Bochner Integral, $\mathbf{x}(t)$ is Bochner μ-integrable. ∎

Corollary. *Let T be a bounded linear operator on a B-space $\mathfrak{X}$ into a B-space $\mathfrak{Y}$, $T \in \mathfrak{E}(\mathfrak{X}, \mathfrak{Y})$. If $\mathbf{x}(t)$ is an $\mathfrak{X}$-valued Bochner μ-integrable function, then $T[\mathbf{x}(t)]$ is a $\mathfrak{Y}$-valued Bochner μ-integrable function, moreover*

$$\int T[\mathbf{x}(t)]\, d\mu = T[\int \mathbf{x}(t)\, d\mu]. \tag{7.5.36}$$

Proof. Let a sequence of simple functions $\{y_n(t)\}$ satisfy

$$\|\mathbf{y}_n(t)\| \leqslant \|\mathbf{x}(t)\|\left(1 + \frac{1}{n}\right) \quad \text{and} \quad \lim_{n\to\infty} \|\mathbf{y}_n(t) - \mathbf{x}(t)\| = 0 \qquad \mu\text{-a.e.}$$

It follows from the linearity and continuity of T that

$$\int T[\mathbf{y}_n(t)]\, d\mu = T[\int \mathbf{y}_n(t)\, d\mu].$$

Furthermore, by the continuity of T,

$$\|T[\mathbf{y}_n(t)]\| \leqslant \|T\|\ \|\mathbf{y}_n(t)\| \leqslant 2\|T\|\ \|\mathbf{x}(t)\|,$$
$$\lim \|T[\mathbf{y}_n(t)] - T[\mathbf{x}(t)]\| = 0 \qquad \mu\text{-a.e.}$$

Thus $T[\mathbf{x}(t)]$ is Bochner μ-integrable and

$$\int T[\mathbf{x}(t)]\,d\mu \overset{s}{=} \lim_{n\to\infty} \int T[\mathbf{y}_n(t)]\,d\mu$$

$$\overset{s}{=} \lim_{n\to\infty} T\left[\int \mathbf{y}_n(t)\,d\mu\right] = T\left[\int \mathbf{x}(t)\,d\mu\right],$$

as asserted.

EXERCISE 7.5

1. Let (S, A, μ) be an abstract measure space. Let F be the class of all real-valued A-measurable functions defined on S. Define what should be meant by saying that an element f of F is quadratically μ-integrable over S. State some properties of such functions.
2. With the measure space as in Problem 1, let $\mathfrak{X}$ be a complex B-space and let F be the class of all $\mathfrak{X}$-valued strongly A-measurable functions defined on S. Define what is to be meant by saying that an element $\mathbf{f}$ of F is quadratically μ-integrable.
3. Let $\mathfrak{B}^2 = \mathfrak{B}^2(S, \mathsf{A}, \mu; \mathfrak{X})$ be the subset of F for which $[\int \|\mathbf{f}(s)\|^2\,d\mu]^{\frac{1}{2}}$ is finite. Prove that this expression is a norm for the space $\mathfrak{B}^2$ and that $\mathfrak{B}^2$ is complete under this norm.
4. Define $\mathfrak{B}^p = \mathfrak{B}^p(S, \mathsf{A}, \mu; \mathfrak{X})$ for any p, $1 \leqslant p < \infty$. Define also $\mathfrak{B}^\infty$.
5. Prove Lemma 7.5.6. [*Hint:* Let $\{\mathbf{x}_n;\ n = 1, 2, \ldots\}$ be a countable set dense in $\mathfrak{X}$. Consider a mapping $x^* \to f_n(x^*) = [x^*(\mathbf{x}_1), x^*(\mathbf{x}_2), \ldots, x^*(\mathbf{x}_n)]$ of the unit sphere of $\mathfrak{X}^*$ into the n-dim Hilbert space $l^2(n)$.]

COLLATERAL READING

The reader may consult:

HILLE, E. and R. S. PHILLIPS, *Functional Analysis and Semi-groups,* American Mathematical Society Colloquium Publications, Vol. 31, revised edition, Providence, R.I. (1957). See, in particular, Chapters 2 and 3. There are numerous references there to the literature and a brief account is given of the large number of alternate definitions of integrability due to G. Birkhoff, N. Dunford, I. M. Gelfand, L. M. Graves, B. J. Pettis, and others. See also:

DUNFORD, N. and J. SCHWARTZ, *Linear Operators,* Part I. *General Theory,* Interscience, New York (1958).

8 COMPLEX ANALYSIS IN LINEAR SPACES

In this chapter we shall be concerned with analytical mappings. We start with analytical functions on complex numbers to vectors. Here it will be shown that a substantial portion of the elements of complex function theory extends to B-spaces. Analytic functions of infinitely many variables were considered by D. Hilbert (1909) and F. Riesz (1913). The basic extensions to abstract spaces were given by Norbert Wiener in 1923, but much was added from different points of view and along various lines by Nelson Dunford, L. Fantappiè (mainly analytic functionals), I. Gelfand, and A. E. Taylor in the 1930's.

There is also a theory of analyticity in commutative B-algebras from the algebra into itself due to E. R. Lorch (1943). There is a much older theory of analytic functions on vectors to vectors started by Maurice Fréchet in 1909 and developed under weaker assumptions by R. Gâteaux (published in 1919–22).

In the present chapter we shall restrict ourselves, on the one hand, to the generalization of the Cauchy theory to B-spaces, and, on the other, to the elements of the Fréchet theory.

There are four sections: Abstract holomorphic functions; Theorem of Vitali; Functions analytic in the sense of Fréchet; and Some properties of (F)-analytic functions.

8.1 ABSTRACT HOLOMORPHIC FUNCTIONS

We consider two B-spaces $\mathfrak{X}$ and $\mathfrak{Y}$ over the complex field, which may coincide, and their corresponding duals, $\mathfrak{X}^*$ and $\mathfrak{Y}^*$. Let s be a complex variable, D a domain in the complex s-plane, and $\mathbf{x} = \mathbf{f}(s)$ a mapping from D into $\mathfrak{X}$. This is a vector-valued function. We shall also have occasion to consider operator-valued functions. Let $T(s) \in \mathfrak{E}(\mathfrak{X}, \mathfrak{Y})$ be defined for s in D. The basic question is: What is to be understood by saying that $\mathbf{f}(s)$ or $T(s)$ is holomorphic in D?

In the Cauchy theory $f(s)$ is holomorphic in D if the difference quotient has a limit everywhere in D. This means that there exists a complex-valued function $f'(s)$ such that

$$\lim_{h \to 0} \left| \frac{1}{h} [f(s+h) - f(s)] - f'(s) \right| = 0. \tag{8.1.1}$$

Let Γ be a simple closed rectifiable oriented curve in D which, together with its interior, is a subset of D. Then it is known that for s interior to Γ we have the Cauchy integral representation

$$f(s) = \frac{1}{2\pi i}\int_{\Gamma}\frac{f(t)\,dt}{t-s}. \tag{8.1.2}$$

Here the integral is a Riemann–Stieltjes integral for the line of integration is a rectifiable curve, i.e. it admits of a representation in the form $t = t(\sigma)$, $0 \leqslant \sigma \leqslant \omega$ in terms of the arclength σ. The integral is then shorthand notation for

$$\frac{1}{2\pi i}\int_{0}^{\omega}\frac{f[t(\sigma)]}{t(\sigma)-s}\,dt(\sigma). \tag{8.1.3}$$

The integral exists since the integrand is a continuous function of σ. We shall see that these formulas extend to the abstract case.

Formula (8.1.1) states that the difference quotient tends to the derivative everywhere in D. Actually the convergence holds uniformly on compact subsets of D. This fact plays a basic role for the extensions, so we shall state and prove it in a form that will meet our needs.

Theorem 8.1.1. *For any function $f(s)$ holomorphic in the sense of Cauchy in the simply connected domain D and for any compact subset S of D there is a finite quantity $M(f; S)$ such that for every s, $s + h$ and $s + k$ in S*

$$\left|\frac{1}{h-k}\left\{\frac{1}{h}[f(s+h)-f(s)] - \frac{1}{k}[f(s+k)-f(s)]\right\}\right| \leqslant M(f; S). \tag{8.1.4}$$

This implies that the difference quotient tends uniformly to its limit in S.

Proof. Let Γ be a simple closed rectifiable oriented curve in D which contains S in its interior and which has a positive distance both from S and from the boundary of D. An elementary calculation based on (8.1.2) shows that the expression inside the absolute value sign in the left member of (8.1.4) is represented by the Cauchy integral

$$\frac{1}{2\pi i}\int_{\Gamma}\frac{f(t)\,dt}{(t-s)(t-s-h)(t-s-k)}. \tag{8.1.5}$$

Here $f(t)$ is bounded on Γ and the denominator is bounded away from zero uniformly in s, $s + h$, and $s + k$, and Γ is of finite length. This shows the existence of an $M(f; S)$ with the stated properties.

Actually (8.1.4) is a stronger assertion than what we need for uniform convergence. The latter is obtained as follows. We first let $k \to 0$, obtaining after multiplying the result by h,

$$\left|\frac{1}{h}[f(s+h)-f(s)] - f'(s)\right| \leqslant |h|\,M(f; S) \tag{8.1.6}$$

from which the uniform convergence of the difference quotient to the derivative follows upon letting $h \to 0$. ∎

Here the restriction to a simply connected domain is superfluous. There are classical extensions of Cauchy's integral to multiply connected domains which may be used in (8.1.5).

We now proceed to definitions of holomorphic vector and operator functions. In analogy with the situation for continuity we would expect weakly and strongly holomorphic vector functions and three varieties of holomorphic operator functions. Actually there is only one of each kind: the weakest property implies the strongest.

Definition 8.1.1. *With the notation as above, we say that* $\mathbf{f}(s)$ *and* $T(s)$ *are holomorphic in D if* $x^*[\mathbf{f}](s)$ *and* $y^*\{T(s)[\mathbf{x}]\}$ *are holomorphic in D in the sense of Cauchy for every choice of* $\mathbf{x} \in \mathfrak{X}$, $x^* \in \mathfrak{X}^*$, *and* $y^* \in \mathfrak{Y}^*$.

This is weaker than weak differentiability. Nevertheless, we shall see that the strongest form of differentiability follows. This is an elegant application of the uniform boundedness theorem.

Theorem 8.1.2. (1) *If* $\mathbf{f}(s)$ *is holomorphic in D, then* $\mathbf{f}(s)$ *is strongly continuous and strongly differentiable in D, uniformly with respect to s in any compact subset of D.* (2) *If* $T(s)$ *is holomorphic in D, then* $T(s)$ *is continuous and differentiable in the uniform operator topology for s in D, uniformly with respect to s in any compact subset of D.*

Proof. We shall give the proof for part (2), which involves two applications of the uniform boundedness principle. To simplify the notation we write

$$\frac{1}{h-k}\left\{\frac{1}{h}[T(s+h) - T(s)] - \frac{1}{k}[T(s+k) - T(s)]\right\} = T(s; h, k).$$

By Theorem 8.1.1 the assumptions in case (2) imply for the complex-valued function $y^*\{[T(s)[\mathbf{x}]\}$ the inequality

$$|y^*\{T(s; h, k)[\mathbf{x}]\}| \leqslant M(y^*, \mathbf{x}, T; S) \tag{8.1.7}$$

for every choice of s, $s + h$, $s + k$ in S. Here, as before, S is any chosen compact subset of D. By Theorem 7.2.3 this implies the existence of a finite $M(\mathbf{x}, T; S)$ such that

$$\|T(s; h, k)[\mathbf{x}]\| \leqslant M(\mathbf{x}, T; S). \tag{8.1.8}$$

Here we can apply Theorem 7.1.5, which in this case asserts the existence of a finite $M(T; S)$ such that

$$\|T(s; h, k)\| \leqslant M(T; S). \tag{8.1.9}$$

From now on we can proceed as in the proof of Theorem 8.1.1. We can let $h, k \to 0$ and use completeness of $\mathfrak{E}(\mathfrak{X}, \mathfrak{Y})$. If we multiply $T(s; h, k)$ by $h - k$ before passing to the limit, we see that the difference quotient has a limit $T'(s) \in \mathfrak{E}(\mathfrak{X}, \mathfrak{Y})$. For $k \to 0$ we obtain

$$\left\| \frac{1}{h}[T(s+h) - T(s)] - T'(s) \right\| \leqslant |h|\, M(T; S) \tag{8.1.10}$$

for all s and $s + h$ in S. Thus the difference quotient tends uniformly to its limit, the derivative, on compact subsets of D just as in the complex-valued case. This, of course, implies continuity of $T(s)$ in the uniform operator topology. We have also differentiability in the strong and the weak topologies.

The proof of part (1) is left to the reader. We shall refer to $\mathbf{f}(s)$ as $\mathfrak{X}$-*holomorphic* since it is by assumption a holomorphic function with values in the B-space $\mathfrak{X}$, with similar terminology in other cases. ∎

We have generalized the first approach to the notion of $\mathfrak{X}$-holomorphic of functions via differentiability of functionals. The same device leads to abstract Cauchy integrals, more generally integrals of the Cauchy type. Our point of departure is Theorem 7.4.2, integrals of the first kind

$$\int_a^b \mathbf{f}(t)\, dg(t). \tag{8.1.11}$$

Here $\mathbf{f}(t)$ is strongly continuous for t in $[a, b]$ and $g(t)$ is numerically valued and of bounded variation. The variable t is real in this formula, while we want to define integrals of the form

$$\int_\Gamma \mathbf{f}(t)\, dt \tag{8.1.12}$$

taken along a rectifiable oriented arc in the complex t-plane, where $\mathbf{f}(t)$ is strongly continuous for t on Γ. The reduction to type (8.1.11) is immediate. Again, using arclength as parameter, we have a representation of Γ by

$$t = t(\sigma), \qquad 0 \leqslant \sigma \leqslant \omega, \tag{8.1.13}$$

where $t(\sigma)$ is a continuous function of bounded variation, ω is the total length of Γ. We have then

$$\mathbf{J} = \int_0^\omega \mathbf{f}[t(\sigma)]\, dt(\sigma), \tag{8.1.14}$$

which is of type (8.1.11) since $\mathbf{f}[t(\sigma)]$ is strongly continuous in σ and has values in $\mathfrak{X}$. This is the sense to be attached to (8.1.12). Once we have made this convention we shall feel free to use (8.1.12) rather than the more elaborate formula (8.1.14).

The integral (8.1.12) has properties which again are analogous to those of the classical Riemann–Stieltjes integral. Thus the passage from $\mathbf{f}$ to $\mathbf{J}[\mathbf{f}]$ is a *linear operation*:

$$\int_\Gamma [\alpha_1 \mathbf{f}_1(t) + \alpha_2 \mathbf{f}_2(t)]\, dt = \alpha_1 \int_\Gamma \mathbf{f}_1(t)\, dt + \alpha_2 \int_\Gamma \mathbf{f}_2(t)\, dt. \tag{8.1.15}$$

It is also a *bounded operation*

$$\left\| \int_\Gamma \mathbf{f}(t)\,dt \right\| \leqslant \max_{t\in\Gamma} \|\mathbf{f}(t)\|\,\omega, \tag{8.1.16}$$

which is the natural generalization of the classical estimate. The integral is also an *additive set function of the path of integration.* If Γ is the union, suitably oriented, of two oriented subarcs Γ_1 and Γ_2 which have only endpoints in common, then

$$\int_\Gamma \mathbf{f}(t)\,dt = \int_{\Gamma_1} \mathbf{f}(t)\,dt + \int_{\Gamma_2} \mathbf{f}(t)\,dt. \tag{8.1.17}$$

We recall that bounded linear operators may be taken under the sign of integration so that

$$T\left\{\int_\Gamma \mathbf{f}(t)\,dt\right\} = \int_\Gamma T[\mathbf{f}](t)\,dt, \qquad T \in \mathfrak{E}(\mathfrak{X}, \mathfrak{Y}). \tag{8.1.18}$$

In particular, this holds for $T = x^* \in \mathfrak{X}^*$

$$x^*\left\{\int_\Gamma \mathbf{f}(t)\,dt\right\} = \int_\Gamma x^*[\mathbf{f}(t)]\,dt. \tag{8.1.19}$$

So far $\mathbf{f}$ was merely strongly continuous for t on Γ. For $\mathfrak{X}$-holomorphic functions we can prove Cauchy's theorem.

Theorem 8.1.3. *Suppose that* $\mathbf{f}(s)$ *is* $\mathfrak{X}$*-holomorphic in the sense of Definition* 8.1.1 *in a domain D of the complex plane. Suppose that* Γ *is a simple closed rectifiable oriented curve in D such that* $\mathbf{f}(s)$ *is holomorphic inside and on* Γ. *Then*

$$\int_\Gamma \mathbf{f}(t)\,dt = \mathbf{0}. \tag{8.1.20}$$

Remark. A curve $\Gamma = t(\sigma)$, $0 \leqslant \sigma \leqslant \omega$, is a simple closed curve iff

$$t(\sigma_1) = t(\sigma_2), \sigma_1 < \sigma_2 \quad \text{implies} \quad \sigma_1 = 0, \sigma_2 = \omega. \tag{8.1.21}$$

For such a curve the *Jordan curve theorem* assigns a definite meaning to the phrase "inside of Γ".

Proof. Apply (8.1.19) to the left member of (8.1.20). Then

$$x^*\left\{\int_\Gamma \mathbf{f}(t)\,dt\right\} = \int_\Gamma x^*[\mathbf{f}(t)]\,dt$$

for every $x^* \in \mathfrak{X}^*$. Since $x^*[\mathbf{f}(t)]$ is a complex-valued function holomorphic in the sense of Cauchy inside and on Γ, by the classical Cauchy theorem its integral along Γ is zero. Thus the integral in (8.1.20) is an element of $\mathfrak{X}$ which annihilates all functionals in $\mathfrak{X}^*$. Since the zero element of $\mathfrak{X}$ is the only one with this property, (8.1.20) holds. ∎

The same type of argument leads to Cauchy's integral.

Theorem 8.1.4. *If* $\mathbf{f}(s)$ *is* $\mathfrak{X}$*-holomorphic inside and on the simple closed rectifiable oriented curve* Γ, *then for* s *inside* Γ

$$\mathbf{f}(s) = \frac{1}{2\pi i}\int_\Gamma \frac{\mathbf{f}(t)\,dt}{t-s}. \tag{8.1.22}$$

Proof. Under the stated assumptions Cauchy's formula is valid for each of the complex-valid functions $x^*[\mathbf{f}(s)]$, $x^* \in \mathfrak{X}^*$, i.e.

$$x^*[\mathbf{f}(s)] = \frac{1}{2\pi i}\int_\Gamma \frac{x^*[\mathbf{f}(t)]\,dt}{t-s} = x^*\left\{\frac{1}{2\pi i}\int_\Gamma \frac{\mathbf{f}(t)\,dt}{t-s}\right\},$$

so that

$$x^*\left\{\mathbf{f}(s) - \frac{1}{2\pi i}\int_\Gamma \frac{\mathbf{f}(t)\,dt}{t-s}\right\} = 0, \qquad x^* \in \mathfrak{X}^*.$$

Thus (8.1.22) is valid. ∎

Theorem 8.1.5. *An* $\mathfrak{X}$*-holomorphic function* $\mathbf{f}(s)$ *has derivatives of all orders and*

$$\mathbf{f}^{(n)}(s) = \frac{n!}{2\pi i}\int_\Gamma \frac{\mathbf{f}(t)\,dt}{(t-s)^{n+1}}, \qquad \forall\, n. \tag{8.1.23}$$

Proof. If $\mathbf{f}(s)$ *is* $\mathfrak{X}$-holomorphic, then it has a derivative $\mathbf{f}'(s)$, and $x^*[\mathbf{f}'(s)]$ is holomorphic in the sense of Cauchy, $x^* \in \mathfrak{X}^*$. This, by Definition 8.1.1, makes $\mathbf{f}'(s)$ $\mathfrak{X}$-holomorphic and ensures the existence of $\mathbf{f}''(s)$. Since $x^*[\mathbf{f}''(s)]$ is Cauchy holomorphic, $\mathbf{f}''(s)$ is $\mathfrak{X}$-holomorphic, and so on. Thus the existence of all derivatives is ensured and each $\mathbf{f}^{(n)}(s)$ is $\mathfrak{X}$-holomorphic in the domain D where $\mathbf{f}(s)$ is supposed to have this property. Further, the representation (8.1.23) must hold, for both sides are meaningful as elements in $\mathfrak{X}$ and their difference annihilates every functional $x^* \in \mathfrak{X}^*$, and this requires that the difference is zero. ∎

Once we have Cauchy's integral at our disposal for $\mathfrak{X}$-holomorphic functions, then one of the main tools for developing a function theory for $\mathfrak{X}$-analytic functions is available. We shall give a few samples of results obtainable in this manner. We start with the *estimates of Cauchy*.

Theorem 8.1.6. *If* $\mathbf{f}(s)$ *is* $\mathfrak{X}$*-holomorphic in the closed disk* $|t - s_0| \leq r$ *and if* $\max_\theta \|\mathbf{f}(s_0 + re^{i\theta})\| = M$, *then*

$$\|\mathbf{f}^{(n)}(s_0)\| \leq Mr^{-n}n!. \tag{8.1.24}$$

Proof. In (8.1.23) set $s = s_0$, $t = s_0 + re^{i\theta}$, take Γ as the circumference of the disk and apply (8.1.16). ∎

The Cauchy estimates lead immediately to Taylor expansions for $\mathfrak{X}$-holomorphic functions.

Theorem 8.1.7. *If* $\mathbf{f}(s)$ *is* $\mathfrak{X}$*-holomorphic in a domain D, if* s_0 *is a point in D at a distance R from the boundary* ∂D *of D, then*

$$\mathbf{f}(s) = \sum_{n=0}^{\infty} \frac{1}{n!} \mathbf{f}^{(n)}(s_0)\,(s - s_0)^n, \tag{8.1.25}$$

where the series is absolutely convergent for $|s - s_0| < R$.

Proof. That the series is absolutely convergent for $|s - s_0| < R$ follows from (8.1.24). Here $\mathbf{f}(s)$ is $\mathfrak{X}$-holomorphic in the disk and for each $x^* \in \mathfrak{X}^*$ we have

$$x^*[\mathbf{f}(s)] = \sum_{n=0}^{\infty} \frac{1}{n!} x^*[\mathbf{f}^{(n)}(s_0)]\,(s - s_0)^n$$

in the same disk. By the usual argument this requires that (8.1.25) holds. ∎

In a similar manner we can prove Laurent expansions for our functions

$$\mathbf{f}(s) = \sum_{-\infty}^{\infty} \mathbf{a}_n(s - s_0)^n, \qquad \mathbf{a}_n \in \mathfrak{X}, \tag{8.1.26}$$

valid in any annulus $0 \leqslant R_1 < |s - s_0| < R_2 \leqslant \infty$, where $\mathbf{f}(s)$ is $\mathfrak{X}$-holomorphic. Here the coefficients $\mathbf{a}_n$ are given by analogues of the classical integral formulas. In particular, if $R_1 = 0$ there is an *isolated singularity* of $f(s)$ at $s = s_0$, namely a *pole of order m* if $\mathbf{a}_{-k} = \mathbf{0}$ for $k > m$ but $\mathbf{a}_{-m} \neq \mathbf{0}$. If no such m exists, we are dealing with an *essential singular point*.

We have also the *uniqueness* or *identity theorem* the proof of which is left to the reader.

Theorem 8.1.8. *If* $\mathbf{f}(s)$ *and* $\mathbf{g}(s)$ *are* $\mathfrak{X}$*-holomorphic in D and if a sequence* $\{s_n\}$ *exists with distinct points and at least one cluster point in D such that* $\mathbf{f}(s_n) = \mathbf{g}(s_n)$, $\forall\, n$, *then* $\mathbf{f}(s) \equiv \mathbf{g}(s)$.

Let us return to formula (8.1.23). Here we set $s = s_0$, $t = s_0 + re^{i\theta}$ and obtain

$$\frac{r^n}{n!} \mathbf{f}^{(n)}(s_0) = \frac{1}{2\pi} \int_0^{2\pi} \mathbf{f}(s_0 + re^{i\theta})\, e^{-ni\theta} d\theta. \tag{8.1.27}$$

This states that the left member is the nth trigonometric Fourier coefficient of $\mathbf{f}(s_0 + re^{i\theta})$ considered as a function of θ. We observe that all Fourier coefficients with negative subscript are zero. The case $n = 0$ is particularly interesting. Here

$$\mathbf{f}(s_0) = \frac{1}{2\pi} \int_0^{2\pi} \mathbf{f}(s_0 + re^{i\theta})\, d\theta. \tag{8.1.28}$$

This shows that $\mathbf{f}(s)$ has the *mean value property* which pertains to Cauchy holomorphic functions as well as to harmonic functions. The formula has an important consequence, namely

$$\|\mathbf{f}(s_0)\| \leqslant \frac{1}{2\pi} \int_0^{2\pi} \|\mathbf{f}(s_0 + re^{i\theta})\|\, d\theta. \tag{8.1.29}$$

This inequality shows that $\|\mathbf{f}(s)\|$ is a so-called *subharmonic function* in D. Such a function has the important property of *having no local maximum in D unless it is identically constant.* We shall give a direct proof of this property rather than appeal to the theory of subharmonic functions which cannot be supposed to be familiar to the reader. The *Principle of the Maximum* now reads:

Theorem 8.1.9. *If* $\mathbf{f}(s)$ *is* $\mathfrak{X}$*-holomorphic inside and on the simple closed rectifiable curve* Γ *and if*

$$\max_{t \in \Gamma} \|\mathbf{f}(t)\| = M, \tag{8.1.30}$$

then for every s inside Γ

$$\|\mathbf{f}(s)\| \leqslant M \tag{8.1.31}$$

with equality for one such s iff there is equality for all.

Proof. Again we use the analyticity of the functionals when applied to $\mathbf{f}(s)$. Here, $x^*[\mathbf{f}(s)]$ is Cauchy holomorphic inside and on Γ, and by the principle of the maximum for such functions

$$|x^*[\mathbf{f}(s)]| \leqslant \max_{t \in \Gamma} |x^*[\mathbf{f}(t)]| \equiv M(\mathbf{f}: x^*)$$

with equality iff $x^*[\mathbf{f}(s)]$ is a constant of absolute value $M(\mathbf{f}; x^*)$. Here $M(\mathbf{f}; x^*) \leqslant M\|x^*\|$. On the other hand,

$$\|\mathbf{f}(s)\| = \sup_{\|x^*\|=1} |x^*[\mathbf{f}(s)]| \leqslant \sup_{\|x^*\|=1} M(\mathbf{f}; x^*) \leqslant M$$

so that (8.1.31) holds everywhere inside and on Γ.

Suppose now that there is a point s_0 inside Γ where $\|\mathbf{f}(s_0)\| = M$. We then appeal to (8.1.29) and obtain

$$M = \|\mathbf{f}(s_0)\| \leqslant \frac{1}{2\pi} \int_0^{2\pi} \|\mathbf{f}(s_0 + re^{i\theta})\| \, d\theta, \tag{8.1.32}$$

where the integrand is everywhere $\leqslant M$ and $r \leqslant d(s_0, \Gamma)$. But a function with such an upper bound cannot have an average value $\geqslant M$ unless it equals M everywhere. Hence $\|\mathbf{f}(s)\| \equiv M$ for s anywhere in the disk $|t - s_0| \leqslant d(s_0, \Gamma)$.

To extend this to the rest of the interior of Γ, we argue as follows (essentially an analytic continuation argument). If s_1 is a point inside Γ but outside the disk just mentioned, we can draw a polygonal line joining s_0 to s_1 with vertices at

$$t_0 = s_0, t_1, t_2, \ldots, t_n = s_1$$

such that for each j

$$|t_{j+1} - t_j| < d(t_j, \Gamma), \qquad j = 0, 1, \ldots, n-1.$$

In the first disk, $|s - t_0| < d(t_0, \Gamma)$, the identity $\|\mathbf{f}(s)\| \equiv M$ holds. In particular it holds at $s = t_1$. We can then apply (8.1.32) again, replacing s_0 by t_1, and find that $\|\mathbf{f}(s)\| \equiv M$ is true also in the second disk $|s - t_1| < d(t_1, \Gamma)$ and so on. Since s_1

is arbitrary, the identity holds everywhere inside Γ. It also holds on Γ by the continuity of the norm. ∎

In Theorem 8.1.9 $\mathfrak{X}$-holomorphism of $\mathbf{f}(s)$ on Γ is more than enough; continuity on and inside Γ plus $\mathfrak{X}$-holomorphism inside Γ is sufficient for the desired conclusion.

The conclusion in the case that there is a point $s = s_0$ inside Γ where $\|\mathbf{f}(s_0)\| = M$ differs from that in the classical case where we simply have $\mathbf{f}(s)$ equal to a constant of absolute value M. No sharper conclusion is possible in the $\mathfrak{X}$-valued case, however. This may be concluded from the situation in the space $\mathfrak{M}_2$ with norm

$$\|\mathcal{A}\| = \max_j (|a_{j1}| + |a_{j2}|).$$

Here we take the matrix

$$\mathcal{A}(s) = \begin{bmatrix} 1 & 0 \\ 0 & s \end{bmatrix} = \mathcal{E}_{11} + \mathcal{E}_{22}\, s. \tag{8.1.33}$$

This is a polynomial in s and certainly not a constant. Since

$$\|\mathcal{A}(s)\| = \max(1, |s|)$$

it is seen that $\|\mathcal{A}(s)\| \equiv 1$ for $|s| \leqslant 1$.

EXERCISE 8.1

1. Verify (8.1.5).
2. Verify (8.1.15) to (8.1.17).
3. Show that the series (8.1.25) converges absolutely for $|s - s_0| < R$. Why is the sum of the series an $\mathfrak{X}$-holomorphic function of s in this disk?
4. Give a representation of the coefficients $\mathbf{a}_n$ of formula (8.1.26).
5. [Theorem of Liouville.] If $\mathbf{f}(s)$ is $\mathfrak{X}$-holomorphic throughout the finite plane and if $\|\mathbf{f}(re^{i\theta})\| \leqslant Mr^{\alpha}$, $0 \leqslant \theta \leqslant 2\pi$, $1 \leqslant r$, M and α fixed, $0 \leqslant \alpha$, show that $\mathbf{f}$ is a polynomial in s of degree $\leqslant \alpha$. [What is a polynomial of degree $\leqslant \frac{1}{2}$?]
6. If $\mathbf{f}$ is $\mathfrak{X}$-holomorphic in $0 < |s - s_0| < R$ and if $\|\mathbf{f}(s_0 + re^{i\theta})\| \leqslant Mr^{-\alpha}$, $0 \leqslant \theta < 2\pi$, $0 < r < \min(1, R)$, α fixed, $0 < \alpha$, show that $s = s_0$ is a pole of $\mathbf{f}$ of order $\leqslant \alpha$. What conclusion may be drawn if the inequality holds for $\alpha = 0$?
7. [Schwarz's Lemma.] If $\mathbf{f}$ is $\mathfrak{X}$-holomorphic for $|s| < 1$, $\mathbf{f}(0) = \mathbf{0}$ and $\|\mathbf{f}(s)\| \leqslant M$, show that $\|\mathbf{f}(s)\| \leqslant M|s|$, $|s| < 1$. What is the condition for equality at some point $s = s_0$, $s_0 \neq 0$?

8. Referring to formula (8.1.33) for notation, we want to construct a linear bounded functional x^* on $\mathfrak{M}_2$ such that $x^*[\mathcal{A}(s)]$ is (i) a constant on $|s| \leqslant 1$, (ii) not a constant. Show that in the first case the functional must be a constant for all s and not merely in the unit disk.

9. Let the sequence $\{\mathbf{f}_n\}$ be made up of functions which are $\mathfrak{X}$-holomorphic in a domain D. Suppose that the sequence converges uniformly on compact subsets of D to a limit $\mathbf{f}(s)$. Show that $\mathbf{f}$ is $\mathfrak{X}$-holomorphic in D.

10. With the same notation as in Problem 9, suppose that D is simply connected and its boundary ∂D is a simple closed rectifiable oriented curve Γ. Suppose that the functions $\mathbf{f}_n$ are strongly continuous in the closure of D and $\mathfrak{X}$-holomorphic in D. Suppose that the sequence $\{\mathbf{f}_n\}$ converges uniformly on Γ. Show that there is uniform convergence (in the strong topology of $\mathfrak{X}$) everywhere in the closure of D and the limit function $\mathbf{f}$ is $\mathfrak{X}$-holomorphic in D.

11. With the same assumptions, show the sequence $\{\mathbf{f}_n^{(p)}(s)\}$, for any fixed positive integer p, converges uniformly to $\mathbf{f}^{(p)}(s)$ on compact subsets of D.

12. [Theorem of Morera.] If $\mathbf{f}(s)$ is strongly continuous in a simply connected domain D and if (8.1.20) holds for any choice of Γ as the perimeter of a triangle which together with its interior is a subset of D, then $\mathbf{f}$ is $\mathfrak{X}$-holomorphic in D.

13. Prove Theorem 8.1.8.

14. If $\mathbf{f}$ is $\mathfrak{X}$-holomorphic in D, prove that its zeros, if any, can have no cluster point in D.

15. A function $\mathbf{f}$ is $\mathfrak{M}_n$-holomorphic in a domain D. Show that $\mathbf{f}$ is either algebraically singular for all s or the algebraic singularities have no cluster point in D.

16. $\mathfrak{F}(s)$ is a polynomial in s of degree $m > 0$ with coefficients in $\mathfrak{M}_n$. Does $\mathfrak{F}(s)$ necessarily have (1) zeros, (2) algebraic singularities?

8.2 THEOREM OF VITALI

We shall give a brief exposition of a result relating to "induced convergence" of sequences of $\mathfrak{X}$-holomorphic functions. The property of being $\mathfrak{X}$-holomorphic is not necessarily preserved under convergence, but it is so strongly adherent that it is preserved if the functions and the mode of convergence satisfy rather mild restrictions. Problem 10 of Exercise 8.1 (with $\mathfrak{X} = C$) is one of the earliest instances of *propagation of convergence*, in this case inward propagation from the boundary to the interior. For complex-valued functions Carl Runge (1836–1927) found this theorem in 1884. Examples of outward propagation, from a subdomain to a containing domain, are known from the 1890's. The problem of whittling down the assumptions was attacked and brought to a satisfactory conclusion in 1903 by Guiseppe Vitali (1875–1932). The result is known as *Vitali's Theorem.* It was rediscovered by the brilliant but little-known American mathematician Milton Brockett Porter (1869–1960) in 1904. Actually there are two Vitali theorems of which only one lends itself readily to extensions for $\mathfrak{X}$-valued functions. The other requires compactness assumptions which are but rarely satisfied in B-spaces. The generalized Vitali theorem reads as follows:

Theorem 8.2.1. *Suppose that* $\{\mathbf{f}_n\}$ *is a sequence of functions* $\mathfrak{X}$*-holomorphic in a domain D where* $\|\mathbf{f}_n(s)\| \leqslant M$ *for all n and all s. Suppose there exists a sequence* $\{s_k\}$ *in D with a cluster point* s_0 *in D such that* $\lim_{n\to\infty} \mathbf{f}_n(s_k)$ *exists for all k. Then* $\lim_{n\to\infty} \mathbf{f}_n(s) \equiv \mathbf{f}(s)$ *exists for all s in D, the limit* $\mathbf{f}(s)$ *is* $\mathfrak{X}$*-holomorphic in D, and the convergence is uniform on compact subsets.*

The original proof of Vitali does not generalize, but an alternate proof found by Ernst Lindelöf in 1913 does extend to abstract spaces. It deals with the special case where D is the open unit disk, $|s| < 1$, and $s_0 = 0$. From this the general result is obtained by routine analysis using the same type of argument as in the proof of the Principle of the Maximum. There is no restriction in taking $M = 1$. Lindelöf's theorem reads:

Theorem 8.2.2. *Suppose that*

$$\mathbf{f}_n(s) = \sum_{j=0}^{\infty} \mathbf{a}_{nj} s^j, \qquad n = 1, 2, 3, \ldots \tag{8.2.1}$$

where the series are absolutely convergent for $|s| < 1$ *and* $\|\mathbf{f}_n(s)\| < 1$, $\forall\, n, s$. *Suppose there is a sequence* $\{s_k\}$ *converging to* 0 *such that*

$$\lim_{n\to\infty} \mathbf{f}_n(s_k) \tag{8.2.2}$$

exists for all k. Then

$$\lim_{n\to\infty} \mathbf{f}_n(s) \equiv \mathbf{f}(s) \tag{8.2.3}$$

exists for $|s| < 1$, *the convergence is uniform for* $|s| \leqslant R < 1$, *any R, and the limit is* $\mathfrak{X}$*-holomorphic for* $|s| < 1$.

Proof. The gist of the proof is to show that all sequences $\{\mathbf{a}_{nj};\, n = 1, 2, 3, \ldots\}$, $j = 0, 1, 2, \ldots$, are Cauchy sequences. Once this has been done for $j = 0$, it will be seen that the same result holds for $j > 0$ by a suitable induction process.

We start by observing that

$$\|\mathbf{a}_{nj}\| \leqslant 1, \qquad \forall\, j, n \tag{8.2.4}$$

by Cauchy's inequalities. Next we use Schwarz's Lemma (Problem 7, Exercise 8.1). Since $\mathbf{f}_n(s) - \mathbf{a}_{n0} = \mathbf{0}$ for $s = 0$ and $\|\mathbf{f}_n(s) - \mathbf{a}_{n0}\| \leqslant 2$, it follows that

$$\|\mathbf{f}_n(s) - \mathbf{a}_{n0}\| \leqslant 2|s|, \qquad \forall\, n. \tag{8.2.5}$$

This we use to prove that $\{\mathbf{a}_{n0}\}$ is a Cauchy sequence. Now

$$\|\mathbf{a}_{n0} - \mathbf{a}_{p0}\| = \|\mathbf{f}_n(s_k) - \mathbf{a}_{n0} - \mathbf{f}_n(s_k) + \mathbf{f}_p(s_k) - \mathbf{f}_p(s_k) + \mathbf{a}_{p0}\|$$

for any choice of s_k in the given sequence. Here the second member is at most equal to

$$\|\mathbf{f}_n(s_k) - \mathbf{a}_{n0}\| + \|\mathbf{f}_n(s_k) - \mathbf{f}_p(s_k)\| + \|\mathbf{f}_p(s_k) - \mathbf{a}_{p0}\| \leqslant 4|s_k| + \|\mathbf{f}_n(s_k) - \mathbf{f}_p(s_k)\|$$

by (8.2.5). Here we can choose k so large that

$$|s_k| < \tfrac{1}{5}\varepsilon,$$

where ε is a preassigned, arbitrarily small positive number. Further, for each fixed k the sequence $\{\mathbf{f}_n(s_k)\}$ is Cauchy so there is an N depending upon ε and k such that for $n, p > N$

$$\|\mathbf{f}_n(s_k) - \mathbf{f}_p(s_k)\| < \tfrac{1}{5}\varepsilon.$$

Thus for $n, p > N$

$$\|\mathbf{a}_{n0} - \mathbf{a}_{p0}\| < \varepsilon.$$

Since ε is arbitrary, this shows that $\{\mathbf{a}_{n0}\}$ is a Cauchy sequence in $\mathfrak{X}$. We set

$$\lim_{n\to\infty} \mathbf{a}_{n0} = \mathbf{a}_0 \tag{8.2.6}$$

and note that

$$\|\mathbf{a}_0\| \leqslant 1 \tag{8.2.7}$$

by the continuity of the norm.

Suppose that we have proved, for $j \leqslant m$, the existence of

$$\lim_{n\to\infty} \mathbf{a}_{nj} = \mathbf{a}_j \tag{8.2.8}$$

and the inequality

$$\|\mathbf{a}_j\| \leqslant 1. \tag{8.2.9}$$

We now define a sequence of functions $\{\mathbf{f}_{n,m+1}(s)\}$ by

$$\mathbf{f}_n(s) - \sum_{j=0}^{m} \mathbf{a}_{nj}s^j = s^{m+1}\mathbf{f}_{n,m+1}(s). \tag{8.2.10}$$

Our next aim is to show that this sequence has the same properties as the original sequence $\{\mathbf{f}_n\}$, i.e. convergence for $s = s_k$, $\forall\, k$, and uniform boundedness in $|s| < 1$, the only difference being that we get a common bound equal to $m + 2$ instead of 1 which does not affect the conclusion that $\{\mathbf{f}_{n,m+1}(0)\} = \{\mathbf{a}_{n,m+1}\}$ is a Cauchy sequence for fixed m.

It is clear that $\lim_{n\to\infty} \mathbf{f}_{n,m+1}(s_k)$ exists for each k. To prove uniform boundedness is more laborious. We choose two numbers r and R, $0 < r < R < 1$. Consider

$$s^{-m-1}\,[\mathbf{f}_n(s) - \sum_{j=0}^{m} \mathbf{a}_{nj}s^j]$$

for $|s| = R$. Since $\|\mathbf{f}_n(s)\| \leqslant 1$ and $\|\mathbf{a}_{nj}\| \leqslant 1$, $\forall\, j, n$, the quantity displayed cannot exceed

$$(m + 2)\, R^{-m-1}$$

in norm. By the Principle of the Maximum,

$$\max_{\theta} \|\mathbf{f}_{n,m+1}(re^{i\theta})\| \leqslant (m+2)\, R^{-m-1}, \tag{8.2.11}$$

and this holds for any fixed $r < R$. Now the left member is independent of R, so we may let $R \to 1$ and obtain

$$\max_{\theta} \|\mathbf{f}_{n,m+1}(re^{i\theta})\| \leqslant m+2$$

or, simply,

$$\|\mathbf{f}_{n,m+1}(s)\| \leqslant m+2 \tag{8.2.12}$$

for all s with $|s| < 1$. This proves the uniform boundedness.

A new application of Schwarz's Lemma gives

$$\|\mathbf{f}_{n,m+1}(s) - \mathbf{a}_{n,m+1}\| \leqslant (m+3)\,|s|$$

true for all n and all m. From this point on the argument proceeds as above and we conclude that (8.2.8) and (8.2.9) hold also for $j = m+1$ and hence for all m.

With the aid of the coefficients $\mathbf{a}_m$ we now form the power series

$$\mathbf{f}(s) = \sum_{m=0}^{\infty} \mathbf{a}_m s^m \tag{8.2.13}$$

and aim to show that

$$\lim_{n\to\infty} \|\mathbf{f}_n(s) - \mathbf{f}(s)\| = 0, \tag{8.2.14}$$

the convergence being uniform for $|s| \leqslant R < 1$. The series (8.2.13) is obviously absolutely convergent for $|s| < 1$, and its sum is $\mathfrak{X}$-holomorphic in this disk. We have now

$$\|\mathbf{f}_n(s) - \mathbf{f}(s)\| \leqslant \sum_{j=0}^{m} \|\mathbf{a}_{nj} - \mathbf{a}_j\|\ |s|^j + \sum_{j=m+1}^{\infty} \|\mathbf{a}_{nj}\|\ |s|^j + \sum_{j=m+1}^{\infty} \|\mathbf{a}_j\|\ |s|^j,$$

where $|s| \leqslant R$. Since the coefficients $\mathbf{a}_{nj}$ and $\mathbf{a}_j$ are at most of norm 1,

$$\sum_{m+1}^{\infty} \|\mathbf{a}_{nj}\|\ |s|^j \leqslant \sum_{m+1}^{\infty} R^j = (1-R)^{-1} R^{m+1}, \qquad \forall\, n,$$

and the same estimate is valid for the second infinite series.

Here we can choose m so large that

$$2\,(1-R)^{-1} R^{m+1} < \tfrac{1}{2}\varepsilon$$

for a given fixed ε. Since $\lim_{n\to\infty} \mathbf{a}_{nj} = \mathbf{a}_j$ for each fixed j, we can choose N so large that for fixed m and $n > N$

$$\sum_{j=0}^{m} \|\mathbf{a}_{nj} - \mathbf{a}_j\|\ R^j < \tfrac{1}{2}\varepsilon.$$

Combining, we get

$$\|\mathbf{f}_n(s) - \mathbf{f}(s)\| < \varepsilon$$

for $n > N$ and $|s| \leqslant R$. Here ε is arbitrary, so (8.2.14) follows. ∎

Proof of Theorem 8.2.1. The proof is essentially an exercise in the use of analytic continuation together with the identity theorem. Without restricting the generality we may assume $s_0 = 0$ and that the disk $\{s; |s| < 1\}$ belongs to D. We have then seen that $\lim_{n\to\infty} \mathbf{f}_n(s) \equiv \mathbf{f}(s)$ exists in the disk and that the convergence is uniform in any smaller concentric disk, say $|s| \leqslant R$. Suppose that $s_1 \in D$ but $|s_1| > R$. We can now join s_1 to the origin by a polygonal line in D with vertices

$$t_0 = 0,\ t_1,\ t_2,\ \ldots,\ t_n = s_1,$$

chosen in such a manner that for each j

$$|t_{j+1} - t_j| < Rd(t_j, \partial D). \tag{8.2.15}$$

We have $\lim_{n\to\infty} \mathbf{f}_n(s) \equiv \mathbf{f}(s)$ in the disk $\{s; |s| < 1\}$ with uniform convergence in the concentric disk $\{s; |s| < R\}$. The latter contains the point $s = t_1$ together with a small neighborhood of this point. In this neighborhood $\mathbf{f}_n(s) \to \mathbf{f}(s)$ everywhere. Further, $\|\mathbf{f}_n(s)\| \leqslant 1$ for all s in the disk

$$|s - t_1| < d(t_1, \partial D) \tag{8.2.16}$$

if we take $M = 1$, which is no restriction. In this disk we have power series expansions

$$\mathbf{f}_n(s) = \sum_{j=0}^{\infty} \mathbf{b}_{nj}(s - t_1)^j$$

analogous to (8.2.1). The radius of convergence of these series is at least $d(t_1, \partial D)$. If this distance is <1, the coefficients are not necessarily bounded uniformly with respect to j, but the argument used in proving Theorem 8.2.2 holds with minor modifications. Thus we conclude that $\lim_{n\to\infty} \mathbf{f}_n(s) \equiv \mathbf{g}(s)$ exists in the disk (8.2.16) and the limit exists uniformly in the smaller disk

$$|s - t_1| < Rd(t_1, \partial D).$$

But in a small neighborhood of the point $s = t_1$ the limit is known to be $\mathbf{f}(s)$. Now both $\mathbf{f}(s)$ and $\mathbf{g}(s)$ are known to be locally $\mathfrak{X}$-holomorphic so that the identity theorem asserts that

$$\mathbf{g}(s) \equiv \mathbf{f}(s).$$

More precisely expressed, $\mathbf{g}(s)$ is the analytic continuation of $\mathbf{f}(s)$ in (8.2.16) and there is one and only one $\mathfrak{X}$-holomorphic function, hereinafter denoted by $\mathbf{f}(s)$, that is the limit of the sequence $\{\mathbf{f}_n(s)\}$ in the union of $\{s; |s| < 1\}$ with the disk (8.2.16). By (8.2.15) the smaller disks of uniform convergence overlap.

In this manner we can proceed from one vertex to the next. After n steps we reach s_1. Thus the polygonal line joining s_1 to 0 is covered by a chain of a finite number of overlapping disks in which the sequence $\{\mathbf{f}_n(s)\}$ converges uniformly to the function $\mathbf{f}(s)$. Here s_1 is any point of D so the limit exists everywhere in D and every compact subset may be included in a region of uniform convergence. ∎

We note that the validity of the identity theorem for $\mathfrak{X}$-holomorphic functions implies that *the Weierstrass principle of analytic continuation holds* for such functions. We have also the *Theorem of Monodromy*:

Theorem 8.2.3. *If a locally $\mathfrak{X}$-holomorphic function* $\mathbf{f}$ *can be continued analytically along every path in a simply connected domain D, then* $\mathbf{f}$ *is $\mathfrak{X}$-holomorphic in D.*

For the classical theorem of monodromy holds for the functionals $x^*[\mathbf{f}]$. Details are left to the reader.

We have based the theory of $\mathfrak{X}$-analyticity upon that of C-analyticity via the functionals. Thus $\mathbf{f}$ is $\mathfrak{X}$-holomorphic in a domain D iff $x^*[\mathbf{f}]$ is C-holomorphic in D for every $x^* \in \mathfrak{X}^*$. This implies that a point $s = s_0$ on ∂D where $\mathbf{f}$ is not $\mathfrak{X}$-holomorphic must be a singular point of at least one (and hence infinitely many) of the functions $x^*[\mathbf{f}]$. It is by no means necessary that all functions $x^*[\mathbf{f}]$ have a singularity at s_0. Thus we see that for a single-valued function $\mathbf{f}$ *the singular points of* $\mathbf{f}$ *are the union of the singularities of* $x^*[\mathbf{f}]$, *where x^* ranges over $\mathfrak{X}^*$. Similarly, the maximal domain of holomorphy of* $\mathbf{f}$ *is the intersection of all the corresponding domains for* $x^*[\mathbf{f}]$.

EXERCISE 8.2

1. How should the proof of Theorem 8.2.2 be modified if the series converges for $|s| < r$ instead of $|s| < 1$ and $\|\mathbf{f}_n(s)\| \leqslant M$?

2. Prove Theorem 8.2.3.

3. [Theorem of the Phragmén–Lindelöf type.] Suppose that $\mathbf{g}$ is $\mathfrak{X}$-holomorphic and bounded in the right half-plane. Suppose that $\lim_{r \to +\infty} \mathbf{g}(r) = \mathbf{a}$. Show that $\lim_{r \to \infty} \mathbf{g}(re^{i\theta}) = \mathbf{a}$ for $-\frac{1}{2}\pi < \theta < \frac{1}{2}\pi$, uniformly in any closed interior sector. [*Hint:* Set $\mathbf{f}_n(s) = \mathbf{g}(ns)$ for $|s| < 1$ and apply Vitali's theorem.]

4. If $\mathbf{g}$ is holomorphic in the right half-plane and there are two rays $\arg s = \theta_1$, $\arg s = \theta_2$ in this half-plane along which $\mathbf{g}$ tends to distinct limits $\mathbf{a}$ and $\mathbf{b}$, show that $\mathbf{g}$ cannot be bounded in any sector containing the rays in its interior.

5. [Theorem of F. Carlson, H. Cramér, S. Wigert type.] If $\mathbf{f}$ is $\mathfrak{X}$-holomorphic and bounded in the right half-plane and vanishes at the positive integers, show that $\mathbf{f}(s) \equiv \mathbf{0}$. [*Hint*: Consider the sequence $\{\mathbf{f}(2^n s)\}$ and construct a sequence $\{s_k\}$, $\lim s_k = 1$ such that $\lim_{n\to\infty} \mathbf{f}(2^n s_k) = \mathbf{0}, \forall k$.]

8.3 FUNCTIONS ANALYTIC IN THE SENSE OF FRÉCHET

The theory of analytic functions on vectors to vectors goes back to Maurice Fréchet, who in 1909 introduced *abstract differentials*, *polynomials*, and *power series*. Among later contributors to this field we note R. Gâteaux, L. M. Graves, I. E. Highberg, R. S. Martin, A. D. Michal, J. Sebastião e Silva, A. E. Taylor, and Max Zorn.

Consider two B-spaces $\mathfrak{X}$ and $\mathfrak{Y}$ over the complex field and a mapping $\mathbf{y} = \mathbf{f}(\mathbf{x})$ from $\mathfrak{X}$ to $\mathfrak{Y}$ defined in a domain $\mathfrak{D} \subset \mathfrak{X}$. Here "domain" means an *open connected set*: open in the strong topology of $\mathfrak{X}$ and connected is understood to mean that any two points of $\mathfrak{D}$ may be joined in $\mathfrak{D}$ by a polygonal line consisting of a finite number of line segments and hence of finite length. We shall be concerned with the properties of $\mathbf{f}(\mathbf{x} + \alpha\mathbf{h})$ where $\mathbf{x} \in \mathfrak{D}$, $\mathbf{h} \in \mathfrak{X}$, α is a complex variable and $|\alpha|$ is so small that $\mathbf{x} + \alpha\mathbf{h} \in \mathfrak{D}$. We plan to examine the differentiability properties of $\mathbf{f}(\mathbf{x} + \alpha\mathbf{h})$ with respect to α.

A simple example will make the problem setting clearer and serve to motivate the terminology to be used. Take $\mathfrak{X} = L_2(-\pi, \pi]$, $\mathfrak{Y} = C[-\pi, \pi]$, $\mathfrak{D} = \mathfrak{X}$, and

$$\mathbf{f}(\mathbf{x})[t] = \int_0^t [\mathbf{x}(s)]^2\, ds, \qquad -\pi \leqslant t \leqslant \pi. \tag{8.3.1}$$

Since

$$\mathbf{f}(\alpha\mathbf{x}) = \alpha^2 \mathbf{f}(\mathbf{x}),$$

we refer to $\mathbf{f}(\mathbf{x})$ as a *homogeneous quadratic polynomial* in $\mathbf{x}$. Here

$$\begin{aligned}\mathbf{f}(\mathbf{x} + \alpha\mathbf{h}) &= \int_0^t [\mathbf{x}(s) + \alpha\mathbf{h}(s)]^2\, ds \\ &= \int_0^t [\mathbf{x}(s)]^2\, ds + 2\alpha \int_0^t \mathbf{x}(s)\,\mathbf{h}(s)\, ds + \alpha^2 \int_0^t [\mathbf{h}(s)]^2\, ds.\end{aligned}$$

The difference quotients

$$\frac{1}{\alpha}[\mathbf{f}(\mathbf{x} + \alpha\mathbf{h}) - \mathbf{f}(\mathbf{x})] \quad \text{and} \quad \frac{1}{\alpha^2}[\mathbf{f}(\mathbf{x} + 2\alpha\mathbf{h}) - 2f(\mathbf{x} + \alpha\mathbf{h}) + \mathbf{f}(\mathbf{x})]$$

tend to limits as $\alpha \to 0$, namely

$$2\int_0^t \mathbf{h}(s)\,\mathbf{x}(s)\, ds \quad \text{and} \quad 2\int_0^t [\mathbf{h}(s)]^2\, ds$$

respectively. They are denoted by

$$\delta\mathbf{f}(\mathbf{x};\mathbf{h}) \quad \text{and} \quad \delta^{(2)}\mathbf{f}(\mathbf{x};\mathbf{h}), \tag{8.3.2}$$

and are known as the *first and second Fréchet differentials of* **f** *at the point* **x** *with respect to the increment* **h**. The term "variations" is also used. There are, of course, also differentials of higher order, but they are all zero in the present case. These differentials have remarkable properties. Thus $\delta\mathbf{f}(\mathbf{x};\mathbf{h})$ is homogeneous of the first order both in **x** and in **h**, and for fixed **x** we see that $\delta\mathbf{f}(\mathbf{x};\cdot) \in \mathfrak{E}(\mathfrak{X}, \mathfrak{Y})$. The second differential is homogeneous of the second order in **h**. Moreover,

$$\|\mathbf{f}(\mathbf{x}+\mathbf{h}) - \mathbf{f}(\mathbf{x}) - \delta\mathbf{f}(\mathbf{x};\mathbf{h})\| \leqslant C(\|\mathbf{h}\|)^2, \tag{8.3.3}$$

where, on the left, we have the sup-norm, i.e. the norm in $\mathfrak{Y}$, and on the right the L_2-norm of $\mathfrak{X}$.

The reader may object that these properties are obvious if not trivial, but the important and not so obvious thing is that they generalize to much more complicated situations.

Some definitions are in order at this point.

Definition 8.3.1. *A function* **f** *from* $\mathfrak{X}$ *to* $\mathfrak{Y}$ *with* $\mathfrak{D} = \mathfrak{X}$ *is said to be homogeneous of degree n if*

$$\mathbf{f}(\alpha\mathbf{x}) = \alpha^n\mathbf{f}(\mathbf{x}), \qquad \forall\, \alpha \in C. \tag{8.3.4}$$

Such a function is said to be bounded if there is a finite M such that

$$\|\mathbf{f}(\mathbf{x})\| \leqslant M\|\mathbf{x}\|^n, \qquad \forall\, \mathbf{x}. \tag{8.3.5}$$

The situation covered by this definition may be regarded as an extension of the linear case. If $T \in \mathfrak{E}(\mathfrak{X}, \mathfrak{Y})$, then $T(\mathbf{x})$ is homogeneous of degree one in **x** and is bounded, both properties in the sense of Definition 8.3.1 with $n = 1$.

Definition 8.3.2. *A function* **f** *defined on a domain* $\mathfrak{D} \subset \mathfrak{X}$ *with range in* $\mathfrak{Y}$ *is said to be Fréchet differentiable at* $\mathbf{x}_0 \in \mathfrak{D}$ *if* (1)

$$\lim_{\beta\to 0} \frac{1}{\beta}[\mathbf{f}(\mathbf{x}_0 + \beta\mathbf{h}) - \mathbf{f}(\mathbf{x}_0)] \equiv \delta\mathbf{f}(\mathbf{x}_0;\mathbf{h}) \tag{8.3.6}$$

exists for all $\mathbf{h} \in \mathfrak{X}$, *and* (2) $\delta\mathbf{f}(\mathbf{x}_0;\mathbf{h})$ *is a linear bounded mapping from* $\mathfrak{X}$ *to* $\mathfrak{Y}$, *i.e.* $\delta\mathbf{f}(\mathbf{x}_0;\cdot) \in \mathfrak{E}(\mathfrak{X}, \mathfrak{Y})$.

For the limit (8.3.6) we use the term *first Fréchet differential or first variation of* **f** *at* $\mathbf{x} = \mathbf{x}_0$ *with respect to the increment* **h** and we say that **f** *is* (F)-*differentiable* for short. There is also the weaker property of being (G)-*differentiable*, G for Gâteaux, but this concept will not be discussed here.

We note that the differential is additive in **f** for if $\mathbf{f} = \mathbf{f}_1 + \mathbf{f}_2$ and $\delta\mathbf{f}_1$ and $\delta\mathbf{f}_2$ exist, then by (8.3.6)

$$\delta[\mathbf{f}_1 + \mathbf{f}_2] = \delta\mathbf{f}_1 + \delta\mathbf{f}_2. \tag{8.3.7}$$

The properties of the differential *qua* function of $\mathbf{h}$ lie much deeper. Our point of departure is the observation that (8.3.6) is equivalent to saying that

$$\left\{\frac{d}{d\alpha}\mathbf{f}(\mathbf{x}+\alpha\mathbf{h})\right\}_{\alpha=0} \equiv \delta\mathbf{f}(\mathbf{x};\mathbf{h}), \tag{8.3.8}$$

where the derivative is the strong derivative in the space $\mathfrak{Y}$. This suggests connections with the theory of $\mathfrak{Y}$-holomorphic functions of α. Our first result is

Theorem 8.3.1. *Suppose that* $\mathbf{f}$ *is a mapping from* $\mathfrak{X}$ *into* $\mathfrak{Y}$ *defined in a sphere* $\mathfrak{S}: \{\mathbf{x};\ \|\mathbf{x}-\mathbf{a}\| < \rho\}$ *where* $\|\mathbf{f}(\mathbf{x})\| \leqslant M$, $\forall\ \mathbf{x}$. *Let* $\mathbf{f}$ *be such that for fixed* $\mathbf{x} \in \mathfrak{S}$ *and fixed* $\mathbf{h} \in \mathfrak{X}$ *the mapping* $\alpha \to \mathbf{f}(\mathbf{x}+\alpha\mathbf{h})$ *is* $\mathfrak{Y}$*-holomorphic in* α *for*

$$|\alpha| < [\rho - \|\mathbf{x}-\mathbf{a}\|]\ \|\mathbf{h}\|^{-1} \equiv \sigma. \tag{8.3.9}$$

Then

$$\left\{\frac{d^n}{d\alpha^n}\mathbf{f}(\mathbf{x}+\alpha\mathbf{h})\right\}_{\alpha=0} \equiv \delta^{(n)}\mathbf{f}(\mathbf{x};\mathbf{h}) \tag{8.3.10}$$

exists for all n *and all* $\mathbf{x} \in \mathfrak{S}$. *There exists a concentric sphere*

$$\mathfrak{S}_0: \{\mathbf{x};\ \|\mathbf{x}-\mathbf{a}\| < \rho_0\}, \qquad \rho_0 \leqslant \rho,$$

in which $\mathbf{f}$ *is strongly continuous together with all its variations in their dependence upon* $\mathbf{x}$. *Further, for fixed* $\mathbf{x}$ *in* $\mathfrak{S}_0$, *the nth variation is homogeneous in* $\mathbf{h}$ *of degree n and* $\delta^{(n)}\mathbf{f}(\mathbf{x};\mathbf{h})$ *is bounded in the sense of* (8.3.5) *and strongly continuous in* $\mathbf{h}$.

Proof. Since $\mathbf{f}(\mathbf{x}+\alpha\mathbf{h})$ is $\mathfrak{Y}$-holomorphic in α for fixed $\mathbf{x}$ and $\mathbf{h}$, it admits of a Maclaurin expansion in powers of α, say

$$\mathbf{f}(\mathbf{x}+\alpha\mathbf{h}) = \mathbf{f}(\mathbf{x}) + \sum_{n=1}^{\infty} \mathbf{f}_n(\mathbf{x},\mathbf{h})\,\alpha^n. \tag{8.3.11}$$

Here $\mathbf{x} \in \mathfrak{S}$, $\mathbf{h} \in \mathfrak{X}$, both fixed, and α satisfies (8.3.9). The series converges in norm for such values of the arguments. Using formulas (8.1.25) and (8.3.10) we can now rewrite (8.3.11) as

$$f(\mathbf{x}+\alpha\mathbf{h}) = \mathbf{f}(\mathbf{x}) + \sum_{n=1}^{\infty} \frac{1}{n!}\,\delta^{(n)}\mathbf{f}(\mathbf{x};\mathbf{h})\,\alpha^n. \tag{8.3.12}$$

By Cauchy's formulas for the derivatives

$$\delta^{(n)}\mathbf{f}(\mathbf{x};\mathbf{h}) = \frac{n!}{2\pi i}\int_{\Gamma} \mathbf{f}(\mathbf{x}+\beta\mathbf{h})\ \beta^{-n-1}\ d\beta, \tag{8.3.13}$$

where Γ, which is at our disposal, may be taken as the circle $|\beta| = \sigma_0 < \sigma$. Since $\|\mathbf{f}(\mathbf{x})\| \leqslant M$ for all $\mathbf{x}$ in $\mathfrak{S}$

$$\|\delta^{(n)}\mathbf{f}(\mathbf{x};\mathbf{h})\| \leqslant Mn!\sigma_0^{\ -n}.$$

Since this holds for all $\sigma_0 < \sigma$, we get

$$\|\delta^{(n)}\mathbf{f}(\mathbf{x};\mathbf{h})\| \leqslant Mn!\sigma^{-n}, \qquad \forall\ n, \tag{8.3.14}$$

and the series (8.3.12) converges in norm for $|\alpha| < \sigma$. Here $\mathbf{x}$ is any point in $\mathfrak{S}$ and $\mathbf{h}$ is arbitrary in $\mathfrak{X}$. It should be noted that σ depends upon the choice of $\mathbf{x}$ and of $\mathbf{h}$. We now choose $\mathfrak{S}_0$ as the sphere $\{\mathbf{x}; \|\mathbf{x} - \mathbf{a}\| < \frac{1}{2}\rho\}$ and restrict $\mathbf{h}$ temporarily to satisfy $\|\mathbf{h}\| \leqslant \frac{1}{2}\rho$. We can then take $\sigma = 1$ and (8.3.14) gives

$$\|\delta^{(n)}\mathbf{f}(\mathbf{x}; \mathbf{h})\| \leqslant Mn!, \qquad \mathbf{x} \in \mathfrak{S}_0, \|\mathbf{h}\| \leqslant \tfrac{1}{2}\rho. \tag{8.3.15}$$

The series (8.3.12) now converges for $|\alpha| \leqslant 1$.

The homogeneity and boundedness properties of $\delta^{(n)}\mathbf{f}(\mathbf{x}; \mathbf{h})$ with respect to $\mathbf{h}$ in the sense of Definition 8.3.1 also follow from (8.3.13) if we observe that the path of integration is at our disposal as long as it surrounds $\beta = 0$ once in the positive sense and $\mathbf{x} + \beta\mathbf{h}$ stays in $\mathfrak{S}$ along the path. Replace $\mathbf{h}$ by $\gamma\mathbf{h}$, where γ is any complex number $\neq 0$, and take $|\gamma\beta| = \sigma_0$ as the new path of integration. A simple calculation gives

$$\delta^{(n)}\mathbf{f}(\mathbf{x}; \gamma\mathbf{h}) = \gamma^n \delta^{(n)}\mathbf{f}(\mathbf{x}; \mathbf{h}), \qquad \forall\, n. \tag{8.3.16}$$

Thus $\delta^{(n)}\mathbf{f}(\mathbf{x}; \mathbf{h})$ is homogeneous in $\mathbf{h}$ of degree n as asserted.

A particular choice of γ gives the boundedness property. Restrict $\mathbf{x}$ to the sphere $\mathfrak{S}_0$ and take

$$\gamma = \tfrac{1}{2}\rho(\|\mathbf{h}\|)^{-1}.$$

Then $\|\gamma\mathbf{h}\| = \frac{1}{2}\rho$ and

$$\delta^{(n)}\mathbf{f}(\mathbf{x}; \gamma\mathbf{h}) = (\tfrac{1}{2}\rho)^n \|\mathbf{h}\|^{-n} \delta^{(n)}\mathbf{f}(\mathbf{x}; \mathbf{h}).$$

By (8.3.15) the left member is at most $Mn!$ so that

$$\|\delta^{(n)}\mathbf{f}(\mathbf{x}; \mathbf{h})\| < M \left(\frac{2}{\rho}\right)^n n!\, \|\mathbf{h}\|^n \tag{8.3.17}$$

for all n, all $\mathbf{x}$ in $\mathfrak{S}_0$, and all $\mathbf{h} \in \mathfrak{X}$.

We shall use these inequalities to estimate various approximations of $\mathbf{f}(\mathbf{x} + \alpha\mathbf{h}) - \mathbf{f}(\mathbf{x})$ by the Maclaurin series (8.3.12). Here $\mathbf{x} \in \mathfrak{S}_0$, $\|\mathbf{h}\| < \frac{1}{2}\rho$, and $|\alpha| \leqslant 1$. For such values

$$\|\mathbf{f}(\mathbf{x} + \alpha\mathbf{h}) - \mathbf{f}(\mathbf{x})\| \leqslant M \sum_{n=1}^{\infty} \left(\frac{2|\alpha|}{\rho} \|\mathbf{h}\|\right)^n.$$

Summing the geometric series and simplifying, we get

$$\|\mathbf{f}(\mathbf{x} + \alpha\mathbf{h}) - \mathbf{f}(\mathbf{x})\| \leqslant M \frac{2|\alpha|\, \|\mathbf{h}\|}{\rho - 2|\alpha|\, \|\mathbf{h}\|}. \tag{8.3.18}$$

In the same manner we get

$$\left\| \frac{1}{\alpha} [\mathbf{f}(\mathbf{x} + \alpha\mathbf{h}) - \mathbf{f}(\mathbf{x})] - \delta\mathbf{f}(\mathbf{x}; \mathbf{h}) \right\| \leqslant \frac{M}{\rho} \frac{4|\alpha|\, \|\mathbf{h}\|^2}{\rho - 2|\alpha|\, \|\mathbf{h}\|}. \tag{8.3.19}$$

The first of these inequalities is a Lipschitz condition for $\mathbf{f}$ and shows that $\mathbf{f}$ is continuous for $\mathbf{x} \in \mathfrak{S}_0$. This has important implications. It is now seen that the integrand $\mathbf{f}(\mathbf{x} + \beta\mathbf{h})$ in (8.3.13) is not merely bounded, but strongly continuous in $\mathbf{x}$ and in $\mathbf{h}$. This holds with respect to $\mathbf{x}$ in $\mathfrak{S}_0$ for any fixed $\mathbf{h} \in \mathfrak{X}$ and with respect to $\mathbf{h}$

for any fixed $\mathbf{x}$ in $\mathfrak{S}_0$. All we have to do is to shrink the radius of Γ to take care of the shifting needs. The continuity of the integrand now implies the continuity of the integral as a function of $\mathbf{x}$ and $\mathbf{h}$, i.e. $\delta^{(n)}\mathbf{f}(\mathbf{x};\mathbf{h})$ is strongly continuous in $\mathbf{x}$ and $\mathbf{h}$.

The significance of (8.3.19) will become evident below. We observe here that it implies that

$$\lim_{\|\mathbf{h}\|\to 0} \frac{1}{\|\mathbf{h}\|} \|\mathbf{f}(\mathbf{x}+\mathbf{h}) - \mathbf{f}(\mathbf{x}) - \delta\mathbf{f}(\mathbf{x};\mathbf{h})\| = 0 \tag{8.3.20}$$

for all $\mathbf{x} \in \mathfrak{S}_0$. ∎

Theorem 8.3.1 tells us very much about the properties of $\mathbf{f}$ and its variations which are consequences of local boundedness and $\mathfrak{Y}$-holomorphy with respect to α of the mapping $\alpha \to \mathbf{f}(\mathbf{x}+\alpha\mathbf{h})$. It looks as if $\mathbf{f}$ is (F)-differentiable in the sense of Definition 8.3.2 at all points of the sphere $\mathfrak{S}_0$, but the proof is not complete: $\delta\mathbf{f}(\mathbf{x};\mathbf{h})$ has been shown to be bounded and homogeneous of degree one as a function of $\mathbf{h}$, but we still have to prove additivity in $\mathbf{h}$. This also follows from the theory of $\mathfrak{Y}$-holomorphic functions, but in this case of two complex variables rather than one. Since we are not planning to go deep into the theory, this difference is not essential. Let $\mathbf{h}_1$ and $\mathbf{h}_2$ be arbitrary elements of $\mathfrak{X}$ and α_1 and α_2 small complex numbers. The assumption that $\alpha \to \mathbf{f}(\mathbf{x}+\alpha\mathbf{h})$ is a $\mathfrak{Y}$-holomorphic mapping then implies that the mapping

$$(\alpha_1, \alpha_2) \to \mathbf{f}(\mathbf{x} + \alpha_1\mathbf{h}_1 + \alpha_2\mathbf{h}_2) \tag{8.3.21}$$

is $\mathfrak{Y}$-holomorphic in α_1 as well as in α_2. We take this as definition of the meaning to be assigned to the statement that $\mathbf{f}(\mathbf{x}+\alpha_1\mathbf{h}_1+\alpha_2\mathbf{h}_2)$ is $\mathfrak{Y}$-holomorphic in (α_1, α_2) in some neighborhood of the origin of the space C^2. In analogy with the situation in C^1 we have now a Maclaurin series expansion

$$\mathbf{f}(\mathbf{x}+\alpha_1\mathbf{h}_1+\alpha_2\mathbf{h}_2) = \sum_{j=0}^{\infty}\sum_{k=0}^{\infty} \alpha_1{}^j\alpha_2{}^k\mathbf{f}_{jk}(\mathbf{x};\mathbf{h}_1,\mathbf{h}_2). \tag{8.3.22}$$

Here the coefficients $\mathbf{f}_{jk}$ are rational multiples of partial derivatives of the left member with respect to α_1 and α_2, evaluated at (0, 0). This is the analogue of (8.3.10). We have also analogues of (8.3.13) with double integrals rather than single ones. We have

$$\mathbf{f}_{00}(\cdot) = \mathbf{f}(\mathbf{x}), \quad \mathbf{f}_{10}(\cdot) = \delta\mathbf{f}(\mathbf{x};\mathbf{h}_1), \quad \mathbf{f}_{01}(\cdot) = \delta\mathbf{f}(\mathbf{h};\mathbf{x}_2). \tag{8.3.23}$$

The other coefficients are mixed variations and will not enter into our discussion. We now set $\alpha_1 = \alpha_2 = \alpha$ in (8.3.22). On the one hand, we get

$$\mathbf{f}[\mathbf{x}+\alpha(\mathbf{h}_1+\mathbf{h}_2)] = \mathbf{f}(\mathbf{x}) + \sum_{n=1}^{\infty}\frac{1}{n!}\delta^{(n)}\mathbf{f}(\mathbf{x};\mathbf{h}_1+\mathbf{h}_2)\,\alpha^n \tag{8.3.24}$$

by virtue of (8.3.13). On the other hand,

$$\mathbf{f}[\mathbf{x}+\alpha(\mathbf{h}_1+\mathbf{h}_2)] = \mathbf{f}(\mathbf{x}) + \alpha[\delta\mathbf{f}(\mathbf{x};\mathbf{h}_1) + \delta\mathbf{f}(\mathbf{x};\mathbf{h}_2)] + \cdots$$

where the ellipsis $(\cdots)$ indicates a power series in α starting with terms of the second degree. By the identity theorem for power series the coefficients of equal powers must be equal. Hence we have

$$\delta\mathbf{f}(\mathbf{x};\mathbf{h}_1+\mathbf{h}_2)=\delta\mathbf{f}(\mathbf{x};\mathbf{h}_1)+\delta\mathbf{f}(\mathbf{x};\mathbf{h}_2). \tag{8.3.25}$$

This is the required relation and we have proved

Theorem 8.3.2. *Under the assumptions of the preceding theorem $\delta\mathbf{f}(\mathbf{x};\mathbf{h})$ is a linear bounded function of $\mathbf{h}$. For fixed $\mathbf{x}\in\mathfrak{S}_0$ the first variation of $\mathbf{f}$ is a linear bounded transformation on $\mathfrak{X}$ to $\mathfrak{Y}$, i.e. $\delta\mathbf{f}(\mathbf{x};\cdot)\in\mathfrak{E}(\mathfrak{X},\mathfrak{Y})$.*

Corollary. $\mathbf{f}$ *is* (F)*-differentiable in the sense of Definition* 8.3.2 *for all* $\mathbf{x}$ *in* $\mathfrak{S}_0$.

A few more definitions will enable us to bring this discussion to a satisfactory conclusion.

Definition 8.3.3. *A mapping $\mathbf{x}\to\mathbf{f}(\mathbf{x})$ defined for $\mathbf{x}\in\mathfrak{D}\subset\mathfrak{X}$ with range in $\mathfrak{Y}$ is said to be locally bounded if for each $\mathbf{a}\in D$ there is a sphere*

$$\mathfrak{S}(\mathbf{a})\colon\{\mathbf{x};\ \|\mathbf{x}-\mathbf{a}\|<\rho(\mathbf{a})\}$$

and a positive number $M(\mathbf{a})$ such that $\|\mathbf{f}(\mathbf{x})\|\leqslant M(\mathbf{a})$ for all $\mathbf{x}\in\mathfrak{S}(\mathbf{a})$.

Definition 8.3.4. *The mapping $\mathbf{x}\to\mathbf{f}(\mathbf{x})$ from $\mathfrak{X}$ to $\mathfrak{Y}$ is said to be (F)-analytic in the domain $\mathfrak{D}$ of $\mathfrak{X}$ if it is single-valued, locally bounded, and (F)-differentiable in D.*

We have now

Theorem 8.3.3. *A mapping $\mathbf{x}\to\mathbf{f}(\mathbf{x})$ from $\mathfrak{X}$ to $\mathfrak{Y}$ is (F)-analytic in a domain $\mathfrak{D}\subset\mathfrak{X}$, if for each $\mathbf{a}\subset\mathfrak{D}$ there exists a sphere $\mathfrak{S}(\mathbf{a})$ in which the conditions of Theorem 8.3.1 hold.*

Formula (8.3.20) used to be postulated as one of the conditions for (F)-differentiability. It was shown by Max Zorn in 1946 that this relation is a consequence of the existence, boundedness and linearity of $\delta\mathbf{f}(\mathbf{x};\mathbf{h})$. It implies a type of uniform (F)-differentiability which may be compared with the corresponding situation for $\mathfrak{Y}$-holomorphic vector and operator functions as exemplified by (8.1.10).

Some further properties of (F)-analytic functions will be studied in the next section.

EXERCISE 8.3

1. Verify (8.3.7) and (8.3.8).
2. Verify (8.3.14) and (8.3.15).

3. Verify (8.3.16) and (8.3.17).

4. Verify (8.3.18) to (8.3.20).

5. Express $\mathbf{f}_{jk}(\mathbf{x}; \mathbf{h}_1, \mathbf{h}_2)$ as (1) a partial derivative, (2) a double integral.

6. Use the double integral to find an estimate of $\mathbf{f}_{jk}$ analogous to (8.3.14).

7. What happens to $\mathbf{f}_{jk}(\mathbf{x}; \mathbf{h}_1, \mathbf{h}_2)$ if $\mathbf{h}_1 = \mathbf{h}_2 = \mathbf{h}$?

8. Express $\delta^{(2)}\mathbf{f}(\mathbf{x}; \mathbf{h}_1 + \mathbf{h}_2)$ in terms of $\mathbf{f}_{20}$, $\mathbf{f}_{11}$, and $\mathbf{f}_{02}$.

9. Show that $f_{11}(\mathbf{x}; \mathbf{h}_1, \mathbf{h}_2) = \delta^{(2)} j\mathbf{f}(\mathbf{x}; \mathbf{h}_1, \mathbf{h}_2)$ is the first variation with increment $\mathbf{h}_2$ of the first variation of $\mathbf{f}$ with increment $\mathbf{h}_1$, and vice versa.

10. If $\mathfrak{Y} = C$ and f is a linear bounded functional on $\mathfrak{X}$, prove that f is (F)-analytic for all $\mathbf{x}$ and determine its first variation.

In the remaining problems (No. 22 excepted) $\mathfrak{X} = \mathfrak{Y} = \mathfrak{B}$ is a non-commutative B-algebra with unit element e.

11. If $f(x) = \Sigma_{n=0}^{\infty} \alpha_n x^n$ where $\alpha_n \in C$, $x \subset \mathfrak{B}$, $x^0 = e$, and if the series converges in norm for $\|x\| < \rho$, show that its sum is (F)-analytic in this sphere.

12. Compute $\delta f(x; h)$ and discuss the convergence of the series. Remember that h and x do not normally commute.

13. Determine spheres of absolute convergence for the two binomial series

$$f(x) = \sum_0^\infty \binom{\frac{1}{2}}{n} x^n, \qquad g(x) = \sum_0^\infty \binom{-\frac{1}{2}}{n} x^n.$$

14. Show that when the series converge their product in either order is the unit element of $\mathfrak{B}$ so that they are inverses of each other.

15. With the same notation, find $\delta f(x; h)$ and $\delta g(x; h)$. When do the series expansions for the differentials converge?

16. Find the second differential of f with respect to h.

17. Prove that

$$f(x + \alpha h) = f(x) f[\alpha g(x) h]$$

for suitable domains for α, h, x and determine such domains.

18. Verify that $[f(x)]^2 = e + x$ when the series converges, so that $f(x)$ defines a square root of $e + x$ in its domain of convergence. How can the validity of this relation be extended by analytic continuation?

19. Similarly, $g(-x)$ should be the square root of $(e - x)^{-1} = R(1, x)$. For what values of x is this true?

20. If $\mathfrak{D}$ is a domain in $\mathfrak{B}$ and if for a fixed complex α the resolvent $R(\alpha, x)$ exists for all x in $\mathfrak{B}$, show that $R(\alpha, x)$ is (F)-analytic in $\mathfrak{D}$.

21. Find the first variation of $R(\alpha, x)$ for x in $\mathfrak{D}$.

22. For general B-spaces show that if $\mathbf{f}$ is (F)-analytic in a domain $\mathfrak{D} \subset \mathfrak{X}$, then for each $\mathbf{a} \in \mathfrak{D}$ there is a sphere $\mathfrak{S}(\mathbf{a})$ in which $\mathbf{f}$ satisfies the conditions of Theorem 8.3.1, i.e. prove the converse of Theorem 8.3.3.

8.4 SOME PROPERTIES OF (F)-ANALYTIC FUNCTIONS

In view of the close relationship between (F)-analyticity and $\mathfrak{Y}$-holomorphism it is to be expected that some of the results of Sections 8.1 and 8.2 will carry over to the (F)-analytic case. We have already seen that the Cauchy integral and related tools carry over. There are some essential differences, however. This is demonstrated in a striking manner by the following extension of Theorem 8.1.8.

Theorem 8.4.1. *If* $\mathbf{f}$ *is (F)-analytic in a domain* $\mathfrak{D} \subset \mathfrak{X}$ *and if* $\mathbf{f}(\mathbf{x}) \equiv \mathbf{0}$ *in a sphere* $\mathfrak{S} \subset \mathfrak{D}$, *then* $\mathbf{f}$ *is identically* $\mathbf{0}$ *in* $\mathfrak{D}$. *There are (F)-analytic functions which vanish on a linear subspace of* $\mathfrak{X}$ *but do not vanish identically.*

Proof. Let $\mathfrak{S}$ be the sphere $\{\mathbf{x}; \|\mathbf{x} - \mathbf{a}\| < \rho\}$. Since $\mathbf{f}$ is (F)-analytic in $\mathfrak{D}$ there is a convergent Taylor series

$$\mathbf{f}(\mathbf{a} + \alpha\mathbf{h}) = \mathbf{f}(\mathbf{a}) + \sum_{n=1}^{\infty} \frac{\alpha^n}{n!} \delta^{(n)}\mathbf{f}(\mathbf{a}; \mathbf{h}). \tag{8.4.1}$$

Here $\mathbf{f}(\mathbf{a} + \alpha\mathbf{h})$ is $\mathfrak{Y}$-holomorphic for small values of $|\alpha|$ and of $\|\mathbf{h}\|$. For such values $\mathbf{f}(\mathbf{a} + \alpha\mathbf{h}) \equiv \mathbf{0}$ so by Theorem 8.1.8

$$\mathbf{f}(\mathbf{a}) = \mathbf{0}, \qquad \delta^{(n)}\mathbf{f}(\mathbf{a}; \mathbf{h}) = \mathbf{0}, \qquad \forall n. \tag{8.4.2}$$

This shows that $\mathbf{f}(\mathbf{x}) \equiv \mathbf{0}$ not merely in $\mathfrak{S}$ but in the concentric sphere $\mathfrak{S}_0: \{\mathbf{x}; \|\mathbf{x} - \mathbf{a}\| < d(\mathbf{a}, \partial\mathfrak{D})\}$ since the series (8.4.1) converges trivially in $\mathfrak{S}_0$ to the sum zero and represents $\mathbf{f}$ there.

If $\mathfrak{S}_0$ does not exhaust $\mathfrak{D}$, we use an analytic continuation argument. If $\mathbf{b} \in \mathfrak{D}$ but is not in $\mathfrak{S}_0$, we choose a finite number of points

$$\mathbf{a}_0 = \mathbf{a}, \mathbf{a}_1, \mathbf{a}_2, \ldots, \mathbf{a}_n = \mathbf{b},$$

such that for each j

$$\|\mathbf{a}_{j+1} - \mathbf{a}_j\| < d(\mathbf{a}_j, \partial\mathfrak{D}).$$

We set $\mathfrak{S}_j: \{\mathbf{x}; \|\mathbf{x} - \mathbf{a}_j\| < d(\mathbf{a}_j, \partial\mathfrak{D})\}$. Then $\mathbf{f}(\mathbf{x}) = \mathbf{0}$ everywhere in $\mathfrak{S}_0$. There is a sphere with center at $\mathbf{x} = \mathbf{a}_1$ which is contained in $\mathfrak{S}_0$ and in this sphere $\mathbf{f}(\mathbf{x}) \equiv \mathbf{0}$. It follows by the argument used above that $\mathbf{f}(\mathbf{x}) \equiv \mathbf{0}$ in all of $\mathfrak{S}_1$ and so on. In a finite number of steps it is found that $\mathbf{f}(\mathbf{x}) \equiv \mathbf{0}$ in $\mathfrak{S}_n$ and in particular $\mathbf{f}(\mathbf{b}) = \mathbf{0}$. Since $\mathbf{b}$ is an arbitrary point in $\mathfrak{D}$, it is seen that $\mathbf{f}(\mathbf{x}) \equiv \mathbf{0}$ in $\mathfrak{D}$.

To prove the final assertion of the theorem we recall that a linear bounded functional $f(\mathbf{x}) \in \mathfrak{X}^*$ is (F)-analytic (see Problem 10, Exercise 8.3). If $\mathfrak{X}_0$ is a linear subspace of $\mathfrak{X}$ which is not dense in $\mathfrak{X}$, then there exists a linear bounded functional f such that $f(\mathbf{x}) = 0$ for all $\mathbf{x}$ in $\mathfrak{X}_0$ while $\|f\| = 1$ so that f is not identically zero. ∎

The *Principle of the Maximum* is valid for (F)-analytic functions.

Theorem 8.4.2. *If* $\mathbf{f}$ *is (F)-analytic in the domain* $\mathfrak{D}$ *and if*

$$\sup_{\mathbf{x}\in\mathfrak{D}} \|\mathbf{f}(\mathbf{x})\| = M, \tag{8.4.3}$$

then either

$$\|\mathbf{f}(\mathbf{x})\| < M \tag{8.4.4}$$

or

$$\|\mathbf{f}(\mathbf{x})\| \equiv M \tag{8.4.5}$$

everywhere in $\mathfrak{D}$.

Proof. Suppose that there is an $\mathbf{a} \in \mathfrak{D}$ such that $\|\mathbf{f}(\mathbf{a})\| = M$. Then by (8.3.13)

$$\mathbf{f}(\mathbf{a}) = \frac{1}{2\pi i} \int_\Gamma \mathbf{f}(\mathbf{a} + \beta\mathbf{h})\, \beta^{-1} d\beta,$$

where Γ is any circle such that $\|\beta\mathbf{h}\| < d(\mathbf{a}, \partial\mathfrak{D})$. This gives

$$\mathbf{f}(\mathbf{a}) = \frac{1}{2\pi} \int_0^{2\pi} \mathbf{f}(\mathbf{a} + \rho e^{i\theta}\mathbf{h})\, d\theta \tag{8.4.6}$$

and

$$\|\mathbf{f}(\mathbf{a})\| \leqslant \frac{1}{2\pi} \int_0^{2\pi} \|\mathbf{f}(\mathbf{a} + \rho e^{i\theta}\mathbf{h})\|\, d\theta. \tag{8.4.7}$$

Here the left member equals M while in the right member the integrand is $\leqslant M$ for all θ. If the average of such a function is to be M, then by the continuity of $\|\mathbf{f}(\mathbf{a} + \rho e^{i\theta}\mathbf{h})\|$ in θ, we must have $\|\mathbf{f}(\mathbf{a} + \rho e^{i\theta}\mathbf{h})\| \equiv M$ for all admissible ρ, θ and $\mathbf{h}$. In other words, (8.4.5) holds for all $\mathbf{x}$ with $\|\mathbf{x} - \mathbf{a}\| < d(\mathbf{a}, \partial\mathfrak{D})$. The extension to all of $\mathfrak{D}$ is then obtained by the now-familiar chain argument and is left to the reader. ▌

The theorem of Vitali is also valid for (F)-analytic functions, but both the assumptions and the conclusions have to be modified. In Section 8.2 we spoke of "induced convergence"; here a much more substantial "induction basis" is required.

Definition 8.4.1. *A sequence* $\{\mathbf{f}_k\}$ *of functions (F)-analytic in a fixed domain* $\mathfrak{D} \subset \mathfrak{X}$ *is said to be equilocally bounded in* $\mathfrak{D}$, *if for each* $\mathbf{a} \in \mathfrak{D}$ *there is a sphere* $\mathfrak{S}(\mathbf{a})$ *and a positive number* $M(\mathbf{a})$ *such that*

$$\|\mathbf{f}_k(\mathbf{x})\| \leqslant M(\mathbf{a}), \qquad \forall\, \mathbf{x} \in \mathfrak{S}(\mathbf{a}), \qquad \forall\, k. \tag{8.4.8}$$

We have now the following analogue of Vitali's theorem.

Theorem 8.4.3. *Suppose that the sequence* $\{\mathbf{f}_k\}$ *is made up of functions which are (F)-analytic and equilocally bounded in a fixed domain* $\mathfrak{D} \subset \mathfrak{X}$ *and suppose that*

$$\lim_{k\to\infty} \mathbf{f}_k(\mathbf{x}) \equiv \mathbf{f}(\mathbf{x}) \tag{8.4.9}$$

exists in a sphere $\mathfrak{S} \subset \mathfrak{D}$. *Then the limit exists everywhere in* $\mathfrak{D}$ *and is (F)-analytic. Further, for all* n, *all* $\mathbf{x} \in \mathfrak{D}$ *and all* $\mathbf{h} \in \mathfrak{X}$

$$\lim_{k\to\infty} \delta^{(n)}\mathbf{f}_k(\mathbf{x}; \mathbf{h}) = \delta^{(n)}\mathbf{f}(\mathbf{x}; \mathbf{h}). \tag{8.4.10}$$

Remark. Note that the theorem does not assert uniform convergence as in Theorem 8.2.1.

Proof. Since $\mathbf{f}_k$ is (F)-analytic, $\mathbf{f}_k(\mathbf{x} + \alpha\mathbf{h})$ is $\mathfrak{Y}$-holomorphic in α for $\mathbf{x} \in \mathfrak{D}$ and $\mathbf{h} \in \mathfrak{X}$, both fixed. We then have a sequence $\{\mathbf{f}_k(\mathbf{x} + \alpha\mathbf{h})\}$ to which Theorem 8.2.1 may possibly apply. We explore this possibility and play back and forth between (F)-analyticity and $\mathfrak{Y}$-holomorphism to obtain the desired result.

Let $\mathbf{a}$ be the center of the sphere $\mathfrak{S}$. Without restricting the generality we may assume that $\mathfrak{S}$ coincides with the open sphere $\mathfrak{S}(\mathbf{a})$ which exists by assumption. For $\mathbf{x} + \alpha\mathbf{h}$ in $\mathfrak{S}$ we have

$$\lim_{h\to\infty} \mathbf{f}_k(\mathbf{x} + \alpha\mathbf{h}) = \mathbf{f}(\mathbf{x} + \alpha\mathbf{h}),$$

again by assumption. This says that for $\mathbf{x} \in \mathfrak{S}$ and any fixed $\mathbf{h}$ the mapping $\alpha \to \mathbf{f}(\mathbf{x} + \alpha\mathbf{h})$ is $\mathfrak{Y}$-holomorphic in α in some neighborhood of $\alpha = 0$. This implies the existence of $\delta\mathbf{f}(\mathbf{x}; \mathbf{h})$ for $\mathbf{x} \in \mathfrak{S}$, and since $\|\mathbf{f}(\mathbf{x})\| \leqslant M(\mathbf{a})$ in $\mathfrak{S}$ we conclude that $\mathbf{f}$ is (F)-differentiable and hence (F)-analytic in $\mathfrak{S}$.

If $\mathbf{b} \in \mathfrak{D}$ but is not in $\mathfrak{S}$, we can join $\mathbf{a}$ to $\mathbf{b}$ by a polygonal line $\mathfrak{P}$ in $\mathfrak{D}$. Each point $\mathbf{x}$ on $\mathfrak{P}$ is the center of a sphere of equilocal boundedness and since $\mathfrak{P}$ is compact in $\mathfrak{X}$, it follows that a finite number of these spheres will form a covering of $\mathfrak{P}$. We can choose these spheres $\mathfrak{S}_1, \mathfrak{S}_2, \ldots, \mathfrak{S}_n$ so that $\mathfrak{S}_1 = \mathfrak{S}$ and the center $\mathbf{a}_j$ of $\mathfrak{S}_j$ is interior to $\mathfrak{S}_{j-1}$ for $j = 2, 3, \ldots, n$ and $\mathbf{a}_n = \mathbf{b}$. A uniform bound of the sequence $\mathbf{f}_k(\mathbf{x})$ in the union of the spheres is $M = \max_j M(\mathbf{a}_j)$.

We now proceed to the point $\mathbf{x} = \mathbf{a}_2$, which by assumption is interior to $\mathfrak{S}_1 = \mathfrak{S}$. It follows that there is a small sphere concentric to $\mathfrak{S}_2$ in which $\lim_{k\to\infty} \mathbf{f}_k(\mathbf{x})$ exists and equals $\mathbf{f}(\mathbf{x})$. Take any $\mathbf{h} \in \mathfrak{X}$ and consider the functions $\mathbf{f}_k(\mathbf{a}_2 + \alpha\mathbf{h})$. If the radius of $\mathfrak{S}_2$ is r_2, then each mapping $\alpha \to \mathbf{f}_k(\mathbf{a}_2 + \alpha\mathbf{h})$ is $\mathfrak{Y}$-holomorphic in α for $|\alpha| \leqslant r_2 \|\mathbf{h}\|^{-1}$ and uniformly bounded in $\mathfrak{S}_2$. It follows that $\lim_{k\to\infty} \mathbf{f}_k(\mathbf{a}_2 + \alpha\mathbf{h})$ exists and must define $\mathbf{f}(\mathbf{a}_2 + \alpha\mathbf{h})$ for such values of α. Here $\mathbf{f}(\mathbf{a}_2 + \alpha\mathbf{h})$ is $\mathfrak{Y}$-holomorphic as the limit of a uniformly convergent sequence of such functions. Since $\mathbf{h}$ is arbitrary, $\mathbf{f}(\mathbf{x})$ exists everywhere in $\mathfrak{S}_2$. It is (F)-differentiable because the existence of $\lim_{k\to\infty} \mathbf{f}_k(\mathbf{a}_2 + \alpha\mathbf{h})$ implies also the uniform convergence of the partial derivatives with respect to α. This shows the existence of $\delta\mathbf{f}(\mathbf{x}; \mathbf{h})$ as well as the boundedness and linearity of the variation. Since $\|\mathbf{f}(\mathbf{x})\| \leqslant M$ in $\mathfrak{S}_2$, we conclude that $\mathbf{f}$ is (F)-analytic in $\mathfrak{S}_2$. Since the spheres $\mathfrak{S}_1$ and $\mathfrak{S}_2$ overlap, $\mathbf{f}$ in $\mathfrak{S}_1$ and $\mathbf{f}$ in $\mathfrak{S}_2$ are local representatives of the same (F)-analytic function.

This argument can now be repeated for the following spheres $\mathfrak{S}_3, \mathfrak{S}_4, \ldots, \mathfrak{S}_n$ and shows the existence of $\mathbf{f}$ as an (F)-analytic function in the union of the spheres. This conclusion then extends to all of $\mathfrak{D}$. We recall that

$$\lim_{k\to\infty} \mathbf{f}_k(\mathbf{x} + \alpha\mathbf{h}) = \mathbf{f}(\mathbf{x} + \alpha\mathbf{h})$$

also implies

$$\lim_{k\to\infty} \frac{\partial^n}{\partial\alpha^n} \mathbf{f}_k(\mathbf{x} + \alpha\mathbf{h}) = \frac{\partial^n}{\partial\alpha^n} \mathbf{f}(\mathbf{x} + \alpha\mathbf{h}), \qquad \forall\, n, \tag{8.4.11}$$

in some neighborhood of $\alpha = 0$. In particular this is true at $\alpha = 0$, which is (8.4.10). ∎

EXERCISE 8.4

1. Suppose that $\{\mathbf{f}_k\}$ is a sequence of functions (F)-analytic in some domain $\mathfrak{D} \subset \mathfrak{X}$ but not necessarily equilocally bounded. Suppose that $\lim_{k\to\infty} \mathbf{f}_k(\mathbf{x}) \equiv \mathbf{f}(\mathbf{x})$ exists everywhere in $\mathfrak{D}$. Show that $\mathbf{f}(\mathbf{x})$ is piecewise (F)-analytic in the following sense. Every sphere in $\mathfrak{D}$ contains a subsphere in which $\mathbf{f}_k(\mathbf{x})$ is uniformly bounded and in which consequently $\mathbf{f}(\mathbf{x})$ is (F)-analytic.

   A mapping $\mathbf{x} \to \mathbf{P}(\mathbf{x})$ from $\mathfrak{X}$ to $\mathfrak{Y}$, defined for all $\mathbf{x}$, is said to be a *polynomial in* $\mathbf{x}$ *of exact degree* n if, for all $\mathbf{a}, \mathbf{h} \in \mathfrak{X}$ and all complex numbers α,

   $$\mathbf{P}(\mathbf{a} + \alpha\mathbf{h}) = \sum_{k=0}^{n} \mathbf{P}_k(\mathbf{a}, \mathbf{h})\, \alpha^k, \quad \mathbf{P}_n(\mathbf{a}, \mathbf{h}) \not\equiv \mathbf{0},$$

   where the $\mathbf{P}_k(\mathbf{a}, \mathbf{h})$ are independent of α. $\mathbf{P}(\mathbf{x})$ is homogeneous of degree n if $\mathbf{P}(\beta\mathbf{x}) = \beta^n\, \mathbf{P}(\mathbf{x})$.

2. If $\mathbf{P}$ is homogeneous of degree n as well as a polynomial of exact degree n, what are the homogeneity properties of the $\mathbf{P}_k(\mathbf{a}, \mathbf{h})$?

3. Show that $\mathbf{P}_0(\mathbf{a}, \mathbf{h}) = \mathbf{P}(\mathbf{a})$, $\mathbf{P}_n(\mathbf{a}, \mathbf{h}) = \mathbf{P}(\mathbf{h})$. Show that $\delta^{(k)}\,\mathbf{P}(\mathbf{a}; \mathbf{h})$ exists for all $\mathbf{a}, \mathbf{h}$ and k with all variations being zero for $k > n$. The variations are not necessarily bounded functions of $\mathbf{a}$ and $\mathbf{h}$ in any sphere. Show that a necessary condition for the boundedness of the nth variation is that $\mathbf{P}(\mathbf{h})$ is bounded in the sense of Definition 8.3.1.

4. Let ω be a primitive $(n + 1)$th root of unity. "Primitive" means that no power of ω with an exponent $< n + 1$ can equal unity. Form the system of equations

   $$\mathbf{P}(\mathbf{a} + \omega^j\, \mathbf{h}) = \sum_{k=0}^{n} \mathbf{P}_k(\mathbf{a}, \mathbf{h})\, \omega^{jk}, \qquad j = 0, 1, 2, \ldots, n.$$

   Show that this system has a determinant different from zero and may be solved for the $\mathbf{P}_k(\mathbf{a}, \mathbf{h})$ in terms of the $\mathbf{P}(\mathbf{a} + \omega^j\, \mathbf{h})$ linearly and with constant coefficients.

5. Use this result to show that the boundedness of $\mathbf{P}(\mathbf{x})$ in some sphere $\|\mathbf{x}\| \leqslant r$ implies the boundedness of all variations. Show that this implies that $\mathbf{P}$ is strongly continuous and (F)-analytic.

6. Suppose that $\mathbf{P}$ is homogeneous of degree n and bounded in the sense of Definition 8.3.1. Show that $\mathbf{P}$ is continuous and (F)-analytic.

   A mapping $x \to f(x)$ from a commutative B-algebra with unit element into itself is said to be (L)-analytic ("L" for E. R. Lorch) in a domain $\mathfrak{D}$ if for each $x \in \mathfrak{D}$ there is an element $f'(x)$ of the algebra such that for $\|h\|$ small

$$\|f(x+h) - f(x) - hf'(x)\| = o(\|h\|).$$

7. Prove that an (L)-analytic function f is (F)-analytic and $\delta f(x; h) = hf'(x)$.
8. Let $x \in \mathfrak{B}$, $a_n \in \mathfrak{B}$, $\forall\, n$. Suppose that

$$f(x) = \sum_{n=0}^{\infty} a_n x^n$$

converges in norm for $\|x\| < r$. Show that f is (L)-analytic and find $f'(x)$.
9. If $\mathfrak{X} = \mathfrak{B}$ is the B-algebra $C[0, 1]$ with the sup-norm, give examples of (L)-analytic functions in this algebra.

COLLATERAL READING

Most of the material in this chapter is based upon:

HILLE, E. and R. S. PHILLIPS, *Functional Analysis and Semi-groups,* American Mathematical Society Colloquium Publications, Vol. 31, revised edition, Providence, R.I. (1957). See, in particular, Chapter 3. Additional material on analytic functions on vectors to vectors is to be found in Chapter 26. This treatise contains numerous references to the literature as well as more general results, especially a discussion of (G)-differentiability.

9 BANACH ALGEBRAS

In this chapter we give a more elaborate discussion of Banach algebras, a topic outlined in Section 2.6. We are chiefly, but not exclusively, concerned with non-commutative B-algebras with unit element. Most of what is to be done may be characterized as analysis in a B-algebra. Here the theory of the resolvent is fundamental, and we study in some detail its properties, the related functional equations and the nature of its isolated singularities. The resolvent is basic for the theory of a wide class of Fréchet analytic functions from $\mathfrak{B}$ to $\mathfrak{B}$ where it plays a role analogous in importance to that of the Cauchy kernel in classical analytic function theory.

Part of this work is extended to B-algebras without unit element where the role of the resolvent is taken over by the so-called dissolvent and the circle product $x \circ y = x + y - xy$ plays an essential role. We also include a treatment of Gelfand's representation theorem for commutative B-algebras with unit element via the maximal ideals along traditional lines.

There are four sections; Review; Resolvents and dissolvents; Gelfand's representation theorem; and (F)-Analytic functions and the operational calculus.

9.1 REVIEW

The theory of Banach algebras, B-algebras for short, was sketched in Section 2.6. Here we shall give a brief review of what has already been given to serve as the foundation for further development. To this is added a brief discussion of the basic modifications required if the algebra has no unit element.

An algebra, as we use the term, is a collection of elements which is closed under the operations of addition, scalar multiplication, and multiplication, subject to the A-, S-, and M-postulates of Sections 2.1 and 2.6. A B-algebra $\mathfrak{B}$ is an algebra which is also a Banach space and where the norm satisfies

$$\|xy\| \leqslant \|x\| \, \|y\|. \tag{9.1.1}$$

The algebra is said to be *commutative* if multiplication is commutative

$$xy = yx \tag{9.1.2}$$

for all x, y. A non-commutative algebra may contain a set of elements z such that

$$xz = zx \tag{9.1.3}$$

for all $x \in \mathfrak{B}$. This set $\{z; xz = zx, \forall x \in \mathfrak{B}\}$ is called the *center* of the algebra. An element e such that

$$xe = ex = x, \qquad \forall x, \tag{9.1.4}$$

is called the *unit element* of $\mathfrak{B}$. There is at most one such element. We recall that the normed metric of $\mathfrak{B}$ may be supposed to satisfy

$$\|e\| = 1. \tag{9.1.5}$$

If $\mathfrak{B}$ has a unit element, then some elements x of $\mathfrak{B}$ have *inverses*, i.e. there exists a $y \in \mathfrak{B}$ such that

$$xy = yx = e. \tag{9.1.6}$$

To each x there is at most one y satisfying this relation. If such a y exists, we write

$$y = x^{-1} \tag{9.1.7}$$

and refer to x as an *invertible* or *regular element* of $\mathfrak{B}$. It is clear that x^{-1} is also regular and $(x^{-1})^{-1} = x$. The set $\mathfrak{G}$ of regular elements of $\mathfrak{B}$ is algebraically a *group*, and topologically an *open set* in $\mathfrak{B}$. In particular, the open sphere

$$\|x - e\| < 1 \tag{9.1.8}$$

is contained in $\mathfrak{G}$. The set $\mathfrak{G}$ is not bounded and need not be connected.

A non-regular element x of $\mathfrak{B}$ is called *singular*. The set $\mathfrak{S}$ of singular elements is unbounded but connected via the origin since constant multiples of singular elements are also singular. Moreover, $x \in \mathfrak{S}$, $y \in \mathfrak{B}$ implies xy and yx in $\mathfrak{S}$. If the product xy or yx should be zero while neither x nor y is zero, then x and y are called *divisors of zero*. Such elements are in $\mathfrak{S}$. The element x is *idempotent* if

$$x^2 = x. \tag{9.1.9}$$

This equation is satisfied by $x = e$. Any other solution is in $\mathfrak{S}$. An x in $\mathfrak{S}$ is said to be *nilpotent of order p* if

$$x^p = 0 \tag{9.1.10}$$

and p is the least integer with this property. It is *quasi-nilpotent* or *topologically nilpotent* if

$$r(x) \equiv \lim_{n\to\infty} \|x^n\|^{1/n} = 0. \tag{9.1.11}$$

With respect to a given element x of $\mathfrak{B}$, the complex numbers fall into two disjoint complementary classes: the *resolvent set* $\rho(x)$ and the *spectrum* $\sigma(x)$, according as $\lambda e - x$ is regular or singular. The former set is unbounded and open, the latter bounded and closed. Neither set need be connected, and $\sigma(x)$ may separate the plane.

Here the quantity $r(x)$ of formula (9.1.11) plays a decisive role. The existence of the limit was asserted in Theorem 2.6.4 but without a proof. We shall supply a proof now.

Proof of Theorem 2.6.4. The existence of the limit is trivial if x is nilpotent since then (9.1.10) holds for any exponent $\geqslant$ the order of nilpotency. Suppose now that $x^n \neq 0$ for all n. Since

$$\|x^{m+n}\| \leqslant \|x^m\| \, \|x^n\|,$$

we see that

$$a_n \equiv \log \|x^n\| \tag{9.1.12}$$

defines a so-called *subadditive sequence*, i.e.

$$a_{m+n} \leqslant a_m + a_n \tag{9.1.13}$$

for all m and n. The property to be proved is the existence of

$$\lim_{n\to\infty} \frac{a_n}{n} \equiv b, \tag{9.1.14}$$

where $-\infty \leqslant b < \infty$. Intrinsically this fact has nothing to do with powers of x in a B-algebra. It is simply a property of subadditive sequences as such. See Section 13.1. Suppose that a quantity b is defined by

$$\inf_n \frac{a_n}{n} = b. \tag{9.1.15}$$

The infimum exists but may possibly be $-\infty$. Here we assume $b > -\infty$ and prove that (9.1.14) holds. By the definition of the infimum there is an integer m such that for given $\varepsilon > 0$

$$\frac{a_m}{m} < b + \varepsilon.$$

Suppose now that $n = km + p$ with $0 \leqslant p < m$. Then

$$b \leqslant \frac{a_n}{n} = \frac{a_{km+p}}{km+p} \leqslant \frac{ka_m + a_p}{km+p} = \frac{km}{km+p} \cdot \frac{a_m}{m} + \frac{a_p}{km+p}.$$

Here the last member has a limit $\leqslant b + \varepsilon$ at $k \to \infty$ since $a_1, a_2, \ldots, a_{m-1}$ are fixed numbers. Since ε is arbitrary, (9.1.14) holds. The case where $b = -\infty$ is left to the reader.

We now return to (9.1.12) and see that

$$\lim_{n\to\infty} \frac{1}{n} \log \|x^n\| = \lim_{n\to\infty} \log \|x^n\|^{1/n} = b,$$

so that

$$r(x) = \lim_{n\to\infty} \|x^n\|^{1/n} = \exp(b),$$

where, if $b = -\infty$, we replace $\exp(b)$ by 0. ∎

We note that

$$0 \leqslant r(x) \leqslant \|x\|. \tag{9.1.16}$$

This quantity $r(x)$ is known as the *spectral radius* of x because, as was shown in

Section 2.6, the spectrum $\sigma(x)$ is confined to the disk

$$|\lambda| \leqslant r(x) \tag{9.1.17}$$

and there is at least one point of $\sigma(x)$ on the rim of the disk. This implies, among other things, that the resolvent set $\rho(x)$ contains $|\lambda| > r(x)$ and for such values of λ

$$R(\lambda, x) = (\lambda e - x)^{-1} = \frac{e}{\lambda} + \sum_{n=1}^{\infty} \frac{x^n}{\lambda^{n+1}}, \tag{9.1.18}$$

where the series converges in norm and $r(x)$ is the precise radius of convergence.

This brings an end to our review. We add a short survey of the case where the algebra $\mathfrak{B}$ lacks a unit element. While a unit element may be adjoined to the algebra, we shall not use this device. Instead we base our discussion on the composition

$$x \circ y = x + y - xy \tag{9.1.19}$$

[S. Perlis (1942), N. Jacobson (1945), with many followers]. The composition is associative, distributive over addition, and is commutative iff $\mathfrak{B}$ is commutative. Here $x = 0$ plays the role of "neutral" element

$$x \circ 0 = 0 \circ x = x, \qquad \forall\, x. \tag{9.1.20}$$

We can now ask about inverses with respect to the circle operation. If $x \in \mathfrak{B}$ and there exists a (necessarily unique) $y \in \mathfrak{B}$ such that

$$x \circ y = y \circ x = 0 \tag{9.1.21}$$

we shall call y the *reverse* of x (*quasi-inverse* and *adverse* are also used) and x is *reversible*. We write x^- for the reverse of x when it exists. The set of reversible elements forms a group $\mathfrak{R}$ in $\mathfrak{B}$ under the circle operation since

$$x \circ y \circ y^- \circ x^- = 0.$$

Further, $\mathfrak{R}$ is an open point set in $\mathfrak{B}$ containing the unit sphere

$$\|x\| < 1 \tag{9.1.22}$$

and for $\|x\| < 1$

$$x^- = -\sum_{n=1}^{\infty} x^n, \tag{9.1.23}$$

since the series converges in norm and

$$x + \left[-\sum_{n=1}^{\infty} x^n\right] - x\left[-\sum_{n=1}^{\infty} x^n\right] = 0.$$

Again we make a disjunction of the complex numbers according as x/λ is reversible or not. Here $\lambda = 0$ plays an exceptional role and requires special conventions, for which see the Exercise below. If $x/\lambda \in \mathfrak{R}$, then λ is said to belong to the *dissolvent set* $\delta(x)$, otherwise to the *spectrum* $\alpha(x)$. The domain

$$r(x) < |\lambda| \tag{9.1.24}$$

is a subset of $\delta(x)$ and for such values of λ the reverse of x/λ, denoted by $D(\lambda, x)$ and called the *dissolvent* of x, is given by

$$D(\lambda, x) = -\sum_{n=1}^{\infty} x^n \lambda^{-n}. \tag{9.1.25}$$

Whenever $D(\lambda, x)$ exists we have

$$x + \lambda D(\lambda, x) = xD(\lambda, x) = D(\lambda, x)\, x. \tag{9.1.26}$$

The circle operation is definable in any B-algebra. If the algebra has a unit element e, then we can consider simultaneously inverses and reverses. Note that

$$(e - x)(e - y) = e \quad \text{iff} \quad x + y - xy = 0. \tag{9.1.27}$$

Thus $e - x$ is invertible iff x is reversible. Applying this to x/λ, $\lambda \neq 0$, we get

$$D(\lambda, x) = -\, xR(\lambda, x). \tag{9.1.28}$$

The relation between the spectral sets we express by writing

$$\alpha(x) \equiv \sigma(x)\, [\text{mod}\, 0], \tag{9.1.29}$$

meaning that any λ, $\lambda \neq 0$, which belongs to one of the sets also belongs to the other, while $\lambda = 0$ may belong to either set without necessarily belonging to the other.

EXERCISE 9.1

1. Show that $b = -\infty$ is a possibility for a subadditive sequence.
2. If $\mathfrak{B} = C[a, b]$ and $f \in \mathfrak{B}$, describe $\rho[f]$ and $\sigma[f]$ and determine $r(f)$.
3. Give an example of an element of $C[a, b]$ whose spectrum separates the plane.
4. Find the dissolvent of an element f of $C[a, b]$ and discuss the role of $\lambda = 0$. Find necessary and sufficient conditions for $\lambda = 0$ to be in the dissolvent set.
5. Give an example of divisors of zero in $C[a, b]$.
6. Are there any nontrivial idempotents in $C[a, b]$?
7. Can the matrix algebra $\mathfrak{M}_n$ contain a quasi-nilpotent element which is not nilpotent?
8. $L_2(-\pi, \pi)$ is a commutative B-algebra under convolution and without unit element (see Theorem 4.3.11). In this algebra define the circle product by $f \circ g = f + g - f * g$ and determine the corresponding spectrum of f in terms of its Fourier coefficients

   $$f_n = \frac{1}{2\pi} \int_{-\pi}^{\pi} f(s) \exp(-ins)\, ds.$$

   Find idempotent elements of this algebra.
9. Find the Fourier series of the dissolvent of f.

10. Prove that the circle product is associative, distributive over addition, and commutative iff $\mathfrak{B}$ is commutative.

11. Show that nilpotent and quasinilpotent elements are reversible. Find the spectrum $\alpha(x)$ of such an element.

12. If $j \in \mathfrak{B}$, show that $j \circ j = j$.

13. Prove that $\mathfrak{R}$ is an open set in $\mathfrak{B}$.

14. The convention with regard to $\lambda = 0$ is as follows. The origin belongs to $\delta(x)$ iff there exist elements j and y of $\mathfrak{B}$ such that $j^2 = j$, $jx = xj = x$, $jy = yj = y$, $xy = yx = j$. If this is the case, $D(0, x) = j$ by definition. Show that under these assumptions

$$D(\lambda, x) = j + \sum_{n=1}^{\infty} y^n \lambda^n$$

satisfies (9.1.26) for $|\lambda|\, r(y) < 1$.

15. Show that $\delta(x)$ is an open set in the complex plane.

16. If $\lambda_0 \in \delta(x)$, $\lambda_0 \neq 0$, obtain an expansion for $D(\lambda, x)$ in terms of powers of $(\lambda - \lambda_0)/\lambda$ and discuss its domain of convergence.

9.2 RESOLVENTS AND DISSOLVENTS

We start with a discussion of resolvents assuming the B-algebra to have a unit element. If now $\lambda \in \rho(x)$, then the resolvent of x is defined as the inverse of $\lambda e - x$:

$$R(\lambda, x) = (\lambda e - x)^{-1}, \qquad \lambda \in \rho(x). \tag{9.2.1}$$

It follows that

$$(\lambda e - x)\, R(\lambda, x) = R(\lambda, x)\, (\lambda e - x) = e. \tag{9.2.2}$$

Suppose now that λ and μ belong to $\rho(x)$ though not necessarily to the same component of the open set $\rho(x)$. We have then

$$(\lambda e - x)\, R(\lambda, x) = e, \qquad R(\mu, x)\, (\mu e - x) = e.$$

Multiply the first relation on the left by $R(\mu, x)$, the second on the right by $R(\lambda, x)$, and subtract. The result is

$$(\mu - \lambda)\, R(\mu, x)\, R(\lambda, x) = R(\lambda, x) - R(\mu, x), \tag{9.2.3}$$

known as the *first resolvent equation*. Similarly, if $\lambda \in \rho(x) \cap \rho(y)$, then

$$(\lambda e - x)\, R(\lambda, x) = e, \qquad R(\lambda, y)\, (\lambda e - y) = e.$$

Multiply the first equation on the left by $R(\lambda, y)$, the second on the right by $R(\lambda, x)$, and subtract. The result is the *second resolvent equation*

$$R(\lambda, y)\, (y - x)\, R(\lambda, x) = R(\lambda, y) - R(\lambda, x). \tag{9.2.4}$$

In the first equation λ and μ are the essential variables and x plays a subordinate role. In the second the situation is reversed: now x and y are essential and λ is secondary.

Let us now consider the functional equations

$$(\mu - \lambda) R(\mu) R(\lambda) = R(\lambda) - R(\mu), \tag{9.2.5}$$

$$S(y)(y - x) S(x) = S(y) - S(x), \tag{9.2.6}$$

per se as functional equations. In the first case $R(\lambda)$ is a mapping from C into $\mathfrak{B}$ defined in some domain Δ of the complex plane with values in the B-algebra $\mathfrak{B}$, and (9.2.5) is assumed to hold for all λ, μ in Δ. In the second case the mapping goes from $\mathfrak{B}$ into itself and (9.2.6) is assumed to hold for all x, y in some domain $\mathfrak{D}$ (= open connected set) of $\mathfrak{B}$. We ask: What are the *a priori* properties of these functions *qua* solutions of the functional equations? For this general type of question, see Section 14.1.

We start with (9.2.5). $\mathfrak{B}$ may be non-commutative but (9.2.5) shows that $R(\lambda)$ and $R(\mu)$ must commute whenever they are defined. We started out with resolvents in deriving (9.2.5). Now it is clear that (9.2.5) will be satisfied by resolvents, but we have no right to assume that all solutions will be algebraically regular everywhere in the domain of existence or even anywhere. This suggests that the obvious crude method of solving (9.2.5) for $R(\lambda)$ is likely to fail. But we can use a method of successive substitutions which together with a boundedness assumption leads to results. It will be shown that $R(\lambda)$ is locally $\mathfrak{B}$-holomorphic.

Suppose that $R(\lambda)$ is defined in some domain Δ of the complex plane where it is bounded, $\|R(\lambda)\| \leqslant M$, and satisfies (9.2.5). Suppose $\alpha \in \Delta$ and set $R(\alpha) = a$. Here $a \in \mathfrak{B}$ is arbitrary and serves as an initial value which will determine $R(\lambda)$ uniquely. Set $\mu = \alpha$ in (9.2.5) and note that

$$R(\lambda) = a + (\alpha - \lambda) aR(\lambda).$$

By repeated substitution

$$R(\lambda) = \sum_{k=0}^{n} (\alpha - \lambda)^k a^{k+1} + (\alpha - \lambda)^{n+1} a^{n+1} R(\lambda).$$

If now

$$|\alpha - \lambda| \, r(a) < 1, \tag{9.2.7}$$

we can pass to the limit with n since the remainder goes to zero and the expansion

$$R(\lambda) = \sum_{k=0}^{\infty} (\alpha - \lambda)^k a^{k+1} \tag{9.2.8}$$

is obtained. This series defines a $\mathfrak{B}$-holomorphic function of λ. Since Δ is connected, the extension of $R(\lambda)$ as a $\mathfrak{B}$-holomorphic mapping to all of Δ is routine analysis.

As stated above, $a \in \mathfrak{B}$ is arbitrary and may be regular or singular. If $a \in \mathfrak{S}$, then $R(\lambda) \in \mathfrak{S}$ for all λ satisfying (9.2.7) for

$$R(\lambda) = a \sum_{k=0}^{\infty} (\alpha - \lambda)^k a^k$$

is the product of $a \in \mathfrak{S}$ with an element of $\mathfrak{B}$. Hence $R(\lambda)$ is singular for all λ in the disk (9.2.7). By analytic continuation with respect to λ this will hold all through Δ. For any continuation will be of the form

$$R(\lambda) = \sum_{k=0}^{\infty} (\beta - \lambda)^k b^{k+1}, \qquad b = R(\beta), \tag{9.2.9}$$

where $b \in \mathfrak{S}$. This implies that $R(\lambda)$ is singular in the domain of convergence of the series (9.2.9) and hence everywhere in Δ. A similar argument shows that if $a \in \mathfrak{G}$ instead, then so does $R(\lambda)$ for all $\lambda \in \Delta$.

A peculiar feature of solutions of (9.2.5) which are resolvents is that $R(\lambda, x)$ is $\mathfrak{B}$-holomorphic in each of the components of the open set $\rho(x)$. There may be countably infinitely many components. Nevertheless, the values of $R(\lambda, x)$ in distinct components are related by (9.2.5). Thus we have the remarkable phenomenon that distinct $\mathfrak{B}$-holomorphic functions may satisfy the same functional equation and their values in distinct components are still linked by the very same equation.

We shall now turn to the second functional equation (9.2.6). Here we assume the existence of a mapping $x \to S(x)$ from $\mathfrak{B}$ to $\mathfrak{B}$ defined in a domain $\mathfrak{D}$ of $\mathfrak{B}$ where it satisfies (9.2.6) and is locally bounded in norm in the sense of Definition 8.3.3. We shall show that $S(x)$ is (F)-analytic in $\mathfrak{D}$ in the sense of Definition 8.3.4. Set $y = a$ in (9.2.6) so that

$$S(x) = S(a) + S(a)\,(x - a)\,S(x). \tag{9.2.10}$$

By repeated substitution

$$S(x) = S(a) + \sum_{k=1}^{n} [(x - a)\,S(a)]^k + [S(a)\,(x - a)]^{n+1} S(x).$$

Since in general $\mathfrak{B}$ is noncommutative, we cannot group together the powers of $(x - a)$ or the powers of $S(a)$. Assuming

$$\|x - a\|\ \|S(a)\| < 1, \tag{9.2.11}$$

we conclude that the remainder goes to zero by the local boundedness of $S(x)$ and we obtain the expansion

$$S(x) = S(a) + S(a) \sum_{k=1}^{\infty} [(x - a)\,S(a)]^k. \tag{9.2.12}$$

This shows that S is (F)-analytic. To start with, this holds for x in the sphere defined by (9.2.11) and then by analytic continuation in $\mathfrak{D}$. Note that if a point

$x = b$ has been reached by this process, then we have

$$S(x) = S(b) + S(b) \sum_{k=1}^{\infty} [(x - b) S(b)]^k \tag{9.2.13}$$

in some neighborhood of $x = b$ and this shows that the (F)-analytic character is preserved. The formulas show, in addition, that if the initial value $S(a)$ of $S(x)$ is algebraically singular, then $S(x)$ has this property everywhere in $\mathfrak{D}$. On the other hand, if $S(a)$ is regular, so is $S(x)$ in $\mathfrak{D}$.

An (F)-analytic function has differentials of all orders. In the present case

$$\delta S(x; h) = S(x)\, hS(x) \tag{9.2.14}$$

and the expressions for the differentials of higher order may be read off from (9.2.12) where $x - a$ is replaced by h.

The preceding discussion was based on the assumption that $\mathfrak{B}$ has a unit element and all statements regarding the algebraic character, regular or singular, of the solutions become meaningful iff there is a unit element. But the functional equations (9.2.5) and (9.2.6) do not involve the unit element, nor does our discussion of these equations employ the existence of a unit element. Thus all non-algebraic results stated above hold for any B-algebra with unit element or not.

We return to equation (9.2.5). We have found the character of the locally holomorphic solutions of the equation in the neighborhood of a finite value $\lambda = \alpha$. The existence of solutions $\mathfrak{B}$-holomorphic at infinity is also known. Thus

$$R(\lambda; a) = e\lambda^{-1} + \sum_{n=1}^{\infty} a^n \lambda^{-n-1}, \tag{9.2.15}$$

convergent for $|\lambda| > r(a)$. We have no assurance, however, that all solutions $\mathfrak{B}$-holomorphic at infinity are of this type and, actually, this is not true. We can replace e by j, an arbitrary singular idempotent, provided

$$aj = ja = a. \tag{9.2.16}$$

This condition, of course, forces a to be singular. Moreover, we can add a constant term q provided

$$q^2 = 0 \quad \text{and} \quad jq = qj = 0. \tag{9.2.17}$$

Again, a solution of (9.2.5) may exist in a neighborhood of infinity without being $\mathfrak{B}$-holomorphic at infinity. We can have polynomial solutions. Suppose that q is a nilpotent in $\mathfrak{B}$ of order p. Then

$$R(\lambda) = q - \lambda q^2 + \lambda^2 q^3 + \cdots + (-\lambda)^{p-2} q^{p-1} \tag{9.2.18}$$

is a solution. We can generalize this by letting q be quasi-nilpotent rather than nilpotent. Thus

$$R(\lambda) = \sum_{n=0}^{\infty} (-\lambda)^n q^{n+1} \tag{9.2.19}$$

is a solution having an essential singular point at infinity. Note that the last solution is an entire function of λ. All these solutions, (9.2.15) excepted, are algebraically singular for all values of λ. The last two solutions are simply (9.2.8) for $\alpha = 0$ and $a = q$.

We come now to the problem of finite singularities of $R(\lambda)$. Since the spectrum of $a \in \mathfrak{B}$ may be any bounded closed set, we realize that a finite singularity of a solution $R(\lambda)$ of (9.2.5)—all resolvents being solutions together with non-resolvents without end—need not be isolated at all. Non-isolated singularities are practically inaccessible to the methods of analysis, with very rare exceptions, and this is all that we can and will say in this case. For isolated singularities there is an elegant theorem of M. Nagumo (1936), and this will now be discussed. At an isolated point in the neighborhood of which R is single-valued, we have a convergent Laurent expansion with coefficients of simple structure. Since equation (9.2.5) is invariant under a shift

$$\lambda \to \lambda + \alpha, \qquad \mu \to \mu + \alpha,$$

the singularity may be placed at the origin. Then

$$R(\lambda) = \sum_{-\infty}^{\infty} a_n \lambda^n = R^+(\lambda) + R^-(\lambda), \tag{9.2.20}$$

where R^- contains the terms with negative exponents and R^+ the rest. We have now

Theorem 9.2.1. *If $\lambda = 0$ is an isolated singularity of a solution R of* (9.2.5), *then all coefficients in* (9.2.20) *commute, coefficients with negative subscripts annihilate those with non-negative subscripts*

$$a_n a_{-k} = 0, \qquad n \geqslant 0, \quad k > 0, \tag{9.2.21}$$

so that

$$R^+(\lambda)\, R^-(\mu) = R^-(\mu)\, R^+(\lambda) = 0 \tag{9.2.22}$$

wherever these functions are defined. Further, there exist an element a, an idempotent j and a quasinilpotent q, all in $\mathfrak{B}$, such that

$$a_n = (-1)^n a^{n+1}, \quad n \geqslant 0, \qquad a_{-1} = j,\, a_{-k} = q^{k-1}, \quad k > 1, \tag{9.2.23}$$

and

$$aj = ja = 0, \qquad qj = jq = q. \tag{9.2.24}$$

Here $R^+(\lambda)$ converges in the disk $r(a)|\lambda| < 1$ which has a finite radius unless a is also a quasinilpotent. $R^-(\lambda)$ converges for all λ with $|\lambda| > 0$.

Proof. We shall give an argument somewhat different from that of Nagumo. The algebra is simpler and the required analysis is elementary function theory. By assumption

$$(\lambda - \mu) \sum_{-\infty}^{\infty} a_m \lambda^m \sum_{-\infty}^{\infty} a_n \mu^n = \sum_{-\infty}^{\infty} a_n \mu^n - \sum_{-\infty}^{\infty} a_m \lambda^m \tag{9.2.25}$$

with absolutely convergent series. Since (9.2.5) is unchanged when λ and μ are interchanged, we conclude that the coefficients commute. Compare coefficients of $\lambda^m\mu^n$ on both sides. If

$$m \neq 0, n \neq 0, \qquad a_{m-1}a_n - a_m a_{n-1} = 0, \tag{9.2.26}$$

$$m = 0, n \neq 0, \qquad a_{-1}a_n - a_0 a_{n-1} = a_n, \tag{9.2.27}$$

$$m = 0, n = 0, \qquad a_{-1}a_0 - a_0 a_{-1} = 0.$$

Since we already know that the coefficients commute, the last equation gives no new information, nor does the case $m \neq 0$, $n = 0$ add to our information.

Since $R^+(\lambda)$ must have a positive radius of convergence, there exists an A, $0 < A < \infty$, such that

$$\|a_n\| \leqslant A^n, \qquad n \geqslant 0. \tag{9.2.28}$$

On the other hand, the power series for R^- must converge for all λ, $0 < |\lambda|$. This requires

$$\lim_{k\to\infty} \|a_{-k}\|^{1/k} = 0. \tag{9.2.29}$$

We now consider (9.2.26) where we replace m by $m + 1$, $0 \leqslant m$, and take $n = -k$, $0 < k$. The result is

$$a_m a_{-k} = a_{m+1}a_{-k-1} = a_{m+2}a_{-k-2} = \cdots = a_{m+p}a_{-k-p}$$

for all p. As $p \to \infty$ the last member goes to zero by (9.2.28) and (9.2.29). It follows that (9.2.21) holds and hence also (9.2.22).

The rest is easy. For $n > 0$ formula (9.2.27) gives

$$a_n = -a_0 a_{n-1}.$$

This shows that we can take $a_0 = a$ and obtain the first of the formulas under (9.2.23).

Next we look at (9.2.27) again, now for $n < 0$. We get first

$$(a_{-1})^2 = a_{-1}$$

so that a_{-1} is an idempotent. This is our j. For $k > 1$

$$a_{-1}a_{-k} = a_{-k} \quad \text{or} \quad ja_{-k} = a_{-k}.$$

We now use (9.2.26) with $m = -k$, $n = -1$ and obtain

$$a_{-k}a_{-2} = a_{-k-1}a_{-1} = a_{-k-1}j = a_{-k-1}.$$

Hence setting $a_{-2} = q$ we get

$$a_{-k} = q^{k-1}, \qquad 1 < k.$$

Since the series for R^- must converge for all $\lambda \neq 0$, it is seen that q is either a nilpotent or a quasinilpotent. If q is a nilpotent of order p, then R^- has a pole of order p at the origin, otherwise $\lambda = 0$ is an essential singular point. ∎

It is worth observing that a solution $R(\lambda)$ of (9.2.5) with an isolated singularity at $\lambda = 0$ is determined completely by three elements a, j, and q which act as coefficients of λ^0, λ^{-1}, λ^{-2}, respectively. They must satisfy (9.2.24) and, in addition, j must be an idempotent, q a nilpotent or quasinilpotent. Otherwise they are completely arbitrary.

It is appropriate at this juncture to devote some attention to the case of a B-algebra without a unit element where we are dealing with dissolvents rather than resolvents. We recall the basic relation (9.1.26)

$$x + \lambda D(\lambda, x) = xD(\lambda, x) = D(\lambda, x)\, x. \tag{9.2.30}$$

We can subject this relation to the same type of analysis as applied to the resolvent relation (9.2.1). If λ and μ are in $\delta(x)$, we can eliminate x and obtain the *first dissolvent equation*

$$(\lambda - \mu)\, D(\mu)\, D(\lambda) = \lambda D(\lambda) - \mu D(\mu), \tag{9.2.31}$$

where we have suppressed x, which plays no further role for this equation. Similarly, if $\lambda \in \delta(x) \cap \delta(y)$, we can eliminate λ and obtain the *second dissolvent equation*

$$E(y)\,(y - x)\, E(x) = E(y)\, x - yE(x), \tag{9.2.32}$$

where λ has been omitted. In the first equation the mapping is from C into $\mathfrak{B}$, in the second from $\mathfrak{B}$ into itself.

Suppose that $D(\lambda)$ is defined in a domain Δ which is at a positive distance from the origin and that $\alpha \in \Delta$. Suppose that $\|D(\lambda)\| \leqslant M$ in Δ and set $D(\alpha) = a$. Here a is the initial value of the solution and may be any element of $\mathfrak{B}$, reversible or not. From (9.2.31) we obtain

$$D(\lambda) = \frac{\alpha}{\lambda} a + \frac{\lambda - \alpha}{\lambda} aD(\lambda)$$

and by repeated substitution

$$D(\lambda) = \frac{\alpha}{\lambda} a + \frac{\alpha}{\lambda}\frac{\lambda - \alpha}{\lambda} a^2 + \cdots + \frac{\alpha}{\lambda}\left(\frac{\lambda - \alpha}{\lambda}\right)^{n-1} a^n + \left(\frac{\lambda - \alpha}{\lambda}\right)^n a^n D(\lambda).$$

Hence, if

$$\left|\frac{\lambda - \alpha}{\lambda}\right| r(a) < 1, \tag{9.2.33}$$

the remainder goes to 0 and we get the expansion

$$D(\lambda) = \frac{\alpha}{\lambda} \sum_{n=0}^{\infty} \left(\frac{\lambda - \alpha}{\lambda}\right)^n a^{n+1}. \tag{9.2.34}$$

The domain of convergence given by (9.2.33) is a circular disk or a half-plane or the exterior of a circle according as $r(a)$ is >1, $=1$ or <1.

A solution of (9.2.31) may be holomorphic at infinity and vanish there as shown by (9.1.25) with

$$D(\lambda, x) = -\sum_{n=1}^{\infty} x^n \lambda^{-n}. \tag{9.2.35}$$

Again a solution of (9.2.31) need not be holomorphic at infinity. The technique used in proving Theorem 9.2.1 will also give

Theorem 9.2.2. *Every solution of* (9.2.31) *which is* $\mathfrak{B}$*-holomorphic in* $0 \leqslant R < |\lambda| < \infty$ *is of the form*

$$D(\lambda) = \sum_{n=0}^{\infty} q^n \lambda^n + j + \sum_{n=1}^{\infty} (-1)^{n-1} a^n \lambda^{-n} \tag{9.2.36}$$

with

$$aj = ja = 0, \quad j^2 = j, \quad qj = jq = q, \quad \|q^n\|^{1/n} \to 0. \tag{9.2.37}$$

Proof. Set

$$D(\lambda) = \sum_{-\infty}^{\infty} a_n \lambda^n.$$

The convergence assumptions for this Laurent series imply that

$$\lim_{n\to\infty} \|a_n\|^{1/n} = 0, \quad \|a_{-k}\| < A^k, \qquad \forall\, k,\, k > 0 \tag{9.2.38}$$

for some A, $R < A < \infty$. Substitution of the series in (9.2.31) shows that the coefficients commute and leads to the relations

$$a_{m-1}a_n = a_m a_{n-1}, \qquad m \neq 0,\, n \neq 0, \tag{9.2.39}$$

$$a_{m-1}a_0 - a_m a_{-1} = a_{m-1}, \quad m \neq 0,\, n = 0. \tag{9.2.40}$$

If in (9.2.39) we replace m by $m+1$, $m \geqslant 0$, and set $n = -k$, $0 < k$, we get

$$a_m a_{-k} = a_{m+1}a_{-k-1} = a_{m+2}a_{-k-2} = \cdots = a_{m+p}a_{-k-p}.$$

This goes to zero as $p \to \infty$ by (9.2.38). Thus the coefficients with negative subscripts annihilate the coefficients with non-negative subscripts.

Next, from (9.2.40) with $m = 1$,

$$a_0^2 = a_0 \quad \text{so} \quad a_0 = j,$$

an idempotent. Again, from (9.2.40),

$$a_{m-1}a_0 = a_{m-1} \quad \text{or} \quad ja_m = a_m j = a_m,\ m \geqslant 0.$$

We set $a_1 = q$ where the notation indicates that q is expected to be a quasinilpotent. Then from (9.2.39) with $m = 1$, $n \geqslant 1$, we get $a_0 a_n = a_1 a_{n-1}$ so that

$$a_n = ja_n = qa_{n-1} \quad \text{or} \quad a_n = q^n.$$

It remains to analyze the coefficients with negative subscripts. It is already known that a_{-k} is annihilated by a_m, $0 < k$, $0 \leqslant m$, and this condition implies and is implied by

$$ja_{-k} = a_{-k}j = 0, \qquad k > 0,$$

since $q = jq = qj$. We again resort to (9.2.40) where we set $m = -k$ and obtain

$$a_{-k-1} = -a_{-1}a_{-k}.$$

We set $a_{-1} = a$ and obtain, finally,

$$a_{-k} = (-1)^{k-1}a^k.$$

This is the last identity to be proved. The convergence assumptions show that q is a quasinilpotent. If q is a nilpotent, then $D(\lambda)$ has a pole at infinity, otherwise an essential singularity. ∎

Note that (9.2.35) is a special case of (9.2.36) for $j = 0$, $q = 0$, $a = -x$.

There is obviously a problem concerning the nature of finite isolated singularities of solutions of (9.2.31). This appears more difficult than either the infinite case or the finite case for (9.2.5). This question is left to the reader for consideration.

Equation (9.2.32) presents further complications which are largely due to the quirks of the underlying B-algebra which are apt to interfere with the proposed argument. We refer to the Exercise below for some of these difficulties. Here we note merely that if E is (F)-analytic in some domain $\mathfrak{D}$ of $\mathfrak{B}$, then its first differential satisfies

$$x\delta E(x; h) = E(x)\,h - E(x)\,hE(x). \tag{9.2.41}$$

EXERCISE 9.2

1. Why is (9.2.9) the form of the continuation of $R(\lambda)$ in Δ? What is the first step in the argument?
2. Same question for (9.2.13).
3. What is the expression for $\delta^{(n)}\, S(x;\, h)$, $n > 1$?
4. Justify (9.2.16) and (9.2.17).
5. If R_1 and R_2 both satisfy (9.2.5) and if

$$R_1(\lambda)\, R_2(\mu) = R_2(\mu)\, R_1(\lambda) = 0,$$

   for all relevant λ and μ, show that $R_1 + R_2$ is also a solution.
6. Show that R^+ and R^- are both solutions of (9.2.5).
7. If R is a solution of (9.2.5), show that $R'(\lambda) = -[R(\lambda)]^2$. Express the higher derivatives in terms of $R(\lambda)$. The differential equation $w'(z) = -[w(z)]^2$, where z and w are complex numbers, is a special case of *Riccati's equation.*
8. Derive (9.2.31).

9. Derive (9.2.32).

10. If D is a solution of (9.2.31), show that

$$D'(\lambda) = [D(\lambda)]^2 - D(\lambda).$$

The classical analogue is also a special case of Riccati's equation.

11. In (9.2.34) set $a = j$, an idempotent, and sum the series in closed form. Can this expression be the dissolvent of some $x \in \mathfrak{B}$?

12. Verify (9.2.41).

13. Consider the commutative convolution algebra $L_2(-\pi, \pi)$ without unit element and consider (9.3.32) in this setting,

$$G*(f - g)*F = G*f - g*F, \qquad F = E(f), \qquad G = E(g).$$

Let f, g, F, G have the Fourier coefficients $\{f_n\}, \{g_n\}, \{F_n\}, \{G_n\}$ respectively, where each sequence is an element of l_2. Find what relations hold between the Fourier coefficients, if the equation is satisfied. Suppose that $\{f_n\}$, $\{g_n\}$ and $\{G_n\}$ are given subject to the conditions

$$|G_n| \leqslant k|g_n|, \quad k \text{ fixed}, \quad 0 < k < 1, \quad |f_n - g_n| \leqslant 1, \qquad \forall\, n.$$

Show that $\{F_n\}$ is uniquely determined in l_2 and $|F_n| \leqslant [k/(1 - k)]\,|f_n|, \forall n$.

14. Let $\mathfrak{Q}$ be a 3 by 3 matrix with $\mathfrak{Q}^2 \neq 0, \mathfrak{Q}^3 = 0$ and consider all matrices of the form $\alpha\mathfrak{Q} + \beta\mathfrak{Q}^2$, where α and β are complex numbers. These matrices form a normed subalgebra $\mathfrak{B}$ of $\mathfrak{M}_3$ without unit element. Equation (9.2.32) is to be studied in $\mathfrak{B}$. Let $(\alpha, \beta) \to f(\alpha, \beta)$ be an arbitrary mapping of $C \times C$ into C and set $\mathfrak{E}(\mathfrak{X}) = f(\alpha, \beta)\,\mathfrak{X}^2$ for all $\mathfrak{X} \in \mathfrak{B}$. Show that this is a solution of (9.2.32) which may be wildly discontinuous. If it is locally bounded, show that $\mathfrak{X}\mathfrak{E}(\mathfrak{X})$ will be continuous. If $\mathfrak{E}(\mathfrak{X})$ is continuous, show that $\mathfrak{X}\mathfrak{E}(\mathfrak{X})$ is (F)-analytic even though $\mathfrak{E}(\mathfrak{X})$ is not.

9.3 GELFAND'S REPRESENTATION THEOREM

In 1941 I. M. Gelfand presented an elegant method of representing a commutative B-algebra with unit element by means of residue class algebras modulo a maximal ideal. Such algebras have important properties. In the following, algebras will be commutative, with unit element, until further notice.

Any B-algebra contains linear subspaces known as ideals. A set $\mathfrak{i}$ is called an *ideal* of $\mathfrak{B}$, if $x, y \in \mathfrak{i}$ imply (1) $x + y \in \mathfrak{i}$ and (2) $x\mathfrak{B} \subset \mathfrak{i}$. The ideal may be *trivial*: the *null ideal* containing only the zero element and the *unit ideal* which is all of $\mathfrak{B}$. An algebra is said to be *simple* if these two are the only ideals admitted. An example of a simple algebra is furnished by C, the field of complex numbers. If $\mathfrak{B}$ is not simple, then there are other ideals called *proper*. Thus if $\mathfrak{B}$ contains a singular element $a \neq 0$, then the set $a\mathfrak{B}$ is a proper ideal. It is an ideal for if $x = ab$, $y = ac$ are in the set, so is $x + y$ and $x\mathfrak{B} \subset a\mathfrak{B}$. The ideal is proper for it contains $a \neq 0$, so it cannot be the null ideal, and if it were all of $\mathfrak{B}$, then there would exist a b such that $ab = ba = e$. This is absurd since a is singular by assumption.

An ideal $\mathfrak{m}$ is said to be *maximal* if $\mathfrak{m}$ is a proper ideal and properly contained in no other proper ideal.

Lemma 9.3.1. *An element x of $\mathfrak{B}$ is regular iff it belongs to no maximal ideal.*

Proof. Suppose that x is regular and belongs to a maximal ideal. Then $xx^{-1} = e$ belongs to the ideal and the latter must be the unit ideal. This is a contradiction. Hence a regular element can belong to no maximal ideal or for that matter to any ideal. Conversely, if x belongs to no maximal ideal, then the assumption that x be singular would lead to a contradiction. For then $x\mathfrak{B}$ would be an ideal containing x, and this ideal is either maximal or a proper subset of a maximal ideal. Again we have a contradiction. ∎

Corollary. *If $\mathfrak{B}$ is simple, then all elements are regular excepting the zero element.*

For there are no proper ideals, hence no singular elements besides zero.

This result may be strengthened, but we shall first insert a remark concerning the topological structure of a maximal ideal.

Lemma 9.3.2. *A maximal ideal in a B-algebra is a closed point set.*

Proof. For $\mathfrak{m} \subset \overline{\mathfrak{m}}$ and $\overline{\mathfrak{m}}$ is also an ideal (why?). There are two possibilities: $\overline{\mathfrak{m}} = \mathfrak{m}$ or $\overline{\mathfrak{m}} = \mathfrak{B}$. The second possibility must be rejected for it would imply that $\mathfrak{m}$ is dense in $\mathfrak{B}$ in particular, in the sphere $\|x - a\| < 1$, all the elements of which are regular. As we have seen, no element of a maximal ideal can be regular. Hence $\overline{\mathfrak{m}} = \mathfrak{m}$ or $\mathfrak{m}$ is closed. ∎

We can strengthen the corollary under Lemma 9.3.1 as follows.

Theorem 9.3.1. *A simple commutative B-algebra with unit element over the complex field is isomorphic to the complex field.*

Proof. If $x \in \mathfrak{B}$, then $\sigma(x)$ cannot be void and there is at least one value λ_0 for which $\lambda_0 e - x$ is singular. For the assumption that $\sigma(x)$ is void would lead to the conclusion that the resolvent is $\mathfrak{B}$-holomorphic in the extended plane. Only a constant can have this property and, since $R(\lambda, x)$ is 0 at infinity, $R(\lambda, x)$ would be identically zero and this is absurd. Hence there is a λ_0 such that $\lambda_0 e - x$ is singular. Since the zero element is the only singular element of a simple algebra, we have $\lambda_0 e - x = 0$ or $x = \lambda_0 e$. Note that the spectrum of x contains one and only one point. Thus there is a 1–1 correspondence between $\mathfrak{B}$ and the complex field C and this correspondence is an isomorphism (why?). ∎

We recall that an isomorphism takes sums into sums and products into products.

The existence of proper ideals in non-simple commutative algebras $\mathfrak{B}$ was shown above: for any singular element $x \in \mathfrak{B}$ the set $x\mathfrak{B}$ is an ideal. The existence of maximal ideals is another matter. Here Zorn's Lemma is commonly invoked. If a given ideal $\mathfrak{i}$ is not maximal, then it is properly contained in a larger ideal $\mathfrak{i}_1$. In this manner a partial ordering by inclusion is defined for the ideals of $\mathfrak{B}$ and Zorn's Lemma establishes the existence of a maximal element of such a partially ordered collection which is the maximal ideal containing $\mathfrak{i}$.

Example 1. We shall give an instance where the ideal under discussion turns out to be maximal. We take the commutative B-algebra $C[0, 1]$ of functions f continuous in $[0, 1]$. Let t_0 be a point of this interval, $0 \leqslant t_0 \leqslant 1$, and set

$$\mathfrak{N} = [f; f(t_0) = 0]. \tag{9.3.1}$$

This is an ideal since $f, g \in \mathfrak{N}$, $h \in \mathfrak{B}$ implies

$$f + g \in \mathfrak{N}, \qquad fh \in \mathfrak{N}.$$

Moreover, $\mathfrak{N}$ is a maximal ideal. For if $\mathfrak{N}$ would be a subset of a proper ideal $\mathfrak{i}$ and if $g \in \mathfrak{i} \ominus \mathfrak{N}$, then $g(t_0) = \lambda_0$, some number $\neq 0$. Let $h(t) = \lambda_0 - g(t)$, then $h \in \mathfrak{N}$ and

$$\lambda_0 = [\lambda_0 - g] + g \in \mathfrak{i}.$$

This says that some constant multiple of the unit element is in $\mathfrak{i}$ and this implies $\mathfrak{i} = \mathfrak{B}$ so that $\mathfrak{N}$ is maximal.

Now suppose that $\mathfrak{B}$ is a commutative B-algebra with unit element e, not simple, and let $\mathfrak{m}$ be a maximal ideal of $\mathfrak{B}$. The elements of $\mathfrak{B}$ fall into equivalence classes with respect to $\mathfrak{m}$. Here x_1 and x_2 belong to the same equivalence class iff

$$x_1 - x_2 \in \mathfrak{m}.$$

In this connection the equivalence classes are called *residue classes of* $\mathfrak{B}$ *modulo* $\mathfrak{m}$. Each class X is of the form

$$X = x + \mathfrak{m}, \tag{9.3.2}$$

i.e. X is the set obtained by adding the fixed element x to the elements of $\mathfrak{m}$. The set of all residue classes X will form the so-called *quotient algebra* $\mathfrak{B}/\mathfrak{m}$. This set is an algebra, actually a B-algebra, under suitable definitions of the algebraic operations and the norm. If

$$X = x + \mathfrak{m}, \qquad Y = y + \mathfrak{m}, \qquad \alpha \in C,$$

we set

$$X + Y = x + y + \mathfrak{m}, \tag{9.3.3}$$

$$\alpha X = \alpha x + \mathfrak{m}, \tag{9.3.4}$$

$$XY = xy + \mathfrak{m}, \tag{9.3.5}$$

and define

$$\|X\| = \{\inf \|s\|; s \in X\}. \tag{9.3.6}$$

The reader should verify that these definitions satisfy the A-, S-, and M-postulates of Sections 2.1 and 2.6, that (9.3.6) defines a norm satisfying

$$\|XY\| \leqslant \|X\| \, \|Y\| \tag{9.3.7}$$

and that, moreover, $\mathfrak{B}/\mathfrak{m}$ is complete in this metric. We have now the important

Theorem 9.3.2. *If* $\mathfrak{m}$ *is a maximal ideal in* $\mathfrak{B}$, *then the quotient algebra* $\mathfrak{B}/\mathfrak{m}$ *is isomorphic to the algebra of complex numbers.*

Proof. It is to be shown that $\mathfrak{B}/\mathfrak{m}$ has no proper ideals. Suppose, contrariwise, that $\mathfrak{I}$ is a proper ideal in $\mathfrak{B}/\mathfrak{m}$, $\mathfrak{I} \neq \{0\}$. Then $\mathfrak{I}$ contains residue classes $X = x + \mathfrak{m}$ and not merely the zero class $0 + \mathfrak{m}$. Let $\mathfrak{M}$ be the union of all the elements of $\mathfrak{B}$ which belong to the residue classes that together make up the ideal $\mathfrak{I}$. It is claimed that $\mathfrak{M}$ is an ideal of $\mathfrak{B}$ which contains $\mathfrak{m}$ which implies that $\mathfrak{M} = \mathfrak{B}$, and this in turn gives $\mathfrak{I} = \mathfrak{B}/\mathfrak{m}$ so that $\mathfrak{I}$ is not proper. To see all this, suppose that $x \in X$ and $y \in Y$ where X and Y belong to $\mathfrak{I}$. Then x and y belong to $\mathfrak{M}$ and $x + y \in X + Y$ implies $x + y \in \mathfrak{M}$ since $X + Y \in \mathfrak{I}$. Further, by assumption, $X(\mathfrak{B}/\mathfrak{m}) \in \mathfrak{I}$ and this says that $x \in \mathfrak{M}$ implies that $x\mathfrak{B} \subset \mathfrak{M}$. Now the set $\mathfrak{m}$ is the zero element of $\mathfrak{B}/\mathfrak{m}$ and hence necessarily an element of $\mathfrak{I}$ and this gives $\mathfrak{m} \subset \mathfrak{M}$. But $\mathfrak{m}$ is maximal, so $\mathfrak{M} = \mathfrak{B}$ and $\mathfrak{I} = \mathfrak{B}/\mathfrak{m}$. Hence $\mathfrak{B}/\mathfrak{m}$ is simple and by Theorem 9.3.1 this implies that $\mathfrak{B}/\mathfrak{m}$ is isomorphic to C. ∎

In this manner every maximal ideal $\mathfrak{m}$ of $\mathfrak{B}$ defines a correspondence between $\mathfrak{B}$ and C in which the element x is mapped on the complex number α. Here α is the number such that x belongs to the residue class $\alpha e + \mathfrak{m}$,which is α times $e + \mathfrak{m}$. Inspection of (9.3.5) shows that the residue class $e + \mathfrak{m} \equiv E$ is the unit element of the algebra $\mathfrak{B}/\mathfrak{m}$. Hence x belongs to the residue class αE modulo $\mathfrak{m}$ and the mapping is $x \to \alpha$. We now define a functional $x(\mathfrak{m})$ on the space $\mathfrak{M}$ of all maximal ideals $\mathfrak{m}$ of $\mathfrak{B}$ by the convention that $x(\mathfrak{m})$ is the complex number uniquely determined by the condition

$$x - x(\mathfrak{m})\, e \in \mathfrak{m}. \tag{9.3.8}$$

This functional now has the following properties:

Theorem 9.3.3. *For each* $\mathfrak{m}$

i) $(x + y)(\mathfrak{m}) = x(\mathfrak{m}) + y(\mathfrak{m}), \quad \forall x \in \mathfrak{B}, \quad \forall y \in \mathfrak{B}$;

ii) $(\alpha x)(\mathfrak{m}) = \alpha x(\mathfrak{m}), \quad \forall \alpha \in C, \quad \forall x \in \mathfrak{B}$;

iii) $(xy)(\mathfrak{m}) = x(\mathfrak{m})\, y(\mathfrak{m}), \quad \forall x \in \mathfrak{B}, \quad \forall y \in \mathfrak{B}$;

iv) $e(\mathfrak{m}) = 1$;

v) $0(\mathfrak{m}) = 0$;

vi) *If* x *is regular, then* $x(\mathfrak{m}) \neq 0$ *and* $(x^{-1})(\mathfrak{m}) = [x(\mathfrak{m})]^{-1}$;

vii) $x(\mathfrak{m}) \in \sigma(x)$;

viii) $r(x) = \sup |x(\mathfrak{m})|$;

ix) *If* $\lambda_0 \in \sigma(x)$, *then there exists a maximal ideal* $\mathfrak{m}_0$ *such that* $x(\mathfrak{m}_0) = \lambda_0$;

x) *If* $x(\mathfrak{m}) = 0$ *for all* $\mathfrak{m}$, *then* x *is nilpotent or quasinilpotent.*

Proof. The first three properties are simply a restatement of the isomorphic properties of the mapping from x to $x(\mathfrak{m})$ implied by Theorem 9.3.2. Properties

(iv) and (v) are implied by e and 0 being idempotents. For if

$$j^2 = j, \quad \text{then} \quad [j(\mathfrak{m})]^2 = j(\mathfrak{m}),$$

so that

$$j(\mathfrak{m}) = 0 \quad \text{or} \quad 1. \tag{9.3.9}$$

Since $0 \cdot x = 0$ and $ex = x$, the choice is categorical: $0(\mathfrak{m}) = 0$, $e(\mathfrak{m}) = 1$.

If x is regular, then $xx^{-1} = e$ gives $x(\mathfrak{m})\,(x^{-1})\,(\mathfrak{m}) = 1$ and (vi) follows. For (vii) we note that $x(\mathfrak{m}) = \alpha$ for a given $\mathfrak{m}$ implies that $x \equiv \alpha e \pmod{\mathfrak{m}}$ or $x - \alpha e \in \mathfrak{m}$ or $x - \alpha e$ is singular. Hence $\alpha \in \sigma(x)$. Since $r(x)$ is the maximum of the absolute value of any element of $\sigma(x)$, it is seen that (vii) implies $|x(m)| \leqslant r(x)$. The stronger assertion made in (viii) follows from (ix), which we now proceed to prove. Suppose that $\lambda_0 \in \sigma(x)$ so that $\lambda_0 e - x$ is singular. We can then form the ideal $(\lambda_0 e - x)\,\mathfrak{B}$ which contains $\lambda_0 e - x$. If this ideal is maximal, we are through; if not, then we can embed it in a maximal ideal $\mathfrak{m}_0$ and $x \equiv \lambda_0 e \pmod{\mathfrak{m}_0}$ and $x(\mathfrak{m}_0) = \lambda_0$. This implies (ix) and hence also (viii). Finally, if $x(\mathfrak{m}) = 0$ for all $\mathfrak{m}$, then by (ix) $\sigma(x) = \{0\}$ and the resolvent

$$R(\lambda, x) = \sum_{n=0}^{\infty} x^n \lambda^{-n-1}$$

converges for all $\lambda \neq 0$. If the series breaks off, then x is a nilpotent, otherwise the convergency condition requires that x is a quasinilpotent. ∎

We return to Example 1 for illustration and inspiration. If t_0 is given, $0 \leqslant t_0 \leqslant 1$, and if

$$\mathfrak{m} = \{f; f(t_0) = 0\},$$

the residue classes modulo $\mathfrak{m}$ are now simply

$$\alpha E = \{f; f(t_0) = \alpha\}.$$

Hence we have $x(\mathfrak{m}) = f(\mathfrak{m}) = f(t_0)$ and the reader will have no difficulty in verifying that the mapping $f \to f(t_0)$ has all the properties (i) to (x). Incidentally, the zero element of $C[0, 1]$ is the only nilpotent and there are no proper quasinilpotents.

In this case the maximal ideals and the residue classes are determined by the value of the element f at a preassigned point $t = t_0$. Now the map from f to $f(t_0)$ determines a linear, multiplicative, and bounded functional on $C[0, 1]$ and this raises the question of the relations between such functionals, on the one hand, and maximal ideals, on the other. Here we note

Theorem 9.3.4. *Let $\mathfrak{B}$ be a complex commutative non-simple B-algebra with unit element and let $\mu(x)$ be a linear, multiplicative bounded functional defined on $\mathfrak{B}$. Let $\mathfrak{N}$ be the nullspace of μ. Then $\mathfrak{N}$ is a maximal ideal. Here the set of all $x \in \mathfrak{B}$ for which $\mu(x)$ has a fixed value α is the residue class αE modulo $\mathfrak{N}$. Conversely, the elements of a maximal ideal in $\mathfrak{B}$ form the null space $\mathfrak{N}$ of a linear multiplicative and bounded functional μ on $\mathfrak{B}$ which takes the value α on the residue class αE modulo $\mathfrak{m}$.*

Proof. Given a non-trivial linear multiplicative bounded functional μ on $\mathfrak{B}$, its nullspace $\mathfrak{N}$ is certainly a proper ideal since

$$\mu(x+y) = \mu(x) + \mu(y),\ \mu(\alpha x) = \alpha\mu(x),$$
$$\mu(xy) = \mu(x)\,\mu(y),\ \mu(e) = 1. \tag{9.3.10}$$

If $\mathfrak{N}$ is not maximal, then there exists a maximal ideal $\mathfrak{m}$ containing $\mathfrak{N}$. Suppose that $x_0 \in \mathfrak{m}$ but is not in $\mathfrak{N}$, so that $\mu(x_0) = \lambda_0 \neq 0$. Then $\lambda_0 e - x_0 \in \mathfrak{N}$ and *a fortiori* belongs to $\mathfrak{m}$. Hence

$$\lambda_0 e = (\lambda_0 e - x_0) + x_0 \in \mathfrak{m}$$

or $\mathfrak{m}$ contains regular elements. This is impossible, so $\mathfrak{N}$ must be maximal and the residue class αE (mod $\mathfrak{N}$) is precisely the set of elements x with $\mu(x) = \alpha$. Conversely, if $\mathfrak{m}$ is a maximal ideal in $\mathfrak{B}$, then the mapping $\mathfrak{B}/\mathfrak{m} \to C$ is an isomorphism and hence determines a linear, multiplicative and bounded functional

$$\mu(x) = x(\mathfrak{m}) \tag{9.3.11}$$

by Theorem 9.3.3. ∎

An important application of Gelfand's theorem is to *spectral analysis.* Properties (vii) and (ix) of Theorem 9.3.3 state that the value of $x(\mathfrak{m}) = \mu(x)$ belongs to the spectrum $\sigma(x)$ of x and, moreover, that all the spectral values of x can be reached by varying the maximal ideal $\mathfrak{m}$. There is one obnoxious restriction, however: the underlying algebra must be commutative. Gelfand has shown how to get around this difficulty. We embed the element or elements under consideration in a commutative subalgebra of $\mathfrak{B}$ which is large enough so that spectra are unchanged. This means that the subalgebra must be closed under inversion so that if a regular element belongs to the subalgebra, so does its inverse.

Let $x_1, x_2, \ldots, x_n$ be elements of $\mathfrak{B}$ which commute and form the subalgebra $\mathfrak{A}$ generated by the elements

$$e, x_1, x_2, \ldots, x_n.$$

Then $\mathfrak{A}$ consists of all polynomials in these elements. $\mathfrak{A}$ is obviously Abelian (= commutative). Next we form $\mathfrak{A}^c$, the *first commutant* of $\mathfrak{A}$. This is the set of all elements of $\mathfrak{B}$ which commute with $x_1, x_2, \ldots, x_n$ and hence with all elements of $\mathfrak{A}$. We have, of course, $\mathfrak{A} \subset \mathfrak{A}^c \subset \mathfrak{B}$. It is easy to see that $\mathfrak{A}^c$ is also an algebra with unit element e, but $\mathfrak{A}^c$ need not be commutative. To get out of this difficulty we repeat the procedure and form

$$(\mathfrak{A}^c)^c \equiv \mathfrak{A}^{cc}, \tag{9.3.12}$$

known as the *second commutant* of $\mathfrak{A}$. Let us note that if $\mathfrak{A}_1$ and $\mathfrak{A}_2$ are subalgebras of $\mathfrak{B}$ and if $\mathfrak{A}_1 \subset \mathfrak{A}_2$, then $\mathfrak{A}_2{}^c \subset \mathfrak{A}_1{}^c$, for the elements of $\mathfrak{B}$ which commute with those of $\mathfrak{A}_2$ must commute with those of $\mathfrak{A}_1$ but the converse does not necessarily hold. From $\mathfrak{A} \subset A^c$ we thus conclude that $\mathfrak{A}^{cc} \subset \mathfrak{A}^c$. Again, $\mathfrak{A}^{cc}$ is an algebra with unit element. It is closed in the normed metric and so is $\mathfrak{A}^c$. Now $\mathfrak{A}^{cc}$ has the added property of being Abelian. This is seen as follows. If x and

y belong to $\mathfrak{A}^{cc}$ and if z is any element of $\mathfrak{A}^c$, then $xz = zx$ by the definition of $\mathfrak{A}^{cc}$. Since $\mathfrak{A}^{cc} \subset \mathfrak{A}^c$ we can take $z = y$ and see that x and y commute, so that $\mathfrak{A}^{cc}$ is commutative. Furthermore, if x is a regular element of $\mathfrak{B}$ and belongs to $\mathfrak{A}^{cc}$ and if z is any element of $\mathfrak{A}^c$, then

$$xz = zx \quad \text{implies} \quad zx^{-1} = x^{-1}z,$$

so that $x^{-1} \in \mathfrak{A}^{cc}$. Thus $\mathfrak{A}^{cc}$ is closed under inversion of regular elements. It follows that if $x \in \mathfrak{A}^{cc}$ and $\alpha e - x$ has no inverse in $\mathfrak{B}$, then it can have no inverse in $\mathfrak{A}^{cc}$, and vice versa. Thus the spectral properties of x are the same with respect to $\mathfrak{B}$ as with respect to the commutative subalgebra $\mathfrak{A}^{cc}$. Gelfand's representation theorem applies to the latter and thus gives the required information about spectra for any $x \in \mathfrak{A}^{cc}$.

EXERCISE 9.3

1. Verify the postulates (A, S, M) for the residue classes.
2. Verify that (9.3.6) defines a norm and that (9.3.7) holds.
3. Verify that for an idempotent $j \neq 0, e$ both 0 and 1 belong to the spectrum.
4. If $\mathfrak{A}$ is the subalgebra generated by e and j, characterize the commutants.
5. Let $P(\lambda)$ be a scalar polynomial with complex coefficients. Let
$$P(a) = \alpha_0 e + \alpha_1 a + \alpha_2 a^2 + \cdots + \alpha_n a^n$$
be the result of replacing λ by a, an element of a non-commutative B-algebra with unit element e. Show that the spectrum of $P(a)$ is $\{P(\lambda); \lambda \in \sigma(a)\}$. [Special case of the *spectral mapping theorem*.]
6. In a B-algebra $\mathfrak{B}$ there are two operations $\mathbf{L}_a$ and $\mathbf{R}_a$ which are particularly important for purposes of representation. Here $\mathbf{L}_a x = ax$ and $\mathbf{R}_a x = xa$. Show that the spectra of $\mathbf{L}_a$ and $\mathbf{R}_a$ in the operator algebra $\mathfrak{E}(\mathfrak{B})$ of linear bounded transformations from $\mathfrak{B}$ into itself are identical with the spectrum of a as an element of $\mathfrak{B}$.
7. The *Jordan operator* $\mathbf{J}_a = \frac{1}{2}[\mathbf{L}_a + \mathbf{R}_a]$ with $\mathbf{J}_a x = \frac{1}{2}[ax + xa]$ has a spectrum relative to $\mathfrak{E}(\mathfrak{B})$. Show that $\sigma(\mathbf{J}_a)$ is contained in the set of mean values
$$\tfrac{1}{2}[\sigma(a) + \sigma(a)] = \{\tfrac{1}{2}(\alpha + \beta); \alpha, \beta \in \sigma(a)\}.$$
8. The *commutator* $\mathbf{C}_a = \mathbf{L}_a - \mathbf{R}_a$, i.e. $\mathbf{C}_a x = ax - xa$ has its spectrum contained in the difference set $\sigma(a) - \sigma(a)$, i.e. $\sigma(\mathbf{C}_a) \subset \{\alpha - \beta; \alpha, \beta \in \sigma(a)\}$.
9. Let a and b be two elements of $\mathfrak{B}$, distinct or not, and consider the operator $\mathbf{S}_{a,b} = \mathbf{L}_a \mathbf{R}_b$ which maps x onto axb. Show that the spectrum of $\mathbf{S}_{a,b}$ is contained in the product set of $\sigma(a)$ and $\sigma(b)$ so that $\sigma[\mathbf{S}_{a,b}] \subset \{\alpha\beta; \alpha \in \sigma(a), \beta \in \sigma(b)\}$.

9.4 (F)-ANALYTIC FUNCTIONS AND THE OPERATIONAL CALCULUS

Let $\mathfrak{B}$ be a B-algebra, not necessarily commutative, with unit element e. By this time we have a fairly clear idea of what is meant by a polynomial in x, by $\exp x$ or by sin x, etc., where x is any element of $\mathfrak{B}$. We simply take the corresponding scalar power series

$$f(\lambda) = \alpha_0 + \alpha_1 \lambda + \cdots + \alpha_n \lambda^n + \cdots \tag{9.4.1}$$

and replace λ by x, writing $\alpha_0 e$ for α_0. Thus

$$f(x) = \alpha_0 e + \alpha_1 x + \cdots + \alpha_n x^n + \cdots. \tag{9.4.2}$$

If f is an entire function, no restriction need be put on x, but if (9.4.1) has a finite radius of convergence R, we require a limitation on the spectral radius of x:

$$r(x) < R. \tag{9.4.3}$$

It is not difficult to show that (9.4.2) defines an (F)-analytic function; this will be proved in a more general context below.

We can look at the problem of defining analytic mappings from $\mathfrak{B}$ to $\mathfrak{B}$ as an *extension problem*. The complex plane can be embedded in $\mathfrak{B}$ since $\mathfrak{B}$ has a unit element and all complex multiples of e belong to $\mathfrak{B}$. Suppose now that on some domain Δ in the intersection of $\mathfrak{B}$ with the complex plane we have defined a function $f(\lambda)e$ where f is Cauchy analytic for $\lambda \in \Delta$. This function is to be extended to adjacent parts of $\mathfrak{B}$ in such a manner that some type of analyticity is preserved. The passage from (9.4.1) to (9.4.2) illustrates such an extension preserving Fréchet analyticity. This can be generalized. We start with a local point of view.

Suppose that $\lambda_0 \in \Delta$ and that

$$f(\lambda) = \sum_{n=0}^{\infty} \frac{1}{n!} f^{(n)}(\lambda_0)\,(\lambda - \lambda_0)^n \tag{9.4.4}$$

for $|\lambda - \lambda_0| < \rho$. We then set

$$f(x) = \sum_{n=1}^{\infty} \frac{1}{n!} f^{(n)}(\lambda_0)\,(x - \lambda_0 e)^n \tag{9.4.5}$$

for x in the set

$$\mathfrak{S}: \{x; r(x - \lambda_0 e) < \rho\}. \tag{9.4.6}$$

An equivalent condition is that $\sigma(x)$ should be located in the interior of the circle of convergence of (9.4.4). We refer to $f(x)$ as the *principle extension* of $f(\lambda)$ from the disk $|\lambda - \lambda_0| < \rho$ to the set $\mathfrak{S}$. We shall sum the series (9.4.4) into a form which suggests further extensions.

Let $0 < \rho_1 < \rho_2 < \rho$. Then

$$\frac{1}{n!} f^{(n)}(\lambda_0) = \frac{1}{2\pi i} \int_\Gamma \frac{f(\mu)\,d\mu}{(\mu - \lambda_0)^{n+1}}, \tag{9.4.7}$$

where Γ is the oriented circle $|\mu - \lambda_0| = \rho_2$ so that for $r(x - \lambda_0 e) < \rho_1$

$$f(x) = \frac{1}{2\pi i} \int_\Gamma f(\mu) \left\{ \sum_{n=0}^{\infty} \frac{(x - \lambda_0 e)^n}{(\mu - \lambda_0)^{n+1}} \right\} d\mu. \tag{9.4.8}$$

The series converges in norm for such μ and x to the sum $R(\mu, x)$ so that

$$f(x) = \frac{1}{2\pi i} \int_{\Gamma} f(\mu)\, R(\mu, x)\, d\mu. \tag{9.4.9}$$

By Cauchy's integral

$$f(\lambda) = \frac{1}{2\pi i} \int_{\Gamma} f(\mu)\, (\mu - \lambda)^{-1}\, d\mu. \tag{9.4.10}$$

Thus the extension of $f(\lambda)$ from the disk $|\lambda - \lambda_0| < \rho$ to the set $\mathfrak{S}$ is obtained by replacing the Cauchy kernel by the resolvent $R(\mu, x)$. The latter is, of course, the principal extension of the Cauchy kernel to the B-algebra. In this discussion the restrictions imposed on Γ and on x may be substantially modified. This leads to

Definition 9.4.1. *Suppose that $f(\lambda)$ is Cauchy holomorphic in a simply connected domain Δ and is continuous inside and on its oriented simple rectifiable boundary Γ. Suppose that $x \in \mathfrak{B}$ is such that $\sigma(x) \subset \Delta$. For such f and such values of x we define the principal extension of $f(\lambda)$ to $\mathfrak{B}$ by formula* (9.4.9) *where now Γ is the boundary of Δ.*

Actually we can relax the assumptions. By using more general forms of Cauchy's integral we can dispense with the assumption that Δ is simply connected. In some applications it is convenient to allow open sets Δ which are not connected but have a finite number of connected components in each of which $f(\lambda)$ is supposed to be holomorphic, though f is only piecewise analytic in Δ. In this context the boundary Γ consists of a finite number of simple rectifiable curves suitably oriented. These refinements are just mentioned for the reader's information; we shall find it convenient to stick to Definition 9.4.1 in the following.

The definition establishes a 1–1 correspondence between two classes of functions, F and H. Here F is the set of functions f given by Cauchy's integral (9.4.10) with Γ taken as the boundary of the domain Δ. They are mappings from C to C, Cauchy holomorphic in Δ and continuous in $\Delta \cup \Gamma$. H, on the other hand, consists of mappings from $\mathfrak{B}$ to $\mathfrak{B}$ defined by (9.4.9) for values of x such that $\sigma(x) \subset \Delta$. This correspondence and these mappings have a number of interesting properties, some of which will be discussed in the following.

The function $f(\lambda)$ is Cauchy holomorphic in Δ, its image $f(x)$ is locally (F)-analytic in the set

$$\mathfrak{S}_{\Delta}: \{x; \sigma(x) \subset \Delta\}.$$

Now Cauchy's integral (9.4.10) defines a function holomorphic in Δ, and for this to be true it is sufficient that f is given as a continuous function of arc length on Γ. No assumptions are needed in the interior of Δ. The analyticity of the integral with respect to λ resides in the analyticity of the Cauchy kernel $(\mu - \lambda)^{-1}$ with respect to λ. In the same manner, the analytical properties of integrals of the form (9.4.9) reside in those of the resolvent $R(\mu, x)$ as a function of x. The discussion in

Section 9.2 shows that $R(\mu, x)$ is locally (F)-analytic. Now we can rewrite (9.2.12) as

$$R(\mu, x + \alpha h) = R(\mu, x) + R(\mu, x) \sum_{k=1}^{\infty} [\alpha h\, R(\mu, x)]^k \tag{9.4.11}$$

and have to examine what limitations have to be put on the arguments α, μ, h, and x for the formula to hold in the present context. Here the main requirement is that the series converge in norm for we can then show by direct substitution that the second resolvent equation is satisfied in the form

$$\alpha R(\mu, x + \alpha h)\, hR(\mu, x) = R(\mu, x + \alpha h) - R(\mu, x). \tag{9.4.12}$$

Since $x \in \mathfrak{S}_\Delta$ we have $\sigma(x) \subset \Delta$. We keep x fixed. Then $R(\mu, x)$ is a $\mathfrak{B}$-holomorphic function of μ in the complement $\mathsf{C}(\Delta)$ of Δ and this function tends to zero as $\mu \to \infty$. Hence there exists a finite $M = M(x)$ such that

$$\|R(\mu, x)\| \leqslant M, \qquad \mu \in C(\Delta). \tag{9.4.13}$$

Suppose now that $\alpha \in C$, $h \in \mathfrak{B}$ are any quantities such that $|\alpha|\, M\, \|h\| < \gamma < 1$. It is then seen that the series (9.4.11) converges in norm, uniformly in the arguments for

$$\mu \in \mathsf{C}(\Delta), \qquad |\alpha| \leqslant 1, \qquad \|h\| \leqslant \gamma M^{-1}, \tag{9.4.14}$$

and for such values the sum of the series is dominated in norm by

$$M(1 - \gamma)^{-1}. \tag{9.4.15}$$

Thus the sum of the series represents a solution of (9.4.12) for the stated values of the arguments and the sum of the series is the resolvent of $x + \alpha h$ at $\lambda = \mu$. Thus $R(\mu, x + \alpha h)$ exists and is bounded in $\mathsf{C}(\Delta)$ for α, h and x as indicated. We can now prove

Theorem 9.4.1. *$\mathfrak{S}_\Delta$ is an open set in $\mathfrak{B}$, possibly not connected. For any $f \in \mathsf{F}$, any $x \in \mathfrak{S}_\Delta$, $|\alpha| \leqslant 1$, $h \in \mathfrak{B}$ with $\|h\| \leqslant \gamma[M(x)]^{-1}$, the corresponding function $f(x)$ is locally (F)-analytic and*

$$f(x + \alpha h) = f(x) + \sum_{k=1}^{\infty} \alpha^k \frac{1}{2\pi i} \int_\Gamma f(\mu)\, R(\mu, x)\, hR(\mu, x) \ldots hR(\mu, x)\, d\mu. \tag{9.4.16}$$

There are $k + 1$ factors $R(\mu, x)$ in the coefficient of α^k alternating with k factors h. For fixed x the series converges in norm and uniformly with respect to α and h, restricted as indicated.

Proof. By the preceding argument $R(\mu, x + \alpha h)$ exists for $\mu \in \mathsf{C}(\Delta)$ when α, h and x are restricted as stated. This asserts that the spectrum of $x + \alpha h$ is confined to Δ so that $x + \alpha h \in \mathfrak{S}_\Delta$. This implies the existence of a sphere in $\mathfrak{B}$ with center at x all the points of which are in $\mathfrak{S}_\Delta$, so that $\mathfrak{S}_\Delta$ is an open set. One of the components of this open set contains the set of elements $\{\lambda e;\, \lambda \in \Delta\}$. However, we have no assurance that this is the only component.

Now since $x + \alpha h \in \mathfrak{S}_\Delta$, we have by definition

$$f(x + \alpha h) = \frac{1}{2\pi i} \int_\Gamma f(\mu)\, R(\mu, x + \alpha h)\, d\mu.$$

Here we can substitute the series (9.4.11) and integrate termwise with respect to μ since the series converges absolutely and uniformly. This gives (9.4.16). ∎

Definition 9.4.2. *A linear mapping T of the class* F *into a class* G *of functions g defined and locally* (F)*-analytic in $\mathfrak{S}_\Delta$ shall be said to be continuous if $g = T[f]$ and*

$$\lim_{n\to\infty} \sup_{\lambda\in\Delta} |f_n(\lambda) - f(\lambda)| = 0 \tag{9.4.17}$$

implies

$$\lim_{n\to\infty} T[f_n](x) = T[f](x) \tag{9.4.18}$$

locally uniformly in x.

Here "locally uniformly" refers to the possibility of $\mathfrak{S}_\Delta$ having infinitely many disjoint components. The convergence is to be uniform in each of them. We now have Gelfand's Uniqueness Theorem.

Theorem 9.4.2. *The mapping T of* F *into* H, *defined by* (9.4.9) *and* (9.4.10), *is a continuous isomorphism taking* 1 *and λ into e and x, respectively, and it is the only linear mapping with these properties.*

Proof. It is clear that T maps 1 and λ into e and x, respectively. The mapping is 1–1 for if the elements f_1 and f_2 of F are distinct, then $T[f_1](x)$ and $T[f_2](x)$ are distinct for $x = \lambda e$ for λ in an open non-void subset of Δ. Hence $T[f_1] \neq T[f_2]$. It is clear that the mapping is linear so sums go into sums, but it is not clear that products go into products. To prove this form

$$f_1(x) f_2(x) = \frac{1}{2\pi i} \int_{\Gamma_1} f_1(\mu)\, R(\mu, x)\, d\mu \cdot \frac{1}{2\pi i} \int_{\Gamma_2} f_2(\nu)\, R(\nu, x)\, d\nu.$$

Here Γ_1 and Γ_2 are at our disposal, subject to two conditions: they must surround $\sigma(x)$ once in the positive sense and they should be confined to $\Delta \cup \Gamma$. We can use this freedom of choice to take $\Gamma_2 = \Gamma$ and Γ_1 interior to Γ. We can then write the product of the two integrals as a double integral

$$\frac{1}{(2\pi i)^2} \int_{\Gamma_1} \int_\Gamma f_1(\mu) f_2(\nu)\, R(\mu, x)\, R(\nu, x)\, d\mu\, d\nu$$

$$= \frac{1}{(2\pi i)^2} \int_{\Gamma_1} \int_\Gamma f_1(\mu) f_2(\nu) \frac{R(\nu, x) - R(\mu, x)}{\mu - \nu}\, d\mu\, d\nu$$

by (9.2.3). This can be written as the difference of two iterated integrals where the order of integration, which is immaterial, is arranged to suit our convenience:

$$\frac{1}{2\pi i}\int_{\Gamma} f_2(v)\,R(v,x)\left\{\frac{1}{2\pi i}\int_{\Gamma_1}\frac{f_1(\mu)}{\mu - v}\,d\mu\right\}dv$$

$$-\frac{1}{2\pi i}\int_{\Gamma_1} f_1(\mu)\,R(\mu,x)\left\{\frac{1}{2\pi i}\int_{\Gamma}\frac{f_2(v)}{\mu - v}\,dv\right\}d\mu.$$

Here the inner integrals are numerical Cauchy integrals. The first one is 0 since v lies outside of Γ_1, the contour of integration, while the second one equals $-f_2(\mu)$. Hence we are left with

$$\frac{1}{2\pi i}\int_{\Gamma_1} f_1(\mu) f_2(\mu)\,R(\mu,x)\,d\mu,$$

which is T operating on the product of f_1 and f_2 in F. Thus

$$T[f_1 f_2] = T[f_1]\,T[f_2] \tag{9.4.19}$$

and products go into products.

As a byproduct of this argument we note that

$$f_1(x) f_2(x) = f_2(x) f_1(x), \tag{9.4.20}$$

so that f_1 and f_2 commute. On the other hand, no information is available whether or not $f_1(x_1)$ and $f_2(x_2)$ commute for distinct values of x_1 and x_2.

Thus T is 1–1, linear, and maps products into products, so it is an isomorphism. It should also be proved that T is continuous in the sense of Definition 9.4.2. We form

$$f(x) - f_n(x) = \frac{1}{2\pi i}\int_{\Gamma}[f(\mu) - f_n(\mu)]\,R(\mu,x)\,d\mu.$$

Suppose that for a given $\varepsilon > 0$ and for $n > k$ we have

$$|f(\lambda) - f_n(\lambda)| \leqslant \varepsilon, \qquad \lambda \in \Delta.$$

By continuity this estimate also holds on Γ and for μ on Γ we also have $\|R(\mu,x)\| \leqslant M(x)$ by (9.4.13). Hence

$$\|f(x) - f_n(x)\| \leqslant \varepsilon l(\Gamma)\,M(x),$$

where $l(\Gamma)$ is the length of Γ. Moreover, if (9.4.14) holds, we have

$$\|f(x + \alpha h) - f_n(x + \alpha h)\| \leqslant \varepsilon(1 - \gamma)^{-1}\,l(\Gamma)\,M(x). \tag{9.4.21}$$

This shows that (9.4.18) holds locally uniformly in $\mathfrak{S}_\Delta$.

Finally uniqueness of the mapping should be proved. It is to be shown that (9.4.9) is the only way of defining a mapping from F into a set of functions from $\mathfrak{B}$ to $\mathfrak{B}$ defined and locally (F)-analytic in $\mathfrak{S}_\Delta$ if the mapping is to be a continuous isomorphism taking 1, λ into e, x. Suppose that T_0 is such a mapping. Since T_0 is an isomorphism, a polynomial in λ goes into the corresponding polynomial in x,

$$P(\lambda) = \alpha_0 + \sum_{k=1}^{m} \alpha_k \lambda^k \to P(x) = \alpha_0 e + \sum_{k=1}^{m} \alpha_k x^k.$$

Since

$$(\lambda - \lambda_0)^n \to (x - \lambda_0 e)^n,$$

this implies, by the continuity property, that formula (9.4.5) is a consequence of (9.4.4), which is equivalent to saying that (9.4.10) implies (9.4.9). To start with, this holds when $\sigma(x)$ is interior to the circle $|\mu - \lambda_0| = \rho$, which is taken as the contour Γ. But if T_0 coincides with T for arbitrary circular disks, then, again by continuity, $T_0 \equiv T$ for arbitrary domains Δ. ∎

These are the main theorems in the Gelfand Operational Calculus. After the appearance of Gelfand's work numerous additions and modifications have been added to the theory. To round out the discussion we shall also prove Gelfand's Spectral Mapping Theorem.

Theorem 9.4.3. *If $f \in \mathsf{F}$ and $x \in \mathfrak{S}_\Delta$ and if $f(x) = T[f](x)$, then*

$$\sigma[f(x)] = f[\sigma(x)]. \tag{9.4.22}$$

Proof. Suppose that $\alpha \in \sigma(x)$. It is required to prove that $f(\alpha) \in \sigma[f(x)]$ and, moreover, that all spectral values of $f(x)$ are of this form. To this end, consider

$$q(\lambda, \alpha) = [f(\lambda) - f(\alpha)](\lambda - \alpha)^{-1},$$

where for $\lambda = \alpha$ the indeterminate form is to be replaced by its limit $f'(\alpha)$. Then $q(\lambda, \alpha)$ is a Cauchy holomorphic function of λ in Δ and belongs to F since f is continuous on Γ. From

$$f(\lambda) - f(\alpha) = (\lambda - \alpha)\, q(\lambda, \alpha)$$

we obtain by the mapping T

$$f(x) - f(\alpha)\, e = (x - \alpha e)\, q(x, \alpha).$$

Here $x - \alpha e$ is singular by assumption. This makes $f(x) - f(\alpha)e$ singular or $f(\alpha) \in \sigma[f(x)]$.

The proof of the converse requires a more cautious approach. Suppose that $\beta \in \sigma[f(x)]$ but not to $f[\sigma(x)]$. Then there exists a domain Δ_0 with oriented rectifiable boundary Γ_0 such that $\sigma(x) \subset \Delta_0 \subset \Delta$ and the equation $f(\lambda) = \beta$ has no roots in Δ_0. Set $h(\lambda) = \beta - f(\lambda)$ and note that both $h(\lambda)$ and $[h(\lambda)]^{-1}$ are Cauchy holomorphic in Δ_0 and continuous in Δ_0. They then belong to the class $\mathsf{F}(\Delta_0)$ which contains the restriction of $\mathsf{F} = \mathsf{F}(\Delta)$ to Δ_0. We write $\mathfrak{S}_{\Delta_0}$ for the set of elements of $\mathfrak{B}$ such that $\sigma(x) \subset \Delta_0$. This set is open and $\mathfrak{S}_{\Delta_0} \subset \mathfrak{S}_\Delta$. Formula (9.4.9) with Γ replaced by Γ_0 now defines the mapping of $\mathsf{F}(\Delta_0)$ into the corresponding class $\mathsf{H}(\Delta_0)$. Hence $h(\lambda) \to h(x)$ and $[h(\lambda)]^{-1} \to [h(x)]^{-1} = [\beta e - f(x)]^{-1}$. This is impossible for $\beta e - f(x)$ is singular by assumption. It follows that no domain Δ_0 can exist with the stated properties. Hence the equation

$$f(\alpha) = \beta$$

has at least one root in Δ. ∎

EXERCISE 9.4

1. Under what conditions is $\exp(x + y) = \exp(x)\exp(y)$?
2. If j is an idempotent commuting with x, show that
$$\exp(x + 2\pi ij) = \exp(x).$$
3. There are two ways of defining a square root of $e - x$ which reduces to e for $x = 0$. One is by the binomial series, the other by (9.4.9) for a suitable choice of f. Give these representations and discuss their validity.
4. Suppose that $\sigma(x)$ lies in the open right half-plane. Define a square root of x using (9.4.9). Show that the root is locally (F)-analytic (where?) and represent $\delta f(x; h)$ by a contour integral. What is the spectrum of the root?
5. Under what conditions on x can (9.4.9) be used to define a logarithm of x?
6. If q is a quasinilpotent, show that
$$\lambda^{\frac{1}{2}}\left\{e\lambda^{-1} + \sum_{n=1}^{\infty} \binom{\frac{1}{2}}{n} q^n \lambda^{-n-1}\right\}$$
is a square root of $R(\lambda, q)$. For what values of λ does the series converge?
7. The Cauchy integral of the product of two Cauchy holomorphic functions should be the product of the Cauchy integrals of the two factors. Verify this!

The remaining problems deal with the operational calculus in a B-algebra without unit element. $D(\lambda, x)$ denotes the dissolvent of x and $\alpha(x)$ its spectrum. Let Δ be a simply connected domain in the complex plane, containing the origin and bounded by an oriented rectifiable contour Γ. F is the class of all functions f Cauchy holomorphic in Δ and continuous in $\bar{\Delta}$ such that $f(0) = 0$; $\mathfrak{S}_\Delta$ is the set of all $x \in \mathfrak{B}$ such that $\alpha(x) \subset \Delta$. Let T be the mapping defined by

$$T[f](x) = f(x) = \frac{1}{2\pi i}\int_\Gamma f(\mu)\, D(\mu, x)\,\frac{d\mu}{\mu}.$$

8. Show that $\mathfrak{S}_\Delta$ is an open set in $\mathfrak{B}$.
9. Show that T defines a continuous isomorphism in which 0 and λ correspond to 0 and x, respectively. Here the first zero is that of C, the second that of $\mathfrak{B}$.

10. Show that T is the only mapping with these properties.
11. If $\mathfrak{B}$ is the convolution algebra $L_2(-\pi, \pi)$, how are the Fourier coefficients of $T[f]$ obtained from those of f?

COLLATERAL READING

DUNFORD, N. and J. T. SCHWARTZ, *Linear Operators,* Part II, *Spectral Theory,* Interscience, New York (1963), Chapter IX.

HILLE, E. and R. S. PHILLIPS, *Functional Analysis and Semi-groups,* American Mathematical Society Colloquium Series, Vol. 31, Providence, R.I. (1957), Chapters 4, 5, and 24.

NAĬMARK, M. A., *Normed Rings,* Noordhoff, Groningen (1959).

RICKHART, C. E., *General Theory of Banach Algebras,* Van Nostrand, Princeton, N.J. (1960).

10 LINEAR TRANSFORMATIONS

Linear operators have been with us almost from the beginning of this treatise. The time has come to tie various loose ends and supplement the earlier discussion, especially that of Chapter 2. Extensions of linear operators, the Banach–Steinhaus theorem, and the closed graph theorem are among the neglected topics which will now be considered. So far we have restricted ourselves to linear bounded operators. In this chapter unbounded operators will also be admitted provided they are closed. Various linear bounded functionals have occurred earlier. Here the missing existence proofs will be provided.

While inverse operators have been considered in many places, we shall now also deal with adjoint operators. They will play an important and more spectacular role in the next chapter. In a Hilbert space, adjoint operators are much more manageable and easier to visualize than, in general, Banach spaces. Finally, we must specialize the discussion of B-algebras in Chapter 9 to the case of operator algebras of the type $\mathfrak{E}(\mathfrak{X})$. Since we shall allow unbounded operators, it will be necessary to mention briefly how spectra and resolvents may change in the new setting.

There are five sections: Boundedness; Closure; Linear functionals; Inverses and adjoints; and Spectra and resolvents.

10.1 BOUNDEDNESS

We start by recalling the basic properties of linear transformations as defined in Chapter 2. The reader is advised to review this chapter.

We restrict ourselves to linear mappings from one B-space $\mathfrak{X}$ to another $\mathfrak{Y}$, both over the complex field. The *domain* $\mathfrak{D}(T)$ of T is a linear subspace of $\mathfrak{X}$, possibly $\mathfrak{X}$ itself, and the *range* $\mathfrak{R}(T)$ of T is a linear subspace of $\mathfrak{Y}$, possibly $\mathfrak{Y}$ itself.

Definition 10.1.1. *T is a linear mapping (equivalently a linear transformation or a linear operator) from $\mathfrak{X}$ to $\mathfrak{Y}$ if $\mathfrak{D}(T)$ and $\mathfrak{R}(T)$ are linear subspaces of $\mathfrak{X}$ and $\mathfrak{Y}$, respectively, and if for all $\mathbf{x}_1, \mathbf{x}_2 \in \mathfrak{D}(T)$ and all $\alpha, \beta \in C$*

$$T(\alpha\mathbf{x}_1 + \beta\mathbf{x}_2) = \alpha T(\mathbf{x}_1) + \beta T(\mathbf{x}_2). \qquad (10.1.1)$$

This is a generalization of Definition 2.2.2.

A linear transformation T maps the zero element of $\mathfrak{X}$ on the zero element of $\mathfrak{Y}$,

$$T(\mathbf{0}) = \mathbf{0}. \tag{10.1.2}$$

There are possibly other elements $\mathbf{x}$ of $\mathfrak{X}$ that are annihilated by T. The set of all such elements $\mathbf{x}$ is the *nullspace* $\mathfrak{N}[T]$ or *kernel* of T

$$\mathfrak{N}[T] = \{\mathbf{x};\, T(\mathbf{x}) = \mathbf{0}\}. \tag{10.1.3}$$

It is clear that $\mathfrak{N}[T] \subset \mathfrak{D}(T)$ is also a linear subspace of $\mathfrak{X}$.

Definition 10.1.2. *A linear transformation T from $\mathfrak{X}$ to $\mathfrak{Y}$ is bounded on $\mathfrak{D}(T)$, if there exists a constant M such that for all $\mathbf{x} \in \mathfrak{D}(T)$*

$$\|T(\mathbf{x})\| \leqslant M\|\mathbf{x}\|. \tag{10.1.4}$$

The least value of M for which the inequality is true is called the *norm* of T on $\mathfrak{D}(T)$, written $\|T\|$.

The norm when it exists may be defined directly by

$$\|T\| = \sup\,[\|T(\mathbf{x})\|;\, \|\mathbf{x}\| = 1,\, \mathbf{x} \in \mathfrak{D}(T)]. \tag{10.1.5}$$

Note that all elements of $\mathfrak{X}$ not $= \mathbf{0}$ are positive multiples of elements on the unit sphere $\|\mathbf{x}\| = 1$.

All linear bounded transformations with $\mathfrak{D}(T) = \mathfrak{X}$ and $\mathfrak{R}(T) \subset \mathfrak{Y}$ form the set $\mathfrak{E}(\mathfrak{X}, \mathfrak{Y})$ with $\mathfrak{E}(\mathfrak{X})$ written instead of $\mathfrak{E}(\mathfrak{X}, \mathfrak{X})$. We recall (see Section 2.4) that these are linear vector spaces which become B-spaces under the norm $\|T\|$.

Definition 10.1.3. *Let $\mathfrak{X} \times \mathfrak{Y}$ be the Cartesian product of $\mathfrak{X}$ and $\mathfrak{Y}$, i.e. the set of all ordered pairs*

$$\{(\mathbf{x}, \mathbf{y});\, \mathbf{x} \in \mathfrak{X},\, \mathbf{y} \in \mathfrak{Y}\}. \tag{10.1.6}$$

The graph $\mathfrak{G}$ of a mapping from $\mathfrak{X}$ to $\mathfrak{Y}$ is the set of all ordered pairs

$$\{[\mathbf{x}, T(\mathbf{x})];\, \mathbf{x} \in \mathfrak{D}(T)\} \subset \mathfrak{X} \times \mathfrak{Y}. \tag{10.1.7}$$

We define a metric on $\mathfrak{X} \times \mathfrak{Y}$ by setting

$$\|(\mathbf{x}, \mathbf{y})\| = \|\mathbf{x}\| + \|\mathbf{y}\|. \tag{10.1.8}$$

Note that the norm symbol $\|\,.\,\|$ refers successively to the spaces $\mathfrak{X} \times \mathfrak{Y}$, $\mathfrak{X}$, and $\mathfrak{Y}$.

Theorem 10.1.1. *The product space $\mathfrak{X} \times \mathfrak{Y}$ is complete in the metric defined by* (10.1.8).

The proof is left to the reader. Actually the space is a B-space if we define

$$\begin{aligned} (\mathbf{x}_1, \mathbf{y}_1) + (\mathbf{x}_2, \mathbf{y}_2) &= (\mathbf{x}_1 + \mathbf{x}_2, \mathbf{y}_1 + \mathbf{y}_2), \\ \alpha(\mathbf{x}, \mathbf{y}) &= (\alpha\mathbf{x}, \alpha\mathbf{y}). \end{aligned} \tag{10.1.9}$$

Definition 10.1.4. *A transformation* $\mathbf{y} = T(\mathbf{x})$ *from* $\mathfrak{X}$ *to* $\mathfrak{Y}$ *is said to be closed if its graph is a closed subset of* $\mathfrak{X} \times \mathfrak{Y}$.

Thus T is closed iff

$$\{\mathbf{x}_n\} \subset \mathfrak{D}, \mathbf{x}_n \to \mathbf{x}_0 \quad \text{together with} \quad \mathbf{y}_n = T(\mathbf{x}_n) \to \mathbf{y}_0 \tag{10.1.10}$$

implies that

$$\mathbf{x}_0 \in \mathfrak{D} \quad \text{and} \quad T(\mathbf{x}_0) = \mathbf{y}_0. \tag{10.1.11}$$

Definition 10.1.5. *If* $T_1(\mathbf{x})$ *and* $T_2(\mathbf{x})$ *are mappings from* $\mathfrak{X}$ *to* $\mathfrak{Y}$ *with domains* $\mathfrak{D}_1$ *and* $\mathfrak{D}_2$, *respectively, if* $\mathfrak{D}_1 \subset \mathfrak{D}_2$ *and if* $T_2(\mathbf{x}) = T_1(\mathbf{x})$ *on* $\mathfrak{D}_1$, *then* T_2 *is called an extension of* T_1 *written* $T_1 \subset T_2$.

Theorem 10.1.2. *Let* $\mathbf{y} = T(\mathbf{x})$ *be a linear transformation from* $\mathfrak{D} \subset \mathfrak{X}$ *to* $\mathfrak{Y}$ *with graph* $\mathfrak{G}$. *Then* T *has a closed extension, iff no vector pair* $(\mathbf{0}, \mathbf{y})$ *with* $\mathbf{y} \neq \mathbf{0}$ *belongs to* $\overline{\mathfrak{G}}$, *the closure of* $\mathfrak{G}$. *If this is the case, then* $\overline{\mathfrak{G}}$ *is the graph of the smallest closed linear extension of* T.

Proof. $\overline{\mathfrak{G}}$ is linear in $\mathfrak{X} \times \mathfrak{Y}$ since $\mathfrak{G}$ has this property. Now $\overline{\mathfrak{G}}$ is the graph of some transformation, say T^0. This transformation is linear and closed since the subspace $\overline{\mathfrak{G}}$ is linear and closed. Further, suppose that $(\mathbf{x}, \mathbf{y}_1)$ and $(\mathbf{x}, \mathbf{y}_2)$ belong to $\overline{\mathfrak{G}}$. Then so does $(\mathbf{x}, \mathbf{y}_1) - (\mathbf{x}, \mathbf{y}_2) = (\mathbf{0}, \mathbf{y}_1 - \mathbf{y}_2)$ and by hypothesis this requires that $\mathbf{y}_1 = \mathbf{y}_2$. It follows that T^0 is single-valued. It is clear that $\overline{\mathfrak{G}}$ is the graph of the smallest closed linear extension of T. Conversely, if T is 1–1 on $\mathfrak{D}$, no proper extension of T can take $\mathbf{0}$ into any element but $\mathbf{0}$. ▮

We return to linear bounded transformations.

Theorem 10.1.3. *If* $T(\mathbf{x})$ *is a linear bounded transformation from* $\mathfrak{D} \subset \mathfrak{X}$ *to* $\mathfrak{Y}$ *with norm* $\|T\|$, *then* T *has a unique linear bounded extension* T^0 *defined on* $\overline{\mathfrak{D}}$ *and* $\|T^0\| = \|T\|$.

Proof. If $\mathfrak{D}$ is a closed subset of $\mathfrak{X}$, then the theorem is trivially true for $T^0 = T$. If $\mathfrak{D}$ is not closed, the extension is obtained as follows. Let $\mathbf{x}_0 \in \overline{\mathfrak{D}} \ominus \mathfrak{D}$, let $\{\mathbf{x}_n\} \subset \mathfrak{D}$ and $\lim \mathbf{x}_n = \mathbf{x}_0$. Then, if $\mathbf{y}_n = T(\mathbf{x}_n)$,

$$\|\mathbf{y}_m - \mathbf{y}_n\| = \|T(\mathbf{x}_m) - T(\mathbf{x}_n)\| = \|T(\mathbf{x}_m - \mathbf{x}_n)\| \leqslant \|T\| \, \|\mathbf{x}_m - \mathbf{x}_n\|.$$

Here $\{\mathbf{x}_n\}$ is a Cauchy sequence in $\mathfrak{X}$, whence it follows that $\{\mathbf{y}_n\}$ is a Cauchy sequence in $\mathfrak{Y}$. Let its limit be $\mathbf{y}_0$.

We have to show that $\mathbf{y}_0$ is independent of the choice of a Cauchy sequence in $\mathfrak{X}$. Suppose that $\{\mathbf{z}_n\} \subset \mathfrak{D}$, $\lim \mathbf{z}_n = \mathbf{x}_0$ and $\mathbf{w}_n = T(\mathbf{z}_n)$. Since $\mathfrak{D}$ is linear, $\mathbf{x}_n - \mathbf{z}_n \in \mathfrak{D}$ and is a Cauchy sequence in $\mathfrak{X}$ converging to zero. Hence

$$\|\mathbf{w}_n - \mathbf{y}_n\| = \|T(\mathbf{z}_n) - T(\mathbf{x}_n)\| = \|T(\mathbf{z}_n - \mathbf{x}_n)\| \leqslant \|T\| \, \|\mathbf{z}_n - \mathbf{x}_n\|,$$

so that $\{\mathbf{w}_n - \mathbf{y}_n\}$ is a Cauchy sequence in $\mathfrak{Y}$, converging to zero. Here $\lim \mathbf{y}_n = \mathbf{y}_0$, so we must have $\lim \mathbf{w}_n = \mathbf{y}_0$, i.e. $\mathbf{y}_0$ is independent of our choice of Cauchy sequence in $\mathfrak{X}$ converging to $\mathbf{x}_0$.

We can then define

$$T^0(\mathbf{x}_0) = \lim T(\mathbf{x}_n), \quad \mathbf{x}_0 \in \overline{\mathfrak{D}} \ominus \mathfrak{D}, \qquad T^0(\mathbf{x}) = T(\mathbf{x}), \quad \mathbf{x} \in \mathfrak{D}.$$

Further, note that

$$\|\mathbf{y}_0\| = \|T^0(\mathbf{x}_0)\| = \lim \|T(\mathbf{x}_n)\| \leqslant \|T\| \lim \|\mathbf{x}_n\| = \|T\| \, \|\mathbf{x}_0\|,$$

so that $\|T^0\| \leqslant \|T\|$. Here we must have equality, since $T^0(\mathbf{x}) = T(\mathbf{x})$ on $\mathfrak{D}$. The uniqueness of the extension follows directly from the facts that $\mathfrak{D}(T)$ is dense and $\mathfrak{Y}$ is a Hausdorff space. ∎

Corollary. *A linear transformation from $\mathfrak{X}$ to $\mathfrak{Y}$ whose domain $\mathfrak{D}$ is dense in $\mathfrak{X}$ admits of a unique extension to all of $\mathfrak{X}$ with norm unchanged.*

Somewhat similar considerations lead to the Banach–Steinhaus theorem.

Theorem 10.1.4. *Let $\{T_n\}$ be a sequence of linear bounded transformations in $\mathfrak{E}(\mathfrak{X}, \mathfrak{Y})$ such that* (1) $\|T_n\| \leqslant M$ *for all n, and* (2) $\lim T_n(\mathbf{x})$ *exists for every $\mathbf{x}$ in a set $\mathfrak{E}$, dense in a sphere $\mathfrak{S}$. Then $\lim T_n(\mathbf{x})$ exists for all $\mathbf{x}$, and the limit defines an element T of $\mathfrak{E}(\mathfrak{X}, \mathfrak{Y})$ with $\|T\| \leqslant \liminf_{n\to\infty} \|T_n\|$.*

Proof. We start by proving convergence in $\mathfrak{S}$. Let $\mathbf{x}_0 \in \mathfrak{S} \ominus \mathfrak{E}$, let $\{\mathbf{x}_n\} \subset \mathfrak{E}$, and let $\lim \mathbf{x}_n = \mathbf{x}_0$. Then for any choice of positive integers j, k, n we have

$$\|T_j(\mathbf{x}_0) - T_k(\mathbf{x}_0)\| \leqslant \|T_j(\mathbf{x}_0) - T_j(\mathbf{x}_n)\| + \|T_j(\mathbf{x}_n) - T_k(\mathbf{x}_n)\| + \|T_k(\mathbf{x}_n) - T_k(\mathbf{x}_0)\|,$$

so that

$$\limsup_{j,k\to\infty} \|T_j(\mathbf{x}_0) - T_k(\mathbf{x}_0)\| \leqslant 2M \, \|\mathbf{x}_0 - \mathbf{x}_n\|.$$

This holds for all n and $\lim \mathbf{x}_n = \mathbf{x}_0$. Hence the sequence $\{T_j(\mathbf{x}_0)\}$ converges and we have convergence everywhere in $\mathfrak{S}$. Suppose that $\mathfrak{S}$ is the sphere $\{\mathbf{x}; \|\mathbf{x} - \mathbf{s}_0\| < r\}$. Suppose that $\mathbf{x}$ is a point of $\mathfrak{X}$ exterior to $\mathfrak{S}$. We can then find a point $\mathbf{y}$ in $\mathfrak{S}$ and a positive number $\rho < 1$ such that

$$\mathbf{y} - \mathbf{s}_0 = \rho(\mathbf{x} - \mathbf{s}_0) \quad \text{or} \quad \mathbf{x} = \left(1 - \frac{1}{\rho}\right)\mathbf{s}_0 + \frac{1}{\rho}\mathbf{y}.$$

Since T_n is linear,

$$T_n(\mathbf{x}) = \left(1 - \frac{1}{\rho}\right) T_n(\mathbf{s}_0) + \frac{1}{\rho} T_n(\mathbf{y}).$$

Here $\mathbf{s}_0$ and $\mathbf{y}$ are in $\mathfrak{S}$ and it follows that the sequence $\{T_n(\mathbf{x})\}$ converges everywhere in $\mathfrak{X}$ and

$$T(\mathbf{x}) \equiv \lim T_n(\mathbf{x})$$

is defined for all $\mathbf{x}$. Here T is obviously linear. It is also bounded for $\|T_n\| \leqslant M$ implies $\|T\| \leqslant M$. Thus $T \in \mathfrak{E}(\mathfrak{X}, \mathfrak{Y})$. Now the convergence of $\{T_n\}$ in $\mathfrak{E}(\mathfrak{X}, \mathfrak{Y})$ implies convergence of any subsequence. Choosing a suitable subsequence, the estimate $\|T\| \leqslant M$ may be improved to $\|T\| \leqslant \liminf_{n\to\infty} \|T_n\|$. ∎

We end this discussion by adducing three related examples which illustrate various points in the preceding discussion. In all three $\mathfrak{X} = \mathfrak{Y} = C[0, 1]$ with the sup norm.

Example 1. Define

$$T[f](t) = \int_0^t f(s)\,ds, \qquad 0 \leqslant t \leqslant 1. \tag{10.1.12}$$

Here $T \in \mathfrak{E}(\mathfrak{X})$ and $\|T\| = 1$. The range $\mathfrak{R}(T)$ is a proper subset of $\mathfrak{Y} = \mathfrak{X}$, namely that subset of $\mathfrak{X}$ whose elements satisfy the conditions (1) $g(0) = 0$, and (2) $g'(t)$ exists and belongs to $\mathfrak{X}$. The first condition implies that $\mathfrak{R}(T)$ is non-dense in $\mathfrak{X}$. For if $f \in \mathfrak{X}$ and $f(0) = a$, $|a| > 0$, then the distance of f from $\mathfrak{R}(T)$ is $\geqslant |a|$.

Example 2. Take

$$U[f](t) = f'(t). \tag{10.1.13}$$

Here $\mathfrak{D}(T)$ is a proper subset of $\mathfrak{X}$ since continuous functions are "normally" not differentiable. On the other hand, every continuous function can be approximated uniformly by differentiable functions—in fact, even by polynomials. This expresses that $\mathfrak{D}(U)$ is dense in $\mathfrak{X}$. Further, U is unbounded in $\mathfrak{D}(T)$. To show this we can consider

$$f(t) = \sin(n\pi t) \quad \text{with} \quad f'(t) = n\pi \cos(n\pi t).$$

Here $\|f\| = 1$ identically in n while $\|f'\| = \pi n$. Thus U is unbounded.

Example 3. With U as in the preceding example, let us note that the unbounded operator U is closed in the sense of Definition 10.1.4. We have to show that $\{f_n\} \subset \mathfrak{D}$, $f_n \to f_0 \in \mathfrak{X}$ together with $f_n' \to g_0 \subset \mathfrak{X}$ implies that $f_0 \in \mathfrak{D}$ and $f_0' = g_0$. The first assumption implies in particular pointwise convergence, so that $\lim f_n(0) = f_0(0)$. Further,

$$f_n(t) = f_n(0) + \int_0^t f_n'(s)\,ds \to f_0(0) + \int_0^t g_0(s)\,ds$$

by the uniform convergence of the sequences $\{f_n(t)\}$ and $\{f_n'(t)\}$. On the other hand, the third member equals $f_0(t)$ by assumption. This shows that $f_0(t)$ is differentiable, i.e. $f_0 \in \mathfrak{D}(U)$, $f_0'(t) \equiv g_0(t) \in \mathfrak{X}$ and U is closed.

These examples illustrate the general observation that differential operators tend to be unbounded but closed. This observation will be made more precise later. The importance of closed operators will be brought out in the next section.

EXERCISE 10.1

1. Take $\mathfrak{X} = l_1$ and the shift operator $T\{x_n\} = \{x_{n+1}\}$. What is $\|T\|$? What is $\mathfrak{R}(T)$? Is $\mathfrak{R}(T)$ dense in l_1?

2. With $\mathfrak{X} = l_1$ and $U\{x_n\} = \{nx_n\}$, determine $\mathfrak{D}(U)$ and $\mathfrak{R}(U)$. Is U bounded? Is $\mathfrak{D}(U)$ dense in l_1?
3. With the same conventions, is U closed?
4. Take $\mathfrak{X} = C[0, 1]$ and $T[f](t) = \int_0^t (t - s) f(s)\, ds$ for $0 \leqslant t \leqslant 1$. Find norm and range of T. Is $\mathfrak{R}(T)$ dense in $\mathfrak{X}$?
5. With $\mathfrak{X} = C[0, 1]$, take $U[f](t) = f''(t)$. Is U bounded? Find $\mathfrak{D}(U)$ and $\mathfrak{R}(U)$. Are they dense in $\mathfrak{X}$?
6. Show that U is closed.
7. Take $\mathfrak{X} = l_\infty$ and consider the vector-valued function
$$\mathbf{F}(t) = \{e^{-\lambda_n t}\}.$$
Here $\{\lambda_n\}$ is an increasing sequence of positive numbers. Show that $\mathbf{F}(t) \in l_\infty$ for all $t \geqslant 0$. What is $\|\mathbf{F}(t)\|$? What additional restrictions are needed on $\{\lambda_n\}$ in order that $\mathbf{F}(t)$ belong to l_∞ for all real t? How should the problem be modified to make sense in l_1?
8. Let the functions $f_n(t)$ be continuous and differentiable on $(0, \infty)$. Is the operator $U\{f_n(t)\} = \{f_n'(t)\}$ bounded on l_1, if $\{f_n(t)\} \in l_1$ for all $t > 0$?

10.2 CLOSURE

The notion of a closed linear transformation was introduced in Definition 10.1.4. Such transformations form a natural generalization of bounded transformations and, while unbounded, have enough properties in common with the bounded ones to be manageable. Most linear transformations encountered in analysis are closed. As a consequence of the definition we get

Lemma 10.2.1. *A linear bounded transformation from $\mathfrak{X}$ to $\mathfrak{Y}$ with $\mathfrak{D}(T) = \mathfrak{X}$ is closed.*

The proof is left to the reader.

We denote by $\mathfrak{C}(\mathfrak{X}, \mathfrak{Y})$ the set of all linear closed transformations from $\mathfrak{X}$ to $\mathfrak{Y}$ and write $\mathfrak{C}(\mathfrak{X})$ for $\mathfrak{C}(\mathfrak{X}, \mathfrak{X})$. Here $\mathfrak{C}(\mathfrak{X}, \mathfrak{Y})$ is in general not a linear system. On the other hand, a closed transformation brings with it an infinite set of "satellites" together with which it forms an equivalence class modulo "boundedness".

Lemma 10.2.2. *If $U \in \mathfrak{C}(\mathfrak{X}, \mathfrak{Y})$ and if T is a linear bounded transformation with $\mathfrak{D}(T) \supset \mathfrak{D}(U)$, then $U + T$ is defined on $\mathfrak{D}(U)$ and is a closed linear transformation and as such is a member of $\mathfrak{C}(\mathfrak{X}, \mathfrak{Y})$.*

Again the proof is left to the reader.

It follows that $\mathfrak{C}(\mathfrak{X}, \mathfrak{Y})$ splits up into equivalence classes where two transformations U_1 and U_2 belong to the same class if $\mathfrak{D}(U_1) = \mathfrak{D}(U_2) = \mathfrak{D}$ and $U_1 - U_2$ is bounded on $\mathfrak{D}$.

We proceed to proving some theorems which are natural extensions of theorems valid for bounded operators. The first is an extension of a theorem due to S. Banach.

Theorem 10.2.1. *Let* $\mathbf{y} = U(\mathbf{x})$ *be a linear closed transformation from* $\mathfrak{D} \subset \mathfrak{X}$ *onto* $\mathfrak{R} \subset \mathfrak{Y}$ *and let* $\mathfrak{R}$ *be a set of the second category in* $\mathfrak{Y}$. *Then*

(1) $\mathfrak{R} = \mathfrak{Y}$;

(2) *there exists an* $m > 0$ *such that for every* $\mathbf{y} \in \mathfrak{Y}$ *there is an* $\mathbf{x} \in \mathfrak{D}$ *with* $\mathbf{y} = U(\mathbf{x})$ *and* $\|\mathbf{x}\| \geqslant m\,\|\mathbf{y}\|$.

Proof. This involves a number of steps. The first one aims at showing that $\mathfrak{R}$ is dense in $\mathfrak{Y}$. Using the assumption that $\mathfrak{R}$ is of the second category we shall show that $\mathfrak{R}$ is dense in a sphere. Set for $n = 1, 2, 3, \ldots$

$$D_n = \{\mathbf{x};\, \mathbf{x} \in \mathfrak{D},\, \|\mathbf{x}\| < n\}. \tag{10.2.1}$$

Then

$$\mathfrak{D} = \bigcup_n (D_n) \quad \text{and} \quad \mathfrak{R} = \bigcup_n U(D_n). \tag{10.2.2}$$

By assumption, $\mathfrak{R}$ is of the second category in $\mathfrak{Y}$, hence one of the image sets $U(D_j)$ is of the second category in $\mathfrak{Y}$ and

$$\operatorname{Int}\left(\overline{U(D_j)}\right) \neq \emptyset.$$

A sphere $\mathfrak{S} = \{\mathbf{y};\, \|\mathbf{y} - \mathbf{y}_0\| < r_0\}$ is therefore contained in $\overline{U(D_j)}$. It follows from $\mathbf{y}_0 \in \overline{U(D_j)}$ that there exists $\mathbf{x}_1 \in D_j$ with $\mathbf{y}_1 = U(\mathbf{x}_1)$ and $\|\mathbf{y}_1 - \mathbf{y}_0\| < \tfrac{1}{2}r_0$. Let

$$\mathfrak{S}_0 = \{\mathbf{y};\, \|\mathbf{y}\| < \tfrac{1}{2}r_0\},$$

then, by the continuity of the functions $(\mathbf{y}, \mathbf{z}) \to \mathbf{y} + \mathbf{z}$ and $(\lambda, \mathbf{y}) \to \lambda\mathbf{y}$,

$$\begin{aligned} \mathfrak{S}_0 \subset \mathfrak{S} - \mathbf{y}_1 \subset \overline{U(D_j)} - \mathbf{y}_1 &\subset \overline{U(D_j)} - \overline{U(D_j)} \\ &\subset \overline{U(D_j) - U(D_j)} = \overline{U(D_{2j})}. \end{aligned} \tag{10.2.3}$$

Thus

$$\mathfrak{Y} = \bigcup_{m=1}^{\infty} m\,\mathfrak{S}_0 \subset \bigcup_{m=1}^{\infty} m\,\overline{U(D_{2j})} \subset \bigcup_{m=1}^{\infty} \overline{U(m\,D_{2j})} \subset \overline{\mathfrak{R}} \tag{10.2.4}$$

and $\mathfrak{R}$ is dense in $\mathfrak{Y}$.

The second step involves showing that $\mathfrak{R}$ contains a sphere. Here is where the property of U being closed comes into play. It will be shown that the sphere

$$\mathfrak{S}_1 = \left\{\mathbf{y};\, \|\mathbf{y}\| < r_1 = \frac{r_0}{4j}\right\}$$

is contained in $\mathfrak{R}$. Formula (10.2.3) shows that the closure of the set $U(2j\,D_1)$ contains the sphere $\mathfrak{S}_0$. By an obvious contraction process we see that implies

$$\overline{U(2^{-k}D_1)} \supset 2^{-k}\,\mathfrak{S}_1; \qquad k = 0, 1, 2, \ldots. \tag{10.2.5}$$

These inclusions relations are the basic tool for the following argument.

Let $\mathbf{y}$ be an arbitrary point in $\mathfrak{S}_1$. Using (10.2.5) with $k = 0$, we can find a point $\mathbf{x}_1 \in D_1$ such that

$$\|\mathbf{y} - U(\mathbf{x}_1)\| < 2^{-1} r_1,$$

where $\|\mathbf{x}_1\| < 1$. Since $\mathbf{y} - U(\mathbf{x}_1) \in \frac{1}{2}\mathfrak{S}_1$, we can go back to (10.2.5), now with $k = 1$, to find a point $\mathbf{x}_2 \in \frac{1}{2}\mathfrak{S}_1$ such that

$$\|\mathbf{y} - U(\mathbf{x}_1) - U(\mathbf{x}_2)\| < \tfrac{1}{4} r_1$$

and here $\|\mathbf{x}_2\| < \frac{1}{2}$. Repeating the process, we find a sequence of points $\{\mathbf{x}_n\}$ all in D_1 such that for each n

$$\left\|\mathbf{y} - \sum_{k=1}^{n} U(\mathbf{x}_k)\right\| = \left\|\mathbf{y} - U\left[\sum_{k=1}^{n} \mathbf{x}_k\right]\right\| < 2^{-n} r_1$$

and now $\|\mathbf{x}_n\| < 2^{-n-1}$. If $\mathbf{s}_n = \sum_{k=1}^{n} \mathbf{x}_k$ and $U(\mathbf{s}_n) = \mathbf{t}_n$, we see that $\lim \mathbf{s}_n = \mathbf{x}_0$ exists, $\|\mathbf{x}_0\| < 2$, and $\lim \mathbf{t}_n = \mathbf{y}$. Since U is closed, this implies that $\mathbf{x}_0 \in \mathfrak{D}(U)$ and $U(\mathbf{x}_0) = \mathbf{y}$. Now $\mathbf{y}$ was an arbitrary point of the sphere $\mathfrak{S}_1$, whence it follows that $\mathfrak{R} \supset \mathfrak{S}_1$. Again, by the homogeneity of U, this implies that $\mathfrak{R} = \mathfrak{Y}$.

The last step is to show property (2) asserted in the theorem. This also comes out of the preceding discussion. We can take $m = 4/r_1$. If $\mathbf{y}$ is any element of $\mathfrak{Y}$, we can find a contraction which will pull $\mathbf{y}$ inside the sphere $\mathfrak{S}_1$ for which we have bounded inverse images. Set

$$\mathbf{y}_1 = r_1(2\|\mathbf{y}\|)^{-1}\mathbf{y}.$$

Then $\mathbf{y}_1 \in \mathfrak{S}_1$ and there is an $\mathbf{x}_0$ with $\|\mathbf{x}_0\| < 2$ such that $U(\mathbf{x}_0) = \mathbf{y}_1$. We have now

$$\|U(\mathbf{x}_0)\| = \|\mathbf{y}_1\| = \tfrac{1}{4} r_1 2 > \tfrac{1}{4} r_1 \|\mathbf{x}_0\|$$

or

$$\|\mathbf{x}_0\| > \frac{4}{r_1}\|U(\mathbf{x}_0)\|$$

as asserted. ∎

Corollary 1. *If U^{-1} exists, then it is bounded.*

Proof. By Theorem 2.2.2 properties (1) and (2) are necessary and sufficient for the boundedness of U^{-1} when it exists, i.e. when the mapping is 1–1. ∎

Corollary 2. *If $\mathbf{y} = T(\mathbf{x})$ is a linear bounded transformation belonging to $\mathfrak{E}(\mathfrak{X}, \mathfrak{Y})$ and if T is 1–1, then T^{-1} is bounded.*

Proof. By Lemma 10.2.1 T is closed, so Theorem 10.2.1 applies. ∎

We come now to the *closed graph theorem* of S. Banach.

Theorem 10.2.2. *If $\mathbf{y} = U(\mathbf{x})$ is a closed linear transformation with domain $\mathfrak{D}$ of the second category in $\mathfrak{X}$, then $\mathfrak{D} = \mathfrak{X}$ and U is bounded.*

Proof. We define a mapping V of the graph $\mathfrak{G}$ of U onto $\mathfrak{D}$ by the convention

$$V[\mathbf{x}, U(\mathbf{x})] = \mathbf{x}. \tag{10.2.6}$$

Since $\mathfrak{X} \times \mathfrak{Y}$ is a B-space, this is a mapping from one B-space $\mathfrak{X} \times \mathfrak{Y}$ to another $\mathfrak{X}$. It is clearly linear. Moreover, V is closed for if

$$(\mathbf{x}_n, \mathbf{y}_n) \to (\mathbf{x}_0 \mathbf{y}_0), \quad \text{then} \quad \mathbf{y}_0 = U(\mathbf{x}_0) \quad \text{and} \quad \mathbf{x}_n \to \mathbf{x}_0$$

by the closure of U. The mapping is 1–1, for the nullspace of V contains only the point $[\mathbf{0}, U(\mathbf{0})] = (\mathbf{0}, \mathbf{0})$. The range of V is $\mathfrak{D}$, by assumption of the second category in $\mathfrak{X}$. We can then apply Theorem 10.2.1 with Corollary 1 and conclude that the range of V is all of $\mathfrak{X}$. Further, V^{-1} exists and is bounded so that

$$\|[\mathbf{x}, U(\mathbf{x})]\| = \|V^{-1}(\mathbf{x})\| \leqslant M \|\mathbf{x}\|$$

or

$$\|\mathbf{x}\| + \|U(\mathbf{x})\| \leqslant M \|\mathbf{x}\|$$

by the definition of the norm in $\mathfrak{X} \times \mathfrak{Y}$. This shows that U is bounded and is an element of $\mathfrak{E}(\mathfrak{X}, \mathfrak{Y})$. ∎

As an application of linear closed transformations we give an extension of Theorem 7.4.3. There it was shown that a linear bounded transformation can be taken under the sign of integration. Here we want to examine under what circumstances this may be done when the operator is just closed. In the bounded case the integrand (or the integrator) is a vector function which assumes values in the domain of the operator. If f is strongly continuous, so is $T[f]$; if g is of bounded variation, so is $T[g]$. In the closed unbounded case the situation is different. It is necessary to postulate that the values of f (or of g) are in $\mathfrak{D}(U)$ and that $U[f]$ (or $U[g]$) has the properties that will ensure the existence of the Stieltjes integral. Since we are more interested in simple results, easy to remember and to apply, than in the utmost generality, we impose stronger restrictions on f and g than necessary.

Theorem 10.2.3. *Given $U \in \mathfrak{C}(\mathfrak{X}, \mathfrak{Y})$. Let $\mathbf{f}$ be a function defined on $[a, b]$ with $\mathbf{f}(s) \in \mathfrak{D}(U)$ for $a \leqslant s \leqslant b$. If both $\mathbf{f}(s)$ and $U[\mathbf{f}](s)$ are strongly measurable in $[a, b]$, further, if $g(s)$ is a numerically valued function of bounded variation in $[a, b]$, then*

$$U\left[\int_a^b \mathbf{f}(s)\, dg(s)\right]$$

exists and

$$U\left[\int_a^b \mathbf{f}(s)\, dg(s)\right] = \int_a^b U[\mathbf{f}](s)\, dg(s). \tag{10.2.7}$$

Instead, if both $\mathbf{f}(s)$ and $U[\mathbf{f}](s)$ are of strongly bounded variation on $[a, b]$, *and if $g(s)$ is a numerically valued continuous function on* $[a, b]$, *then*

$$\int_a^b g(s)\, d\mathbf{f}(s) \in \mathfrak{D}(U)$$

and

$$U\left[\int_a^b g(s)\, d\mathbf{f}(s)\right] = \int_a^b g(s)\, d[U[\mathbf{f}](s)]. \tag{10.2.8}$$

Proof. In the first case consider a partition of the interval $[a, b]$, say

$$a = s_1 < s_2 < \cdots < s_n = b,$$

a set of intermediary points $\{t_j\}$, $s_{j-1} \leqslant t_j \leqslant s_j$, and the corresponding Riemann–Stieltjes sum

$$\sum_2^n \mathbf{f}(t_j)\, [g(s_j) - g(s_{j-1})]. \tag{10.2.9}$$

By assumption each term in this finite sum belongs to $\mathfrak{D}(U)$, hence also the sum, and by the linearity of U

$$U\left\{\sum_2^n \mathbf{f}(t_j)\, [g(s_j) - g(s_{j-1})]\right\} = \sum_2^n U[\mathbf{f}(t_j)]\, [g(s_j) - g(s_{j-1})].$$

As the partitions become finer and finer, the sequence of sums of type (10.2.9) and the corresponding sequence with typical element

$$\sum_2^n U[\mathbf{f}(t_j)]\, [g(s_j) - g(s_{j-1})]$$

converge to limits

$$\int_a^b \mathbf{f}(s)\, dg(s) \quad \text{and} \quad \int_a^b U[\mathbf{f}](s)\, dg(s), \tag{10.2.10}$$

respectively. Here we have used that $\mathbf{f}(s)$ and $U[\mathbf{f}](s)$ are strongly continuous and $g \in BV[a, b]$. Because U is closed, it follows that the first integral in (10.2.10) belongs to $\mathfrak{D}(U)$ and (10.2.7) holds.

The same type of argument applies in the second case. From the equality of the Riemann–Stieltjes sums

$$U\left\{\sum_2^n g(t_j)\, [\mathbf{f}(s_j) - \mathbf{f}(s_{j-1})]\right\} = \sum_2^n g(t_j)\, \{U[\mathbf{f}(s_j)] - U[\mathbf{f}(s_{j-1})]\}$$

and the closure of U we get (10.2.8). ∎

We shall give an example of a large class of linear closed operators which play an important role in analysis, particularly in the theory of linear differential equations. The assumptions imposed on the operator may strike the reader as artificial, but they are satisfied at least in the special case mentioned.

Theorem 10.2.4. *Given* $\mathfrak{X} = C[0, b]$ *with the sup-norm. Let* U *be a linear operator from* $\mathfrak{X}$ *to itself. Suppose that a kernel function* $K(s, t)$ *is defined and continuous in* $[0, b] \times [0, b]$ *such that any function* $s \to z(s)$ *satisfying*

$$z(s) = y(s) + \int_0^s K(s, t) f(t)\, dt \tag{10.2.11}$$

with $0 \leqslant s \leqslant b$, $U[y](s) \equiv 0$ *and* $f \in \mathfrak{X}$ *is a solution of the functional equation*

$$U[z](s) = f(s). \tag{10.2.12}$$

Suppose further that any solution to (10.2.12) *is of the form* (10.2.11). *Then* U *is a closed linear transformation, if* $\mathfrak{N}[U]$, *the nullspace of* U, *is closed.*

Remark. The same result holds if the upper limit in the integral is fixed and equal to b.

Proof. It is to be shown that $\{z_n\} \subset \mathfrak{D}(U)$, $z_n \to z_0$ together with $f_n = U[z_n] \to f_0$, both in the sup-norm of $\mathfrak{X}$, implies that $z_0 \in \mathfrak{D}(U)$ and $U[z_0] = f_0$. By assumption

$$z_n(s) = y_n(s) + \int_0^s K(s, t) f_n(t)\, dt \tag{10.2.13}$$

for some choice of y_n in $\mathfrak{N}[U]$. The assumptions on f_n imply that

$$\int_0^s K(s, t) f_n(t)\, dt \to \int_0^s K(s, t) f_0(t)\, dt, \tag{10.2.14}$$

to start with in the sense of point-wise convergence. Both sides are evidently elements of $\mathfrak{X} = C[0, b]$ and from

$$\left| \int_0^s K(s, t)\,[f_n(t) - f_0(t)]\, dt \right| \leqslant bB \| f_n - f_0 \|,$$

where $B = \max |K(s, t)|$, $0 \leqslant t \leqslant s \leqslant b$, we see that (10.2.14) holds in the sense of the metric.

In (10.2.13) there are three terms. The first and the third converge to limits in $\mathfrak{X}$ as $n \to \infty$. It follows that the sequence $\{y_n\}$ also converges to a limit $y_0 \in \mathfrak{X}$. Now each y_n belongs to the closed nullspace $\mathfrak{N}[U]$, so it follows that $y_0 \in \mathfrak{N}[U]$ and

$$z_0(s) = y_0(s) + \int_0^s K(s, t) f_0(t)\, dt. \tag{10.2.15}$$

By assumption a function of this form satisfies

$$U[z_0](s) = f_0(s),$$

so that U is closed. ∎

The assumptions are satisfied, in particular, for linear nth order differential operators. If

$$L[z](s) = z^{(n)}(s) + a_1(s)\,z^{(n-1)}(s) + \cdots + a_n(s)\,z(s), \tag{10.2.16}$$

where the $a_j(s) \in C[0, b]$, then the solution of

$$L[z](s) = f(s) \tag{10.2.17}$$

is given by a formula of type (10.2.11), where $K(s, t) = G(s, t)$ is the Green's function of the problem and $y(s)$ is a solution of the homogeneous equation

$$L[y](s) = 0. \tag{10.2.18}$$

In this setting formula (10.2.11) is obtained, for instance, by the method of variation of the parameters.

The assumptions of Theorem 10.2.4 are as usual unnecessarily restrictive. We could replace continuous functions by square-integrable ones. It requires a little more work and the result may be desirable for various applications, but the present formulation permits a brief proof, exhibiting the essential ideas.

EXERCISE 10.2

1. Take $\mathfrak{X} = C[0, b]$, let $g(t)$ be a strictly positive fixed element of $\mathfrak{X}$ and let U be the linear operator on $\mathfrak{X}$ to itself defined by $U[f](t) = (d/dt)[f(t)\,g(t)]$. It is not assumed that g is differentiable. If $h(t)$ belongs to the range of U, express f in terms of h and show that the range is all of $\mathfrak{X}$. Show that U is closed.
2. Prove that the domain of U is dense in $\mathfrak{X}$.
3. In formulas (10.2.16) to (10.2.18) take $n = 2$ and let y_1 and y_2 be two linearly independent solutions of (10.2.18), i.e.

$$W(t; y_1, y_2) = \begin{vmatrix} y_1'(t) & y_2'(t) \\ y_1(t) & y_2(t) \end{vmatrix} \neq 0, \quad \forall\, t.$$

Form the function

$$z(t) = C_1 y_1(t) + C_2 y_2(t) + \int_0^t \frac{y_1(t)\,y_2(x) - y_2(t)\,y_1(x)}{W(x; y_1, y_2)} f(x)\,dx,$$

where C_1 and C_2 are arbitrary constants. Show that it is a solution of $L[z](t) = f(t)$ for any f in $\mathfrak{X}$. Verify that the assumptions of Theorem 10.2.4 are satisfied so that L is a closed operator on $\mathfrak{X} = C[0, b]$ to itself.

4. Consider the space l_1 with $\|\{x_n\}\| = \sum_{n=1}^{\infty} |x_n|$ and the linear unbounded operator U which takes $\{x_n\}$ into $\{nx_n\}$. Is U^k, $k \geqslant 1$ closed?
5. Consider the space $L_1(0, 1)$ and the linear operator U which takes $f(t) \in L_1(0, 1)$ into $t^{-\frac{1}{2}} f(t)$. Is U bounded? Is it closed? Is the domain of U dense?
6. What happens if in the preceding problem the space $L_1(0, 1)$ is replaced by $L_1(0, \infty)$, other data being unchanged?

10.3 LINEAR FUNCTIONALS

We shall base the discussion on the properties of Minkowski gauge functions first encountered in Section 1.6. Let $\mathfrak{X}$ be a B-space to start with over the reals.

Definition 10.3.1. *A function* $p(\mathbf{x})$ *from* $\mathfrak{X}$ *to* R *is called a gauge function for the space if* (1) $p(\mathbf{0}) = 0$, (2) $p(\alpha\mathbf{x}) = \alpha p(\mathbf{x})$, $\forall \alpha$, $\alpha > 0$, $\forall \mathbf{x}$, (3) $p(\mathbf{x}) \geqslant 0$, *and* (4)

$$p(\mathbf{x} + \mathbf{y}) \leqslant p(\mathbf{x}) + p(\mathbf{y}). \tag{10.3.1}$$

Thus $p(\mathbf{x})$ is positive-homogeneous and subadditive. In a normed linear space there are such functions, e.g. $\beta\|\mathbf{x}\|$, where β is any positive number. On the other hand, if a linear Hausdorff space contains a convex set K containing the origin but not the whole space, then K has a gauge function and a normed topology may be introduced which is equivalent to the Hausdorff topology [Kolmogorov (1934)].

Lemma 10.3.1. *A gauge function* $p(\mathbf{x})$ *is continuous for all* $\mathbf{x}$.

Proof. For $\alpha > 0$ and any $\mathbf{y} \in \mathfrak{X}$, $\mathbf{y} \neq 0$,

$$p(\mathbf{x}) \leqslant p(\mathbf{x} - \alpha\mathbf{y}) + p(\alpha\mathbf{y}) \leqslant p(\mathbf{x}) + p(-\alpha\mathbf{y}) + p(\alpha\mathbf{y}) = p(\mathbf{x}) + \alpha[p(\mathbf{y}) + p(-\mathbf{y})].$$

Hence, by (1) and (2), $p(\mathbf{x})$ is continuous at the origin; furthermore,

$$p(\mathbf{x}) \leqslant \liminf_{\alpha\to 0} p(\mathbf{x} - \alpha\mathbf{y}) \leqslant \limsup_{\alpha\to 0} p(\mathbf{x} - \alpha\mathbf{y}) \leqslant p(\mathbf{x}). \tag{10.3.2}$$

This holds for all $\mathbf{x}$ and $\mathbf{y}$ in $\mathfrak{X}$. Thus $p(\mathbf{x})$ is continuous everywhere. ∎

Theorem 10.3.1. *The point set*

$$\mathsf{K} = \{\mathbf{x}; \mathbf{x} \in \mathfrak{X}, p(\mathbf{x}) \leqslant 1\} \tag{10.3.3}$$

is closed and convex.

Proof. If $\{\mathbf{x}_n\} \subset \mathsf{K}$, $\mathbf{x}_n \to \mathbf{x}_0$, then $p(\mathbf{x}_n) \leqslant 1$ implies $p(\mathbf{x}_0) \leqslant 1$ by the continuity of p. Hence K is closed. If $\mathbf{x}_1 \in \mathsf{K}$, $\mathbf{x}_2 \in \mathsf{K}$, and if $0 < \alpha < 1$, then by conditions (2) and (4)

$$p\big(\alpha\mathbf{x}_1 + (1 - \alpha)\mathbf{x}_2\big) \leqslant \alpha p(\mathbf{x}_1) + (1 - \alpha)\, p(\mathbf{x}_2) \leqslant \alpha + (1 - \alpha) = 1,$$

so that $\alpha\mathbf{x}_1 + (1 - \alpha)\mathbf{x}_2 \in \mathsf{K}$. Thus K is convex. ∎

Convex sets made their first appearance in Definition 1.2.3.

A gauge function is a functional, i.e. a mapping from an abstract space into R. It is bounded functional. For boundedness we have given two slightly different definitions, one in Section 5.3, the other in Section 7.1. Here we want boundedness understood in the sense of (7.1.9), which is as made for the present situation.

In Section 2.3 we were faced with the problem of constructing linear bounded functionals on a B-space. We shall now tackle this problem in earnest. For the case of real functionals, the argument is based on the extension theorem of Hahn–Banach, for complex functionals on that of Bohnenblust–Sobczyk.

Theorem 10.3.2. *Let $\mathfrak{X}$ be a B-space over the reals, $\mathfrak{X}_0$ a linear subspace non-dense in $\mathfrak{X}$. Let p be a gauge function defined on $\mathfrak{X}$ with properties as stated in Definition 10.3.1. Let f_0 be a linear bounded functional defined in $\mathfrak{X}_0$ where it satisfies*

$$f_0(\mathbf{x}) \leqslant p(\mathbf{x}), \qquad \forall\, \mathbf{x} \in \mathfrak{X}_0. \tag{10.3.4}$$

Then there exists a linear bounded functional f such that (i) $\mathfrak{D}[f] = \mathfrak{X}$, (ii) $f(\mathbf{x}) \leqslant p(\mathbf{x})$, $\forall\, \mathbf{x} \in \mathfrak{X}$, *and* (iii) $f(\mathbf{x}) = f_0(\mathbf{x})$, $\forall\, \mathbf{x} \in \mathfrak{X}_0$.

Proof. The proof involves a series of successive extensions, partially ordered by inclusion, and an appeal to Zorn's Lemma for the existence of a maximal element, which is the desired functional.

First let $\mathbf{x}_0 \in \mathfrak{X} \ominus \mathfrak{X}_0$ and form the space

$$\mathfrak{X}_1 = \{\mathbf{x};\, \mathbf{x} = \mathbf{y} + \lambda \mathbf{x}_0,\, \lambda \in R,\, \mathbf{y} \in \mathfrak{X}_0\}. \tag{10.3.4a}$$

This is a linear subspace of $\mathfrak{X}$ and it contains $\mathfrak{X}_0$ as a proper subset. Note also that the representation of $\mathbf{x}$ as the sum of an element of $\mathfrak{X}_0$ and a multiple of $\mathbf{x}_0$ is unique. We shall need some inequalities. If $\mathbf{x}_1, \mathbf{x}_2 \in \mathfrak{X}_0$, then

$$f_0(\mathbf{x}_2) - f_0(\mathbf{x}_1) = f_0(\mathbf{x}_2 - \mathbf{x}_1) \leqslant p(\mathbf{x}_2 - \mathbf{x}_1)$$

and

$$p(\mathbf{x}_2 - \mathbf{x}_1) \leqslant p(\mathbf{x}_2 + \mathbf{x}_0) + p(-\mathbf{x}_1 - \mathbf{x}_0).$$

Hence

$$-p(-\mathbf{x}_1 - \mathbf{x}_0) - f_0(\mathbf{x}_1) \leqslant p(\mathbf{x}_2 + \mathbf{x}_0) - f_0(\mathbf{x}_2). \tag{10.3.5}$$

Let us put

$$r_1 = \sup_{\mathbf{x}_1 \in \mathfrak{X}_0} [-p(-\mathbf{x}_1 - \mathbf{x}_0) - f_0(\mathbf{x}_1)], \tag{10.3.6}$$

$$r_2 = \inf_{\mathbf{x}_2 \in \mathfrak{X}_0} [p(\mathbf{x}_2 + \mathbf{x}_0) - f_0(\mathbf{x}_2)]. \tag{10.3.7}$$

These are finite numbers and $r_1 \leqslant r_2$. Now choose r so that

$$r_1 \leqslant r \leqslant r_2,$$

then r separates the two sides in (10.3.5) so that we have

$$-p(-\mathbf{x}_1 - \mathbf{x}_0) - f_0(\mathbf{x}_1) \leqslant r \leqslant p(\mathbf{x}_2 + \mathbf{x}_0) - f_0(\mathbf{x}_2) \tag{10.3.8}$$

for all $\mathbf{x}_1, \mathbf{x}_2$ in $\mathfrak{X}_0$.

We can now define an extension from $\mathfrak{X}_0$ to $\mathfrak{X}_1$ by setting

$$f_1(\mathbf{x}) = f_0(\mathbf{y}) + \lambda r, \qquad \mathbf{x} = \mathbf{y} + \lambda \mathbf{x}_0. \tag{10.3.9}$$

We have clearly $f_1(\mathbf{x}) = f_0(\mathbf{x})$ if $\mathbf{x} = \mathbf{y}$, $\lambda = 0$. Further, for a $\mathbf{y}_0$ in $\mathfrak{X}_0$ which will be disposed over later we have

$$f_1(\mathbf{x}) = f_0(\mathbf{y}) + \lambda r \leqslant f_0(\mathbf{y}) + \lambda[p(\mathbf{y}_0 + \mathbf{x}_0) - f_0(\mathbf{y}_0)]$$
$$= f_0(\mathbf{y} - \lambda\mathbf{y}_0) + p(\lambda\mathbf{y}_0 + \lambda\mathbf{x}_0).$$

Here we set $\lambda\mathbf{y}_0 = \mathbf{y}$ and obtain

$$f(\mathbf{x}) \leqslant p(\mathbf{y} + \lambda\mathbf{x}_0) = p(\mathbf{x}),$$

so that the extension is also bounded above by the gauge function.

Let E be the class of all linear extensions of f_0 which are dominated above by the gauge function on their domains. Partially order E by setting $g \leqslant h$ if h is a linear extension of g, i.e. $\mathfrak{D}(g) \subset \mathfrak{D}(h)$ and $g(\mathbf{x}) = h(\mathbf{x})$, $\forall\, \mathbf{x} \in \mathfrak{D}(g)$. Let C be a chain in E and define a linear functional F by specifying the domain of F to be the union of the domains of the members of C, by assigning $F(\mathbf{x})$ to be $g(\mathbf{x})$ for $\mathbf{x} \in \mathfrak{D}(F) \cap \mathfrak{D}(g)$, where $g \in$ C. Since C is a chain, F is well defined. Moreover, it is clear that F is a linear extension of f_0 and $F(\mathbf{x}) \leqslant p(\mathbf{x})$, $\forall\, \mathbf{x} \in \mathfrak{D}(F)$, so that $F \in$ E and it is an upper bound of C By Zorn's Lemma, E has a maximal element f. Here f must be an extension to $\mathfrak{X}$ since, if its domain were a proper subspace $\mathfrak{M}$ and if $\mathbf{t}_0 \in \mathfrak{X} \ominus \mathfrak{M}$, then the method of constructing f_1 guarantees an extension to $\mathfrak{M}_1 = \{\mathbf{x}; \mathbf{x} = \mathbf{y} + \lambda\mathbf{t}_0, \lambda \in R, \mathbf{y} \in \mathfrak{M}\}$. Clearly f satisfies conditions (i) to (iii). ∎

So far we have dealt with a B-space over R and real functionals. The extension to the complex case was given by H. F. Bohnenblust and A. Sobczyk in 1938. This requires an extension of the notion of a gauge function.

Definition 10.3.2. *A gauge function $c(\mathbf{x})$ is said to be circular if it is subadditive and positive-homogeneous and, in addition, $c(\mathbf{0}) = 0$ and for every complex number α we have*

$$c(\alpha\mathbf{x}) = |\alpha| c(\mathbf{x}). \tag{10.3.10}$$

Theorem 10.3.3. *Given a B-space $\mathfrak{X}$ over the complex field and a circular gauge function $c(\mathbf{x})$. Given a complex-linear functional $F_0(\mathbf{x})$ defined on a complex-linear subspace $\mathfrak{X}_0$ of $\mathfrak{X}$ and such that $|F_0(\mathbf{x})| \leqslant c(\mathbf{x})$, $\forall\, \mathbf{x} \in \mathfrak{X}_0$. Then there exists a complex-linear functional $F(\mathbf{x})$ defined on all of $\mathfrak{X}$ such that $F(\mathbf{x}) = f_0(\mathbf{x})$ in $\mathfrak{X}_0$ and $|f(\mathbf{x})| \leqslant c(\mathbf{x})$, $\forall\, \mathbf{x}$.*

Proof. Set $\mathfrak{R}[F_0(\mathbf{x})] = f_0(\mathbf{x})$. This is a real-linear functional defined on $\mathfrak{X}_0$ and $f_0(\mathbf{x}) \leqslant |F_0(\mathbf{x})| \leqslant c(\mathbf{x})$ on $\mathfrak{X}_0$. Further,

$$F_0(\mathbf{x}) = f_0(\mathbf{x}) - i f_0(i\mathbf{x}). \tag{10.3.11}$$

We can now use the preceding theorem to find a real-linear extension $f(\mathbf{x})$ of $f_0(\mathbf{x})$ to all of $\mathfrak{X}$ such that (i) $f(\mathbf{x}) = f_0(\mathbf{x})$ for $\mathbf{x}$ in $\mathfrak{X}_0$ and (ii) $|f(\mathbf{x})| \leqslant c(\mathbf{x})$ for all $\mathbf{x}$. It is now clear that $F(\mathbf{x}) = f(\mathbf{x}) - i f(i\mathbf{x})$ is a complex-linear extension of $F_0(\mathbf{x})$ to all of $\mathfrak{X}$. Since $\mathfrak{X}_0$ is a complex-linear subspace of $\mathfrak{X}$ we have

$i\,\mathfrak{X}_0 = \mathfrak{X}_0$ with obvious meaning. This implies that $f(i\mathbf{x}) = f_0(i\mathbf{x})$ on $\mathfrak{X}_0$ so that $F(\mathbf{x}) = F_0(\mathbf{x})$ on $\mathfrak{X}_0$. Further, if $\theta = \arg[F(\mathbf{x})]$ we have

$$|F(\mathbf{x})| = e^{-i\theta}F(\mathbf{x}) = F(e^{-i\theta}\mathbf{x}) = f(e^{-i\theta}\mathbf{x}) \leqslant c(e^{-i\theta}\mathbf{x}) = c(\mathbf{x}),$$

so that the extension has all the desired properties. ∎

Theorem 10.3.4. *If $\mathbf{x}_0 \neq \mathbf{0}$ is a point in a B-space $\mathfrak{X}$, there exists a linear bounded functional F defined on $\mathfrak{X}$ such that $F(\mathbf{x}_0) = \|\mathbf{x}_0\|$ and $\|F\| = 1$.*

Proof. We take

$$\mathfrak{X}_0 = \{\mathbf{x};\, \mathbf{x} = \alpha\mathbf{x}_0, \alpha \in C\}, \qquad F_0(\mathbf{x}) = \alpha\|\mathbf{x}_0\|, \qquad \mathbf{x} \in \mathfrak{X}_0.$$

We can take $c(\mathbf{x}) = \|\mathbf{x}\|$ and note that $F_0(\mathbf{x}_0) = \|\mathbf{x}_0\|$. These data satisfy the requirements of Theorem 10.3.3 and the conclusion follows. ∎

Theorem 10.3.5. *Given a B-space $\mathfrak{X}$, a linear subspace $\mathfrak{X}_0$ non-dense in $\mathfrak{X}$, and a point $\mathbf{x}_1$ at a distance $d > 0$ from $\mathfrak{X}_0$. Then there exists a linear bounded functional F on $\mathfrak{X}$ such that* (1) $f(\mathbf{x}) = 0$, $\forall\, \mathbf{x}$ *in* $\mathfrak{X}_0$ *and* (2) $\|f\| = 1/d$.

Proof. We set

$$\mathfrak{X}_1 = \{\mathbf{x};\, \mathbf{x} = \mathbf{x}_0 + \alpha\mathbf{x}_1, \mathbf{x}_0 \in \mathfrak{X}_0, \alpha \in C\}, \qquad F_1(\mathbf{x}) = \alpha.$$

The representation $\mathbf{x} = \mathbf{x}_0 + \alpha\mathbf{x}_1$ is unique and $F_1(\mathbf{x})$ is a complex-linear functional defined on $\mathfrak{X}_1$ and vanishing on $\mathfrak{X}_0$. We take $c(\mathbf{x}) = \beta\|\mathbf{x}\|$, where

$$\beta = \sup\left\{\frac{|\alpha|}{\|\mathbf{x}_0 + \alpha\mathbf{x}_1\|};\, \mathbf{x}_0 \in \mathfrak{X}_0, \alpha \neq 0\right\}. \tag{10.3.12}$$

A simple calculation shows that $\beta = 1/d$. The extension theorem completes the proof, giving us a complex-linear functional F of norm $1/d$ which vanishes on $\mathfrak{X}_0$. ∎

Theorem 10.3.6. *If $\mathbf{x}_1 \neq \mathbf{x}_2$ are two points of a B-space $\mathfrak{X}$, then there exists a linear bounded functional F defined on all of $\mathfrak{X}$ such that $F(\mathbf{x}_1) \neq F(\mathbf{x}_2)$.*

Proof. Since $\mathbf{x}_1 - \mathbf{x}_2 \neq 0$, there exists a functional F defined on $\mathfrak{X}$ such that $F(\mathbf{x}_1 - \mathbf{x}_2) = \|\mathbf{x}_1 - \mathbf{x}_2\| \neq 0$. This F satisfies $F(\mathbf{x}_1) \neq F(\mathbf{x}_2)$. ∎

Actually we can prescribe the values of a functional at two or more points. We shall give the argument for two points.

Theorem 10.3.7. *Let $\mathbf{x}_1$, $\mathbf{x}_2$ be linearly independent elements of $\mathfrak{X}$ and c_1, c_2 two distinct complex numbers. Then there exists a linear bounded functional F on $\mathfrak{X}$ such that*

$$F(\mathbf{x}_1) = c_1, \qquad F(\mathbf{x}_2) = c_2.$$

Proof. Set

$$\mathfrak{X}_0 = \{\mathbf{x}; \mathbf{x} = \alpha_1\mathbf{x}_1 + \alpha_2\mathbf{x}_2, \alpha_1, \alpha_2 \in C\},$$
$$F_0(\mathbf{x}) = \alpha_1 c_1 + \alpha_2 c_2, \mathbf{x} \in \mathfrak{X}_0.$$

This is a linear functional defined on $\mathfrak{X}_0$ and

$$F_0(\mathbf{x}_1) = c_1, \qquad F_0(\mathbf{x}_2) = c_2.$$

We have to prove that F_0 is a bounded functional. This leads to the following question. Suppose

$$\|\alpha_1\mathbf{x}_1 + \alpha_2\mathbf{x}_2\| = 1. \tag{10.3.13}$$

Does this imply that

$$|\alpha_1 c_1 + \alpha_2 c_2| \equiv G(\alpha_1, \alpha_2) \tag{10.3.14}$$

is bounded? If this is the case and the supremum is M, then we can take $c(\mathbf{x}) = M\|\mathbf{x}\|$ and apply the extension theorem.

Suppose that $G(\alpha_1, \alpha_2)$ is unbounded, i.e. $F_0(\mathbf{x})$ is unbounded for $\|\mathbf{x}\| = 1$, $\mathbf{x} \in \mathfrak{X}_0$. This implies the existence of two infinite sequences $\{\alpha_{1n}\}$ and $\{\alpha_{2n}\}$ with the following properties. If

$$\mathbf{x}_n = \alpha_{1n}\mathbf{x}_1 + \alpha_{2n}\mathbf{x}_2,$$

then

$$\|\mathbf{x}_n\| = 1, \qquad |F_0(\mathbf{x}_n)| > 2n.$$

Further,

$$\limsup |\alpha_{1n}| = \limsup |\alpha_{2n}| = +\infty.$$

For if only one of the sequences were unbounded $\limsup \|\mathbf{x}_n\| = +\infty$ instead of 1. We have now

$$\|\mathbf{x}_n\| = |\alpha_{1n}|\,\|\mathbf{x}_1 + \beta_n\mathbf{x}_2\|, \qquad \beta_n = \frac{\alpha_{2n}}{\alpha_{1n}}.$$

Here the first factor in the right member of the first relation is unbounded, so there is at least a subsequence for which the second factor goes to zero or

$$\mathbf{x}_1 = -(\lim \beta_{n_k})\mathbf{x}_2,$$

and this contradicts the assumed linear independence of $\mathbf{x}_1$ and $\mathbf{x}_2$. It follows that

$$\sup\{|F_0(\mathbf{x})|; \mathbf{x} \in \mathfrak{X}_0, \|\mathbf{x}\| = 1\}$$

exists as a finite number M. We can then take $c(\mathbf{x}) = M\|\mathbf{x}\|$ and the extension theorem does the rest. ■

We state without proof

Theorem 10.3.8. *Let $\mathbf{x}_1, \mathbf{x}_2, \ldots, \mathbf{x}_n$ be n linearly independent elements of a B-space $\mathfrak{X}$ and let $c_1, c_2, \ldots, c_n$ be n given constant numbers. Then there exists a linear bounded functional F defined on $\mathfrak{X}$ such that*

$$F(\mathbf{x}_j) = c_j, \qquad j = 1, 2, \ldots, n.$$

More general existence theorems for linear bounded functionals have been studied by several authors (H. Hahn, E. Helly, and F. Riesz among others).

Theorems 10.3.4 to 10.3.6 establish facts stated in Theorem 2.3.2.

Definition 10.3.3. *A set S in a B-space $\mathfrak{X}$ is said to be fundamental if the closed linear extension of S is all of $\mathfrak{X}$.*

In view of Theorem 10.3.5 we have

Theorem 10.3.9. *A necessary and sufficient condition that a set S in a B-space $\mathfrak{X}$ be fundamental is that every linear bounded functional which vanishes on S vanishes identically, i.e. is the zero functional.*

EXERCISE 10.3

1. Prove Theorem 10.3.7.
2. Prove Theorem 10.3.8.
3. Suppose that $F_1, F_2, \ldots, F_n$ are n linear bounded functionals on a B-space $\mathfrak{X}$ of dimension $>n$. Show that the system
$$F_1(\mathbf{x}) = 0, \qquad F_2(\mathbf{x}) = 0, \ldots, \qquad F_n(\mathbf{x}) = 0$$
always has non-trivial solutions.

Three questions concerning fundamental sets.

4. Show that $\{t^n; n = 0, 1, 2, \ldots\}$ is fundamental in $C[a, b]$. If $t = 0$ belongs to $[a, b]$, can the unit element 1 be omitted without affecting the result?
5. The set of functions holomorphic and bounded in the unit disk of the complex plane form a B-space under the sup-norm. Show that $\{z^n; n = 0, 1, 2, \ldots\}$ is a fundamental set. Can any power of z be omitted?
6. Give a fundamental set for $L_2(-\pi, \pi)$.

The next two problems illustrate the question raised in (10.3.14).

7. Take $\mathfrak{X} = L_2(-\pi, \pi)$ and $\mathbf{x}_1 = e^{it}$, $\mathbf{x}_2 = e^{2it}$ and find $\sup |\alpha_1 c_1 + \alpha_2 c_2|$ for the stated choice.
8. Same question for $\mathfrak{X} = C[0, 1]$ with $\mathbf{x}_1 = 1$, $\mathbf{x}_2 = t$.
9. Back to $L_2(-\pi, \pi)$ and the orthonormal system
$$\{(2\pi)^{-\frac{1}{2}} e^{nit}; n = 0, \pm 1, \pm 2, \ldots\}.$$
Is it possible to find a linear bounded functional F which assumes preassigned values c_n for $\mathbf{f} = \mathbf{f}_n$ where $\mathbf{f}_n$ runs through the orthonormal basis?

10.4 INVERSES AND ADJOINTS

Consider now a linear transformation from $\mathfrak{X}$ into $\mathfrak{Y}$, two B-spaces over the complex field. Let $\mathfrak{D} = \mathfrak{D}(T)$ be the domain of T, $\mathfrak{R} = \mathfrak{R}(T)$ the range. We recall that T has an inverse, T^{-1}, iff the mapping of $\mathfrak{D}$ onto $\mathfrak{R}$ is 1–1. An equivalent condition is that the nullspace of T reduces to the zero element

$$\mathfrak{N}[T] = \{\mathbf{0}\}. \tag{10.4.1}$$

We also recall that if T has an inverse, then T^{-1} is linear.

Theorem 10.4.1. *If a closed linear transformation T has an inverse, then T^{-1} is also closed.*

Proof. T is closed iff the set $\{(\mathbf{x}, T(\mathbf{x})); \mathbf{x} \in \mathfrak{D}\}$ is closed in the product space $\mathfrak{X} \times \mathfrak{Y}$. The same set may be written as $\{(T^{-1}(\mathbf{y}), \mathbf{y}); \mathbf{y} \in \mathfrak{R}\}$. Hence, if the graph of T is closed, so is the graph of T^{-1}. ∎

A closed linear transformation which has an inverse may have a bounded inverse. Theorem 2.2.2 asserts that $T^{-1} \in \mathfrak{E}(\mathfrak{Y}, \mathfrak{X})$ iff $\mathfrak{R} = \mathfrak{Y}$ and there is a finite positive m such that

$$\|T(\mathbf{x})\| \geqslant m\|\mathbf{x}\|, \qquad \forall\, \mathbf{x} \in \mathfrak{D}. \tag{10.4.2}$$

The largest permissible value of m is then the reciprocal of the norm of T^{-1}.

Theorem 10.4.2. *If T and its inverse are bounded linear transformations from $\mathfrak{D}$ onto $\mathfrak{R}$ and from $\mathfrak{R}$ onto $\mathfrak{D}$, respectively, and if $\mathfrak{D}$ is closed, so is $\mathfrak{R}$.*

Proof. Suppose that $\{\mathbf{x}_n\} \subset \mathfrak{D}$, $T(\mathbf{x}_n) = \mathbf{y}_n \in \mathfrak{R}$ and that $\mathbf{y}_n \to \mathbf{y}_0$. Then $\|\mathbf{x}_n - \mathbf{x}_m\| \leqslant \|T^{-1}\|\,\|\mathbf{y}_n - \mathbf{y}_m\|$, so that $\lim \mathbf{x}_n \equiv \mathbf{x}_0$ exists. This requires that $T(\mathbf{x}_0)$ exists since T is bounded and that $T(\mathbf{x}_0) = \lim_{n\to\infty} T(\mathbf{x}_n) = \mathbf{y}_0$, or, $\mathbf{y}_0 \in \mathfrak{R}$ and $\mathfrak{R}$ is closed. ∎

We turn now to the adjoint transformation. Here we are dealing with four B-spaces, $\mathfrak{X}$, $\mathfrak{Y}$, $\mathfrak{X}^*$, $\mathfrak{Y}^*$, all over the complex field. As the notation indicates, $\mathfrak{X}^*$ is the adjoint space of $\mathfrak{X}$ and $\mathfrak{Y}^*$ is the adjoint space of $\mathfrak{Y}$.

Definition 10.4.1. *Let T be a linear transformation (not necessarily bounded) from $\mathfrak{X}$ to $\mathfrak{Y}$ with domain $\mathfrak{D}(T)$ dense in $\mathfrak{X}$. The adjoint transformation T^* has domain $\mathfrak{D}(T^*)$ made up of all elements y^* of $\mathfrak{Y}^*$ for which there is an $x^* \in \mathfrak{X}^*$ such that*

$$y^*[T(\mathbf{x})] = x^*(\mathbf{x}), \qquad \mathbf{x} \in \mathfrak{D}(T). \tag{10.4.3}$$

For any such y^* we set

$$T^*(y^*) = x^*. \tag{10.4.4}$$

Some comments are in order. The density of $\mathfrak{D}(T)$ implies that T^*, defined by (10.4.4), is a well-defined mapping, $\mathfrak{D}(T^*)$, the domain of T^* is not empty, since (10.4.3) is at least satisfied by the trivial linear bounded functional which maps every $\mathbf{y} \in \mathfrak{Y}$ to zero. From (10.4.3) it follows that $\mathfrak{D}(T^*)$ is a subspace of $\mathfrak{Y}^*$. From (10.4.4) the linearity of T^* follows. T^* is actually a closed linear transformation. This will be shown in Theorem 10.4.3.

We shall illustrate the definition by an example which is not completely trivial and where the symbols can be unraveled.

Example 1. We take

$$\mathfrak{X} = \mathfrak{X}^* = L_2(0, 1), \qquad \mathfrak{Y} = C[0, 1], \qquad \mathfrak{Y}^* = BV[0, 1]$$

and define T by

$$T[\mathbf{x}](t) = \int_0^t \mathbf{x}(s)\, ds.$$

Here, if $g \in BV[0, 1]$ defines the functional y^*, we have

$$y^*[T(\mathbf{x})] = \int_0^1 \left[\int_0^t \mathbf{x}(s)\, ds\right] dg(t)$$

$$= \int_0^1 [g(1) - g(s)]\, \mathbf{x}(s)\, ds.$$

This is a functional x^* defined on $\mathfrak{X}$ as the inner product of $\mathbf{x}(s)$ and the conjugate of $g(1) - g(s)$. Thus T^* takes the functional y^* defined by g on $\mathfrak{Y}$ into the functional x^* defined on $\mathfrak{X}$ by the conjugate of $g(1) - g(s)$.

The operator T^* has a number of properties, interesting in their own right but also for the light they throw on T and T^{-1}.

Theorem 10.4.3. *Let T be a linear transformation with domain $\mathfrak{D}$ dense in $\mathfrak{X}$ and range $\mathfrak{R}$ in $\mathfrak{Y}$. Then T^* is a closed linear transformation on $\mathfrak{D}(T^*) \subset \mathfrak{Y}^*$ to $\mathfrak{X}^*$. If, in addition, T is bounded, then $T^* \in \mathfrak{E}(\mathfrak{Y}^*, \mathfrak{X}^*)$ and $\|T^*\| = \|T\|$.*

Proof. We refer to Theorem 10.1.3 and its Corollary. It shows that $S(\mathbf{x}) = y^*[T(\mathbf{x})]$ has a unique bounded extension to all of $\mathfrak{X}$ provided T is bounded on $\mathfrak{D}$ which is dense in $\mathfrak{X}$. Thus T^* is well defined and it is linear. Let us prove that T^* is closed. Consider a sequence $\{y_n^*\} \subset \mathfrak{D}(T^*)$ such that $y_n^* \to y_0^*$. Then $T^*(y_n^*) = x_n^* \to x_0^*$. Further, $y_n^*[T(\mathbf{x})] = x_n^*(\mathbf{x})$, $\forall\, \mathbf{x} \in \mathfrak{D}$, implies that $y_0^*[T(\mathbf{x})] = x_0^*(\mathbf{x})$, $\forall\, \mathbf{x} \in \mathfrak{D}$, so that $y_0^* \in \mathfrak{D}(T^*)$ and $T^*(y_0^*) = x_0^*$. This is closure. If, in addition, T is bounded, then $S(\mathbf{x}) = y^*[T(\mathbf{x})]$ defines a linear bounded functional on $\mathfrak{D}$ and we have

$$|S(\mathbf{x})| \leqslant \|y^*\|\, \|T\|\, \|\mathbf{x}\|.$$

Thus $\mathfrak{D}(T^*) = \mathfrak{Y}^*$ and $\|T^*\| \leqslant \|T\|$. To get the reverse inequality we observe that, given $\varepsilon > 0$, we can find an $\mathbf{x}_\varepsilon$ with $\|\mathbf{x}_\varepsilon\| = 1$ such that $\|T(\mathbf{x}_\varepsilon)\| \geqslant \|T\| - \varepsilon$. If $\mathbf{y}_\varepsilon = T(\mathbf{x}_\varepsilon)$, we choose a functional $y_\varepsilon^* \in \mathfrak{Y}^*$ such that $y_\varepsilon^*(\mathbf{y}_\varepsilon) = \|\mathbf{y}_\varepsilon\|$, $\|y_\varepsilon^*\| = 1$; Theorem 10.3.4 guarantees the existence of such a functional. We have now

$$\|T^*\| \geqslant \|T^*(y_\varepsilon^*)\| \geqslant |[T^*(y_\varepsilon^*)]\,(\mathbf{x}_\varepsilon)| = \|\mathbf{y}_\varepsilon\| \geqslant \|T\| - \varepsilon.$$

Hence $\|T^*\| = \|T\|$. ∎

Theorem 10.4.4. *If T is a linear transformation with domain $\mathfrak{D}$ dense in $\mathfrak{X}$, then $(T^*)^{-1}$ exists iff $\mathfrak{R}$ is dense in $\mathfrak{Y}$. More generally, the closure of $\mathfrak{R}$ is made up of all points $\mathbf{y}$ such that $T^*(y^*) = 0$ implies $y^*(\mathbf{y}) = 0$.*

Proof. If $T^*(y_0^*) = 0$, then $[T^*(y_0)](\mathbf{x}) = y_0^*[T(\mathbf{x})] = 0$, $\forall\, \mathbf{x}$ in $\mathfrak{D}$, and hence y_0^* annihilates the closure of $\mathfrak{R}$. If, in particular, $\mathfrak{R}$ is dense in $\mathfrak{Y}$, then y_0^* must be the zero functional and $\mathfrak{R}[T^*] = \{\mathbf{0}\}$ so that T^* has an inverse. On the other hand, if $\mathfrak{R}$ is not dense in $\mathfrak{Y}$ and if $\mathbf{y}_0 \in \mathfrak{Y} \ominus \mathfrak{R}$, then by Theorem 10.4.5 we can find a linear bounded functional $y_0^* \in \mathfrak{Y}^*$ such that $y_0^*(\mathbf{y}_0) = 1$ while y_0^* annihilates $\overline{\mathfrak{R}}$. For this functional $y_0^*[T(\mathbf{x})] = 0$, $\forall\, \mathbf{x} \in \mathfrak{D}$. Hence $y_0^* \in \mathfrak{D}(T^*)$ and $T^*(y_0^*) = 0$ while $y_0^*(\mathbf{y}_0) \neq 0$. It follows that if $\mathfrak{R}$ is not dense in $\mathfrak{Y}$, then T^* cannot have an inverse. ∎

Several other results are listed in the Exercise below.

EXERCISE 10.4

1. In Example 1 functionals $x^* \in \mathfrak{X}^*$ are defined. Are they dense in $\mathfrak{X}^*$?
2. Does T^* as defined in Example 1 have an inverse?
3. Take $\mathfrak{X} = \mathfrak{Y} = l_1$, $\mathfrak{X}^* = \mathfrak{Y}^* = l_\infty$ and let T be the shift operator $T(\mathbf{x}) = (x_2, x_3, \dots, x_{n+1}, \dots)$ if $\mathbf{x} = (x_1, x_2, \dots, x_n, \dots)$. Find T^*.
4. Let T be a linear transformation with domain $\mathfrak{D}$ dense in $\mathfrak{X}$. If $\mathfrak{R}(T^*)$ is dense (weakly* dense suffices) in $\mathfrak{X}^*$, then T has an inverse. Prove!
5. Let T be a linear transformation with an inverse and such that $\mathfrak{D}$ is dense in $\mathfrak{X}$ and $\mathfrak{R}$ dense in $\mathfrak{Y}$. Then $(T^*)^{-1} = (T^{-1})^*$, further T^{-1} is bounded iff $(T^*)^{-1}$ is bounded on $\mathfrak{X}^*$. Prove!
6. Prove that a linear transformation T with domain dense in $\mathfrak{X}$ has a linear bounded inverse iff $\mathfrak{R}(T^*) = \mathfrak{X}^*$.

10.5 SPECTRA AND RESOLVENTS

These concepts have been with us for a long time. For matrices they appeared in Sections 1.4 and 1.5, for operators in Section 2.6, and for elements of a Banach algebra they were the central notions of Chapter 9. In this section we shall consider the special case when the B-algebra is an operator algebra, $\mathfrak{E}(\mathfrak{X})$ in our previous notation. Here there are some new features.

Let $\mathfrak{B} = \mathfrak{E}(\mathfrak{X})$, i.e. the B-algebra of linear bounded operators from $\mathfrak{X}$ to $\mathfrak{X}$, a B-space over the complex field. If $T \in \mathfrak{E}(\mathfrak{X})$ and if $\lambda_0 I - T$ has a bounded inverse with domain dense in $\mathfrak{X}$, we say that $\lambda_0 \in \rho(T)$, the *resolvent set* of T, and write

$$(\lambda_0 I - T)^{-1} = R(\lambda_0, T). \qquad (10.5.1)$$

The resolvent set is open and unbounded; it contains the point at infinity of the

λ-plane and need not be connected. $R(\lambda, T)$ is B-holomorphic in each of the connected components of $\rho(T)$. The complement of $\rho(T)$ is the spectrum of T, denoted by $\sigma(T)$. It is a closed bounded set and may separate the plane. It is confined to the disk

$$|\lambda| \leqslant r(T), \tag{10.5.2}$$

where $r(T)$ is the spectral radius of T,

$$r(T) = \lim_{n\to\infty} \|T^n\|^{1/n}. \tag{10.5.3}$$

There is at least one point of $\sigma(T)$ on the rim of the disk.

The resolvent satisfies, for $\lambda \in \rho(T)$,

$$(\lambda I - T)\, R(\lambda, T) = R(\lambda, T)\,(\lambda I - T) = I \tag{10.5.4}$$

by virtue of (10.5.1). It also satisfies the two resolvent equations.

Theorem 9.2.1 of Nagumo now takes the form

Theorem 10.5.1. *In a neighborhood of an isolated point λ_0 of $\sigma(T)$ there is an expansion*

$$R(\lambda, T) = J \sum_{n=0}^{\infty} Q^n(\lambda - \lambda_0)^{-n-1} + \sum_{n=0}^{\infty} A^{n+1}(\lambda_0 - \lambda)^n. \tag{10.5.5}$$

Here A, J, Q are elements of $\mathfrak{E}(\mathfrak{X})$ uniquely determined by T. J is a projection operator (i.e. idempotent), Q is quasinilpotent (or nilpotent) and

$$AJ = JA = 0, \qquad JQ = QJ = Q. \tag{10.5.6}$$

The first series converges for all $\lambda \neq \lambda_0$, while the second converges in the largest open disk with center at λ_0 containing no point of $\sigma(T)$ except for λ_0.

The proof follows directly from Theorem 9.2.1.

We note that

$$J = \frac{1}{2\pi i}\int_\Gamma R(\mu, T)\, d\mu, \tag{10.5.7}$$

where Γ is any simple closed rectifiable oriented curve ("scroc" for short in the following) surrounding $\mu = \lambda_0$ once and leaving the rest of $\sigma(T)$ on the outside. This is a special case of a more general *spectral resolution*. Suppose that it is possible to split the spectrum into n disjoint *spectral sets*

$$\sigma(T) = \sigma_1 \cup \sigma_2 \cup \cdots \cup \sigma_n, \tag{10.5.8}$$

where σ_j and σ_k have a positive distance from each other. We may assume that Γ_k is a scroc surrounding σ_k and separating it from the rest of the spectrum.

Theorem 10.5.2. *Set*

$$J_k = \frac{1}{2\pi i} \int_{\Gamma_k} R(\mu, T)\, d\mu. \tag{10.5.9}$$

Then each J_k is an idempotent with $J_j J_k = \delta_{jk} J_k$ and the idempotents define a resolution of the identity

$$I = J_1 + J_2 \cdots + J_n. \tag{10.5.10}$$

Proof. This follows from Cauchy's theorem for B-holomorphic functions together with the first resolvent equation

$$R(\lambda, T)\, R(\mu, T) = \frac{R(\mu, T) - R(\lambda, T)}{\lambda - \mu} \tag{10.5.11}$$

and the expansion

$$R(\lambda, T) = \lambda^{-1} I + \lambda^{-2} T + \cdots + \lambda^{-n-1}\, T^n + \cdots \tag{10.5.12}$$

valid for $|\lambda| > r(T)$. The right member of (10.5.10) is the integral of $R(\mu, T)$ taken over all the scrocs Γ_k. By Cauchy's theorem this equals the integral taken over a circle $|\mu| = r > r(T)$. Here we can use (10.5.12) and see that the integral reduces to I, so that (10.5.10) follows. Next we consider the product of two integrals

$$J_j J_k = \frac{1}{(2\pi i)^2} \int_{\Gamma_j} R(\lambda, T)\, d\lambda \int_{\Gamma_k} R(\mu, T)\, d\mu$$

$$= \frac{1}{(2\pi i)^2} \int_{\Gamma_j} \int_{\Gamma_k} R(\lambda, T)\, R(\mu, T)\, d\lambda\, d\mu.$$

Here we use (10.5.11). Suppose first that $j \neq k$. We then obtain the difference of two repeated integrals. One of these can be written

$$\frac{1}{2\pi i} \int_{\Gamma_k} R(\mu, T) \left\{ \frac{1}{2\pi i} \int_{\Gamma_j} \frac{d\lambda}{\lambda - \mu} \right\} d\mu = 0 \tag{10.5.13}$$

since μ is outside of Γ_j. In the same manner, but interchanging the order of integration, we see that the second integral is zero.

If now $j = k$ we let Γ_j denote a contour surrounding Γ_k and still leaving the rest of the spectrum outside. Now μ is inside Γ_j in (10.5.13) so that the numerical integral reduces to unity and the repeated integral equals J_k. The second repeated integral still equals zero. ∎

A point $\lambda_0 \in \sigma(T)$ iff $\lambda_0 I - T$ does not have a bounded inverse with domain dense in $\mathfrak{X}$. There are various possibilities. We write $T_\lambda = \lambda I - T$.

Definition 10.5.1. *The values of λ for which T_λ has an unbounded inverse with domain dense in $\mathfrak{X}$ form the continuous spectrum $C\sigma(T)$. The values of λ for which T_λ has an inverse whose domain is not dense in $\mathfrak{X}$ form the residual spectrum $R\sigma(T)$. The values of λ for which no inverse exists form the point spectrum $P\sigma(T)$. The union of $C\sigma(T)$, $R\sigma(T)$ and $P\sigma(T)$ is the spectrum $\sigma(T)$ of T.*

Theorem 10.5.3. *The four sets $\rho(T)$, $C\sigma(T)$, $R\sigma(T)$, $P\sigma(T)$ are mutually disjoint and their union is the extended complex plane.*

Verification is left to the reader.

The nature of a particular spectral value is usually not evident and calls for a special investigation. The relations between analytical properties of the resolvent and metric or topological properties of the spectrum, on the one hand, and the classification of the spectral values, on the other, are quite baffling. The following is a simple connection, one of the few known to the author.

Theorem 10.5.4. *If $\lambda = \lambda_0$ is a pole of the resolvent, then $\lambda_0 \in P\sigma(T)$.*

Proof. In formula (10.5.5) the operator Q is now a nilpotent. Suppose that Q^m is the zero operator but Q^{m-1} is not. We can then find an element $\mathbf{x}_0 \in \mathfrak{X}$ with $\|\mathbf{x}_0\| = 1$ such that $Q^{m-1}\mathbf{x}_0 \neq 0$. We recall that

$$Q = (T - \lambda_0 I)\, J = J(T - \lambda_0 I). \tag{10.5.14}$$

Cf. formula (1.5.35). This gives

$$[T - \lambda_0 I]\,[Q^{m-1}\mathbf{x}_0] = [T - \lambda_0 I]\,[J\,Q^{m-1}\mathbf{x}_0] = Q^m\mathbf{x}_0 = \mathbf{0}.$$

Thus $\lambda_0 \in P\sigma[T]$ and $Q^{m-1}\mathbf{x}_0$ is a characteristic vector. ∎

Spectral values which are not poles of the resolvent may belong to any one of the three spectral classes. Examples illustrating the various possibilities are to be found in the Exercise below.

In the case of an unbounded linear operator U, Definition 10.5.1 still makes sense. There are, however, new possibilities: (1) the spectrum may be unbounded; (2) the spectrum may coincide with the finite complex plane; (3) the spectrum may be vacuous. The last possibility calls for $R(\lambda, U)$ to be an entire function of λ. It is customary to say that $\lambda = \infty$ belongs to the *extended spectrum*, and this terminology is also used if $\lambda = \infty$ is an isolated singularity of $R(\lambda, U)$, even if this is not an entire function. Various examples to illustrate these possibilities are to be found in Exercise 10.5.

In discussing the nature of spectral points an alternate approach is often useful.

Definition 10.5.2. *The operator T has the property*

P_1 if there exists an element $\mathbf{x}_0$ of $\mathfrak{X}$ such that $\|\mathbf{x}_0\| = 1$ and $T\mathbf{x}_0 = \mathbf{0}$;

P_2 if $\mathfrak{R}(T)$ is non-dense in $\mathfrak{X}$;

P_3 if there exists a sequence $\{\mathbf{x}_n\} \subset \mathfrak{X}$ such that $\|\mathbf{x}_n\| = 1$, $\forall\, n$, and $T\mathbf{x}_n \to \mathbf{0}$.

With this terminology we can reformulate the classification as follows.

Theorem 10.5.5. *A point λ belongs to*
$\sigma(T)$ *if T_λ has at least one of the properties P_ν;*
$P\sigma(T)$ *if T_λ has P_1;*
$R\sigma(T)$ *if T_λ has P_2 but not P_1;*
$C\sigma(T)$ *if T_λ has P_3 but neither P_1 nor P_2; and*
$\rho(T)$ *if T_λ has none of the properties P_ν.*

This is just a paraphrase of the previous classification, but it brings out what may be called the "pecking order"

$$P_1 > P_2 > P_3. \tag{10.5.15}$$

P_1 does not exclude P_2 or P_3 but P_1 decides the classification and so on.

There are various relations between the spectra of T and of its adjoint T^* the most important of which is

$$\rho(T) = \rho(T^*), \quad \text{if} \quad \overline{\mathfrak{D}} = \mathfrak{X}. \tag{10.5.16}$$

See further the Exercise below.

EXERCISE 10.5

In the next three problems show that $\sigma(T) = \{0\}$ and $\lambda = 0$ is an essential singularity of $R(\lambda, T)$ with the spectral classification as stated in the problem.

1. $\mathfrak{X} = l_2, T$ takes $(x_1, x_2, \ldots, x_n, \ldots)$ into $(x_2, x_3/2, \ldots, x_{n+1}/n, \ldots)$. Point spectrum.
2. $\mathfrak{X} = C[0, 1]$, T takes $f(t)$ into $\int_0^t f(s)\,ds$. Residual spectrum.
3. $\mathfrak{X} = C_0[0, 1]$, the subspace of $C[0, 1]$ with $f(0) = 0, T$ as in the preceding problem. Continuous spectrum.

The next problem exhibits a "solid" two-dimensional point spectrum.

4. $\mathfrak{X} = C[0, \infty], T$ takes $f(t)$ into $f(t + h)$, where h is a fixed positive number. Show that $P\sigma(T) = \{\lambda; (|\lambda| < 1) \cup \{1\}\}$. Show that the rest of the unit circle belongs to $C\sigma(T)$.

The next three problems exhibit various spectral possibilities of unbounded operators.

5. $\mathfrak{X} = L_2(-\pi, \pi)$, U takes f into f'. $P\sigma(U) = \{ni, n = 0, \pm 1, \pm 2, \ldots\}$.
6. Take the Kolmogorov matrix $\mathcal{A} = (a_{mn})$ with $a_{11} = -1$, $a_{12} = 1$, $a_{n,n-1} = n - 1$, $a_{nn} = -n$, $a_{n,n+1} = 1$ for $n > 1$, all other a_{mn} being equal to 0. Use $\mathcal{A}$ as an operator, $\mathbf{y} = \mathcal{A}\mathbf{x}$, on a weighted sequence space to itself with

$$\|\mathbf{x}\| = \sum_1^\infty \frac{|x_n|}{(2n)!}.$$

Here the system $\mathcal{A}\mathbf{x} = \lambda\mathbf{x}$ is satisfied by a vector $\mathbf{x}(\lambda)$ where the nth component is a polynomial $P_n(\lambda)$ of degree $n - 1$, $P_1(\lambda) \equiv 1$. Show that $P_n(\lambda)$ has positive integral coefficients, $P_n(0) \equiv 1$ and $|P_n(\lambda)| \leqslant n!|\lambda|^{n-1}$ for $|\lambda| \geqslant 1$. Hence show that the point spectrum of the operator $\mathcal{A}$ covers the finite plane.

7. $\mathfrak{X} = C_0[0, 1]$, U takes $f(t)$ into $-f'(t)$. Prove that $R(\lambda, U)$ exists and

$$R(\lambda, U)[f] = \int_0^t e^{\lambda(s-t)} f(s)\, ds,$$

which is a C_0-valued entire function of λ.

8. Let $\mathfrak{X} = L_2(-\pi, \pi)$ and let T be a shift operator on the Fourier coefficients which takes $\sum_{-\infty}^{\infty} f_n e^{nit}$ into $\sum_{-\infty}^{\infty} f_{n+1} e^{nit}$. If $F_n(\lambda)$ is the nth Fourier coefficient of $R(\lambda, T)[f]$ when it exists, show that

$$F_n(\lambda) = \frac{1}{2\pi} \int_{-\pi}^{\pi} \frac{f(s)}{\lambda - e^{-is}} e^{-ins} ds.$$

Discuss the existence of $F_n(\lambda)$ as a function of λ and determine the spectrum of T.

9. Show that $\{F_n(e^{-i\alpha})\} \in l_2$ iff $f(t)[e^{-i\alpha} - e^{-it}]^{-1} \in L_2(-\pi, \pi)$, α real. What bearing does this have on the nature of the spectrum?

10. Verify Theorem 10.5.1.

11. Fill in missing details in the proof of Theorem 10.5.2.

12. Verify (10.5.14).

13. If J is the idempotent in (10.5.5), if $\lambda = \lambda_0$ belongs to $P\sigma(T)$ and $\mathbf{x}_0$ is a characteristic vector, prove that $J\mathbf{x}_0 = \mathbf{x}_0$. Prove also that $(\lambda - \lambda_0)\, R(\lambda, T)\, \mathbf{x}_0 \equiv \mathbf{x}_0$.

14. Prove (10.5.16). Prove also that $R(\lambda, T^*) = R(\lambda, T)^*$ both under the assumption that $\mathfrak{D}$ is dense in $\mathfrak{X}$.

COLLATERAL READING

The following is a small selection from the available texts:

DUNFORD, N. and J. SCHWARTZ, *Linear Operators,* Part I, *General Theory*, Interscience, New York (1958). See in particular Chapters 2, 5, 6 and 7.

EPSTEIN, B., *Linear Functional Analysis. Introduction to Lebesgue Integration and Infinite-dimensional Problems,* Saunders, Philadelphia (1970), Chapters 5 and 6.

HILLE, E. and R. S. PHILLIPS, *Functional Analysis and Semi-groups,* American Mathematical Society Colloquium Publications, Vol. 31, Providence, R.I. (1957), Chapters 2, 4, and 5.

SCHAEFFER, H. H., *Topological Vector Spaces,* Macmillan, New York (1966), Chapters 3 and 4.

11 INNER PRODUCT SPACES

Many threads went into the fabric displayed in Chapters 1 and 2. Most of these threads have been examined in some detail in preceding chapters, but we still have to give a fuller account of inner-product spaces, introduced as a general concept in Section 2.5. The spaces C^n, l_2, and L_2 are important instances of such spaces with which we are familiar. The linear bounded functionals on these special spaces are defined by inner products and are associated in a natural manner with the theory of quadratic and bilinear forms. These in their turn generalize in a natural manner to the relation between an operator and its adjoint

$$(T\mathbf{x}, \mathbf{y}) = (\mathbf{x}, T^*\mathbf{y}).$$

The case where T is self-adjoint, $T^* = T$, requires that the spectrum of T be real and leads to the notion of the associated spectral function and the resolution of the identity defined by it. These are the main topics in the discussion.

There are four sections: Review; Spectrum and the numerical range; Operational calculus; and The spectral theorem.

11.1 REVIEW

The following is a brief reminder of the basic concepts presented in Section 2.5.

We consider a linear space $\mathfrak{X}$ over the complex field and a mapping taking ordered pairs $\mathbf{x}, \mathbf{y} \in \mathfrak{X}$ into an element $(\mathbf{x}, \mathbf{y})$ of C^1, called the inner product of $\mathbf{x}$ with $\mathbf{y}$. See Definition 2.5.1 for the properties of $(\mathbf{x}, \mathbf{y})$. We recall that $(\mathbf{x}, \mathbf{y})$ is *bilinear* and *skew-symmetric.* We introduce a norm in $\mathfrak{X}$ by setting

$$\|\mathbf{x}\| = [(\mathbf{x}, \mathbf{x})]^{\frac{1}{2}} \tag{11.1.1}$$

and define the metric by

$$d(\mathbf{x}, \mathbf{y}) = \|\mathbf{x} - \mathbf{y}\|. \tag{11.1.2}$$

If $\mathfrak{X}$ is complete in this metric it is called a *Hilbert space*, for which we use the generic notation $\mathfrak{H}$.

For fixed $\mathbf{y} \in \mathfrak{X}$, the mapping $\mathbf{x} \to (\mathbf{x}, \mathbf{y})$ defines a linear bounded functional $l(\mathbf{x})$ on $\mathfrak{X}$. It was shown that the norm of the functional equals $\|\mathbf{y}\|$. If $\mathfrak{X} = \mathfrak{H}$, all linear bounded functionals are given by inner products, i.e. to $l \in \mathfrak{H}^*$ there is a $\mathbf{y} \in \mathfrak{H}$ such that $l(\mathbf{x}) = (\mathbf{x}, \mathbf{y})$. There is a 1–1 correspondence between $\mathfrak{H}$ and $\mathfrak{H}^*$ which is an isometry and a skew symmetry. Thus $\mathfrak{H}$ is its own adjoint space.

For any two vectors of $\mathfrak{X}$ the *Parallelogram Law* holds:

$$\|\mathbf{x}_1 + \mathbf{x}_2\|^2 + \|\mathbf{x}_1 - \mathbf{x}_2\|^2 = 2[\|\mathbf{x}_1\|^2 + \|\mathbf{x}_2\|^2]. \tag{11.1.3}$$

Two vectors $\mathbf{x}_1$ and $\mathbf{x}_2$ are said to be *orthogonal* or *perpendicular*, in symbols $\mathbf{x}_1 \perp \mathbf{x}_2$, if

$$(\mathbf{x}_1, \mathbf{x}_2) = 0. \tag{11.1.4}$$

For such vectors the *Pythagorean Law* holds:

$$\|\mathbf{x}_1 + \mathbf{x}_2\|^2 = \|\mathbf{x}_1\|^2 + \|\mathbf{x}_2\|^2. \tag{11.1.5}$$

Let $\mathfrak{M} \subset \mathfrak{H}$ be a closed linear subspace of $\mathfrak{H}$. To $\mathfrak{M}$ corresponds another subspace denoted by $\mathfrak{M}^\perp$ and known as the *orthogonal complement* of $\mathfrak{M}$. Here

$$(\mathbf{u}, \mathbf{v}) = 0 \quad \text{if} \quad \mathbf{u} \subset \mathfrak{M} \quad \text{and} \quad \mathbf{v} \in \mathfrak{M}^\perp. \tag{11.1.6}$$

Any $\mathbf{x} \in \mathfrak{H}$, $\mathbf{x} \neq \mathbf{0}$ admits a unique representation

$$\mathbf{x} = \mathbf{u} + \mathbf{v}, \qquad \mathbf{u} \in \mathfrak{M}, \qquad \mathbf{v} \in \mathfrak{M}^\perp, \qquad \mathfrak{H} = \mathfrak{M} \oplus \mathfrak{M}^\perp. \tag{11.1.7}$$

This defines a mapping P of $\mathfrak{H}$ into $\mathfrak{M}$:

$$P(\mathbf{x}) = P(\mathbf{u} + \mathbf{v}) = P(\mathbf{u}) = \mathbf{u}, \qquad P(\mathbf{v}) = \mathbf{0}. \tag{11.1.8}$$

P is a *projection operator*, i.e. an idempotent element of the operator algebra $\mathfrak{E}(\mathfrak{H})$. For $\mathfrak{M} \neq \{\mathbf{0}\}$ we have

$$P^2 = P, \qquad \|P\| = 1, \qquad \mathfrak{N}[P] = M^\perp. \tag{11.1.9}$$

The operator $I - P$ is also a projection. It projects $\mathfrak{H}$ onto $\mathfrak{M}^\perp$. Its nullspace is $\mathfrak{M}$ and its norm is 1 unless $\mathfrak{M}^\perp = \{\mathbf{0}\}$.

If $\mathfrak{H}$ is infinite-dimensional, we can find a countable set of elements $\{\mathbf{v}_k\}$, any finite number of which are linearly independent, and by the Gram–Schmidt process we can obtain an orthonormal system $\{\mathbf{u}_k\}$. This leads to the formal Fourier series

$$\mathbf{x} \sim \sum_1^\infty \hat{x}_k \mathbf{u}_k, \qquad \hat{x}_k = (\mathbf{x}, \mathbf{u}_k). \tag{11.1.10}$$

The partial sums of this series form a Cauchy sequence in $\mathfrak{H}$ with the limit $\tilde{\mathbf{x}}$ which may or may not coincide with $\mathbf{x}$. Necessary and sufficient conditions for $\tilde{\mathbf{x}} = \mathbf{x}$ are found in Theorem 2.5.2.

We can generalize (11.1.7) in an obvious manner:

$$\mathfrak{H} = \Sigma\, \mathfrak{M}_j, \qquad \mathfrak{M}_j \cap \mathfrak{M}_k = \{\mathbf{0}\}, \qquad j \neq k, \tag{11.1.11}$$

and

$$\mathbf{x} = \Sigma\, \mathbf{x}_j, \qquad \mathbf{x}_j \in \mathfrak{M}_j, \qquad (\mathbf{x}_j, \mathbf{x}_k) = 0, \qquad j \neq k. \tag{11.1.12}$$

Here the sums are finite or countably infinite. Such a decomposition may be based upon an orthonormal system and the linear manifolds spanned by the elements (singly or grouped) which form closed linear subspaces any two of which are orthogonal.

To the decomposition (11.1.11) corresponds a set of projection operators P_j such that

$$P_j(\mathfrak{H}) = \mathfrak{M}_j, \qquad P_j P_k = 0, \qquad j \neq k, \tag{11.1.13}$$

$$P_j(\mathbf{x}) = \mathbf{x}_j, \qquad \mathbf{x}_j \in \mathfrak{M}_j, \tag{11.1.14}$$

where $\mathbf{x}_j$ is the vector occurring in (11.1.12). If the set of subspaces $\mathfrak{M}_j$ is finite, say m in number, then we write

$$I = P_1 + P_2 + \cdots + P_m, \tag{11.1.15}$$

and speak of this as a *resolution of the identity*. If the set is infinite, we can still write

$$I = \Sigma\, P_j \tag{11.1.16}$$

and note that the partial sums of the operator series converge to I in the strong operator topology. Each P_j is an element of $\mathfrak{E}(\mathfrak{H})$, the space of linear bounded operators from $\mathfrak{H}$ into itself.

If $T \in \mathfrak{E}(\mathfrak{H})$, a basic tool in the study of T is given by the inner products $(\mathbf{Tx}, \mathbf{x})$ and $(\mathbf{Tx}, \mathbf{y})$ which are the *quadratic* and *polar forms*, respectively, that correspond to T. The *adjoint operator* T^* is then defined by

$$(T\mathbf{x}, \mathbf{y}) = (\mathbf{x}, T^*\mathbf{y}) \tag{11.1.17}$$

for all $\mathbf{x}$, $\mathbf{y} \in \mathfrak{H}$. Here $T \in \mathfrak{E}(\mathfrak{H})$ implies and is implied by $T^* \in \mathfrak{E}(\mathfrak{H})$. If, in particular,

$$T^* = T, \tag{11.1.18}$$

then T is said to be *self-adjoint* or *Hermitian* and the corresponding forms $(T\mathbf{x}, \mathbf{x})$ and $(T\mathbf{x}, \mathbf{y})$ are also called Hermitian.

In working with polar forms the analogue of (1.7.8) is a convenient tool. For all $\mathbf{x}, \mathbf{y} \in \mathfrak{H}$ and all complex numbers s, t we have

$$\begin{aligned} &s\bar{t}(T\mathbf{x}, \mathbf{y}) + \bar{s}t(T\mathbf{y}, \mathbf{x}) \\ &\qquad = \big(T(s\mathbf{x} + t\mathbf{y}), s\mathbf{x} + t\mathbf{y}\big) - |s|^2(T\mathbf{x}, \mathbf{x}) - |t|^2(T\mathbf{y}, \mathbf{y}). \end{aligned} \tag{11.1.19}$$

This identity implies that $(T\mathbf{x}, \mathbf{x}) = 0$ for all $\mathbf{x}$ holds iff T is the zero operator in analogy with Theorem 1.7.2. For if $s = t = 1$

$$(T\mathbf{x}, \mathbf{y}) = -(T\mathbf{y}, \mathbf{x})$$

and for $s = 1$, $t = i$ we find that $(T\mathbf{x}, \mathbf{y}) = (T\mathbf{y}, \mathbf{x})$ so that

$$(T\mathbf{x}, \mathbf{y}) = 0, \qquad \forall\, \mathbf{x}, \mathbf{y}.$$

In particular, for $\mathbf{y} = T\mathbf{x}$ we get $(T\mathbf{x}, T\mathbf{x}) = \|T\mathbf{x}\|^2 = 0$ for all $\mathbf{x}$ and hence T is the zero operator.

This fact implies first that T^* is unique, secondly that

$$(T^*)^* = T, \tag{11.1.20}$$

so that the operation of taking the adjoint is an *involution*. Among other consequences we list

Theorem 11.1.1. *If $T \in \mathfrak{E}(\mathfrak{H})$, then $(T\mathbf{x}, \mathbf{x})$ is real for all $\mathbf{x}$ iff T is Hermitian.*

Proof. (1) T Hermitian implies $(T\mathbf{x}, \mathbf{x})$ real, $\forall\, \mathbf{x}$. This was proved in Theorem 2.5.12.

(2) $(T\mathbf{x}, \mathbf{x})$ real for all $\mathbf{x}$ implies that T is Hermitian. For

$$(T\mathbf{x}, \mathbf{x}) = \overline{(T\mathbf{x}, \mathbf{x})} = (\mathbf{x}, T\mathbf{x}) = (\mathbf{x}, T^*\mathbf{x}),$$

so $T^* = T$ by the uniqueness of the adjoint. ∎

We recall that if S and T are Hermitian operators in $\mathfrak{E}(\mathfrak{H})$ and if α is real, then $S + T$ and αT are Hermitian, while ST is Hermitian iff S and T commute. We also recall that an arbitrary operator A in $\mathfrak{E}(\mathfrak{H})$ admits of a unique representation by Hermitian operators, namely

$$A = B + iC, \qquad B = \tfrac{1}{2}(A + A^*), \qquad C = \frac{1}{2i}(A - A^*). \tag{11.1.21}$$

Finally, we recall the definition of *normal* and of *unitary* operators. A is normal iff A and A^* commute, which is the case iff B and C commute in (11.1.21). U is unitary if

$$UU^* = U^*U = I, \quad \text{or, equivalently,} \quad U^{-1} = U^*. \tag{11.1.22}$$

EXERCISE 11.1

1. The norm defined by (11.1.1) is differentiable in the sense that

$$\lim \frac{1}{\alpha}[\|\mathbf{x} + \alpha\mathbf{h}\| - \|\mathbf{x}\|]$$

exists if α decreases to zero. Find the limit. What happens at $\mathbf{x} = \mathbf{0}$? Cf. Section 8.4.

2. The norm is differentiable in this sense for each of the spaces l_p and L_p, $1 \leqslant p < \infty$. Verify this for l_p.
3. Verify (11.1.9) and statements about $I - P$.
4. Verify that the partial sums in (11.1.10) do form a Cauchy sequence.
5. Why is the representation (11.1.12) unique?
6. Why does $T \in \mathfrak{E}(\mathfrak{H})$ imply $T^* \in \mathfrak{E}(\mathfrak{H})$, and vice versa?
7. Show that T^* is unique.
8. Show that $(T_1T_2)^* = T_2{}^*T_1{}^*$.
9. Prove that $S^* = T^*$ implies $S = T$.
10. Verify (11.1.20).

11. Prove that if T is Hermitian, so is T^n, $n = 2, 3, \ldots$, and so is any polynomial in T with real coefficients.
12. If T has an inverse, show that so does T^* and that $(T^{-1})^* = (T^*)^{-1}$.
13. If T is Hermitian, show that $\|T^2\| = \|T\|^2$.
14. Prove that $\|T^*\| = \|T\|$.
15. Let $T \in \mathfrak{E}(\mathfrak{H})$ and let $\lambda = i$ belong to the resolvent set of T. Show that $iI + T$ and $(iI - T)^{-1}$ commute.
16. With T as in the preceding problem, form the Cayley transform

$$\mathbf{C}(T) = (iI + T)(iI - T)^{-1}$$

and show that $\mathbf{C}(T)$ is unitary if T is Hermitian.

11.2 SPECTRUM AND THE NUMERICAL RANGE

We return to the quadratic form $(T\mathbf{x}, x)$ for any T in $\mathfrak{E}(\mathfrak{H})$. If $\mathbf{x}$ is restricted to the unit sphere, the values taken on by $(T\mathbf{x}, \mathbf{x})$ constitute the *numerical range* of T, denoted by $\mathsf{W}(T)$. Here the W stands for the German term "Wertvorrat". Thus

$$\mathsf{W}(T) = \{\alpha;\ (T\mathbf{x}, \mathbf{x}) = \alpha,\ \|\mathbf{x}\| = 1\}. \tag{11.2.1}$$

This is a set of complex numbers which has interesting geometrical properties and is closely related to the spectrum of T. More precisely we shall prove

Theorem 11.2.1. $\mathsf{W}(T)$ *is a bounded convex set.*

Theorem 11.2.2. *The spectrum* $\sigma(T)$ *of* T *is a subset of the closure of* $\mathsf{W}(T)$.

The first result is known as the *Toeplitz–Hausdorff theorem.* Otto Toeplitz (1881–1940) proved in 1918 that the boundary of $\mathsf{W}(T)$ is a convex curve but left open the possibility that there could be holes in the interior not belonging to $\mathsf{W}(T)$. This gap was filled by Felix Hausdorff in 1919. For the proof of this theorem we need the simple

Lemma 11.2.1. *If* s *and* t *are complex numbers, then*

$$\mathsf{W}(sT + tI) = s\mathsf{W}(T) + t, \tag{11.2.2}$$

i.e. the numerical range of an affine transform of T *is the same affine transform of the numerical range of* T.

Proof. Note that

$$\big((sT + tI)\mathbf{x}, \mathbf{x}\big) = s(T\mathbf{x}, \mathbf{x}) + t(\mathbf{x}, \mathbf{x}) \tag{11.2.3}$$

which for $\|\mathbf{x}\| = 1$ implies (11.2.2). ∎

Corollary 1. $\mathsf{W}(T)$ *is convex iff* $\mathsf{W}(sT + tI)$ *has this property for all* s *and* t.

Proof of Theorem 11.2.1. It is given that $\mathsf{W}(T)$ contains two complex numbers α and β and it is desired to prove that the line segment joining α and β is also in $\mathsf{W}(T)$. It is no restriction to assume that $\alpha = 0$ and $\beta = 1$, for we can always use an affine transformation $z \to sz + t$ which sends α into 0 and β into 1 and T into $sT + tI$. Since straight line segments go into straight line segments under an affine transformation, the line segment joining α and β goes into the line segment joining 0 and 1.

In this modified formulation of the question we have two unit vectors $\mathbf{x}$ and $\mathbf{y}$ such that

$$(T\mathbf{x}, \mathbf{x}) = 0, \qquad (T\mathbf{y}, \mathbf{y}) = 1 \tag{11.2.4}$$

and it is required to prove that the line segment $[0, 1]$ belongs to $\mathsf{W}(T)$. In the canonical representation (11.1.24) for T, say

$$T = B + iC,$$

we have

$$(C\mathbf{x}, \mathbf{x}) = 0, \qquad (C\mathbf{y}, \mathbf{y}) = 0 \tag{11.2.5}$$

by virtue of (11.2.4). The unit vectors which satisfy (11.2.4) are not uniquely determined and could be replaced by $\omega_1\mathbf{x}$ and $\omega_2\mathbf{y}$, respectively, where $|\omega_1| = |\omega_2| = 1$ without affecting (11.2.4) and (11.2.5). We can use this added amount of freedom to require that the real part of $(C\mathbf{x}, \mathbf{y})$ is 0,

$$\Re(C\mathbf{x}, \mathbf{y}) = 0. \tag{11.2.6}$$

With $\mathbf{x}$ and $\mathbf{y}$ chosen in this manner we set

$$\mathbf{z}(s) = s\mathbf{x} + (1 - s)\,\mathbf{y}. \tag{11.2.7}$$

Since $\|\mathbf{z}(s)\| \leqslant s\,\|\mathbf{x}\| + (1 - s)\,\|\mathbf{y}\| = 1$, the vector $\mathbf{z}(s)$ is confined to the unit ball. It may possibly reduce to the zero vector for some value of s. This would mean $s\mathbf{x} = (s - 1)\,\mathbf{y}$ and since $\mathbf{x}$ and $\mathbf{y}$ are unit vectors, this can only happen for $s = \frac{1}{2}$. But if $\mathbf{x} = -\mathbf{y}$, then $(T\mathbf{x}, \mathbf{x}) = (T\mathbf{y}, \mathbf{y})$, which contradicts (11.2.4). Thus $\|\mathbf{z}(s)\| > 0$ for $0 \leqslant s \leqslant 1$. Using (11.1.19) with $s = s$, $t = 1 - s$ we get

$$\begin{aligned}\big(T[\mathbf{z}(s)], \mathbf{z}(s)\big) = s^2(T\mathbf{x}, \mathbf{x}) &+ (1 - s)^2\,(T\mathbf{y}, \mathbf{y}) \\ &+ s(1 - s)\,[(T\mathbf{x}, \mathbf{y}) + (T\mathbf{y}, \mathbf{x})].\end{aligned} \tag{11.2.8}$$

Here

$$(T\mathbf{x}, \mathbf{y}) + (T\mathbf{y}, \mathbf{x}) = 2\Re[(B\mathbf{x}, \mathbf{y})] + 2i\Re[(C\mathbf{x}, \mathbf{y})] = 2\Re[(B\mathbf{x}, \mathbf{y})] \tag{11.2.9}$$

by virtue of (11.2.6). It follows that (11.2.8) is real. Since $\|\mathbf{z}(s)\| \leqslant 1$ with equality assured only at the endpoints, we replace $\mathbf{z}(s)$ in (11.2.8) by the unit vector $\mathbf{z}(s)\,[\|\mathbf{z}(s)\|]^{-1}$. The result is an element of $\mathsf{W}(T)$ and a continuous function of s which takes on the value 0 for $s = 1$ and 1 for $s = 0$. It follows that all intermediary values are taken on at least once for s in $[0, 1]$. Hence $\mathsf{W}(T)$ is actually convex. ∎

Corollary 2. *If T is Hermitian, $\mathsf{W}(T)$ is an interval on the real axis, more precisely a subinterval of $[-\|T\|, \|T\|]$.*

Proof. $\mathsf{W}(T)$ is convex and the points of $\mathsf{W}(T)$ are real by Theorem 11.1.1. A real convex set is an interval. This interval must be a subinterval of $[-\|T\|, \|T\|]$ since for $\|\mathbf{x}\| = 1$

$$|(T\mathbf{x}, \mathbf{x})| \leqslant \|T\|. \blacksquare$$

As a preliminary to the proof of Theorem 11.2.2 we recall that the spectrum of a linear bounded operator T in $\mathfrak{E}(\mathfrak{H})$ breaks up into three disjoint parts:

(i) *The point spectrum* $P\sigma(T)$. The linear transformation $T_\lambda = \lambda I - T$ is not 1–1 for $\lambda = \lambda_0 \in P\sigma(T)$ and there is a unit vector $\mathbf{x}_0$ such that $T\mathbf{x}_0 = \lambda_0\mathbf{x}_0$.

(ii) *The continuous spectrum* $C\sigma(T)$. The mapping T_λ is 1–1 for $\lambda = \lambda_0 \in C\sigma(T)$ but the image of the unit sphere is not bounded away from the zero element. There exists a sequence of unit vectors $\{\mathbf{x}_n\}$ such that $\lambda_0\mathbf{x}_n - T\mathbf{x}_n \to 0$.

(iii) *The residual spectrum* $R\sigma(T)$. The mapping T_{λ_0} is 1–1 but the range of T_{λ_0} is not dense in $\mathfrak{H}$.

These conventions recalled, we can now proceed to the

Proof of Theorem 11.2.2. We take the several parts of the spectrum one at a time.

(1) $\lambda_0 \in P\sigma(T)$. Since there is an $\mathbf{x}_0$ with $\|\mathbf{x}_0\| = 1$ and $T\mathbf{x}_0 = \lambda_0\mathbf{x}_0$ we have

$$(T\mathbf{x}_0, \mathbf{x}_0) = (\lambda_0\mathbf{x}_0, \mathbf{x}_0) = \lambda_0(\mathbf{x}_0, \mathbf{x}_0) = \lambda_0$$

and $P\sigma(T)$ belongs to $\mathsf{W}(T)$.

(2) $\lambda_0 \in C\sigma(T)$. By the definition of the continuous spectrum there is a sequence of unit vectors $\{\mathbf{x}_n\}$ such that $\lambda_0\mathbf{x}_n - T\mathbf{x}_n$ goes to zero. Hence

$$(T\mathbf{x}_n, \mathbf{x}_n) \to \lambda_0(\mathbf{x}_n, \mathbf{x}_n) = \lambda_0$$

and λ_0 belongs to the closure of $\mathsf{W}(T)$, if not to $\mathsf{W}(T)$.

(3) $\lambda_0 \in R\sigma(T)$. The range of T_{λ_0} is not dense in $\mathfrak{H}$. Since the range is a linear subspace of $\mathfrak{H}$ there exists a linear functional which annihilates $\mathfrak{R}(T_{\lambda_0})$ without being identically zero. Since it is a functional on $\mathfrak{H}$ we can find a $\mathbf{y} \in \mathfrak{H}$ such that

$$\big(T_{\lambda_0}(\mathbf{x}), \mathbf{y}\big) = 0, \quad \forall\, \mathbf{x} \in \mathfrak{H}, \qquad \|\mathbf{y}\| = 1.$$

This gives

$$(T\mathbf{x}, \mathbf{y}) = \lambda_0(\mathbf{x}, \mathbf{y}), \quad \forall\, \mathbf{x},$$

and for $\mathbf{x} = \mathbf{y}$ we get $(T\mathbf{y}, \mathbf{y}) = \lambda_0$, so λ_0 belongs to $\mathsf{W}(T)$. $\blacksquare$

Corollary 3. *The convex hull of the spectrum is confined to the closure of the numerical range*

$$\operatorname{conv}[\sigma(T)] \subseteq \overline{\mathsf{W}(T)}. \tag{11.2.10}$$

Here the left member may very well be a proper subset of the right. An extreme case is furnished by quasi-nilpotent operators where $\sigma(T) = \{0\}$ while $\mathsf{W}(T) = \{0\}$ iff T is the zero operator. The other extreme, equality in (11.2.10), is furnished by normal operators. To prove this we need some preliminary results which are of independent interest.

Lemma 11.2.2. *If A is normal, then*

$$\|A^n\| = \|A\|^n. \tag{11.2.11}$$

Proof. We note first that for a normal operator A

$$(A\mathbf{x}, A\mathbf{x}) = (A^*A\mathbf{x}, \mathbf{x}) = (AA^*\mathbf{x}, \mathbf{x}) = (A^*\mathbf{x}, A^*\mathbf{x}),$$

so that

$$\|A\mathbf{x}\| = \|A^*\mathbf{x}\|, \qquad \forall\, \mathbf{x}. \tag{11.2.12}$$

Formula (11.2.11) is trivially true for $n = 1$. Suppose that it holds for $n \leqslant k$. Then

$$\begin{aligned}\|A^k\mathbf{x}\|^2 &= (A^k\mathbf{x}, A^k\mathbf{x}) = (A^{k-1}\mathbf{x}, A^*A^k\mathbf{x}) \\ &\leqslant \|A^{k-1}\mathbf{x}\|\;\|A^*A^k\mathbf{x}\| = \|A^{k-1}\mathbf{x}\|\;\|A^{k+1}\mathbf{x}\|\end{aligned}$$

by (11.2.12) with $\mathbf{x}$ replaced by $A^k\mathbf{x}$. Hence

$$\|A^k\mathbf{x}\|^2 \leqslant \|A^{k-1}\|\;\|A^{k+1}\|, \qquad \forall\, \mathbf{x}, \quad \|\mathbf{x}\| \leqslant 1,$$

and this inequality must hold also for the supremum of the left member, so that

$$\|A^k\|^2 \leqslant \|A^{k-1}\|\;\|A^{k+1}\|.$$

By the induction hypothesis this gives

$$\|A\|^{2k} \leqslant \|A\|^{k-1}\;\|A^{k+1}\|$$

or

$$\|A\|^{k+1} \leqslant \|A^{k+1}\|.$$

Since the converse inequality is trivially true for all operators, (11.2.11) holds for $n = k + 1$ and hence for all n. ∎

Corollary 1. *The spectral radius of a normal operator equals its norm,*

$$r(A) = \|A\|. \tag{11.2.13}$$

Proof. Use (11.2.11) and the definition of $r(A)$. ∎

In view of the properties of the spectral radius we have

Corollary 2. *The spectrum of a normal operator A, which is confined in the disk $|\lambda| \leqslant \|A\|$, has at least one point on the circumference of the disk.*

This implies

Corollary 3. *For $\|\mathbf{x}\| = 1$ and A normal*

$$\sup |(A\mathbf{x}, \mathbf{x})| = \|A\|. \tag{11.2.14}$$

Proof. We know that the supremum is at most equal to $\|A\|$, i.e. $\mathsf{W}(A)$ is confined to the disk. Now $\mathsf{W}(A)$ contains conv $[\sigma(A)]$, *a fortiori* it contains $\sigma(A)$ at least one point of which lies on the rim of the disk. This implies (11.2.14). ∎

In particular we get

Corollary 4. *If T is Hermitian, then at least one of the endpoints of the interval $[-\|T\|, \|T\|]$ belongs to $\sigma(T)$.*

There is an interesting consequence of Corollary 3 which is worth stating as

Lemma 11.2.3. *If A is normal and there exist $\mathbf{x}_0$ and λ_0 such that $\|\mathbf{x}_0\| = 1$, $(A\mathbf{x}_0, \mathbf{x}_0) = \lambda_0$ and $|\lambda_0| = \|A\|$, then $A\mathbf{x}_0 = \lambda_0\mathbf{x}_0$, so that $\lambda_0 \in P\sigma(A)$.*

Proof. We have

$$\|A\| = |\lambda_0| = |(A\mathbf{x}_0, \mathbf{x}_0)| \leqslant \|A\mathbf{x}_0\| \, \|\mathbf{x}_0\| \leqslant \|A\|. \tag{11.2.15}$$

This requires that equality holds throughout the relation. Now by Theorem 2.5.1 equality requires that there is a constant γ such that $A\mathbf{x}_0 = \gamma\mathbf{x}_0$ and since $(A\mathbf{x}_0, \mathbf{x}_0) = \lambda_0$ we must have $\gamma = \lambda_0$, so that $\lambda_0 \in P\sigma(A)$ as asserted. ∎

We can sharpen Corollary 4 in a direction which is important for the following.

Lemma 11.2.4. *For T Hermitian set*

$$m = \inf\{\lambda; \lambda \in \mathsf{W}(T)\}, \qquad M = \sup\{\lambda; \lambda \in \mathsf{W}(T)\}. \tag{11.2.16}$$

Then $\lambda = m$ and $\lambda = M$ belong to $\sigma(T)$ and $\sigma(T)$ is a subset of the interval $[m, M]$.

Proof. That $\sigma(T) \subset [m, M]$ follows from (11.2.10). We have only to prove that the endpoints of this interval belong to the spectrum. To prove this we use suitable affine transformations on T and the induced transformations on the numerical range and spectrum. Formula (12.2.2) shows how the numerical range is affected by an affine transformation. We have, similarly,

$$\sigma(sT + tI) = s\sigma(T) + t, \tag{11.2.17}$$

which holds for any $T \in \mathfrak{E}(\mathfrak{H})$. In the particular case where T is normal, $sT + tI$ is also normal and its spectral radius equals its norm. We now take $s = 1$, $t = -m$, so that

$$\mathsf{W}(T - mI) = \mathsf{W}(T) - m \subset [0, M - m].$$

Here $M \in \overline{\mathsf{W}(T)}$, so

$$M - m = \sup\{\lambda; \lambda \in \mathsf{W}(T - mI)\}.$$

In the other direction we have

$$0 \leqslant \inf\{\lambda; \lambda \in \mathsf{W}(T - mI)\}.$$

These two relations together with (11.2.14) give

$$\|T - mI\| = M - m,$$

and by Corollary 4 of Lemma 11.2.2 this implies

$$M - m \in \sigma(T - mI),$$

whence

$$M \in \sigma(T - mI) + m = \sigma(T).$$

Choosing $s = 1$, $t = -M$, we prove in the same manner that $m \in \sigma(T)$. ∎

Definition 11.2.1. *A Hermitian operator T is said to be positive if $0 \leqslant m$, positive definite if $0 < m$, negative if $M \leqslant 0$, and negative definite if $M < 0$. If T_1 and T_2 are two Hermitian operators, then $T_1 \leqslant T_2$ shall mean that $T_2 - T_1$ is positive.*

Lemma 11.2.5. *If A is normal and $A = B + iC$ where B and C are Hermitian, then*

$$\sigma(A) \subseteq \{\beta + i\gamma;\ \beta \in \sigma(B),\ \gamma \in \sigma(C)\}, \tag{11.2.18}$$

$$\mathsf{W}(A) \subseteq \mathsf{W}(B) + i\mathsf{W}(C). \tag{11.2.19}$$

Proof. Since B and C commute, I, B, C may be embedded in a commutative B-algebra $\mathfrak{A}^{cc}$ to which Gelfand's Representation Theorem 9.3.3 applies. If μ is a linear, multiplicative, and bounded functional on $\mathfrak{A}^{cc}$, then

$$\mu(A) = \mu(B) + i\mu(C) \tag{11.2.20}$$

by the linearity of the functional. Here $\mu(A)$, $\mu(B)$, $\mu(C)$ are spectral values of A, B, C, respectively. Moreover, the functional can be varied so that every spectral value is represented. This, however, does not mean that there is identity between the two sides of (11.2.18); inclusion is the best that can be asserted.

The inclusion (11.2.19) follows from the identity

$$(A\mathbf{x}, \mathbf{x}) = (B\mathbf{x}, \mathbf{x}) + i(C\mathbf{x}, \mathbf{x}). \ ∎$$

We can now prove

Theorem 11.2.3. *For normal operators the closure of the numerical range coincides with the convex hull of the spectrum.*

Proof. The convex hull of $\sigma(A)$ is the intersection of all closed half-planes which contain $\sigma(A)$. Let h be such a half-plane. Since an affine transformation obviously leaves inclusion properties invariant and also preserves normality of operators, we may assume that h is the closed right half-plane: $\mu \geqslant 0$ where $\lambda = \mu + i\nu$. If $A = B + iC$ is the canonical representation of A, we see that B is a positive operator. For B is Hermitian and if the spectrum of A lies in the closed right half-plane, then $\sigma(B)$ must lie on the positive real axis including the origin and this makes B positive.

It follows that the numerical range $\mathsf{W}(B)$ is real non-negative and this means that $\mathsf{W}(A)$ is located in the closed right half-plane. From this fact the general assertion follows. ▮

EXERCISE 11.2

1. A linear transformation from C^2 into itself is defined by the matrix $\mathcal{E}_{12} \in \mathfrak{M}_2$ which has a one in the place (1, 2) and zero elsewhere. Use the Euclidean metric in C^2 and find $\sigma(T)$ and $\mathsf{W}(T)$.
2. Let $\{\lambda_n; n = 0, \pm 1, \pm 2, \ldots\}$ be given and consider an operator T on $L_2(-\pi, \pi)$ defined by
$$T[f] \sim \sum_{-\infty}^{\infty} \lambda_n f_n e^{nit}, \qquad f \sim \sum_{-\infty}^{\infty} f_n e^{nit}.$$
How should the λ's be chosen so that $T \in \mathfrak{E}(L_2)$? If this condition is satisfied, find $\sigma(T)$ and $\mathsf{W}(T)$.
3. Prove that the operator is normal if it is in $\mathfrak{E}(L_2)$ and find its representation in terms of Hermitian operators.
4. Prove that an affine transformation V really has the two properties stated in the text: (1) If S_1 and S_2 are two sets in the complex plane and $S_1 \subseteq S_2$, then $V(S_1) \subseteq V(S_2)$. (2) If A is normal, so is $V(A)$.
5. Suppose that A is normal and $A = B + iC$. The spectrum of A is projected vertically on the real axis and horizontally on the imaginary axis. Show that the first projection coincides with $\sigma(B)$ and the second with $\sigma(C)$. Note that (11.2.18) is a weaker statement.
6. Suppose that T is Hermitian and that $\nu > 0$, $\lambda = \mu + i\nu$. Consider the resolvent $R(\lambda, T)$ and prove that it is a normal operator. Find its spectrum and numerical range. Show that both lie in the open lower half-plane. Here λ is fixed.
7. Prove that a Hermitian operator $T \neq 0$ cannot be quasi-nilpotent.
8. Prove that (11.2.12) is sufficient for A to be normal.
9. Prove that if $\mathbf{x}_1$ and $\mathbf{x}_2$ are characteristic vectors of a normal operator A corresponding to distinct characteristic values, then $(\mathbf{x}_1, \mathbf{x}_2) = 0$.
10. Let A be normal, $\mathbf{x}_0$ a characteristic vector with the characteristic value λ_0. Prove that $\mathbf{x}_0$ is also a characteristic vector of A^* corresponding to the characteristic value $\bar{\lambda}_0$.
11. Prove that if U is unitary, then $\sigma(U)$ is a subset of the unit circle $\{\lambda; |\lambda| = 1\}$.
12. Prove that the product of a finite number of unitary operators is unitary.

11.3 OPERATIONAL CALCULUS

The study of Hermitian operators, bounded as well as unbounded, started and had an explosive development around 1930 with work by J. von Neumann (1903–57) and M. H. Stone. Among the forerunners should be mentioned T. Carleman

(1892–1948), E. Hellinger (1883–1950), D. Hilbert, F. Riesz, and H. Weyl (1885–1955). While the work of von Neumann and of Stone can be said to be motivated by the requirements of quantum mechanics (or, equivalently, singular boundary value problems raised by physics), the forerunners were concerned with functions of infinitely many variables—in particular, infinite quadratic forms, singular boundary value problems, and singular integral equations.

Within the framework of this treatise there is no room for this vast theory in satisfactory detail, but we shall give a brief account of some of the basic facts essentially following the ideas developed by R. Riesz in 1930.

Let us revert to the theory of Hermitian matrices given in Section 1.7. In Exercise 1.7 the reader will find some of the problems which can serve as suggestive analogues for what must be done for Hermitian operators on a Hilbert space. Let $\mathfrak{H}$ be a Hermitian matrix in $\mathfrak{M}_n$, $\{\lambda_j\}$ its characteristic values,

$$\lambda_1 < \lambda_2 < \cdots < \lambda_n,$$

$\mathfrak{P}_j$ the idempotent matrix (= projection matrix) corresponding to λ_j. We have then

$$\mathfrak{H} = \lambda_1 \mathfrak{P}_1 + \lambda_2 \mathfrak{P}_2 + \cdots + \lambda_n \mathfrak{P}_n, \tag{11.3.1}$$

$$\mathfrak{R}(\lambda, \mathfrak{H}) = \frac{\mathfrak{P}_1}{\lambda - \lambda_1} + \frac{\mathfrak{P}_2}{\lambda - \lambda_2} + \cdots + \frac{\mathfrak{P}_n}{\lambda - \lambda_n}. \tag{11.3.2}$$

More generally, if $t \to F(t)$ is a function holomorphic on the spectrum of $\mathfrak{H}$, we can define

$$\mathcal{F}(\mathfrak{H}) = F(\lambda_1)\mathfrak{P}_1 + F(\lambda_2)\mathfrak{P}_2 + \cdots + F(\lambda_n)\mathfrak{P}_n. \tag{11.3.3}$$

Compare similar considerations for B-algebras in Section 9.4.

We can write these finite sums as Stieltjes integrals with a matrix-valued step-function as integrator. Set

$$\mathfrak{E}(t) = \begin{cases} \mathfrak{O}, \ -\infty < t \leqslant \lambda_1, \\ \mathfrak{P}_1 + \mathfrak{P}_2 + \cdots + \mathfrak{P}_k, \quad \lambda_k < t \leqslant \lambda_{k+1}, \\ \mathfrak{I}, \ \lambda_n < t < \infty. \end{cases} \tag{11.3.4}$$

Here k goes from 1 to $n - 1$ and $\mathfrak{I}$ denotes the unit matrix of $\mathfrak{M}_n$. Then

$$\mathcal{F}(\mathfrak{H}) = \int_{-\infty}^{\infty} F(t)\, d\mathfrak{E}(t) \tag{11.3.5}$$

or, equivalently, as an ordinary Stieltjes integral representing the Hermitian form

$$(\mathcal{F}(\mathfrak{H})\mathbf{x}, \mathbf{x}) = \int_{-\infty}^{\infty} F(t)\, d_t(\mathfrak{E}(t)\mathbf{x}, \mathbf{x}). \tag{11.3.6}$$

In both formulas as well as in (11.3.3) we can allow $\mathcal{F}$ to be merely continuous. The reader will have discovered that $F(t) = t$ in (11.3.1) and $(\lambda - t)^{-1}$ in (11.3.2).

These are the formulas to be generalized to Hermitian operators T on an arbitrary Hilbert space $\mathfrak{H}$. To each such T corresponds a *spectral function* $E(t)$.

This is an operator from $\mathfrak{H}$ into itself for all real t and belongs to $\mathfrak{E}(\mathfrak{H})$ if T is bounded. It defines a *resolution of the identity* (some authors say that it "is" a resolution of the identity). The operator $E(t)$ is non-decreasing with respect to t in the sense of Definition 11.2.1. It is a projection operator; $E(s)$ commutes with $E(t)$ and with T for all s and t. Further, $E(s)\,E(t) = E(s)$ for $s < t$ and $E(t) - E(s)$ is also a projection. If T is bounded and its spectrum is confined to the interval $[a, b]$, then $E(t) = 0$ for $t \leqslant a$ and $E(t) = I$ for $b < t$. The reader should verify that the operator $E(t)$ defined by (11.3.4) has all these properties.

We have now

$$\big(F(T)\mathbf{x}, \mathbf{x}\big) = \int_{-\infty}^{+\infty} F(t)\, d_t\big(E(t)\mathbf{x}, \mathbf{x}\big). \tag{11.3.7}$$

If T is bounded this holds for all $\mathbf{x} \in \mathfrak{H}$ and all functions F continuous on the spectrum of T. If T is unbounded, $\mathbf{x}$ and F are subject to limitations.

The problem is now to find the spectral function $E(t)$ and to study its properties. The basic idea of F. Riesz was to obtain $F(T)$ directly by a limiting process for positive continuous functions F and bounded Hermitian operators T and to reduce the problem of finding $E(T)$ to that of finding the operator T^+ corresponding to

$$t \to t^+ = \max(0, t). \tag{11.3.8}$$

In the matrix case $\mathcal{H}^+$ would be obtained by summing over the non-negative characteristic values in (11.3.1). The mapping $t \to F(t)$ corresponds to the mapping $T \to F(T)$ and *the mapping* $F(t) \to F(T)$ *has the fundamental property of preserving positivity.* The importance of such mappings has come to our attention repeatedly and this will not be the last time.

The mapping $F(t) \to F(T)$ is easily effected if F is a polynomial: in the expression for the polynomial we simply replace the scalar-valued variable t by the operator-valued T while numerical constants are replaced by I times the constant in question. To a sequence of scalar polynomials $\{F_n(t)\}$ corresponds a sequence of operator polynomials $\{F_n(T)\}$ and, as will be seen, monotonicity of the first sequence implies the same property for the second one in the sense of the partial ordering.

We shall need a number of auxiliary results concerning positive functions and positive operators. We start with a theorem due to F. Hausdorff (1921) concerning polynomials of the form

$$P(t) = \sum_{k=0}^{n} c_k(t-a)^k\,(b-t)^{n-k}, \tag{11.3.9}$$

where $-\infty < a < b < \infty$. For the special case $a = 0$, $b = 1$ we encountered such polynomials in the proof of Theorem 3.2.4 (Bernštein's Approximation Theorem). Let $\mathsf{P}(a, b)$ denote the set of all such polynomials for fixed a and b. Actually this class contains all polynomials and, moreover, any polynomial may be written in the form (11.3.9) in infinitely many ways. This is due to the fact that for all positive integers n we have

$$1 = (b-a)^{-n} \sum_{k=0}^{n} \binom{n}{k} (t-a)^k (b-t)^{n-k}. \tag{11.3.10}$$

It should also be noted that $\mathsf{P}(a, b)$ is an algebra. In particular, the product of two polynomials in $\mathsf{P}(a, b)$ is also in $\mathsf{P}(a, b)$. It is a question here of formal invariance. The product is obtained by termwise multiplication and collection of terms. Hausdorff's theorem is concerned with the subclass $\mathsf{P}^+(a, b)$ containing all polynomials of the form (11.3.9) with non-negative coefficients c_k. This class is closed under addition, multiplication by positive numbers, and element-wise multiplication.

Theorem 11.3.1. *Each polynomial which is positive in* $[a, b]$ *equals some element of* $\mathsf{P}^+(a, b)$. *Conversely, each element of* $\mathsf{P}^+(a, b)$ *is positive in* (a, b).

Proof. The converse part is obvious by inspection. To prove the direct part, we note that a polynomial real on the real axis admits of a unique factorization into linear and irreducible quadratic factors. If the polynomial is positive in $[a, b]$, then each of the factors may be assumed to be positive there. A linear factor is then of the form

$$L(t) = (b-a)^{-1} [C_1(t-a) + C_2(b-t)], \tag{11.3.11}$$

where $C_1 = L(b)$ and $C_2 = L(a)$, both positive numbers, so that $L \in \mathsf{P}^+[a, b]$. Quadratic factors require more elaborate manipulations. We write such a factor in the form

$$Q(t) = A + 2B(t-a) + C(t-a)^2, \qquad A > 0, \quad C > 0, \quad AC - B^2 > 0. \tag{11.3.12}$$

Here we multiply the three summands in $Q(t)$ by the right member of (11.3.10) with $n = p,\ p-1,\ p-2$, respectively, and where p is a positive integer to be determined. Thus

$$\begin{aligned} Q(t) &= A(b-a)^{-p} \sum_{m=0}^{p} \binom{p}{m} (t-a)^m (b-t)^{p-m} \\ &\quad + 2B(b-a)^{-p+1} \sum_{m=1}^{p} \binom{p-1}{m-1} (t-a)^m (b-t)^{p-m} \\ &\quad + C(b-a)^{-p+2} \sum_{m=2}^{p} \binom{p-2}{m-2} (t-a)^m (b-t)^{p-m} \\ &= (b-a)^{-p} \sum_{m=0}^{p} \frac{(p-2)!}{m!(p-m)!} C_m (t-a)^m (b-t)^{p-m}, \end{aligned}$$

where

$$C_m = Ap(p-1) + 2B(b-a)(p-1)m + C(b-a)^2 m(m-1). \tag{11.3.13}$$

This is a quadratic polynomial in m which will be positive for all values of m provided the discriminant is positive, which will obviously be the case for sufficiently large values of p. Thus $Q(t)$ has representatives in $\mathsf{P}^+(a, b)$. Since each factor is representable, so is the product, i.e. each polynomial which is positive in $[a, b]$ is equivalent to an element of $\mathsf{P}^+(a, b)$. ∎

We shall be concerned with sequences of polynomials $\{P_n(t)\}$ which converge uniformly to a given continuous function F for t in some interval $[a, b]$. It is no restriction to assume that the sequence is also monotone increasing (or decreasing). For if this is not the case at the outset, we can construct a new sequence with the desired properties as follows. We first select a rapidly convergent subsequence $\{P_{n_m}\}$ such that

$$|P_{n_m}(t) - F(t)| < 3^{-m}, \qquad \forall\, n, \quad a \leqslant t \leqslant b,$$

and set

$$Q_m(t) = P_{n_m}(t) - 2^{-m}.$$

Then

$$\begin{aligned} Q_{m+1}(t) - Q_m(t) &= P_{n_{m+1}}(t) - P_{n_m}(t) + 2^{-m-1} \\ &> -4(3^{-m-1}) + 2^{-m-1} > 0 \end{aligned} \tag{11.3.14}$$

for $m > 2$. Thus the sequence $\{Q_m(t);\, m > 2\}$ is increasing for all t in $[a, b]$ and obviously converges uniformly to $F(t)$.

We shall take up some properties of positive operators. We start by observing that the square of a Hermitian operator is positive since

$$(T^2\mathbf{x}, \mathbf{x}) = (T\mathbf{x}, T\mathbf{x}) = \|T\mathbf{x}\|^2. \tag{11.3.15}$$

A less obvious fact is that a positive operator has a positive square root as well as a positive pth root for any $p > 1$. The reader can construct a proof of this by carrying out the steps indicated in Problems 12 and 13 in Exercise 11.3 below.

Formula (11.3.15) implies that if T is Hermitian and S is positive and commutes with T, then $ST^2 = TST = T^2S$ is positive for

$$(TST\mathbf{x}, \mathbf{x}) = (ST\mathbf{x}, T\mathbf{x}) = (S\mathbf{y}, \mathbf{y}) \geqslant 0, \qquad \mathbf{y} = T\mathbf{x}. \tag{11.3.16}$$

F. Riesz observed that a positive operator S can be written as a sum of squares. He proved

Theorem 11.3.2. *If S is positive and if, for $\|\mathbf{x}\| = 1$, $\sup\,(S\mathbf{x}, \mathbf{x}) \leqslant 1$, then with*

$$A_1 = S, \qquad A_{n+1} = A_n - A_n^2, \qquad n = 1, 2, \ldots, \tag{11.3.17}$$

it is found that for each n the operators A_n and $I - A_n$ are positive and

$$S = \sum_{n=1}^{\infty} A_n^2 \tag{11.3.18}$$

in the sense that

$$\sum_{n=1}^{\infty} (A_n^2\mathbf{x}, \mathbf{x}) = (S\mathbf{x}, \mathbf{x}), \qquad \forall\, \mathbf{x}. \tag{11.3.19}$$

Remark. The assumption that $\sup (S\mathbf{x}, \mathbf{x}) \leqslant 1$ for $\|\mathbf{x}\| = 1$ simplifies the formulas and implies no restriction of the generality. If it is not satisfied at the outset, we simply divide S by the appropriate supremum and then proceed as indicated.

Proof. It is clear that $A_1 = S$ and $I - A_1$ are positive operators. We then proceed by induction. For $n > 1$

$$A_{n+1} = A_n^2(I - A_n) + A_n(I - A_n)^2, \qquad I - A_{n+1} = (I - A_n) + A_n^2,$$

i.e. by (11.3.16) the operators in the left members are sums of positive operators if A_n and $I - A_n$ are known to be positive. Since

$$S = A_1^2 + A_2^2 + \cdots + A_n^2 + A_{n+1}$$

is the sum of positive operators, we see that

$$\sum_{k=1}^{n} (A_k^2\mathbf{x}, \mathbf{x}) < (S\mathbf{x}, \mathbf{x}) \tag{11.3.20}$$

for all n. Thus the series in the left member of (11.3.17) is convergent for all $\mathbf{x} \in \mathfrak{H}$. That the sum of the series is $(S\mathbf{x}, \mathbf{x})$ follows from the fact that $(A_n^2\mathbf{x}, \mathbf{x}) = \|A_n\mathbf{x}\|^2$ goes to zero. ■

From this result we get

Theorem 11.3.3. *If S and T are positive operators which commute, then ST is also positive.*

Proof. Again supposing $\sup (S\mathbf{x}, \mathbf{x}) \leqslant 1$ for $\|\mathbf{x}\| = 1$ we have the representation (11.3.18) and hence also

$$(ST\mathbf{x}, \mathbf{x}) = \sum_{n=1}^{\infty} (A_n^2 T\mathbf{x}, \mathbf{x}), \tag{11.3.21}$$

where every term of the infinite series is non-negative. ■

Let us also remind the reader that if T is Hermitian and

$$a = \inf \{t; t \in \mathsf{W}(T)\} \leqslant \sup \{t; t \in \mathsf{W}(T)\} = b, \tag{11.3.22}$$

then the Hermitian operators

$$T - aI \quad \text{and} \quad bI - T \tag{11.3.23}$$

are positive.

We can now take up the main mapping theorems.

Theorem 11.3.4. *Suppose that the scalar polynomial $P(t)$ is positive in the interval $[a, b]$ and that T is a Hermitian operator satisfying* (11.3.22), *then the operator $P(T)$ is positive.*

Proof. By assumption $P(t)$ has a representation by an element of $\mathsf{P}^+(a, b)$.

Suppose that (11.3.9) is this representation where the coefficients $c_k \geqslant 0$. Then

$$P(T) = \sum_{k=0}^{m} c_k(T - aI)^k (bI - T)^{n-k}. \tag{11.3.24}$$

Here the factors $T - aI$ and $bI - T$ are positive operators which commute, so every term in the representation is a positive operator. Multiplication by positive numbers and adding preserves the positive character. ∎

Corollary. *If P and Q are two scalar polynomials such that $P(t) < Q(t)$ for $a \leqslant t \leqslant b$ and if T is a Hermitian operator satisfying* (11.3.22), *then $Q(T) - P(T)$ is a positive operator or, in other words, $P(T) \leqslant Q(T)$ in the partial ordering.*

This gives

Theorem 11.3.5. *If F is positive and continuous for $a \leqslant t \leqslant b$ and if T is a Hermitian operator satisfying* (11.3.22), *then there exists a unique operator $F(T)$ which is positive.*

Proof. Since F is continuous and positive we can find an increasing sequence of positive polynomials $\{P_n(t)\}$ which converges uniformly to $F(t)$ in $[a, b]$. This follows from Bernštein's approximation theorem together with the discussion leading up to formula (11.3.14). If $\max F(t) = K$, the sequence $\{K - P_n(t)\}$ is also made up of positive polynomials which decrease to $K - F(t)$. We also consider the two sequences $\{P_n{}^2(t)\}$ and $\{K^2 - P_n{}^2(t)\}$. The first increases to the limit $F^2(t)$, the second decreases to $K^2 - F^2(t)$. To these four sequences correspond four sequences of positive operators,

$$\{P_n(T)\}, \quad \{KI - P_n(T)\}, \quad \{P_n{}^2(T)\}, \text{ and } \{K^2I - P_n{}^2(T)\}.$$

The first and the third are increasing, the second and the fourth decreasing in the sense of the partial ordering. This implies corresponding monotonic properties for the associated Hermitian forms. Since

$$0 \leqslant (P_n{}^2(T)\mathbf{x}, \mathbf{x}) \leqslant K^2(\mathbf{x}, \mathbf{x}), \tag{11.3.25}$$

it is seen that the sequence

$$\{(P_n{}^2(T)\mathbf{x}, \mathbf{x})\} \tag{11.3.26}$$

is bounded and non-decreasing for fixed $\mathbf{x}$ with $\|\mathbf{x}\| = 1$. Hence it has a limit $\leqslant K^2$. Further, for $m < n$

$$P_m{}^2(t) < P_m(t)\,P_n(t) < P_n{}^2(t), \qquad \forall\, t, \qquad a \leqslant t \leqslant b,$$

and this implies

$$(P_m{}^2(T)\mathbf{x}, \mathbf{x}) < (P_m(T)\,P_n(T)\mathbf{x}, \mathbf{x}) < (P_n{}^2(T)\mathbf{x}, \mathbf{x}),$$

and, as $m, n \to \infty$, all three members converge to the same limit. This in its turn implies that

$$([P_n(T) - P_m(T)]^2\mathbf{x}, \mathbf{x}) = \|P_n(T)\mathbf{x} - P_m(T)\mathbf{x}\|^2 \to 0.$$

Thus $\{P_n(T)\mathbf{x}\}$ is a Cauchy sequence in $\mathfrak{H}$ for each $\mathbf{x} \in \mathfrak{H}$. Moreover, $\|P_n(T)\|$ is bounded by K for all n. Hence $P_n(T)$ converges strongly to an element $F(T)$ of $\mathfrak{E}(\mathfrak{H})$.

We have shown that starting with a Cauchy sequence $P_n(t)$ in $C[a, b]$ which converges to $F(t)$ we obtain an operator $F(T)$. This operator, however, might conceivably depend upon what particular Cauchy sequence we start out with. That this is not the case is seen as follows. Suppose that $\{Q_n(t)\}$ is another monotone increasing sequence of positive polynomials converging to $F(t)$ uniformly in $[a, b]$. To this sequence corresponds an operator sequence $\{Q_n(T)\}$ and a limit, the operator $G(T)$. It should be shown that $F(T) = G(T)$. To this end we need the following simple observation. For each m we can find an $n > m$ such that

$$\begin{aligned} P_m(t) - 2^{-m} &< Q_n(t), \qquad \forall\, t, \\ Q_m(t) - 2^{-m} &< P_n(t), \qquad \forall\, t. \end{aligned} \tag{11.3.27}$$

Note that if one integer n will do for the first inequality, then any larger n will also do. This implies for each $\mathbf{x}$ with $\|\mathbf{x}\| = 1$

$$((P_m(T) - 2^{-m}I)\mathbf{x}, \mathbf{x}) < (Q_n(T)\mathbf{x}, \mathbf{x})$$

or

$$(P_m(T)\mathbf{x}, \mathbf{x}) - 2^{-m} < (Q_n(T)\mathbf{x}, \mathbf{x})$$

so that for $m \to \infty$

$$(F(T)\mathbf{x}, \mathbf{x}) \leqslant (G(T)\mathbf{x}, \mathbf{x}).$$

The opposite inequality follows in the same manner and from

$$(F(T)\mathbf{x}, \mathbf{x}) = (G(T)\mathbf{x}, \mathbf{x})$$

it follows that $F(T)$ and $G(T)$ are identical operators. Thus the mapping $F(t) \leftrightarrow F(T)$ is unique. ▮

So far $F(t)$ was restricted to be positive on the interval $[a, b]$ but the extension to arbitrary real or complex-valued continuous functions follows the usual pattern. Write

$$F(t) = F_1(t) + iF_2(t) - F_3(t) - iF_4(t), \tag{11.3.28}$$

where each $F_j(t)$ is non-negative. Then $F_j(T)$ is a uniquely defined element of $\mathfrak{E}(\mathfrak{H})$ and we define

$$F(T) = F_1(T) + iF_2(T) - F_3(T) - iF_4(T). \tag{11.3.29}$$

Further extensions to discontinuous functions such as step functions are feasible. A step function $S(t)$ can be approximated by an increasing sequence of continuous functions each of which can be approximated from below by polynomials with any desired degree of accuracy. We can thus select an increasing sequence of polynomials $\{P_n(t)\}$ which converges boundedly to $S(t)$. The corresponding sequence $\{P_n(T)\}$ converges to a bounded operator $S(T)$. In particular, if $S(t)$ is the characteristic function of an interval J, i.e. $S(t)$ is 1 or 0 according as $t \in J$ or not, then $S(T)$ is a projection operator since $S^2(t) = S(t)$. Here $S(T)$ is the zero operator if $J \cap [a, b] = \emptyset$.

To fix the ideas, suppose that

$$a < 0 < b \tag{11.3.30}$$

and consider the characteristic functions of the intervals $[a, 0)$ and $[0, b)$. Denote the corresponding operators by $E^-(T) = E^-$ and $E^+(T) = E^+$, respectively. They are projection operators and add up to the unit operator

$$E^-(T) + E^+(T) = I. \tag{11.3.31}$$

They commute with each other and with T as well as with all $F(T)$ constructed above. We set

$$TE^-(T) = T^-, \qquad TE^+(T) = T^+. \tag{11.3.32}$$

Then

$$T^- + T^+ = T \tag{11.3.33}$$

and the following properties hold.

Theorem 11.3.6. *T^- is a negative operator, T^+ a positive one. Any element $\mathbf{x} \in \mathfrak{H}$ such that $T\mathbf{x} = 0$ satisfies*

$$E^-\mathbf{x} = \mathbf{0}, \qquad E^+\mathbf{x} = \mathbf{x}. \tag{11.3.34}$$

Proof. To prove that T^+ is a positive operator we have to prove that $(T^+\mathbf{x}, \mathbf{x}) \geqslant 0$ for all $\mathbf{x}$. We observe that T^+ is the image of the function $t \to t^+ = \max(0, t)$. This function is nowhere negative. Thus for any given $\delta > 0$, an increasing sequence of polynomials $\{P_n(t)\}$ converging to t^+ must satisfy $P_n(t) > -\delta$ for $n > N_\delta$ and this implies $P_n(T) \geqslant -\delta I$ and

$$(P_n(T)\mathbf{x}, \mathbf{x}) \geqslant -\delta(\mathbf{x}, \mathbf{x}), \qquad N_\delta < n.$$

Hence

$$(T^+\mathbf{x}, \mathbf{x}) \geqslant -\delta(\mathbf{x}, \mathbf{x})$$

and $(T^+\mathbf{x}, \mathbf{x}) \geqslant 0$ since δ is arbitrary. T^- is handled in the same manner. For the proof of (11.3.34) we need a partial generalization of the spectral mapping theorem.

Lemma 11.3.1. *If A is Hermitian and if $A\mathbf{x}_0 = \lambda_0\mathbf{x}_0$ where $\|\mathbf{x}_0\| = 1$, then for any function F, continuous on $[a, b]$ or the limit of continuous functions, we have*

$$F(A)\mathbf{x}_0 = F(\lambda_0)\mathbf{x}_0. \tag{11.3.35}$$

Proof. By the "classical" spectral mapping theorem, more precisely the fine structure theorem, this holds for polynomials, and hence for any F which is the limit of polynomials on $[a, b]$. ∎

Continuation of the proof of Theorem 11.3.6. We now return to (11.3.34). If $T\mathbf{x}_0 = \mathbf{0}$, then $\lambda_0 = 0$ in the preceding lemma. If F^- and F^+ are the scalar functions corresponding to $E^-(T)$ and $E^+(T)$, respectively, then

$$E^-(T)\mathbf{x}_0 = F^-(0)\mathbf{x}_0, \qquad E^+(T)\mathbf{x}_0 = F^+(0)\mathbf{x}_0. \tag{11.3.36}$$

We now note that as long as we are working with bounded Hermitian operators, that is bounded intervals $[a, b]$, then $F^-(0)$ and $F^+(0)$ are independent of the choice of A in the lemma. But if A is the zero operator, then

$$E^-(0) = 0, \qquad E^+(0) = I. \tag{11.3.37}$$

Substituting this in (11.3.36) we see that

$$F^-(0) = 0, \qquad F^+(0) = 1, \tag{11.3.38}$$

and this gives (11.3.34). ∎

F. Riesz refers to Theorem 11.3.6 and the preceding formulas as the "splitting theorem" (= Zerlegungssatz). From it the spectral resolution is obtained by applying the theorem to the family of operators $T - tI$, $-\infty < t < \infty$. Set

$$E(t) = E^-(T - tI). \tag{11.3.39}$$

This defines a family of operators with properties most of which were given in the preview earlier in this section.

Theorem 11.3.7. *The operators $E(t)$ are projections for each fixed t, the zero projection for $t \leqslant a$, the identity operator for $b < t$. Further, $E(s)$ commutes with $E(t)$, with T and with any admissible $F(T)$. For $s < t$*

$$E(s) \leqslant E(t), \qquad E(s)\,E(t) = E(s), \tag{11.3.40}$$

and $E(t) - E(s)$ is a projection. $E(t)$ is left-continuous,

$$\lim_{s \uparrow t} E(s) = E(t). \tag{11.3.41}$$

Finally,

$$(tI - T)\mathbf{x} = \mathbf{0} \text{ implies } E(t)\mathbf{x} = \mathbf{0},\ [I - E(t)]\mathbf{x} = \mathbf{x}. \tag{11.3.42}$$

Proof. That $E(t)$ is a projection follows from the fact that $E^-(A)$ is a projection for any choice of A, in particular for $A = T - tI$. Similarly, (11.3.42) is just a special case of (11.3.34). The commutativity properties are obtained by observing that $E(s)$ commutes with $sI - T$ and hence with $tI - T$, which commutes with $E(t)$. In fact, any two operators $F_1(T)$ and $F_2(T)$ commute since the polynomial operators commute.

The second formula under (11.3.40) is equivalent to

$$[I - E(t)]E(s) = 0.$$

Denote the left member by P. It is obviously a projection. It should be noted that

$$E(s)P = P, \qquad [I - E(t)]P = P.$$

The splitting theorem asserts that for any Hermitian operator A, the operators $AE^{-}(A)$ and $AE^{+}(A)$ are negative and positive operators respectively. We apply this observation to the two operators $T - sI$ and $T - tI$ with $s < t$. Further, for any $\mathbf{x} \in \mathfrak{H}$ we form $\mathbf{y} = P\mathbf{x}$. Then

$$\big(E(s)\,(T - sI)\mathbf{y}, \mathbf{y}\big) \leqslant 0, \qquad \big([I - E(t)](T - tI)\mathbf{y}, \mathbf{y}\big) \geqslant 0.$$

By the special choice of $\mathbf{y}$ these formulas simplify to

$$\big((T - s\,I)\mathbf{y}, \mathbf{y}\big) \leqslant 0, \qquad \big((T - t\,I)\mathbf{y}, \mathbf{y}\big) \geqslant 0$$

or

$$t(\mathbf{y}, \mathbf{y}) \leqslant (T\mathbf{y}, \mathbf{y}) \leqslant s(\mathbf{y}, \mathbf{y}).$$

Since $s < t$ this holds iff $(\mathbf{y}, \mathbf{y}) = \|P\mathbf{x}\|^2 = 0$. Hence $P\mathbf{x} = \mathbf{0}$ for all $\mathbf{x}$ and P is the zero operator.

This gives

$$\begin{aligned}[E(t) - E(s)]^2 &= E^2(t) - 2\,E(s)\,E(t) + E^2(s)\\ &= E(t) - 2\,E(s) + E(s)\\ &= E(t) - E(s),\end{aligned}$$

so $E(t) - E(s)$ is indeed a projection. Since projections are positive operators, $E(s) \leqslant E(t)$ as asserted.

The left hand continuity also follows from the splitting theorem. Since $E(s)$ is non-decreasing,

$$\lim_{s \uparrow t} E(s) \equiv E(t - 0) \tag{11.3.43}$$

exists in the strong sense. Set

$$\Delta = \Delta_{st} = E(t) - E(s), \qquad \Delta_0 = E(t) - E(t - 0).$$

It is clear that Δ_0 is Hermitian and a projection. It should be proved that $\Delta_0 = 0$. From

$$E(t)\,\Delta = \Delta, \qquad E(s)\Delta = 0, \tag{11.3.44}$$

it follows that

$$E(t)\,\Delta_0 = \Delta_0, \qquad E(t - 0)\,\Delta_0 = 0. \tag{11.3.45}$$

The splitting theorem gives

$$\big(E(t)(T - tI)\,\Delta\mathbf{x}, \Delta\mathbf{x}\big) \geqslant 0, \qquad \big([I - E(s)]\,(T - sI)\,\Delta\mathbf{x}, \Delta\mathbf{x}\big) \leqslant 0,$$

so that by (11.3.44)

$$\big((T - tI)\,\Delta\mathbf{x}, \Delta\mathbf{x}\big) \geqslant 0, \qquad \big((T - sI)\,\Delta\mathbf{x}, \Delta\mathbf{x}\big) \leqslant 0.$$

Letting $s \uparrow t$, we see that $\Delta\mathbf{x} \to \Delta_0\mathbf{x}$ and the resulting double inequality implies the equality

$$(T\Delta_0\mathbf{x}, \Delta_0\mathbf{x}) = t(\Delta_0\mathbf{x}, \Delta_0\mathbf{x}).$$

Here we use the Hermitian character of Δ_0 and its being a projection commuting with T to obtain

$$(T\Delta_0\mathbf{x}, \mathbf{x}) = t(\Delta_0\mathbf{x}, \mathbf{x}).$$

Since this holds for all $\mathbf{x}$ we must have

$$T(\Delta_0 \mathbf{x}) = t(\Delta_0 \mathbf{x}).$$

This asserts that $\mathbf{y} = \Delta_0 \mathbf{x}$ is a characteristic vector of the operator T corresponding to the characteristic value $\lambda = t$. We now invoke Theorem 11.3.6 with T replaced by $T - tI$ to obtain

$$0 = E(t)\mathbf{y} = E(t)\,\Delta_0 \mathbf{x} = \Delta_0 \mathbf{x}$$

by (11.3.45). Since this holds for all $\mathbf{x}$ we see that $\Delta_0 = 0$. ∎

So far T was supposed to be a bounded Hermitian operator. F. Riesz considered also the case of unbounded operators. *Suppose that T is a closed operator with domain $\mathfrak{D}(T)$ dense in $\mathfrak{H}$ and $\mathfrak{R}(T) \subset \mathfrak{H}$.* Suppose that

$$(T\mathbf{x}, \mathbf{y}) = (\mathbf{x}, T\mathbf{y}) \tag{11.3.46}$$

for all $\mathbf{x}$ and $\mathbf{y}$ in $\mathfrak{D}(T)$. We still refer to T as Hermitian.

The basic fact in the study of such operators is the observation that the equation

$$(iI - T)\mathbf{y} = \mathbf{x} \tag{11.3.47}$$

has a unique solution $\mathbf{y}$ for each $\mathbf{x} \in \mathfrak{H}$. In other words, $R(i, T)$ exists as an element of $\mathfrak{E}(\mathfrak{H})$. This gives as a necessary and sufficient condition for, $\mathbf{z} \in \mathfrak{D}(T)$, the existence of an $\mathbf{x} \in \mathfrak{H}$ such that

$$\mathbf{z} = R(i, T)\mathbf{x}. \tag{11.3.48}$$

F. Riesz could extend his splitting theorem to such operators T and he showed the existence of a spectral function $E(t)$ with essentially the same properties as in the bounded case. Since the spectrum of T is now an unbounded real point set, the operator $E(t)$ is not constant outside of a finite interval $[a, b]$. Instead we have the limit relations

$$\lim_{t \downarrow -\infty} E(t)\mathbf{x} = \mathbf{0}, \qquad \lim_{t \uparrow \infty} E(t)\mathbf{x} = \mathbf{x} \tag{11.3.49}$$

for all $\mathbf{x}$. Thus $E(t)$ converges strongly to 0 and I as $t \to -\infty$ and $+\infty$, respectively.

The other properties of $E(t)$ are preserved. Thus the spectral function satisfies (1) $E(t)$ is a projection operator for each t, (ii) $E(s)$ commutes with $E(t)$, (iii) for $s < t$, $E(s) < E(t)$, (iv) $E(s)\,E(t) = E(s)$ for $s < t$, (v) $E(t) - E(s)$ is a projection, and $\lim_{s \uparrow t} E(s) = E(t)$. We shall refer to these results as Theorem 11.3.8. It is given without a proof.

EXERCISE 11.3

1. Verify (11.3.13) and that the discriminant of the polynomial in m is positive for sufficiently large values of p.
2. Fill in missing details in the proof of Theorem 11.3.2.

3. If sup $(S\mathbf{x}, \mathbf{x}) > 1$, how should the formulas and the proof be modified in Theorem 11.3.2?

4. Verify (11.3.31) and the commuting properties claimed for these operators.

5. Verify (11.3.37) and show that $F^{-}(0)$ and $F^{+}(0)$ are indeed independent of the choice of A under the stated conditions.

6. Fill in missing details in the proof of Theorem 11.3.7.

7. Show that the equation $(iI - T)\,\mathbf{y} = \mathbf{0}$, T Hermitian, has $\mathbf{y} = \mathbf{0}$ as its only solution, so that the solution of (11.3.47) is unique if it exists.

8. In equation (11.3.47) the domain of $iI - T$ is $\mathfrak{D}(T)$. Show that the range $\mathfrak{R}$ is dense in $\mathfrak{H}$. [*Hint*: If not, there would exist $\mathbf{z} \in \mathfrak{H}$, $\|\mathbf{z}\| = 1$ such that $\mathbf{z}$ is orthogonal to all of $\mathfrak{R}$. This leads to a contradiction since $(T\mathbf{y}, \mathbf{y})$ is real. Details?]

9. If $\{\mathbf{x}_n\} \subset \mathfrak{R}$ is a Cauchy sequence with $\mathbf{x}_n \to \mathbf{x}_0$, show that $\{\mathbf{y}_n\}$ and $\{T\mathbf{y}_n\}$ are also Cauchy sequences, so that by the fact that T is a closed operator $\mathbf{x}_0 \in \mathfrak{R}$ and $\mathfrak{R} = \mathfrak{H}$.

10. Show that (11.3.47) has a unique solution $\mathbf{y}$ for all $\mathbf{x}$ and verify that $\mathbf{y} = R(i, T)\,\mathbf{x}$ and that $R(i, T) \in \mathfrak{E}(\mathfrak{H})$.

11. Prove (11.3.49).

12. Suppose that T is a positive operator, $T \in \mathfrak{E}(\mathfrak{H})$, and that sup $(T\mathbf{x}, \mathbf{x}) = M$ for $\|\mathbf{x}\| = 1$. Define

$$S_2 = \sum_{n=0}^{\infty} (-1)^n \binom{\frac{1}{2}}{n} M^{\frac{1}{2}-n} (MI - T)^n.$$

Show that (1) the series converges in the uniform operator topology, (2) S_2 is positive, (3) sup $(S_2\mathbf{x}, \mathbf{x}) = M^{\frac{1}{2}}$ for $\|\mathbf{x}\| = 1$, and (4) $S_2{}^2 = T$ so that S_2 is a positive square root of T.

13. In the preceding problem replace 2 by p, $\frac{1}{2}$ by $1/p$, where p is a positive integer. Carry through a similar argument with a view of showing that S_p is a positive pth root of T.

14. Let $\{\lambda_n\}$ be any sequence of real numbers and suppose that

$$\sum_{n=1}^{\infty} a_n e^{\lambda_n it} \equiv f(t),$$

where the series is uniformly convergent in $(-\infty, \infty)$. Consider the space of all such functions and adjoin to it the limits of uniformly convergent sequences $\{f_n(t)\}$. Show that the so completed space is a Hilbert space under the inner product

$$(f, g) = \lim_{a\to\infty} (2a)^{-1} \int_{-a}^{a} f(t)\,\overline{g(t)}\,dt.$$

If λ is any real number show that $\exp(\lambda it) \in \mathfrak{H}$ and that $\{\exp(\lambda it);\ -\infty < \lambda < +\infty\}$ is a non-denumerable orthonormal system. Is $\mathfrak{H}$ separable? This is the space of *almost periodic functions* in the sense of Harald Bohr (1887–1951).

11.4 THE SPECTRAL THEOREM

We have now the basic properties of the spectral function $E(t)$ corresponding to a Hermitian operator T which is closed and has its domain $\mathfrak{D}(T)$ dense in $\mathfrak{H}$. We have to define and discuss the Stieltjes integrals which figured in formula (11.3.7) and which form the core of the Spectral Theorem for Hermitian operators.

Our point of departure is the operator

$$\Delta = \Delta_{st} = E(t) - E(s), \qquad s < t \tag{11.4.1}$$

introduced in the proof of Theorem 11.3.7. By Theorem 11.3.8 its properties are essentially unaffected by the passage from a bounded to an unbounded Hermitian operator. We note first that $(T - tI)\Delta$ is a negative bounded Hermitian operator while $(T - sI)\Delta$ is positive and bounded. It follows that

$$\Delta T = T\Delta$$

is also bounded and Hermitian. Further,

$$s(\Delta \mathbf{x}, \mathbf{x}) \leqslant (T\Delta \mathbf{x}, \mathbf{x}) \leqslant t(\Delta \mathbf{x}, \mathbf{x}).$$

If now $s < \lambda < t$, we have

$$|((T - \lambda I)\,\Delta \mathbf{x}, \mathbf{x})| \leqslant (t - s)\,(\Delta \mathbf{x}, \mathbf{x}) = (t - s)\|\Delta \mathbf{x}\|^2,$$

so that

$$\|\lambda \Delta \mathbf{x} - T\Delta \mathbf{x}\| \leqslant (t - s)\|\Delta \mathbf{x}\|. \tag{11.4.2}$$

Next we recall that the operators Δ_{st} are projections, so that

$$(\Delta_{st})^2 = \Delta_{st}. \tag{11.4.3}$$

Moreover, if (s_1, s_2) and (t_1, t_2) are non-overlapping intervals having at most an endpoint in common, say

$$s_1 < s_2 \leqslant t_1 < t_2,$$

and if Δ_1 and Δ_2 are the corresponding projections, then

$$\begin{aligned}\Delta_1\Delta_2 &= [E(s_2) - E(s_1)]\,[E(t_2) - E(t_1)]\\ &= E(s_2)\,E(t_2) - E(s_2)\,E(t_1) - E(s_1)\,E(t_2) + E(s_1)\,E(t_1)\\ &= E(s_2) - E(s_2) - E(s_1) + E(s_1) = 0,\end{aligned}$$

so that the projections are orthogonal.

We now consider a partition of the real axis into a countable number of intervals (s_k, t_k) with no clusterpoint in the finite domain. Here $t_k = s_{k+1}$ for all k. Set

$$\Delta_k = E(t_k) - E(s_k). \tag{11.4.4}$$

The projections $\{\Delta_k\}$ form a complete orthonormal system in the sense that

$$\Delta_k{}^2 = \Delta_k, \qquad \Delta_j\Delta_k = 0, \qquad j \neq k, \tag{11.4.5}$$

and

$$\Sigma\Delta_k = I, \tag{11.4.6}$$

where the series on the left converges in the uniform operator topology. Then if $\mathbf{x} \in \mathfrak{H}$ we set

$$\mathbf{x}_k = \Delta_k \mathbf{x}, \qquad \Sigma \mathbf{x}_k = \mathbf{x} \tag{11.4.7}$$

and note that

$$(\mathbf{x}_j, \mathbf{x}_k) = 0, \qquad \Sigma \|\mathbf{x}_k\|^2 = \|\mathbf{x}\|^2 \tag{11.4.8}$$

since

$$\Sigma \|\mathbf{x}_k\|^2 = \Sigma (\Delta_k \mathbf{x}, \Delta_k \mathbf{x}) = \Sigma (\Delta_k \mathbf{x}, \mathbf{x}) = (\mathbf{x}, \mathbf{x}).$$

If now $\mathbf{x} \in \mathfrak{D}(T)$ we set $T\mathbf{x} = \mathbf{y}$ and note corresponding representations for $\mathbf{y}$. If $\mathbf{y}_k = \Delta_k \mathbf{y}$, then

$$\mathbf{y}_k = \Delta_k \mathbf{y} = \Delta_k T\mathbf{x} = T\Delta_k \mathbf{x} = T\mathbf{x}_k$$

and

$$T\mathbf{x} = \Sigma\, T\mathbf{x}_k, \qquad \Sigma \|T\mathbf{x}_k\|^2 = \|T\mathbf{x}\|^2. \tag{11.4.9}$$

The convergence of the last series is necessary and sufficient for $\mathbf{x} \in \mathfrak{D}(T)$. Suppose this condition is satisfied and let us show that $\mathbf{x} \in \mathfrak{D}(T)$. For $j \neq k$

$$(T\Delta_j)(T\Delta_k) = T(\Delta_j T\Delta_k) = T^2(\Delta_j \Delta_k) = 0.$$

Thus the elements $T\mathbf{x}_k = T\Delta_k \mathbf{x}$ of $\mathfrak{H}$ are orthogonal and

$$\left\| \sum_{-p}^{p} T\mathbf{x}_k - \sum_{-n}^{n} T\mathbf{x}_k \right\|^2 = \left\| \sum_{n+1 \leqslant |k| \leqslant p} T\mathbf{x}_k \right\|^2 = \sum_{n+1 \leqslant |k| \leqslant p} \|T\mathbf{x}_k\|^2,$$

where $n < p$. By the assumed convergence of the second series in (11.4.9) the last display tends to zero as $n \to \infty$. It follows that the first series in (11.4.9) converges to a limit, $\mathbf{z}$ say. If now $\mathbf{u}$ is any element of $\mathfrak{D}(T)$ we have

$$(T\mathbf{u}, \mathbf{x}_1 + \mathbf{x}_2 + \cdots + \mathbf{x}_n) = (\mathbf{u}, T\mathbf{x}_1 + T\mathbf{x}_2 + \cdots + T\mathbf{x}_n).$$

As $n \to \infty$ the left member converges to $(T\mathbf{u}, \mathbf{x})$, the right to $(\mathbf{u}, \mathbf{z})$, so that $(T\mathbf{u}, \mathbf{x}) = (\mathbf{u}, \mathbf{z})$ and $\mathbf{z} = T\mathbf{x}$ since T is Hermitian. Hence $\mathbf{x} \in \mathfrak{D}(T)$.

Suppose that the partition of the real axis is so fine that all intervals are at most δ in length where δ is a given positive number. Let $s_k \leqslant \lambda_k \leqslant t_k$ for all k and let $\mathbf{x} \in \mathfrak{D}(T)$. Then with the notation as above

$$\|T\mathbf{x}_k - \lambda_k \mathbf{x}_k\| = \|T\Delta_k \mathbf{x} - \lambda_k \Delta_k \mathbf{x}\| \leqslant (t_k - s_k)\|\mathbf{x}_k\| \leqslant \delta \|\mathbf{x}_k\|,$$

so that

$$\Sigma \|T\mathbf{x}_k - \lambda_k \mathbf{x}_k\|^2 \leqslant \delta^2 \|\mathbf{x}\|^2. \tag{11.4.10}$$

Consider the three series

$$\Sigma\,[T\mathbf{x}_k - \lambda_k \mathbf{x}_k], \qquad \Sigma\, T\mathbf{x}_k, \qquad \Sigma\, \lambda_k \mathbf{x}_k. \tag{11.4.11}$$

The first and the second series converge in norm to elements of $\mathfrak{H}$, the sum in the second case being $T\mathbf{x}$. It follows that the third series also converges in norm so that the convergence of

$$\Sigma\, \lambda_k^2 \|\mathbf{x}_k\|^2 = \Sigma\, \lambda_k^2 (\Delta_k \mathbf{x}, \mathbf{x}) \tag{11.4.12}$$

is necessary and sufficient for $\mathbf{x} \in \mathfrak{D}(T)$. By (11.4.10) for any $\mathbf{y} \in \mathfrak{H}$

$$\|(T\mathbf{x}, \mathbf{y}) - \Sigma\, \lambda_k(\Delta_k\mathbf{x}, \mathbf{y})\| \leqslant \delta\|\mathbf{x}\|\;\|\mathbf{y}\|. \tag{11.4.13}$$

Now to any positive δ we have Riemann–Stieltjes sums

$$S_1(\delta) = \Sigma\, \lambda_k^{\,2}\big(E(t_k)\mathbf{x} - E(s_k)\mathbf{x}, \mathbf{x}\big),$$
$$S_2(\delta) = \Sigma\, \lambda_k\big(E(t_k)\mathbf{x} - E(s_k)\mathbf{x}, \mathbf{y}\big),$$

and by the preceding estimates (11.4.10) and (11.4.13) any sequence of such sums with $\delta \downarrow 0$ converges to

$$\int_{-\infty}^{\infty} s^2 d\big(E(s)\mathbf{x}, \mathbf{x}\big) \quad \text{and} \quad \int_{-\infty}^{\infty} s d\big(E(s)\mathbf{x}, \mathbf{y}\big),$$

respectively. Thus we have proved

Theorem 11.4.1. *A necessary and sufficient condition for* $\mathbf{x}$ *to belong to* $\mathfrak{D}(T)$ *is the convergence of*

$$\int_{-\infty}^{\infty} s^2 d\big(E(s)\mathbf{x}, \mathbf{x}\big), \tag{11.4.14}$$

and for any such $\mathbf{x}$ *and any* $\mathbf{y} \in \mathfrak{H}$

$$(T\mathbf{x}, \mathbf{y}) = \int_{-\infty}^{\infty} s d\big(E(s)\mathbf{x}, \mathbf{y}\big). \tag{11.4.15}$$

These are ordinary Riemann–Stieltjes integrals. With $\mathfrak{H}$-valued integrals we have

$$T\mathbf{x} = \int_{-\infty}^{\infty} s dE(s)\mathbf{x} \tag{11.4.16}$$

for any $\mathbf{x}$ satisfying (11.4.14). If T is a bounded operator, then this condition is trivially satisfied for all $\mathbf{x} \in \mathfrak{H}$.

Conversely, suppose that an operator function $E(s) \in \mathfrak{E}(\mathfrak{H})$ is defined for all values of s, $-\infty < s < +\infty$, and that $E(s)$ has the properties of a spectral function as stated in Theorem 11.3.8. Then a linear operator T is defined by (11.4.16) for all $\mathbf{x}$ satisfying (11.4.14). It may be shown that T is Hermitian and that $E(s)$ is its spectral function.

We can generalize Theorem 11.4.1 in various ways. A simple extension is the following. Suppose that $F(s) \in C[-\infty, \infty]$, then

$$\int_{-\infty}^{\infty} F(s)\, d\big(E(s)\mathbf{x}, \mathbf{y}\big) \tag{11.4.17}$$

is well defined for all $\mathbf{x}, \mathbf{y} \in \mathfrak{H}$. We set

$$F(T)\mathbf{x} = \int_{-\infty}^{\infty} F(s)\, d\big(E(s)\mathbf{x}\big). \tag{11.4.18}$$

This defines a linear bounded operator $F(T) \in \mathfrak{E}(\mathfrak{H})$ which is Hermitian. This definition represents an extension of the Operational Calculus of Section 11.3 to unbounded operators. It will be proved below that for a bounded operator the two definitions coincide.

Theorem 11.4.2. *If* $F \in C[-\infty, \infty]$ *and* $\mathbf{x} \in \mathfrak{D}(T)$, *then*

$$F(T)T\mathbf{x} = TF(T)\mathbf{x} = \int_{-\infty}^{\infty} sF(s)\, d(E(s)\mathbf{x}). \tag{11.4.19}$$

Proof. Since $\mathbf{x} \in \mathfrak{D}(T)$ and F is bounded by its C-norm $\|F\|_C$ we have

$$\int_{-\infty}^{\infty} s^2 |F(s)|^2\, d(E(x)\mathbf{x}, \mathbf{x}) \leqslant \|F\|_C{}^2 \int_{-\infty}^{\infty} s^2\, d(E(s)\mathbf{x}, \mathbf{x}),$$

which exists by (11.4.14). It follows (by the Bounyakovsky–Schwarz inequality) that the integral in (11.4.19) exists. On the other hand, $F(T)$ is bounded and $\mathbf{x} \in \mathfrak{D}(T)$, so that $F(T)\, T\mathbf{x}$ exists and is the limit of Riemann–Stieltjes sums (with notation as above),

$$\Sigma\, F(\lambda_k)\, [E(t_k)T\mathbf{x} - E(s_k)T\mathbf{x}] = T\{\Sigma\, F(\lambda_k)\, [E(t_k)\mathbf{x} - E(s_k)\mathbf{x}]\}.$$

For a sequence of increasingly refined partitions, the sum on the left converges to $F(T)T\mathbf{x}$ while the sum on the right converges to $F(T)\mathbf{x}$. Since T is closed, this gives

$$[F(T)\, T]\mathbf{x} = T[F(T)\mathbf{x}]$$

for $\mathbf{x} \in \mathfrak{D}(T)$. Thus $F(T)\mathbf{x} \in \mathfrak{D}(T)$ and the operators T and $F(T)$ commute on $\mathfrak{D}(T)$.

Next, note that

$$\Sigma\, F(\lambda_k)[E(t_k) - E(s_k)]\, [T\mathbf{x} - \lambda_k\mathbf{x}] = \Sigma\, F(\lambda_k)\, [T\mathbf{x}_k - \lambda_k\mathbf{x}_k].$$

The square of the norm of the last expression does not exceed

$$\|F\|_C{}^2\, \Sigma\, \|T\mathbf{x}_k - \lambda_k\mathbf{x}_k\|^2 \leqslant (\delta\|F\|_C\, \|\mathbf{x}\|)^2$$

by (11.4.10). It follows that

$$\int_{-\infty}^{\infty} sF(s)\, d(E(s)\mathbf{x}, \mathbf{y}) = \int_{-\infty}^{\infty} F(s)\, d(E(s)T\mathbf{x}, \mathbf{y}), \qquad \forall\, \mathbf{y}.$$

Here the last expression is $(F(T)T\mathbf{x}, \mathbf{y})$ and (11.4.19) holds. ∎

This implies

Theorem 11.4.3. *If T is a bounded Hermitian operator, then the operator $F(T)$ defined in Theorem 11.3.2 for continuous functions $F(t)$ coincides with that defined by* (11.4.18).

Proof. We start by proving the identity for powers of T. If the spectrum of T is confined to the finite interval $[a, b]$, we have by (11.4.16)

$$T\mathbf{x} = \int_a^b s d(E(s)\mathbf{x}).$$

More generally, by (11.4.18),

$$T^n\mathbf{x} = \int_a^b s^n d(E(s)\mathbf{x}). \tag{11.4.20}$$

To justify this we note that for

$$F(s) = \begin{cases} a^n, & s < a, \\ s^n, & a \leqslant s \leqslant b, \\ b^n, & b < s, \end{cases}$$

formula (11.4.18) makes sense and gives precisely (11.4.20). Hence for any polynomial $P(s)$ we have

$$P(T)\mathbf{x} = \int_a^b P(s)\, d(E(s)\mathbf{x}). \tag{11.4.21}$$

Now if $F(s) \in C[a, b]$ we extend the definition by setting $F = F(a)$ for $s < a$ and $F = F(b)$ for $b < s$. The so extended function is an element of $C[-\infty, \infty]$ and formula (11.4.18) applies and gives

$$F(T)\mathbf{x} = \int_a^b F(s)\, d(E(s)\mathbf{x}).$$

On the other hand, we can find a sequence of polynomials $\{P_n(s)\}$ converging uniformly to $F(s)$ in $[a, b]$. From

$$P_n(T)\mathbf{x} = \int_a^b P_n(s)\, d(E(s)\mathbf{x})$$

we get

$$\lim_{n\to\infty} P_n(T)\mathbf{x} = \lim_{n\to\infty} \int_a^b P_n(s)\, d(E(s)\mathbf{x})$$

$$= \int_a^b \lim_{n\to\infty} P_n(s)\, d(E(s)\mathbf{x}) = \int_a^b F(s)\, d(E(s)\mathbf{x}).$$

The first member is $F(T)\mathbf{x}$ as defined by Theorem 11.3.2, while the last member is $F(T)\mathbf{x}$ according to (11.4.18). Thus the two definitions coincide whenever they are both meaningful. ∎

We call attention to two special cases of formula (11.4.18).

Example 1. Let λ be a complex number, $\lambda = \mu + i\nu$, $\gamma \neq 0$, and set

$$F(s) = \frac{1}{\lambda - s}$$

so that

$$F(T)\mathbf{x} = \int_{-\infty}^{\infty} \frac{1}{\lambda - s}\, d(E(s)\mathbf{x}).$$

If now $\mathbf{x} \in \mathfrak{D}(T)$,

$$TF(T)\mathbf{x} = F(T)T\mathbf{x} = \int_{-\infty}^{\infty} \frac{s}{\lambda - s}\, d(E(s)\mathbf{x})$$

and

$$(\lambda I - T)F(T)\mathbf{x} = \int_{-\infty}^{\infty} \frac{\lambda - s}{\lambda - s}\, d(E(s)\mathbf{x}) = \mathbf{x},$$

and this relation holds for all $\mathbf{x}$. On the other hand,

$$F(T)(\lambda I - T)\mathbf{x} = \mathbf{x}$$

holds only for $\mathbf{x} \in \mathfrak{D}(T)$. Thus we see that the resolvent of T exists for all non-real values of λ and

$$R(\lambda, T)\mathbf{x} = \int_{-\infty}^{\infty} \frac{1}{\lambda - s}\, d(E(s)\mathbf{x}). \tag{11.4.22}$$

For the existence of $R(\lambda_0, T)$ for a real λ_0 it is necessary and sufficient that there is an interval $(\lambda_0 - \delta, \lambda_0 + \delta)$ in which the spectral function $E(s)$ keeps a constant value.

Example 2. Take

$$F(s) = \frac{i + s}{i - s}$$

and define

$$F(T)\mathbf{x} \equiv \mathbf{C}(T)\mathbf{x} = \int_{-\infty}^{\infty} \frac{i + s}{i - s}\, d(E(s)\mathbf{x}), \qquad \forall\, \mathbf{x}. \tag{11.4.23}$$

For $\mathbf{x} \in \mathfrak{D}(T)$

$$(iI - T)\, \mathbf{C}(T)\mathbf{x} = i\mathbf{x} + T\mathbf{x},$$

so that $\mathbf{C}(T)$ is the Cayley operator (see Problem 16, Exercise 11.1).

We can generalize still further. Formula (11.4.18) stays meaningful as long as the integral exists. This will be the case if $F(s) \in C(-\infty, \infty)$ and $\mathbf{x}$ is so chosen that

$$\int_{-\infty}^{\infty} |F(s)|^2\, d(E(s)\mathbf{x}, \mathbf{x}) \quad \text{exists.} \tag{11.4.24}$$

EXERCISE 11.4

1. What is the norm of the operator $F(T)$ defined by (11.4.18)?
2. Use (11.4.23) to show that $\mathbf{C}(T)$ is unitary.

3. Take $\mathfrak{H} = L_2(-\pi, \pi)$, set $f(t) \sim \Sigma f_n e^{nit}$, and let $T = -i(d/dt)$. Determine $\mathfrak{D}(T)$. Show that $\sigma(T)$ is the set of all integers and that the spectral function $E(s)$ is the operation of taking partial sums of an unconventional kind so that

$$\big(E(s)f, f\big) = \Sigma\,|f_n|^2 \quad \text{for} \quad n \leqslant s.$$

Find (Tf, g) if $f \in \mathfrak{D}(T)$, $g \in \mathfrak{H}$.

4. With the same choice of $\mathfrak{H}$, take $Tf = -f''$ and carry through a similar discussion.

5. If T is Hermitian with spectral function $E(s)$, show that

$$\exp(-\,itT)\mathbf{x} = \int_{-\infty}^{\infty} e^{-ist} d\big(E(s)\mathbf{x}\big)$$

is meaningful for all $\mathbf{x}$ if $-\infty < t < \infty$. Find its inverse and show that the operator is unitary.

6. Show that for $\lambda = \mu + i\nu$, $\nu > 0$, for all $\mathbf{x}$,

$$-\,i\int_0^{\infty} e^{i\lambda t} \exp(-\,itT)\mathbf{x}\,dt$$

makes sense and equals $R(\lambda, T)\mathbf{x}$. What is the corresponding formula for $\nu < 0$?

7. If $E(s)$ is constant in some interval $[\alpha, \beta)$, show that for $\alpha < \lambda < \beta$ formula (11.4.22) makes sense and represents $R(\lambda, T)\mathbf{x}$. In particular, if T is bounded with spectrum confined to $[a, b]$, show that the formula represents $R(\lambda, T)\mathbf{x}$ for all λ not in $[a, b]$.

8. Take $\mathfrak{H} = L_2(-\infty, \infty)$ and define the *Dirichlet operator* by

$$D_s[f](t) = \int_{-\infty}^{\infty} \frac{\sin su}{u} f(u + t)\,du, \qquad s > 0,$$

and let D_s be the zero operator for $s \leqslant 0$. Show that D_s is a spectral function. It corresponds to the operator T which takes f into the conjugate of its derivative where "conjugate" is taken in the sense of potential theory, so that the conjugate of g is

$$\tilde{g}(t) = -\frac{1}{\pi}\lim_{\delta \downarrow 0} \int_{|u|>\delta} g(u + t)\frac{du}{u}.$$

9. Let $\{\omega_n(t)\}$ be a complete orthonormal system for $L_2(-\infty, \infty)$ and define

$$A[f](t) \sim \sum_0^{\infty} (-i)^n f_n \omega_n(t) \quad \text{if} \quad f(t) \sim \sum_0^{\infty} f_n \omega_n(t).$$

Show that this is a bounded normal operator. Find its spectrum and the canonical decomposition $A = B + iC$ where B and C are Hermitian. Find the spectral functions corresponding to B and to C. If the set $\{\omega_n(t)\}$ is the system of normalized weighted Hermitian polynomials, then $A[f]$ is the Fourier transform of f.

10. Let $\{c_n;\ n = 1, 2, 3, \ldots\}$ be a strictly decreasing sequence of positive numbers such that $\sum_1^\infty n{c_n}^2 < \infty$. Define a transformation T on l_2 by

$$T\mathbf{x} = \{y_1, y_2, \ldots, y_n, \ldots\} \quad \text{where} \quad y_n = \sum_{k=n}^{\infty} c_k x_k.$$

Show that T is linear and bounded and that its spectrum is the set $\{c_n\}$. Is T Hermitian or normal? If T is taken as a mapping from l_2 to l_∞ and if also $\sum_1{}^\infty c_k < \infty$, show that the point spectrum covers the whole plane less the origin.

11. Show that a Hilbert space is separable iff it contains a complete orthonormal system.

12. For a Hilbert space and a Hermitianoperat or A show that the Landau–Kallman–Rota inequality (5.3.22) may be sharpened to $\|A(\mathbf{x})\|^2 \leqslant \|\mathbf{x}\|\ \|A^2(\mathbf{x})\|$.

COLLATERAL READING

Carleman, T., *Sur les équations intégrales singulières à noyau réel et symétrique*, Uppsala Universitets Årsskrift, 1923.

Dunford, N. and J. Schwartz, *Linear Operators*, Part II, *Spectral Theory; Self Adjoint Operators in Hilbert Space*, Interscience, New York (1963).

Epstein, B., *Linear Functional Analysis; Introduction to Lebesgue Integration and Infinite-dimensional Problems,* Saunders, Philadelphia (1970).

Halmos, P., *A Hilbert Space Problem Book*, Van Nostrand, Princeton, N.J. (1967).

Hausdorff, F., Der Wertvorrat einer Bilinearform, *Math. Z.* **3** (1919) 314–16.

Neumann, J. von, Allgemeine Eigenwerttheorie hermitescher Funktionaloperatoren, *Math. Ann.* **102** (1929) 49–131.

Riesz, F., Über die linearen Transformationen des komplexen Hilbertschen Raumes, *Acta Sci. Math.* (*Szeged*) **5** (1930–2) 23–54.

Stone, M. S., *Linear Transformation in Hilbert Spaces and Their Applications to Analysis,* American Mathematical Society Colloquium Publications, Vol. 15, New York (1932).

Toeplitz, O., Das algebraische Analogon zu einem Satze von Fejér, *Math. Z.* **2** (1918) 187–97.

12 FUNCTIONAL INEQUALITIES I FUNCTIONS OF A SINGLE VARIABLE

In several branches of analysis functional inequalities enter in a decisive manner. They underline uniqueness theorems and they characterize important classes of functions. Moreover, it is often possible to describe functions satisfying a given functional inequality in terms of the solution of an associated fixed-point theorem.

In this and the following chapter we shall discuss two different types of functional inequality involving functions of a single variable and functions of two variables, respectively. For the first type we consider the canonical form

$$f(t) \leqslant T[f](t),$$

where T is a mapping from a given metric space of functions f into itself. Here partial ordering is often a useful tool.

There are five sections: Classification; Some determinative inequalities; Use of fixed-point theorems; Some applications; and Remarks on a class of multiplicative inequalities.

12.1 CLASSIFICATION

Let $\mathfrak{X}$ be a complete metric space the elements of which are mappings from some interval $[a, b]$ into R^1. Further properties of $\mathfrak{X}$ and of its elements f will be specified as need arises. Let T be a mapping of $\mathfrak{X}$ into itself and consider the inequality

$$f(t) \leqslant T[f](t), \qquad \forall\, t \in [a, b], \tag{12.1.1}$$

or

$$f \leqslant T[f], \tag{12.1.2}$$

if partial ordering is introduced in $\mathfrak{X}$ in the natural manner, so that

$$f_1 \leqslant f_2 \quad \text{iff} \quad f_1(t) \leqslant f_2(t), \qquad \forall\, t. \tag{12.1.3}$$

There exists a subset $\mathfrak{X}_0$ of $\mathfrak{X}$ the elements of which satisfy (12.1.2). The nature of the subset will lead to a useful classification of the inequality.

(1) $\mathfrak{X}_0 = \mathfrak{X}$. In this case the inequality is *trivial* for $\mathfrak{X}$. As an example, take $\mathfrak{X} = C[a, b]$ and

$$T[f] = \tfrac{1}{4}[3 + f^4],$$

so that the inequality reads

$$f(t) \leqslant \tfrac{1}{4}\{3 + [f(t)]^4\}. \tag{12.1.4}$$

Now the quartic

$$y = x^4 - 4x + 3$$

has a unique minimum for $x = 1$ with $y = 0$ and $y' < 0$ for $x < 1$ and > 0 for $x > 1$. It follows that y is never negative and this implies that (12.1.4) holds for every $f \in C[a, b]$. Thus the inequality is trivial, which does not impair its usefulness.

(2) $\mathfrak{X}_0$ is a non-empty proper subset of $\mathfrak{X}$ which contains more than one element. Then (12.1.2) is said to be *restrictive*.

An example is given by $\mathfrak{X} = C[a, b]$ restricted to the real-valued elements and $T[f] = f^2$, the inequality being

$$f(t) \leqslant [f(t)]^2. \tag{12.1.5}$$

There are obviously elements of $\mathfrak{X}$ which satisfy this inequality. A sufficient but not necessary condition is that the range of f omits the interval (0, 1). On the other hand, there are just as obviously elements of $\mathfrak{X}$ which do not satisfy (12.1.5). Hence $\mathfrak{X}_0$ is a proper non-void subset of $\mathfrak{X}$ containing more than one element (actually infinitely many). Thus (12.1.5) is restrictive in our classification.

(3) If $\mathfrak{X}_0$ reduces to a single element, the inequality is said to be *determinative*.

The next section is devoted to determinative inequalities. Here we give a fairly trivial example. Take $\mathfrak{X} = C^+[0, 1]$ and the inequality

$$f(t) \leqslant t[f(t)]^2, \qquad 0 \leqslant t \leqslant 1. \tag{12.1.6}$$

This has a unique solution in $C^+[0, 1]$, namely $f(t) \equiv 0$. Hence the inequality is determinative.

(4) If $\mathfrak{X}_0$ is void, we say that the inequality is *absurd* (for the given space $\mathfrak{X}$, enlarging the space may change the classification).

Among the possible inequalities of this nature we select

$$f(t) \leqslant -1 - [f(t)]^2, \tag{12.1.7}$$

$\mathfrak{X}$ being the subset of real-valued elements of $C[a, b]$. There is no element of $\mathfrak{X}$ which satisfies the inequality, so (12.1.7) is absurd in this setting.

EXERCISE 12.1

If $\mathfrak{X} = C^+[0, 1]$ or a subset thereof as specified, classify the following inequalities. If a subset is specified, verify that it by itself is a complete metric space in the induced topology.

1. $f(t) \leqslant \{1 - [f(t)]^2\}^{\frac{1}{2}}$, $\|f\| \leqslant 1$.
2. $f(t) \leqslant 2[f(t)]^2 - [f(t)]^3$, $\|f\| \leqslant 1$.
3. $f(t) \leqslant \sin[f(t)]$, $\|f\| \leqslant \pi$.
4. $f(t) \leqslant \tan[f(t)]$, $\|f\| < \frac{1}{2}\pi$.

5. $f(t) \leqslant \dfrac{[f(t)]^2}{1 + f(t)}$.

6. $f(t) \leqslant 1 + [f(t)]^2$.

7. $f(t) \leqslant [f(t)]^2 - 1$ for $T[f](t) \in C^+[0, 1]$.

8. $f(t) \leqslant \dfrac{[f(t)]^2}{1 + f(t)} - \frac{1}{2}$ for $T[f](t) \in C^+[0, 1]$.

9. $f(t) \leqslant f(\frac{1}{2}t)$.

10. $f(t) \leqslant \frac{1}{2}f(\frac{1}{2}t) + \frac{1}{3}f(\frac{1}{3}t)$.

11. $f(t) \leqslant F[f(t)]$ with $\|f\| \leqslant 1$ where

$$F(u) = \sum_0^\infty a_n u^n, \quad 0 \leqslant a_n \ \ \forall n, \quad \sum_0^\infty a_n = b < 1$$

and the radius of convergence of the power series equals 1. There are two cases according as $a_0 = 0$ or $a_0 > 0$.

12.2 SOME DETERMINATIVE INEQUALITIES

There are several inequalities of this type which play a decisive role for uniqueness theorems. In view of their importance we state them as theorems. We also list the resulting uniqueness theorems.

Theorem 12.2.1. *If $f \in C^+[0, b]$ and if*

$$f(t) \leqslant K \int_0^t f(s)\, ds, \qquad 0 < t \leqslant b, \tag{12.2.1}$$

then $f(t) \equiv 0$.

Proof. We shall give two distinct proofs here, and other proofs will be found in Exercise 12.2 or as corollaries of later theorems.

1. *Proof by iteration.* By substitution we get

$$0 \leqslant f(t) \leqslant K^2 \int_0^t \int_0^s f(u)\, du\, ds = K^2 \int_0^t (t - s) f(s)\, ds$$

by a classical reduction formula for multiple integrals. Repeating the procedure, we get

$$0 \leqslant f(t) \leqslant \frac{1}{(n-1)!} K^n \int_0^t (t - s)^{n-1} f(s)\, ds. \tag{12.2.2}$$

Hence the sup-norm satisfies

$$\|f\| \leqslant \frac{1}{n!}(Kb)^n \|f\|, \tag{12.2.3}$$

and for large values of n the multiplier on the right is <1. This requires $\|f\| = 0$ and hence $f = 0$.

2. *Proof by integration.* Set

$$F(t) = \int_0^t f(s)\,ds \tag{12.2.4}$$

and note that

$$F(0) = 0, \qquad F(t) \geqslant 0, \qquad F'(t) = f(t).$$

Hence

$$F'(t) - KF(t) \leqslant 0, \tag{12.2.5}$$

i.e. an integral inequality has been replaced by a differential inequality. Normally the latter type is more difficult to discuss than the former, but in the present case we reach the desired goal after multiplying both members by $\exp(-Kt)$ to obtain an exact derivative

$$\frac{d}{dt}[F(t)\exp(-Kt)] \leqslant 0. \tag{12.2.6}$$

This shows that $F(t)\exp(-Kt)$ is non-increasing in $(0, b)$. Since $F(0) = 0$, this implies $F(t) \leqslant 0$. But we already know that $F(t) \geqslant 0$, so we must have $F(t) \equiv 0$ and hence also $f(t) \equiv 0$. ∎

This type of argument is quite useful in handling various types of functional inequality involving integration.

Let us return for a moment to Theorem 6.4.1, where the uniqueness proof was omitted.

Theorem 12.2.2. *The solution of a differential equation of type* (6.4.5) *with a Lipschitz condition* (6.4.4) *is unique and the same is true for a matrix equation* (6.4.7) *with the Lipschitz condition* (6.4.6).

Proof. This is an immediate consequence of Theorem 12.2.1. To fix the ideas, take the matrix case. If there are two solutions $\mathcal{Y}_1$ and $\mathcal{Y}_2$ of the equation

$$\mathcal{Y}'(x) = \mathcal{F}[x, \mathcal{Y}(x)], \qquad \mathcal{Y}(0) = \mathcal{O}, \tag{12.2.7}$$

valid for $|x| < r_0$, then for $0 < x < r_0$

$$\|\mathcal{Y}_2(x) - \mathcal{Y}_1(x)\| \leqslant \int_0^x \|\mathcal{F}[s, \mathcal{Y}_2(s)] - \mathcal{F}[s, \mathcal{Y}_1(s)]\|\,ds.$$

By the Lipschitz condition

$$\|\mathcal{Y}_2(x) - \mathcal{Y}_1(x)\| \leqslant K\int_0^x \|\mathcal{Y}_2(s) - \mathcal{Y}_1(s)|\,ds.$$

This states that the continuous non-negative function $\|\mathcal{Y}_2(x) - \mathcal{Y}_1(x)\|$ satisfies

(12.2.1) with $b = r_0$ and is consequently identically zero. The extension to $(-r_0, 0)$ is trivial. ∎

We can generalize Theorem 12.2.1, replacing the constant K by an integrable function $K(s)$ under the sign of integration.

Theorem 12.2.3. *If $f \in C^+[0, b]$, if $K \in C^+(0, b] \cap L(0, b)$, and if*

$$f(t) \leqslant \int_0^t K(s) f(s)\, ds, \qquad \forall\, t, \tag{12.2.8}$$

then $f(t) \equiv 0$.

A proof can be given by either of the methods used above in the case $K(t) \equiv K$ and will also follow as a corollary of later theorems. This theorem can be extracted from an existence theorem for differential equations due to the brilliant and colorful Turkish–Greek–German mathematician Constantin Carathéodory (1875–1950) in 1918. His theorem is obtained if in the formulation of Theorem 6.4.1 the boundedness condition (6.4.9) be replaced by

$$\|\mathbf{F}(x, \mathbf{y})\| \leqslant K(|x|) \tag{12.2.9}$$

and the Lipschitz condition by

$$\|\mathbf{F}(x, \mathbf{y}_2) - \mathbf{F}(x, \mathbf{y}_1)\| \leqslant K(|x|)\, \|\mathbf{y}_2 - \mathbf{y}_1\|. \tag{12.2.10}$$

Such conditions permit the consideration of equations where the right member becomes infinite as $x \to 0$, provided the infinitude is integrable. The uniqueness of the solution follows from the fact that if $\mathbf{y}_1(x)$ and $\mathbf{y}_2(x)$ are two solutions with the same initial value, then $\|\mathbf{y}_2(x) - \mathbf{y}_1(x)\|$ satisfies (12.2.8).

A uniqueness theorem found by Mitio Nagumo for differential equations in 1926 is based on the following functional inequality, which is not a special case of Theorem 12.2.3.

Theorem 12.2.4. *Let $\mathfrak{X}$ be the subspace of $C^+[0, b]$, the elements of which satisfy $f(0) = 0$, $\lim_{h \downarrow 0} f(h)/h = 0$. Then, if $f \in \mathfrak{X}$ and if*

$$f(t) \leqslant \int_0^t f(s) \frac{ds}{s}, \tag{12.2.11}$$

it follows that $f(t) \equiv 0$.

A proof may be based upon the integration method. The details are left to the reader.

Nagumo's uniqueness theorem is relevant if in Theorem 6.4.1 the Lipschitz condition is replaced by

$$\|\mathbf{F}(x, \mathbf{y}_2) - \mathbf{F}(x, \mathbf{y}_1)\| \leqslant \frac{1}{|x|}\, \|\mathbf{y}_2 - \mathbf{y}_1\|. \tag{12.2.12}$$

If there were two solutions, then the norm of their difference is a function f

satisfying the conditions of Theorem 12.2.4 and is hence identically zero. Thus the solution is unique.

Other determinative inequalities will be encountered later.

EXERCISE 12.2

1. Verify Theorem 12.2.1 by a step-by-step argument. Start by showing $f(t) \equiv 0$ if $bK < 1$ and extend to longer intervals.

If $\mathfrak{X} = C^+[0, b]$, show that the following inequalities are determinative:

2. $f(t) \leqslant \log[1 + f(t)]$.
3. $f(t) \leqslant 1 - \exp[-f(t)]$.
4. With the same choice of $\mathfrak{X}$, for what values of the positive number a is

$$f(t) \leqslant \int_0^t [f(s)]^a \, ds$$

determinative? Distinguish the two cases $0 < a < 1$ and $1 \leqslant a$ which require different treatment.

5. Prove Theorem 12.2.3.
6. Prove Theorem 12.2.4.
7. Verify that (12.2.12) implies uniqueness.
8. It was said at the end of the proof of Theorem 12.2.2 that the extension to the interval $(-r_0, 0)$ is trivial. If it is trivial, it can be done. Do it!

12.3 USE OF FIXED-POINT THEOREMS

The reader may have noticed that the problems in the preceding part of the chapter frequently led to conditions related to the fixed point of the corresponding transformation T. Thus it turned out that the determinative inequalities could only be satisfied by the invariant element, which in Section 12.2 was $f(t) \equiv 0$. But restrictive inequalities lead to similar considerations. Thus in Problem 4, Exercise 12.2, take $a = \frac{1}{2}$ so that the inequality reads

$$f(t) \leqslant \int_0^t [f(s)]^{\frac{1}{2}} \, ds. \tag{12.3.1}$$

It is found that $f(t)$ must satisfy $0 \leqslant f(t) \leqslant \frac{1}{4}t^2$. In this case the transformation T defined by the right member of (12.3.1) has infinitely many fixed points, the extremal solutions being $f(t) \equiv 0$ and $f(t) \equiv \frac{1}{4}t^2$. In the cases considered the underlying spaces would admit partial ordering. This suggests the existence of general relations between functional inequalities and fixed-point theorems in partially ordered spaces when the transformation T is order preserving. A typical result obtained from this starting point is

Theorem 12.3.1. *Let $\mathfrak{X}$ be a complete metric space which is partially ordered in such a manner that if $\{x_n\}$ is an increasing sequence in $\mathfrak{X}$ so that $x_n \leqslant x_{n+1}$, for all n, and if $\lim_{n\to\infty} x_n \equiv x_0$ exists in the sense of the metric, then $x_n \leqslant x_0$, for all n. Let T be an order-preserving mapping from $\mathfrak{X}$ into $\mathfrak{X}$ such that T^m is a contraction for some m. Let f_0 be the unique fixed point of T. Then*

$$f \leqslant T[f] \quad \textit{implies} \quad f \leqslant f_0. \tag{12.3.2}$$

Proof. We say that T is *order preserving* if for $f_1, f_2 \in \mathfrak{X}$

$$f_1 \leqslant f_2 \quad \text{implies} \quad T[f_1] \leqslant T[f_2]. \tag{12.3.3}$$

Suppose that $f \in \mathfrak{X}_0$, that is $f \leqslant T[f]$. Note that $\mathfrak{X}_0$ is not void since it contains f_0 at least. Then

$$f \leqslant T[f] \leqslant T^2[f] \leqslant \cdots \leqslant T^n[f] \leqslant \cdots. \tag{12.3.4}$$

Here

$$\lim_{k\to\infty} T^{km}[T^j(f)] = f_0, \qquad j = 0, 1, \cdots, m-1,$$

since T^m is a contraction and the limit is the same for all elements of $\mathfrak{X}$. See Section 5.4. It follows that the whole sequence, which is increasing, must converge to f_0. The theorem is proved since order is preserved under the limit operation. ∎

Example 1. We take the inequality of Theorem 12.2.1 with

$$T[f](t) \equiv K \int_a^t f(s)\,ds \tag{12.3.5}$$

and $\mathfrak{X} = C^+[a, b]$. Here (5.4.10) and (5.4.11) show that T^m is a strict contraction for all large values of m. Further, T is a positive operator; it takes positive elements of $C[a, b]$ into positive elements, i.e. $C^+[a, b]$ is mapped into itself. This implies that T is order preserving. It is clear that the zero element is in $\mathfrak{X}$ and is left invariant by T, so it is a fixed point. Theorem 12.3.1 now gives $f \leqslant 0$, which together with $f \geqslant 0$ gives $f = 0$ as the only element of $\mathfrak{X}$ that satisfies (12.3.5).

The same type of argument applies to the Carathéodory inequality (12.2.8). In these cases linear spaces and positive operators are involved. Such cases may be handled by the Volterra fixed point theorem, which gives

Theorem 12.3.2. *Let $\mathfrak{X}$ be a partially ordered B-space such that the positive cone $\mathfrak{X}^+$ is a closed point set. Let S be a positive linear bounded operator from $\mathfrak{X}$ to $\mathfrak{X}$ such that*

$$\sum \|S^n\| < \infty. \tag{12.3.6}$$

Let g be a given element of $\mathfrak{X}^+$, f_0 the unique fixed point of

$$T[f] = g + S[f]. \tag{12.3.7}$$

Then

$$f \leqslant g + S[f] \quad \textit{implies} \quad f \leqslant f_0. \tag{12.3.8}$$

Proof. The iterates

$$T^n[f] = g + S[g] + \cdots + S^{n-1}[g] + S^n[f] \tag{12.3.9}$$

form an increasing sequence since $f \leqslant T[f]$, while g is positive and S is a positive linear operator and hence order preserving. By Theorem 5.6.1

$$\lim_{n\to\infty} T^n[f] = f_0.$$

Combining this with $f \leqslant T[f]$ and the order, preserving properties of limits, we get (12.3.8). ∎

An important special case is the following.

Theorem 12.3.3. *Let $K \in C^+(a, b) \cap L(a, b)$. Let f and g belong to $C^+[a, b]$ and suppose that*

$$f(t) \leqslant g(t) + \int_a^t K(s) f(s)\, ds, \qquad \forall\, t \in [a, b]. \tag{12.3.10}$$

Then

$$f(t) \leqslant g(t) + \int_a^t K(s) \exp\left[\int_s^t K(u)\, du\right] g(s)\, ds. \tag{12.3.11}$$

Proof. Here

$$S[f](t) = \int_a^t K(s) f(s)\, ds, \tag{12.3.12}$$

which clearly exists as an element of $C^+[a, b]$. Further,

$$S^2[f](t) = \int_a^t K(s) \int_a^s K(u) f(u)\, du\, ds$$

$$= \int_a^t K(u) f(u) \left[\int_u^t K(s)\, ds\right] du.$$

It follows that

$$|S^2[f](t)| \leqslant \|f\| \int_a^t K(u) \left[\int_u^t K(s)\,ds \right] du$$

$$= \|f\| \frac{1}{2!} \left[\int_a^t K(s)\,ds \right]^2.$$

Hence

$$\|S^2\| \leqslant \frac{1}{2!} \left[\int_a^b K(s)\,ds \right]^2,$$

and by induction

$$\|S^n\| \leqslant \frac{1}{n!} \left[\int_a^b K(s)\,ds \right]^n. \tag{12.3.13}$$

Hence (12.3.6) holds. Thus

$$S^n[g](t) = \frac{1}{(n-1)!} \int_a^t g(s)\,K(s) \left[\int_s^t K(u)\,du \right]^{n-1} ds.$$

Summing for n, we obtain the right member of (12.3.11). ∎

Corollary. *Theorem 12.2.3.*

For if $g(s) \equiv 0$, so is the right member of (12.3.11) and $f(t) \equiv 0$ results. Let us specialize again. If $K(t) \equiv K$, then (12.3.10) becomes

$$f(t) \leqslant g(t) + K \int_a^t f(s)\,ds \tag{12.3.14}$$

and (12.3.11) reduces to

$$f(t) \leqslant g(t) + K \int_a^t \exp[K(t-s)]\,g(s)\,ds. \tag{12.3.15}$$

If $g(t)$ is also constant, then

$$f(t) \leqslant g + K \int_a^t f(s)\,ds \tag{12.3.16}$$

implies

$$f(t) \leqslant g \exp[K(t-a)]. \tag{12.3.17}$$

The set of the three inequalities (12.3.11), (12.3.15), and (12.3.17) is known as *Gronwall's Lemma* after the Swedish–American mathematician Thomas Hakon Gronwall (1877–1932), who found a special case in 1918. We shall consider some implications of this lemma in the next section.

In this discussion and also in the preceding section we have imposed unnecessary restriction on the space $\mathfrak{X}$. Thus in Theorems 12.2.1, 12.2.3, and

12.3.3 we could replace $C^+[a, b]$ by $L^+(a, b)$ with the inequalities holding for almost all t. We can also generalize in a different direction.

Consider the space $\mathfrak{M}_n C^+[a, b]$ briefly mentioned towards the end of Section 3.2. This is the set of all n by n matrices $\mathcal{F}(t) = (f_{ij}(t))$, where $f_{ij}(t) \in C^+[a, b]$. Let $\mathcal{K} = (k_{ij})$, $k_{ij} \geqslant 0$ be a positive constant matrix. Then Theorem 12.2.1 generalizes to

Theorem 12.3.4. *The zero matrix is the only element of $\mathfrak{M}_n C^+[a, b]$ which satisfies*

$$\mathcal{F}(t) \leqslant \mathcal{K} \int_a^t \mathcal{F}(s)\, ds, \qquad \forall\, t \in [a, b]. \tag{12.3.18}$$

Proof. We recall that partial ordering is available in this space and both sides of (12.3.18) are positive matrices. As the norm in $\mathfrak{X}$ we use

$$\|\mathcal{F}\| = \sup_t \sup_j \sum_{k=1}^{n} |f_{jk}(t)|. \tag{12.3.19}$$

If $T[\mathcal{F}]$ denotes the right member of (12.3.18), then

$$\|T\| \leqslant (b - a)\, \|\mathcal{K}\|,$$

and, by induction,

$$\|T^m\| \leqslant \frac{1}{m!} [(b - a)\, \|\mathcal{K}\|]^m, \tag{12.3.20}$$

so that T^m is a contraction from $\mathfrak{X}$ into $\mathfrak{X}$ for all large m. Thus T has a single fixed point. Now the zero matrix is obviously left invariant by T, so that it is the fixed point. Theorem 12.3.1 now gives $\mathcal{F} \leqslant \mathcal{O}$, which, combined with $\mathcal{F} \geqslant \mathcal{O}$, gives $\mathcal{F} = \mathcal{O}$. ∎

Similar generalizations may be found for the other theorems listed above.

We end this discussion by noting a result for the transformation T^2 where T is the operator of (12.2.1).

Theorem 12.3.5. *If f and $g \in C^+[a, b]$ and if*

$$f(t) \leqslant g(t) + K^2 \int_a^t (t - s) f(s)\, ds, \tag{12.3.21}$$

then

$$f(t) \leqslant g(t) + K \int_a^t \sinh[K(t - s)]\, g(s)\, ds. \tag{12.3.22}$$

The proof is left to the reader.

Let us return briefly to the point of departure, namely Gronwall's Lemma or Theorem 12.3.3. There we are concerned with the positive cone of the

Banach algebra $C[a, b]$ and the operator

$$T[f](t) = \int_a^t K(s) f(s)\, ds, \qquad a \leqslant t \leqslant b. \tag{12.3.23}$$

T is a linear positive quasi-nilpotent element of the operator algebra $\mathfrak{E}\{C[a, b]\}$. J. B. Miller has called my attention to the fact that T is an *antiderivation*. This is a class of linear bounded operators from a B-algebra $\mathfrak{B}$ into itself to which he has given much attention. The basic identity for an antiderivation G is

$$(\mathsf{G}x) \cdot (\mathsf{G}y) = \mathsf{G}[(\mathsf{G}x) \cdot y + x \cdot (\mathsf{G}y)], \tag{12.3.24}$$

where $x, y \in \mathfrak{B}$. Miller has determined the resolvent of G and shown that the spectrum of G reduces to $\{0\}$. It follows that

$$\lim_{n\to\infty} \|\mathsf{G}^n\|^{1/n} = 0. \tag{12.3.25}$$

Suppose now that $\mathfrak{B}$ is partially ordered and G is a positive operator from $\mathfrak{B}$ to $\mathfrak{B}$, which is also an antiderivation. Consider the functional inequality

$$f \leqslant g + \mathsf{G}f, \tag{12.3.26}$$

where $f, g \in \mathfrak{B}^+$ and g is fixed. By Theorem 12.3.2

$$f \leqslant g + \sum_{n=1}^{\infty} \mathsf{G}^n[g], \tag{12.3.27}$$

where the series is absolutely convergent by (12.3.25). This is the analogue of Theorem 12.3.3 for positive antiderivations.

EXERCISE 12.3

1. Verify (12.3.13).
2. Show that Theorem 12.2.1 holds for $L^+(0, b)$. Estimate T^m in this space.
3. Verify (12.3.20).
4. Extend (12.3.16) to the matrix case, i.e. show that if $\mathfrak{X} = \mathfrak{M}_n C^+[a, b]$ and $\mathcal{G}$ and $\mathcal{K}$ are fixed positive matrices, then

   $$\mathcal{F}(t) \leqslant \mathcal{G} + \mathcal{K} \int_a^t \mathcal{F}(s)\, ds$$

   implies that

   $$\mathcal{F}(t) \leqslant \exp[\mathcal{K}(t-a)]\,\mathcal{G}.$$

   Note the order of the factors: $\mathcal{G}$ and $\mathcal{K}$ need not commute.
5. Prove Theorem 12.3.5.

6. It is desired to give an extension of Nagumo's inequality analogous to Theorem 12.3.3. Let $\mathfrak{X}$ be the space of functions f defined on $[0, b]$ with the following properties: (1) $f \in C^+[0, b]$, (2) $\lim_{t \downarrow 0} f(t) = 0$, (3) $\lim_{t \downarrow 0} t^{-1} f(t) = 0$, (4) $t^{-2} f(t) \in L(0, b)$. Show that $\mathfrak{X}$ is a complete metric space under a suitable norm. Use the method of integration [method (2) in the proof of Theorem 12.2.1] or other device to prove that if g and $f \in \mathfrak{X}$, g fixed, and if
$$f(t) \leqslant g(t) + \int_0^t f(s)\, s^{-1}\, ds,$$
then
$$f(t) \leqslant g(t) + t \int_0^t g(s)\, s^{-2}\, ds.$$

7. Extend the preceding result to the matrix case, i.e. let $\mathfrak{X}$ be the subspace of $\mathfrak{M}_n C^+[a, b]$ where the elements satisfy conditions (2) to (4) with f replaced by a matrix $\mathcal{F}$.

8. Let f satisfy (12.3.1) and $f(t) = 0, 0 \leqslant t \leqslant a \leqslant b$. Show that $0 \leqslant f(t) \leqslant \frac{1}{4}(t - a)^2$, $a \leqslant t \leqslant b$. Show that the majorant is also a fixed point of T acting on $C^+[0, b]$.

12.4 SOME APPLICATIONS

Gronwall developed his lemma with a view to applications in the theory of differential equations. To illustrate the power of the lemma we shall discuss two such applications.

The first concerns the change in the solution caused by a small change in the initial condition. To fix the ideas, consider the matrix case (see Theorem 6.4.2)
$$\mathcal{Y}'(x) = \mathcal{F}[x, \mathcal{Y}(x)], \tag{12.4.1}$$
where $\mathcal{F}$ is bounded and satisfies a Lipschitz condition for $|x| \leqslant a$, $\|\mathcal{Y}\| \leqslant b$. We denote by $\mathcal{Y}_1$ and $\mathcal{Y}_2$ the two solutions defined by
$$\mathcal{Y}_1(0) = \mathcal{O}, \qquad \mathcal{Y}_2(0) = \mathcal{Y}_0, \tag{12.4.2}$$
respectively, where for convenience we assume $\|\mathcal{Y}_0\| \leqslant \frac{1}{2}b$.

Theorem 12.4.1. *Under the stated conditions there exists an $r_1 > 0$ such that for $|x| < r_1$*
$$\|\mathcal{Y}_2(x) - \mathcal{Y}_1(x)\| \leqslant \|\mathcal{Y}_0\| \exp[K|x|]. \tag{12.4.3}$$

Proof. Suppose that the solutions exist for $|x| \leqslant r_1 \leqslant a$. Then
$$\mathcal{Y}_1(x) = \int_0^t \mathcal{F}[s, \mathcal{Y}_1(s)]\, ds, \qquad \mathcal{Y}_2(x) = \mathcal{Y}_0 + \int_0^t \mathcal{F}[s, \mathcal{Y}_2(s)]\, ds,$$

and by familiar estimates for $0 < |x| \leqslant r_1$,

$$\|\mathcal{Y}_2(x) - \mathcal{Y}_1(x)\| \leqslant \|\mathcal{Y}_0\| + K\int_0^{|x|} \|\mathcal{Y}_2(s) - \mathcal{Y}_1(s)\|\, ds.$$

This inequality is of type (12.3.10), and (12.4.3) then follows from (12.3.17). ∎

As the second application we take the problem of giving *a priori* estimates of the rate of growth of solutions for approach to a singular point of the equation. This method applies to linear equations and we shall consider first order matrix equations on the interval $(0, b]$.

Theorem 12.4.2. *Let*

$$\mathcal{Y}'(t) = \mathcal{A}(t)\mathcal{Y}(t), \qquad \mathcal{Y}(b) = \mathcal{E}, \tag{12.4.4}$$

where $\mathcal{A}(t) \in \mathfrak{M}_n C(0, b)$ and becomes infinite as $t \downarrow 0$ in such a manner that there is a fixed $c > 0$ for which

$$0 < \sup\,[t^c\|\mathcal{A}(t)\|] \equiv m < \infty. \tag{12.4.5}$$

Then the behavior of $\|\mathcal{Y}(t)\|$ as $t \downarrow 0$ is governed by the value of c as follows. If

(1) $0 < c < 1$, *then $\|\mathcal{Y}(t)\|$ is bounded away from 0 and ∞;*

(2) $c = 1$, *then*

$$(t/b)^m \leqslant \|\mathcal{Y}(t)\| \leqslant (b/t)^m; \tag{12.4.6}$$

(3) $c > 1$, *then there exist positive numbers A and B such that*

$$A \exp\left\{-\frac{m}{c-1}t^{1-c}\right\} \leqslant \|\mathcal{Y}(t)\| \leqslant B \exp\left\{\frac{m}{c-1}t^{1-c}\right\}. \tag{12.4.7}$$

Proof. For $0 < \alpha < \beta < b$

$$\mathcal{Y}(\beta) - \mathcal{Y}(\alpha) = \int_\alpha^\beta \mathcal{A}(s)\,\mathcal{Y}(s)\, ds. \tag{12.4.8}$$

This gives

$$\|\mathcal{Y}(\beta)\| \leqslant \|\mathcal{Y}(\alpha)\| + \int_\alpha^\beta \|\mathcal{A}(s)\|\,\|\mathcal{Y}(s)\|\, ds$$

$$\leqslant \|\mathcal{Y}(\alpha)\| + m\int_\alpha^\beta s^{-c}\|\mathcal{Y}(s)\|\, ds. \tag{12.4.9}$$

Gronwall's Lemma applies to this inequality and gives

$$\|\mathcal{Y}(\beta)\| \leqslant \|\mathcal{Y}(\alpha)\| \exp\left[m\int_\alpha^\beta s^{-c}\, ds\right]. \tag{12.4.10}$$

Here is where the three cases separate. The value of the exponential factor equals

$$\textit{Case (1).} \qquad \exp\left\{\frac{m}{1-c}\,[\beta^{1-c}-\alpha^{1-c}]\right\}; \tag{12.4.11}$$

$$\textit{Case (2).} \qquad (\beta/\alpha)^m; \tag{12.4.12}$$

$$\textit{Case (3).} \qquad \exp\left\{\frac{m}{c-1}\,[\alpha^{1-c}-\beta^{1-c}]\right\}. \tag{12.4.13}$$

Here we set $\alpha = t$, $\beta = b$ and obtain the lower bounds for $\|\mathcal{Y}(t)\|$ stated in the theorem.

To obtain the upper bounds we need the additional inequality

$$\|\mathcal{Y}(\alpha)\| \leqslant \|\mathcal{Y}(\beta)\| \exp\left[m\int_{\alpha}^{\beta} s^{-c}\,ds\right]. \tag{12.4.14}$$

There are various ways of proving this. The simplest is perhaps to make a change of variables. We set

$$t = b - u, \qquad \mathcal{Y}(t) = \mathcal{X}(u), \tag{12.4.15}$$

so that

$$\mathcal{X}'(u) = -\mathcal{A}(b-u)\,\mathcal{X}(u),\ \mathcal{X}(0) = \mathcal{E}. \tag{12.4.16}$$

Treating this equation as (12.4.4) was treated, we get

$$\|\mathcal{X}(u)\| \leqslant \|\mathcal{X}(0)\| \exp\left[m\int_0^u (b-v)^{-c}\,dv\right]$$

or

$$\|\mathcal{Y}(t)\| \leqslant \|\mathcal{Y}(b)\| \exp\left[m\int_t^b s^{-c}\,ds\right]. \tag{12.4.17}$$

Again separating cases, we obtain the upper bounds stated in the theorem. The details are left to the reader. ▮

In Case (1) a sharper statement may be proved, namely, that $\mathcal{Y}(t)$ tends to a limit $\neq 0$ as $t \downarrow 0$.

Cases (2) and (3) will arise if, for instance, $t = z$ is a complex variable and

$$\mathcal{A}(z) = z^{-p-1}\sum_{j=0}^{\infty} \mathcal{A}_j z^j, \tag{12.4.18}$$

where p is a non-negative integer, $\mathcal{A}_0 \neq \mathcal{O}$, and the matrix power series has a positive radius of convergence. Here $p = 0$ gives Case (2) and $z = 0$ is known as a *regular singular point* of the equation, while $p > 0$ leads to case (3) and an *irregular singular point*.

EXERCISE 12.4

1. Make a comparison along the lines of Theorem 12.4.1 between the solutions of
$$\mathcal{Y}'(x) = \mathcal{F}[x, \mathcal{Y}(x)] \quad \text{and} \quad \mathcal{Z}'(x) = \mathcal{G}[x, \mathcal{Z}(x)]$$
with $\mathcal{Y}(0) = \mathcal{Z}(0) = \mathcal{Y}_0$, if $\mathcal{F}$ and $\mathcal{G}$ are bounded and continuous in some neighborhood of $(0, \mathcal{O})$. Let $\mathcal{F}$ satisfy a Lipschitz condition and $\|\mathcal{F}(x, \mathcal{Y}) - \mathcal{G}(x, \mathcal{Y})\| < \delta$ for some fixed δ.
2. Let $\mathcal{A}$ be a constant matrix. Then the equation
$$\mathcal{W}'(z) = \mathcal{A}z^{-\frac{1}{2}}\mathcal{W}(z)$$
is of Case (1) at the origin. Solve the equation explicitly by a power series in $z^{1/2}$ and verify that $\mathcal{W}(z)$ has a limit as $z \to 0$.
3. Prove the general assertion that in Case (1) the solution of (12.4.4) tends to a limit when $t \downarrow 0$.
4. Construct examples of equations of type (12.4.4) where Cases (2) and (3) hold and explicit solutions are available to show that the estimates (12.4.6) and (12.4.7) are the best possible.
5. Supply omitted details in the proof of Theorem 12.4.2.
6. If Case (1) is present, show that the initial value problem $\mathcal{W}'(z) = \mathcal{A}(z)\mathcal{W}(z)$, $\mathcal{W}(0) = \mathcal{E}$ has a unique solution.

12.5 REMARKS ON A CLASS OF MULTIPLICATIVE INEQUALITIES

In the classical theory of analytic functions and of functional equations, functions with a *multiplication theorem* and functions satisfying a so-called *q-difference equation* have played a certain role. In the first case an equation of the form

$$F(qz) = F[f(z)] \tag{12.5.1}$$

is considered where q is a given real or complex number $\neq 1$ and $F(u)$ is a given analytic function of u, say holomorphic in some neighborhood of $u = 0$. The reader has encountered *addition theorems* in the theory of trigonometric functions, exponential functions, possibly also elliptic functions. These are relations of the form

$$f(z_1 + z_2) = F[f(z_1), f(z_2)] \tag{12.5.2}$$

where normally $F(u, v)$ is a symmetric algebraic function of the two arguments. Setting $z_1 = z_2 = z$, we obtain an equation of type (12.5.1) with $q = 2$. By repeated use of (12.5.2) we can obtain other equations of type (12.5.1), where q is any preassigned positive integer. Specific examples are given by

$$f(2z) = 2[f(z)]^2 - 1, \tag{12.5.3}$$

$$g(3z) = \frac{3g(z) - [g(z)]^3}{1 - 3[g(z)]^2}, \tag{12.5.4}$$

satisfied by $\cos az$ and $\tan az$, respectively. Complex multipliers occur in the theory of so-called *complex multiplication for elliptic functions.*

To this class of functional equations corresponds a class of functional inequalities of the form

$$f(qt) \leqslant F[f(t)], \tag{12.5.5}$$

where now q and t are real positive and the coefficients of the Maclaurin series of $F(u)$ are also real.

Another source of functional inequalities is offered by q-difference equations which in the linear case are of the form

$$\sum_{m=0}^{n} a_{n-m}(z) f[q^m z] = 0, \qquad a_0(z) \equiv 1, \tag{12.5.6}$$

where the $a_k(z)$ are given analytic functions and $q \neq 1$ is a complex number. More generally, the multipliers $1, q, q^2, \cdots, q^n$ may be replaced by arbitrary distinct complex numbers $\lambda_0, \lambda_1, \cdots, \lambda_n$. The corresponding functional inequalities will then be of the form

$$f(\lambda_n t) \leqslant \sum_{k=0}^{n-1} b_{n-k}(t) f(\lambda_k t), \tag{12.5.7}$$

where now $1 = \lambda_0 < \lambda_1 < \lambda_2 < \cdots < \lambda_n$ and the b_k's are given real-valued continuous functions.

Neither (12.5.5) nor (12.5.7) seems to have received much attention in the literature. The author is in no position to develop a theory of such inequalities. What follows is just some examples of such inequalities with rambling comments.

Example 1. We start with the simple case

$$f(2t) \leqslant f(t), \qquad 0 \leqslant t, \tag{12.5.8}$$

which is a special case of (12.5.5) as well as of (12.5.7). Any positive decreasing function f satisfies this inequality on R^+. Thus a solution may increase arbitrarily fast as t decreases to 0. Not all solutions are of this form, however. Thus any real-valued periodic function of

$$\frac{\log t}{\log 2}$$

of period 1 will satisfy (12.5.8), actually with equality. Such a function approaches no limit as t tends to 0 or infinity.

Example 2. We take Problem 10, Exercise 12.1, with t replaced by $6t$, i.e.

$$f(6t) \leqslant \tfrac{1}{2} f(3t) + \tfrac{1}{3} f(2t), \tag{12.5.9}$$

which is of type (12.5.7). This inequality turned out to be determinative in $C^+[0, 6]$, actually in $C^+[0, \infty]$, with the unique solution $f = 0$. This does not mean that there are no positive solutions for $t \in R^+$, but they are unbounded as t approaches the origin. Thus any power of t of the form

$$t^{-p} \tag{12.5.10}$$

will satisfy (12.5.9) provided p exceeds the unique positive root of the transcendental equation

$$2^{s-1} + 3^{s-1} = 1, \tag{12.5.11}$$

which, to six decimal places, is 0.212127. Moreover,

$$-t^{-q} \tag{12.5.12}$$

satisfies for $q < s$. Since the inequality is linear in f, any linear combination of solutions with positive constant multipliers will also be a solution.

Example 3. The inequality

$$f(2t) \leqslant 2f(t) \tag{12.5.13}$$

is satisfied by any subadditive function. See Section 13.1 for properties of such functions.

Example 4. Our last example is

$$f(2t) \leqslant 2[f(t)]^2 - 1, \tag{12.5.14}$$

which is of type (12.5.5). If a is any real number, the functions

$$1, \quad \cos at, \quad \cosh at$$

are solutions since they satisfy (12.5.14) with equality. More generally, for any $A \geqslant 1$,

$$A, \quad A\cos at, \quad A\cosh at \tag{12.5.15}$$

are solutions. There are also polynomial solutions. A simple instance is that of the partial sums of the Maclaurin series of $\cosh at$. Thus

$$P_{2n}(t) = \sum_{k=0}^{n} \frac{(at)^{2k}}{(2k)!} \tag{12.5.16}$$

is always a solution. The verification is left to the reader.

With these scanty indications we leave this class of functional inequalities.

EXERCISE 12.5

1. How are the inequalities (12.5.5) and (12.5.7) reduced to the canonical form $f(t) \leqslant T[f](t)$ considered in this chapter?
2. Verify the assertions made in Example 1.
3. Same question for Example 2.

4. For $t \in R^+$ equation (12.5.9) has solutions of the form t^p and $-t^q$. Find the limitations on p and q.
5. Show that if a solution of (12.5.14) has a positive infimum m for $t \in R^+$, then $m \geqslant 1$.
6. Similarly, verify that a finite supremum M of a solution satisfies $M \geqslant 1$.
7. Show that $A \exp(at)$, $a > 0$, is a solution of (12.5.14) for any $A > \frac{1}{2}$.
8. Why are the polynomials (12.5.16) solutions?
9. Are the partial sums of the Maclaurin series of cos at also solutions?

COLLATERAL READING

The reader may consult Section 1.5 of:

HILLE, E., *Lectures on Ordinary Differential Equations,* Addison-Wesley, Reading, Mass. (1969). Sections 3.1 and 6.3 of this treatise have a bearing on Section 12.4 above.

For differential and integral inequalities the basic work is:

WALTER, W., *Differential- und Integralungleichungen,* Springer Tracts on Natural Philosophy, Vol. 2, Springer-Verlag, Berlin (1966).

The original Gronwall Lemma occurs in:

GRONWALL, T. H., Note on the derivative with respect to a parameter of the solutions of a system of differential equations, *Ann. Math.* (2) **20,** (1918) 292–6.

For antiderivations and relevant literature see:

MILLER, J. B., Some properties of Baxter operators, *Acta Math. Acad. Sci. Hung.* **17,** (1966) 387–400.

13 FUNCTIONAL INEQUALITIES II FUNCTIONS ON PRODUCT SPACES

Functional inequalities on product spaces play an important role in analysis. Two striking examples come to mind right away: subadditive functions and convex functions. Actually these are closely related. The first class is intimately connected with the concept of invariance under addition as well as with the notion of convexity. Functions which are piecewise convex or concave are, whether we like it or not, basic in calculus, elementary as well as advanced. Convexity is next to positivity the most fruitful concept in analysis, and its implications are legion. We hope to bring out some of these connections in the present chapter. Convexity is, of course, a geometrical concept, but ever since the days of Minkowski, who was the pioneer and pathfinder in this area, it has become the happy playground of the analyst.

There are four sections: Subadditive and suboperative functions; Semi-modules and subadditive functions in R^m; Convex functions; and Non-Archimedean valuations.

13.1 SUBADDITIVE AND SUBOPERATIVE FUNCTIONS

A general form of functional inequalities on product spaces is the following. Let $\circ$ denote an associative, binary and commutative operation and let $G(u, v)$ be a given symmetric function from $R \times R$ to R (more generally from $R^m \times R^m$ to R) and consider the inequality

$$f(x \circ y) \leqslant G[f(x), f(y)]. \tag{13.1.1}$$

There are a number of interesting special cases.

We shall start with

$$\circ = +, \qquad G(u, v) = u + v. \tag{13.1.2}$$

A function satisfying

$$f(x + y) \leqslant f(x) + f(y) \tag{13.1.3}$$

is said to be *subadditive*. The domain of definition $D = \mathfrak{D}[f]$ of f is for the time being taken to be a subset of $R = R^1$. $\mathfrak{D}[f]$ *must be closed under addition*. The three most important cases are: D is R or R^+ or Z^+ (= the set of positive integers). To get interesting results in the first two cases, we must assume f to be measurable. As will be seen in Section 14.2, there are non-measurable additive functions, every one of which is trivially subadditive.

As simple examples of subadditive functions in R^+ we present

$$x^\alpha,\ 0 \leqslant \alpha \leqslant 1, \quad \text{and} \quad -x^\beta,\ 1 \leqslant \beta < \infty. \tag{13.1.4}$$

The verification is left to the reader.

Theorem 13.1.1. *If $\mathfrak{D}[f] = R^+$, if f is measurable and subadditive and if f assumes the value $+\infty$ at most in a set of measure zero, then f is bounded above on every interval $[\alpha, \beta]$ where $0 < \alpha < \beta < +\infty$.*

Proof. Suppose the assertion is false and there exists an f with the stated properties which is not bounded in an interval $[\alpha, \beta]$. We can then find a sequence of points $\{x_n\}$ such that (1) $\alpha < x_n < \beta$, (2) $\lim_{n\to\infty} x_n \equiv x_0$ exists, (3) $f(x_n) > 2n$.

We have then for $0 < s < x_n$

$$2n < f(x_n) = f(x_n - s + s) \leqslant f(x_n - s) + f(s). \tag{13.1.5}$$

Thus either $f(x_n - s) > n$ or $f(s) > n$ and the measurable set

$$S_n = \{t;\ n < f(t),\ 0 < t < x_n\}$$

has measure

$$m[S_n] \geqslant \tfrac{1}{2}x_n > \tfrac{1}{2}\alpha.$$

Now consider the set

$$T_n = \{t;\ n < f(t),\ 0 < t < \beta\}.$$

It is measurable and contains S_n. Hence

$$m[T_n] > \tfrac{1}{2}\alpha.$$

This holds for every n and $T_n \supset T_{n+1} \supset \bigcap_k T_k \equiv S$. Since every inequality $f(t) > n$ holds in S, $f(t) = +\infty$ for all $t \in S$. Here $m[S] \geqslant \frac{1}{2}\alpha > 0$ so we have a contradiction. Thus for every measurable subadditive f and every interval I whose closure is in R^+ there is an $M(f, I) < \infty$ such that $f(x) \leqslant M(f, I)$ for $x \in I$. ∎

Theorem 13.1.2. *If in addition to the hypotheses of Theorem 13.1.1, f assumes the value $-\infty$ at most in a set of measure zero, then f is bounded in every interval $I = [\alpha, \beta]$, $0 < \alpha < \beta < \infty$.*

Proof. Again suppose there exists an f for which the assertion does not hold. Such an f is bounded above on I. It is to be shown that f is also bounded below in I. If not, there is a sequence $\{s_n\}$ such that (1) $\alpha < s_n < \beta$, (2) $\lim_{n\to\infty} s_n = s_0$ exists, and (3) $f(s_n) < -n$ for each n. Then for $\alpha \leqslant s \leqslant \beta$

$$f(s + s_n) \leqslant f(s) + f(s_n) < M - n, \tag{13.1.6}$$

where $M = M(f, I)$ is the least upper bound for f in I. This says that there is an interval $I_n = [s_n + \alpha, s_n + \beta]$ where

$$f(t) < M - n, \qquad t \in I_n,$$

and the length of I_n is $\beta - \alpha$. Set

$$I_n^* = \bigcup_{k \geq n} I_k.$$

Then $\{I_n^*\}$ is a nested sequence of intervals, each of length at least $\beta - \alpha$ and $f(t) < M - n$ for $t \in I_n^*$. It follows that all the inequalities are valid in $\bigcap_m I_m^*$; this is a measurable set of measure at least $\beta - \alpha > 0$. This requires that f assumes the value $-\infty$ in a set of positive measure and this violates the assumptions. It follows that no f can satisfy the stated hypotheses and be unbounded below in I. Hence there exists a finite $M_0(f, I)$ such that $f(x) \geq M_0(f, I)$ and f is bounded in I. ∎

The restriction to an interval bounded away from 0 and infinity is essential. The function $\exp(1/x)$ is subadditive (why?) and continuous in R^+ but clearly not bounded above as $x \downarrow 0$. The second function under (13.1.4) suggests that a function subadditive on R^+ may tend arbitrarily fast to $-\infty$ as $x \to +\infty$. The reader will have a chance to prove this in Problem 2, Exercise 13.1. The first function under (13.1.4) shows that $f(x)$ may tend to $+\infty$ with x, but here the rate of growth appears to be more moderate. In fact, we have

Theorem 13.1.3. *If f is measurable, finite-valued and subadditive in R^+, then*

$$\lim_{x \to \infty} \frac{f(x)}{x} = \inf_{x > 0} \frac{f(x)}{x} < \infty. \tag{13.1.7}$$

Proof. The infimum in question is either a finite number b or $-\infty$. We shall give the proof for the finite case. By the definition of the infimum, we can find a positive number a such that for a given, arbitrarily small, $\varepsilon > 0$ we have

$$\frac{f(a)}{a} < (1 + \varepsilon)\, b.$$

Suppose now that $x = (n + 1)\, a + c$ where $0 \leq c < a$. Then

$$b \leq \frac{f(x)}{x} = \frac{f(na + a + c)}{(n + 1)\, a + c} \leq \frac{nf(a) + f(a + c)}{(n + 1)\, a + c}$$

$$\leq \frac{na}{(n + 1)\, a + c}(1 + \varepsilon)\, b + \frac{M(f, I)}{(n + 1)\, a + c}. \qquad I = [a, 2a].$$

As $n \to \infty$ the last member tends to $(1 + \varepsilon)\, b$ and (13.1.7) follows. The case where the infimum is $-\infty$ is left to the reader. ∎

If $\mathfrak{D}[f] = R$ instead, the corresponding results are

Theorem 13.1.4. *If f is measurable, finite-valued and subadditive in R, then f is bounded on compact subsets of R.*

Theorem 13.1.5. *Under the same assumptions*

$$\lim_{x\to+\infty}\frac{f(x)}{x}=\inf_{0<x}\frac{f(x)}{x}=b<\infty, \tag{13.1.8}$$

$$\lim_{x\to-\infty}\frac{f(x)}{x}=\sup_{x<0}\frac{f(x)}{x}=a>-\infty. \tag{13.1.9}$$

Here $a\leqslant b$.

The proofs are left to the reader.
From $f(2x)\leqslant 2f(x)$ it follows that

$$\limsup_{x\downarrow 0} f(x)=\limsup_{x\downarrow 0} f(2x)\leqslant 2\limsup_{x\downarrow 0} f(x). \tag{13.1.10}$$

Hence, if $\limsup_{x\downarrow 0} f(x)$ is finite, it is necessarily non-negative. This is also true for $\liminf_{x\downarrow 0} f(x)$. If f is defined and subadditive in R, there are similar inequalities for approach to $x=0$ from the left. A particularly interesting case is that where

$$\lim_{x\to 0} f(x)=0 \tag{13.1.11}$$

either for approach from the right or for two-sides approach.

Theorem 13.1.6. *If f is measurable, finite-valued and subadditive in R^+ and if* (13.1.11) *holds, then for* $x>0$

$$\limsup_{h\downarrow 0} f(x+h)\leqslant f(x)\leqslant \liminf_{h\downarrow 0} f(x-h). \tag{13.1.12}$$

If the same assumptions are valid in R, then f is continuous in R.

Proof. Here (13.1.12) follows from the inequalities

$$f(x+h)\leqslant f(x)+f(h), \qquad f(x)\leqslant f(x-h)+f(h).$$

If (13.1.11) holds for two-sided approach, then h can be replaced by $-h$ in both inequalities to obtain

$$\limsup_{h\downarrow 0} f(x-h)\leqslant f(x)\leqslant \liminf_{h\downarrow 0} f(x+h).$$

This combined with (13.1.12) gives continuity. ∎

Moduli of continuity [see (3.2.11) and Problem 5 of Exercise 3.2] are examples of subadditive functions. So are Minkowski functionals or gauge functions [see (1.6.18) and following]. Subadditive functions also play a basic role in the theory of semi-groups of linear operators.

C. T. Ionescu Tulčea (1960) generalized the inequality (13.1.3) in various directions. One of his inequalities is

$$f(x+y)\leqslant g(x)+h(y), \tag{13.1.13}$$

which has important bearings on semi-group theory. Among other results he proved a generalization of Theorem 13.1.1.

Theorem 13.1.7. *If g and h assume the value $+\infty$ at most on sets of zero measure and if either g or h is measurable, then f is bounded above on compact sets.*

Proof. To fix the ideas, suppose that (13.1.13) holds for x and y in R^+ and that g is measurable. Then f can take the value $+\infty$ at most on a set of measure zero. Suppose now that f is not bounded above in $[\alpha, \beta]$. There is then a sequence of points $\{x_n\}$ in this interval where $f(x_n) > 2n$. Let x_0 satisfy $0 < x_0 < \inf x_n$. We have now

$$2n < f(x_n) \leqslant \begin{cases} g(s) + h(x_n - s), \\ & 0 < s < x_n. \\ g(x_n - s) + h(s), \end{cases}$$

Set

$$S_n = \{s;\, n \leqslant g(s),\, 0 < s < x_0\},$$
$$T_n = \{s;\, n > g(s),\, 0 < s < x_0\}. \tag{13.1.14}$$

These are disjoint measurable sets so that

$$m[S_n] + m[T_n] = x_0.$$

In T_n we have

$$h(x_n - s) \geqslant n.$$

Next note that

$$S_n \supset S_{n+1} \supset \bigcap_k S_k \equiv S.$$

If $m[S] \neq 0$, then $g(s) \geqslant n$ for all n and all $s \in S$. This requires that $g(s) = +\infty$ in S, a set of positive measure, and this contradicts the assumptions.

Suppose now that S is of measure zero. This implies that $\{m[S_n]\}$ is a non-increasing sequence of limit 0. Then $\{m[T_n]\}$ is a non-decreasing sequence with limit x_0. In the sequence $\{x_n\}$ we can find a sub-sequence $\{x_{n_k}\}$ which converges to a limit $x^* > x_0$. Set

$$U_n = \{s;\, s = x_n - t,\, t \in T_n\}. \tag{13.1.15}$$

The sets U_{n_k} are confined to the interval $[x^* - x_0 - \varepsilon, x^* + \varepsilon]$ for all large values of k. These sets are measurable and $m[U_{n_k}]$ increases to x_0. Further, $h(s) > n_k$ for s in U_{n_k}. Set

$$V_p = \bigcup_{p \leqslant k} U_{n_k}. \tag{13.1.16}$$

Then $m[V_p] \geqslant x_0$ for each p and $h(s) \geqslant n_p$ for all $s \in V_p$. Here $V_p \supset V_{p+1} \supset \bigcap_p V_p \neq \emptyset$. It follows that $h(s) = +\infty$ in a set of positive measure and this is again a contradiction. The reader should note that our discussion of sets in which $h(s)$ is large does not involve a tacit assumption that h is measurable. Thus it is seen that f must be bounded on compact sets. ▮

There is an interesting generalization of the subadditive inequality to the case

$$f(x \circ y) \leqslant f(x) + f(y), \tag{13.1.17}$$

where again $\circ$ denotes an associative, binary, commutative operation. A solution of such an equation is said to be *suboperative* (with respect to the operation $\circ$). E. Hille and S. Phillips (1957) have considered special cases of such inequalities and given some discussion of the general case. The latter was treated in detail by C. T. Ionescu Tulčea in 1960. Tulčea also considered the generalization

$$f(x \circ y) \leqslant g(x) + h(y), \tag{13.1.18}$$

which is important in the theory of Lie semi-groups of linear operators in the strong topology.

Subadditive functions on $R^m \times R^m$ to R have been studied by R. A. Rosenbaum (1950). Some related results are discussed in the next section.

Actually subadditivity extends to more general situations. In considering the inequality (13.1.3) it is necessary to specify domain and range of f. The domain $\mathfrak{D}[f]$ in an additive space must be closed under addition and the range $\mathfrak{R}[f]$ must belong to an additive partially ordered space. There is no lack of such sets $\mathfrak{D}$ and $\mathfrak{R}$ and each choice gives rise to a theory of subadditive functions. A particular extension may be trivial but should not be condemned without a fair trial.

EXERCISE 13.1

1. Prove that a positive non-increasing function on R^+ is subadditive. What consequences will this have for the question of growth properties of a subadditive function?
2. If g is positive and non-decreasing on R^+, prove that $-g(x)$ is subadditive on R^+. Growth properties?
3. Prove (13.1.7) if the infimum is $-\infty$.
4. Prove Theorem 13.1.4.
5. Prove Theorem 13.1.5.
6. Verify that x^k, $0 \leqslant k \leqslant 1$, is subadditive in R^+. How about $|x|^k$ in R?
7. Verify that $|\sin x|$ is subadditive in R.
8. Prove the analogue of Theorem 13.2.1 for functions subadditive in an interval (a, ∞) where $0 < a$. The method used above applies to an interval $[\alpha, \beta]$ with $2a < \alpha$.
9. Extend Theorem 13.1.3 under similar assumptions. What infimum enters in the limit relation?
10. Formulate and prove analogues of Theorem 13.1.3 for Z^+, i.e. if $\{a_n\}$ is a sequence such that $a_{m+n} \leqslant a_m + a_n$ for all positive integers, what is $\lim_{n\to\infty} a_n/n$?

11. Does an analogue of Theorem 13.1.1 hold for the suboperative inequality (13.1.17) when $\circ = \cdot$ and $\mathfrak{D}[f] = R^+$?
12. What is the analogue of Theorem 13.1.3 for this case?
13. Suppose that $x \circ y = (x + y)(1 + xy)^{-1}$ and consider the corresponding inequality (13.1.17). Show that the operation has the required properties. Show the existence of a function g such that $f[g(t)]$ is subadditive. Find analogues of Theorems 13.1.1 and 13.1.3.

13.2 SEMI-MODULES AND SUBADDITIVE FUNCTIONS IN R^m

We recall that the domain of definition of a subadditive function f must be invariant under addition. This motivates the following

Definition 13.2.1. *A semi-module $\mathfrak{S}$ is an addition-invariant subset of a vector space $\mathfrak{X}$ such that* (1) *$\mathfrak{X}$ is a commutative group under addition satisfying Postulates* (A_1) *to* (A_5) *of Definition* 2.1.1, (2) *$\mathfrak{X}$ is a Hausdorff space in which the operations $(x, y) \to x + y$ and $x \to -x$ are continuous.*

Thus if $x, y \in \mathfrak{S}$, so does $x + y$. Hence any semi-module $\mathfrak{S}$ can be the domain of definition of a subadditive function and, vice versa, the domain of definition of a subadditive function is a semi-module.

We start by deriving some properties of semi-modules to be used in the following. The first is a lemma the proof of which is left to the reader.

Lemma 13.2.1. *$\mathfrak{S} \subset \mathfrak{X}$ is a semi-module, so is the closure of $\mathfrak{S}$, as well as the interior of $\mathfrak{S}$.*

The zero element plays a peculiar role in our theory, as is shown already on the line. The following class of semi-modules is particularly important.

Definition 13.2.2. *An angular semi-module is one where* (1) *the zero element belongs to $\overline{\mathfrak{S}}$ and* (2) *$\mathfrak{S}$ is open.*

The reason for the name will become clear later. On the line there are only three angular semi-modules, namely, R^+, R^-, and R itself. Already in the plane there is a continuum of possibilities, and the angular character becomes evident. In R^m, $m > 1$, there is a peculiar alternating role of semi-modules and subadditive functions. To define a subadditive function on R^m we need an angular semi-module, but the boundary of the latter is defined by a subadditive function on R^{m-1} and so on until we get down to R^1. We shall get a glimpse of this interplay in the following. The next result still holds for an additive Hausdorff group $\mathfrak{X}$.

Lemma 13.2.2. *If $\mathfrak{S} \subset \mathfrak{X}$ is an angular semi-module, then* $\mathfrak{S} = \operatorname{Int}(\overline{\mathfrak{S}})$.

Proof. Since $\mathfrak{S}$ is open by assumption, $\mathfrak{S} = \operatorname{Int}(\mathfrak{S}) \subseteq \operatorname{Int}(\overline{\mathfrak{S}})$. To show that the inclusion is actually an equality, we have to prove that no point x of the complement of $\mathfrak{S}$ can be in Int $(\bar{\mathfrak{S}})$. Suppose that $x \notin \mathfrak{S}$ and let N_x be a neighborhood of x, arbitrarily small. It is enough to show that N_x contains an open set G such that $G \cap \mathfrak{S} = \emptyset$. To this end, take a neighborhood N_0 of the origin so small that $x - N_0 \subset N_x$. By assumption (1) of Definition 13.2.2, the zero element of $\mathfrak{X}$ belongs to $\bar{\mathfrak{S}}$. Since $\mathfrak{S}$ is open, we can find an element y which together with a full neighborhood N_y belongs to $\mathfrak{S} \cap N_0$. Then $G = x - N_y$ is an open set belonging to N_x. No point u of G can belong to $\mathfrak{S}$, for if it did, then $x = (x - u) + u$ would belong to $\mathfrak{S}$ since $x - u$ already has this property (why?). It follows that $x \notin$ Int $(\bar{\mathfrak{S}})$. ∎

Corollary. *Under the same assumptions* $\mathfrak{S} + \bar{\mathfrak{S}} \subset \mathfrak{S}$.

Proof. Let $x \in \mathfrak{S}$, $y \in \bar{\mathfrak{S}}$. Since $\mathfrak{S}$ is open, there is a neighborhood N_x of x also in $\mathfrak{S}$ and hence in $\bar{\mathfrak{S}}$. This says that the open set $y + N_x \subset \bar{\mathfrak{S}}$ and hence it belongs to Int $(\bar{\mathfrak{S}}) = \mathfrak{S}$. ∎

From this point on we specialize and take $\mathfrak{X} = R^m$.

Lemma 13.2.3. *If $\mathfrak{S}$ is a semi-module in R^m and if every neighborhood N_0 of* $\mathbf{0}$ *contains an element of $\mathfrak{S}$ distinct from* $\mathbf{0}$, *then there exists at least one non-zero vector* $\mathbf{b}$ *such that $\rho\mathbf{b} \in \bar{\mathfrak{S}}$ for all* $\rho > 0$.

Proof. The assumptions imply the existence of a sequence $\{\mathbf{x}_k\} \subset \mathfrak{S}$ with $\mathbf{x}_k \neq \mathbf{0}$ and $\lim_{k\to\infty} \mathbf{x}_k = \mathbf{0}$. The sequence of unit vectors $\{\mathbf{x}_k \|\mathbf{x}_k\|^{-1}\}$ admits of at least one vector which is a cluster point of the sequence and without loss of generality we may assume that the whole sequence converges to a limit $\mathbf{b}$ with $\|\mathbf{b}\| = 1$. It is to be shown that all vectors $\rho\mathbf{b}$, $0 < \rho$, are in $\bar{\mathfrak{S}}$. To this end, set $n_k = [\rho\|\mathbf{x}_k\|^{-1}] + 1$ where $[u]$ is the largest integer less than or equal to u. Now $\{n_k\mathbf{x}_k\}$ is a sequence of vectors in $\mathfrak{S}$, for if $\mathbf{x} \in \mathfrak{S}$, so does $n\mathbf{x}$ for any natural number n. Here

$$n_k\mathbf{x}_k = \{[\rho\|\mathbf{x}_k\|^{-1}] + 1\}\,\mathbf{x}_k,$$

with the limit $\rho\mathbf{b}$ as $k \to \infty$. It follows that $\rho\mathbf{b} \in \bar{\mathfrak{S}}$. ∎

Corollary. *The conclusion of Lemma 13.2.3 holds for any angular semi-module in R^m.*

It may very well happen that there is one and only one vector $\mathbf{b}$ with the stated properties. This is illustrated by

Example 1. Take the plane R^2 and define $\mathfrak{S}$ as the set of points (x, y) such that $0 < x$, $0 < y < x^2$. This is seen to be an angular semi-module and the only unit vector $\mathbf{b}$ such that $\rho\mathbf{b} \in \bar{\mathfrak{S}}$ is $\mathbf{b} = (1, 0)$ and this vector is in $\bar{\mathfrak{S}}$ but not in $\mathfrak{S}$. There is no vector in $\mathfrak{S}$ with the stated properties.

Lemma 13.2.4. *If $\mathfrak{S}$ is an angular semi-module in R^m, then there is at least one unit vector* $\mathbf{b}$ *such that* $\mathbf{x} \in \mathfrak{S}$ *implies* $\mathbf{x} + \rho\mathbf{b} \in \mathfrak{S}$, $0 \leqslant \rho$.

Proof. We take $\mathbf{b}$ as in the preceding lemma and combine with the corollary of Lemma 13.2.2. Here $\mathbf{x} \in \mathfrak{S}$, $\rho\mathbf{b} \in \overline{\mathfrak{S}}$, so that $\mathbf{x} + \rho\mathbf{b} \in \mathfrak{S} + \overline{\mathfrak{S}} \subset \mathfrak{S}$. ∎

This says that an angular semi-module $\mathfrak{S}$ in R^m is the union of a system of parallel line segments extending to infinity in one direction. The problem is now to characterize the other endpoints. What is the greatest lower bound for ρ in order that $\mathbf{x} + \rho\mathbf{b}$ shall belong to $\mathfrak{S}$? In Example 1 there is a simple answer to this question: for a given $y > 0$ we must have $x > y^{\frac{1}{2}}$. It is no accident that $f(y) = y^{\frac{1}{2}}$ is actually a subadditive function of y for $y > 0$. We shall carry out the characterization of the boundary of angular semi-modules in R^2, where it is easy to visualize what is going on. The reader will find it instructive to carry out the analogous description in R^3. The basis for the discussion is the observation that the projection of an angular semi-module in R^m on the hyperplane perpendicular to the distinguished vector $\mathbf{b}$ is an angular semi-module in R^{m-1}. To this new semi-module the same observation applies, which gives the step-by-step reduction.

We now take an angular semi-module in R^2 represented by the complex plane. Since a rotation about the origin does not change the character of $\mathfrak{S}$ being a semi-module, we may assume that the distinguished vector $\mathbf{b}$ is $(1, 0)$. Then the points of $\mathfrak{S}$ are of the form $z = x + iy$ and if $z_0 \in \mathfrak{S}$, so does $z_0 + \rho$, $0 < \rho$. We denote the projection of $\mathfrak{S}$ on the imaginary axis by $\mathfrak{P}$, so that

$$\mathfrak{P} = \{y;\, x + iy \in \mathfrak{S}\}. \tag{13.2.1}$$

Lemma 13.2.5. *$\mathfrak{P}$ is an angular semi-module on R.*

Proof. Let $z_1 \in \mathfrak{S}$, $z_2 \in \mathfrak{S}$ where $z_1 = x_1 + iy_1$, $z_2 = x_2 + iy_2$. This implies that y_1 and y_2 belong to $\mathfrak{P}$ as well as their sum $y_1 + y_2$. Since $\mathfrak{S}$ is open, so is $\mathfrak{P}$, and since $0 \in \overline{\mathfrak{S}}$ we have $0 \in \overline{\mathfrak{P}}$ and this proves the assertion. ∎

There are only three possibilities for $\mathfrak{P}$, namely R, R^+, and R^-. We can discard the last case since it is symmetric to R^+ and does not give rise to a different theory. We now define

$$f(y) = \inf\{x;\, x + iy \in \mathfrak{S}\}. \tag{13.2.2}$$

Theorem 13.2.1. *$f(y)$ is a subadditive function of y on $\mathfrak{P}$, upper semi-continuous and with* $\liminf_{y \to 0} f(y) = 0$ *or* $-\infty$. *In the latter case* $f(y) \equiv -\infty$ *in* $\mathfrak{P}$.

Proof. Suppose $y_1 \in \mathfrak{P}$, $y_2 \in \mathfrak{P}$. By the property of an infimum, for every $\delta > 0$ there is an x_1, with $x_1 + iy_1 \in \mathfrak{S}$, such that $x_1 < f(y_1) + \delta$. There is also an x_2, with $x_2 + iy_2 \in \mathfrak{S}$, such that $x_2 < f(y_2) + \delta$. Set $z_1 = x_1 + iy_1$, $z_2 = x_2 + iy_2$. By the continuity of the mapping $(z_1, z_2) \to z_1 + z_2$, for every neighborhood N of $z_1 + z_2$ there are neighborhoods N_1 of z_1 and N_2 of z_2 such that $N_1 + N_2 \subset N$

where $N_1 = \{x + iy; |x - x_1| < \delta, |y - y_1| < \delta\}$ and $N_2 = \{x + iy; |x - x_2| < \delta, |y - y_2| < \delta\}$. From $x_1 < f(y_1) + \delta$ and $x_2 < f(y_2) + \delta$ we infer that

$$f(y_1 + y_2) < x_1 + x_2 < f(y_1) + f(y_2) + 2\delta.$$

Here δ can be arbitrarily small so that

$$f(y_1 + y_2) \leqslant f(y_1) + f(y_2),$$

which is the subadditive property.

To prove the upper semi-continuity of f, let $y_1 \in \mathfrak{P}$. By the property of the infimum, for every $\delta > 0$ there is an x_1 with $z_1 = x_1 + iy_1 \in \mathfrak{S}$ such that $x_1 < f(y_1) + \frac{1}{2}\delta$. Since $\mathfrak{S}$ is open, there is a neighborhood N_1 of z_1 such that $N_1 \subset \mathfrak{S}$. Now $x_1 < f(y_1) + \frac{1}{2}\delta$ implies $f(y) < x < x_1 + \frac{1}{2}\delta < f(y_1) + \delta$ for every

$$x + iy \in N_1 = \{x + iy; |x - x_1| < \tfrac{1}{2}\delta, |y - y_1| < \tfrac{1}{2}\delta\}.$$

This is the upper semi-continuity. Such a function, in particular, is measurable and the set where $f(y) = +\infty$ is void. Finally, since $0 \in \bar{\mathfrak{S}}$, the inequality

$$f(2y) \leqslant 2f(y)$$

implies

$$\liminf_{y \to 0} f(2y) = \liminf_{y \to 0} f(y) \leqslant 2 \liminf_{y \to 0} f(y)$$

with the two alternatives $\liminf_{y \to 0} f(y) = 0$ or $-\infty$. In the second case

$$\liminf_{h \to 0} f(y + h) \leqslant f(y) + \liminf_{h \to 0} f(h) = -\infty$$

for all $y \in \mathfrak{P}$. ∎

The second alternative cannot be excluded. Thus if $\mathfrak{S} = C$, the whole complex plane, $f(y) = -\infty$ for all $y \in R$. Similarly for $\mathfrak{S}$ equal to the upper or the lower half-plane where $f(y) = -\infty$ in R^+ or in R^-, respectively.

Lemma 13.2.6. *An angular semi-module in R^2 is a simply connected point set. It is either the whole plane or a subset of a half-plane.*

Proof. We can join two points z_1 and z_2 in $\mathfrak{S}$ by a broken line $[z_1, z_3, z_4, z_2]$ where $z_3 = x_3 + iy_1$, $z_4 = x_3 + iy_2$, and x_3 exceeds the finite least upper bound of $f(y)$ in the interval $[y_1, y_2]$. If $\mathfrak{P} = R^+$ or R^-, then $\mathfrak{S}$ is confined to a half-plane, the upper or the lower, respectively. If $\mathfrak{P} = R$, then the positive real axis belongs to $\mathfrak{S}$. Suppose now that there are vectors z_1 and z_2 in $\mathfrak{S}$ where $\arg z_k = \theta_k$ and $-\pi < \theta_1 < 0 < \theta_2 < \pi$. Then by the additivity all points $z = re^{i\theta}$ with $\theta_1 - \varepsilon < \theta < \theta_2 + \varepsilon$ and large values of r are in $\mathfrak{S}$. If now $\theta_2 - \theta_1 \geqslant \pi$, then all the distant points in the complementary sector would also be in $\mathfrak{S}$ and this can be true iff $\mathfrak{S}$ is the whole plane. It follows that if $\mathfrak{S} \neq C$, then $\mathfrak{S}$ will occupy an angle at the origin of opening at most π. ∎

Similarly, we see that an angular semi-module in R^m, $m > 2$, is either R^m itself or is confined to a half-space. If $\mathbf{v}$ denotes a vector in R^m and $\mathfrak{S}$ is a semi-module in R^m, we shall consider a subadditive function $F(\mathbf{v})$ defined in $\mathfrak{S}$. Thus

$$F(\mathbf{v}_1 + \mathbf{v}_2) \leqslant F(\mathbf{v}_1) + F(\mathbf{v}_2) \tag{13.2.3}$$

for all $\mathbf{v}_1$, $\mathbf{v}_2$ in $\mathfrak{S}$. F shall be finite-valued and measurable. Here measurability is taken in a rather sweeping sense and is meant to include k-dimensional measurability of the restrictions of F to k-dimensional manifolds for all $k \leqslant m$. In particular, we shall need *radial measurability*.

Definition 13.2.3. *A function F from R^m to R^1 defined in an angular semi-module $\mathfrak{S}$ is said to be radially measurable if for each fixed $\mathbf{v} \in \mathfrak{S}$ the function $F(r\mathbf{v})$ is a measurable function of r for all large values of r.*

Since $\mathfrak{S}$ is open and $\mathbf{v} \in \mathfrak{S}$ by assumption, then $r\mathbf{v} \in \mathfrak{S}$ for all values of r in a sequence of intervals

$$[n(1 - \delta), n(1 + \delta)], \qquad n = 1, 2, 3, \ldots$$

for a small value of δ. These intervals ultimately overlap. Thus there exists for each $\mathbf{v} \in \mathfrak{S}$ a number $a(\mathbf{v}) \geqslant 0$ such that $F(r\mathbf{v})$ is defined for all $r > a(\mathbf{v})$.

Theorem 13.2.2. *Let F be defined, finite-valued, radially measurable and subadditive in an angular semi-module $\mathfrak{S} \in R^m$. Then*

$$\lim_{r \to \infty} \frac{F(r\,\mathbf{v})}{r} \equiv G(\mathbf{v}) \tag{13.2.4}$$

exists for all $\mathbf{v} \in \mathfrak{S}$. Moreover,

$$-\infty \leqslant G(\mathbf{v}) = \inf \left\{ \frac{F(r\,\mathbf{v})}{r};\, \mathbf{v} \in \mathfrak{S},\, r > 2\,a(\mathbf{v}) \right\}. \tag{13.2.5}$$

Proof. For fixed $\mathbf{v} \in \mathfrak{S}$ the function $F(r\mathbf{v})$ is defined for $r > a(\mathbf{v})$ where $a(\mathbf{v})$ is the infimum of the numbers a such that $r\mathbf{v} \in \mathfrak{S}$ for all $r > a$. Further, $F(r\mathbf{v})$ is a subadditive function of r for $r > a(\mathbf{v})$. Now Theorem 13.2.3 may be extended to functions $f(r)$ which are subadditive on an interval (a, ∞). Again the limit of $f(r)/r$ exists and equals the infimum of the ratio for $r > 2a$. See Problem 9, Exercise 13.1. ∎

To illustrate we shall work out some examples in R^2, again represented by the complex plane. As we shall see, it is enough to work out the results for unit vectors. We set

$$G(e^{i\theta}) = g(\theta). \tag{13.2.6}$$

Example 2. Take $\mathfrak{S}$ as the right half-plane, $x > 0$, and set

$$F(z) = F(x + iy) = |y|^{\frac{1}{2}} + x. \tag{13.2.7}$$

This is the sum of an additive and a subadditive function, each of one variable, so the sum is subadditive in z. Here

$$g(\theta) = \cos\theta.$$

Example 3. Again take $\mathfrak{S}$ as the right half-plane and set

$$F(x + iy) = y^2 x^{-1}. \tag{13.2.8}$$

The subadditivity follows from the identity

$$x_2(x_1 + x_2)y_1{}^2 + x_1(x_1 + x_2)y_2{}^2 - x_1x_2(y_1 + y_2)^2 = (x_1y_2 - x_2y_1)^2 \geqslant 0.$$

Here

$$g(\theta) = \sin^2\theta \sec\theta, \qquad |\theta| < \tfrac{1}{2}\pi. \tag{13.2.9}$$

If $\mathfrak{S} = C$, then $g(\theta)$ is defined for all θ. In the general case there exists an angle of opening $\leqslant \pi$ in which g is defined. This includes the possibility that g is identically $-\infty$ in part or all of the sector. The following example illustrates this phenomenon.

Example 4. Take $\mathfrak{S}$ to be the right half-plane and set

$$F(x + iy) = -x^2 = -r^2\cos^2\theta, \qquad |\theta| < \tfrac{1}{2}\pi, \tag{13.2.10}$$

with $g(\theta) = -\infty$, for all θ.

Following a terminology introduced by G. Pólya (1929) for functions typified by $g(\theta)$, we shall refer to the function G defined by (13.2.4) as the *radial growth indicator* or simply as the *indicator* of the subadditive function F defined on the semi-module $\mathfrak{S}$. Such indicators were first introduced by E. Phragmén (1863–1937) and Lindelöf in their epoch-making memoir on the principle of the maximum of 1908. If f is (Cauchy) holomorphic in a sector $\mathfrak{S}$, say $|\arg z| < \alpha$, and is of exponential growth in $\mathfrak{S}$, i.e.

$$|f(z)| \leqslant M \exp(B|z|), \qquad z \in \mathfrak{S},$$

then

$$\limsup_{r\to\infty} \frac{1}{r} \log |f(re^{i\theta})| \equiv g(\theta), \qquad |\theta| < \alpha, \tag{13.2.11}$$

exists and $-\infty \leqslant g(\theta) \leqslant B$. Phragmén and Lindelöf studied the properties of this function and observed its convexity properties. A more detailed study was made in 1929 by Pólya, who was able to utilize the advances made in the theory of convex domains and convex functions in the meantime. Pólya based his study on a three-membered functional inequality satisfied by g, our formula (13.2.23) below.

So much for the early history. The indicator G has a number of interesting properties. We expect it to have connections with convexity and our expectations will be amply fulfilled. The basic properties follow more or less directly from the definition.

Theorem 13.2.3. *G is positive-homogeneous, i.e.*

$$G(a\,\mathbf{v}) = a\,G(\mathbf{v}), \qquad \mathbf{v} \in \mathfrak{S}, \qquad 0 < a. \tag{13.2.12}$$

Proof by inspection. ∎

Actually we see that the domain of definition of G is not $\mathfrak{S}$ but $\mathsf{C}[\mathfrak{S}]$, the least cone with vertex at the origin which contains $\mathfrak{S}$, in other words, the *convex hull* of $\mathfrak{S}$. Thus, while $\mathfrak{S}$ is not necessarily a convex domain in R^m, $\mathsf{C}[\mathfrak{S}]$ has this property.

The subadditivity of F in $\mathfrak{S}$ implies subadditivity of G in $\mathsf{C}[\mathfrak{S}]$.

Theorem 13.2.4. *If* $\mathbf{v}_1, \mathbf{v}_2 \in \mathsf{C}[\mathfrak{S}]$, *then*

$$G(\mathbf{v}_1 + \mathbf{v}_2) \leqslant G(\mathbf{v}_1) + G(\mathbf{v}_2). \tag{13.2.13}$$

Proof. For large values of r the vectors $r\mathbf{v}_1$ and $r\mathbf{v}_2 \in \mathfrak{S}$ if $\mathbf{v}_1, \mathbf{v}_2 \in \mathsf{C}[\mathfrak{S}]$. Hence

$$F[r(\mathbf{v}_1 + \mathbf{v}_2)] \leqslant F(r\mathbf{v}_1) + F(r\mathbf{v}_2).$$

Divide by r and pass to the limit to obtain (13.2.13). ∎

Combination of (13.2.12) and (13.2.13) yields convexity.

Theorem 13.2.5. *If* $\mathbf{x}_1, \mathbf{x}_2 \in \mathsf{C}[\mathfrak{S}]$, *so does* $\frac{1}{2}(\mathbf{x}_1 + \mathbf{x}_2)$ *and*

$$G[\tfrac{1}{2}(\mathbf{x}_1 + \mathbf{x}_2)] \leqslant \tfrac{1}{2}[G(\mathbf{x}_1) + G(\mathbf{x}_2)]. \tag{13.2.14}$$

Proof. The first assertion follows from the convexity of $\mathsf{C}[\mathfrak{S}]$, the second from

$$G[\tfrac{1}{2}(\mathbf{x}_1 + \mathbf{x}_2)] \leqslant G(\tfrac{1}{2}\mathbf{x}_1) + G(\tfrac{1}{2}\mathbf{x}_2) = \tfrac{1}{2}G(\mathbf{x}_1) + \tfrac{1}{2}G(\mathbf{x}_2). \ ∎$$

For the properties of convex functions on R^m to be used in the following, see the end of the next section. It is shown there that (13.2.14) implies

$$G[a_1\mathbf{x}_1 + a_2\mathbf{x}_2 + \cdots + a_n\mathbf{x}_n] \leqslant a_1G(\mathbf{x}_1) + a_2G(\mathbf{x}_2) + \cdots + a_nG(\mathbf{x}_n) \tag{13.2.15}$$

for any choice of points $\mathbf{x}_1, \mathbf{x}_2, \ldots, \mathbf{x}_n$ in the domain of definition of G, $\mathsf{C}[\mathfrak{S}]$ in our case, and for any choice of non-negative weights a_j of sum unity. Actually in Section 13.3 the argument is given only for weights which are rational numbers. In general, irrational weights require continuity of G, a property which we want to prove rather than assume. But in the present case (13.2.15) holds for arbitrary real non-negative weights. We need merely to use subadditivity and imitate the derivation of (13.2.14) from (13.2.13). Thus we have

Corollary of Theorem 13.2.5. *Formula* (13.2.15) *holds for any choice of points in* $\mathsf{C}[\mathfrak{S}]$ *and any non-negative weights of sum unity.*

We have seen that an indicator G can take on the value $-\infty$. Formula (13.2.15) throws further light on this situation. Take $n = m$, the dimension of the space and of $\mathfrak{S}$ and $\mathsf{C}[\mathfrak{S}]$, and choose m linearly independent vectors $\{\mathbf{x}_j\}$ in $\mathsf{C}[\mathfrak{S}]$. If $G(\mathbf{x}_j) = -\infty$ for some j, then the inequality shows that $G(\mathbf{x}) = -\infty$ for all vectors $\mathbf{x}$ which are linear combinations of the basis vectors with positive weights adding up to unity. This is an $(m-1)$-dimensional variety, $\mathfrak{B}_0$ say. But the conclusion also holds for any vector $\mathbf{y} = a\mathbf{x}$ such that $\mathbf{x} \in \mathfrak{B}_0$, $0 < a$, and this is a cone V, a conical subset of C[$\mathfrak{S}$]. By varying the basis vectors we extend the conclusion to all of C[$\mathfrak{S}$]. Hence we have

Theorem 13.2.6. *G takes on the value $-\infty$ either everywhere in* C[$\mathfrak{S}$] *or nowhere.*

Ultimately this result is based on Theorem 13.1.3, which states that for a measurable finite-valued function, subadditive on R^+, the ratio $f(r)/r$ tends to a limit when r becomes infinite and this limit may be $-\infty$. This possibility is excluded if f is subadditive on R. Consequently, if F is defined, finite-valued, measurable (including radially measurable) and subadditive on all of R^m, then for $\mathbf{x}$ in a suitable half-space

$$-\infty < G(-x) \leqslant G(x) < \infty. \tag{13.2.16}$$

G is bounded above in C[$\mathfrak{S}$] in the following sense:

Theorem 13.2.7. *If G is finite-valued in* C[$\mathfrak{S}$], *then for each proper sub-cone* V *there is a number M*[V] *such that*

$$G(\mathbf{x}) \leqslant M[\mathsf{V}]\,\|\mathbf{x}\|, \qquad \forall\, \mathbf{x} \in \mathsf{V}. \tag{13.2.17}$$

Proof. As above, choose a basis $\{\mathbf{u}_j\}$ of vectors belonging to C[$\mathfrak{S}$]. This time each $\mathbf{u}_j$ shall be a unit vector. The $(m-1)$-dimensional polytope $\mathfrak{B}_0$ now consists of all vectors of the form

$$\mathbf{x} = \sum_{j=1}^{m} a_j \mathbf{u}_j, \qquad a_j \geqslant 0, \qquad \sum_{j=1}^{m} a_j = 1, \tag{13.2.18}$$

and

$$G(\mathbf{x}) \leqslant \sum_{j=1}^{m} a_j G(\mathbf{u}_j) \leqslant \max_j G(\mathbf{u}_j). \tag{13.2.19}$$

Here the distance of $\mathfrak{B}_0$ from the origin is a positive number d which depends upon the geometrical configuration and is, in any case, <1. If V is the cone swept out by $\mathfrak{B}_0$ we have, for any $\mathbf{x} \in$ V,

$$G(\mathbf{x}) \leqslant d^{-1}\left[\max_j G(\mathbf{u}_j)\right] \|\mathbf{x}\|. \tag{13.2.20}$$

Thus we can take $M[\mathsf{V}] = d^{-1} \max G(\mathbf{u}_j)$. ∎

An adaptation of the technique used in proving Theorem 13.2.2 could be used to prove that G is also bounded below on V, but convexity gives a much stronger result.

Theorem 13.2.8. *A finite-valued indicator G defined on the cone* $\mathsf{C}[\mathfrak{S}]$ *in* R^m *is continuous on any closed proper sub-cone* V *and uniformly continuous on compact subsets of* V.

Proof. We appeal to Theorem 13.3.7 of the next section, where it is proved that a function which is convex, measurable and finite-valued in a convex domain D is continuous in D. These properties hold for a finite-valued indicator. ∎

Suppose again that $\{\mathbf{u}_j\}$ is a set of m linearly independent vectors, all in $\mathsf{C}[\mathfrak{S}]$. Let $\mathfrak{B}_0$ and V be defined as above. Then the vectors in V are linear combinations of the basis vectors with non-negative constant multipliers. By Cramer's rule we can write such a combination as follows:

$$\Delta \mathbf{x} = \sum_{j=1}^{m} \Delta_j(\mathbf{x})\, \mathbf{u}_j. \tag{13.2.21}$$

Here the coefficients are certain determinants. If

$$\mathbf{u}_j = (u_{j1}, u_{j2}, \ldots, u_{jm}), \qquad \mathbf{x} = (x_1, x_2, \ldots, x_m),$$

then $\Delta = \det(u_{jk})$ and $\Delta_j(\mathbf{x})$ is the determinant obtained by replacing the jth row of Δ by $x_1, x_2, \ldots, x_m$. Here $\Delta \neq 0$ and we can choose the numbering of the basis so that $\Delta > 0$. For $\mathbf{x} \in \mathsf{V}$ the $\Delta_j(\mathbf{x})$ are non-negative. The basic inequality now becomes

$$\Delta G(\mathbf{x}) \leqslant \sum_{j=1}^{m} \Delta_j(\mathbf{x})\, G(\mathbf{u}_j). \tag{13.2.22}$$

The special case $m = 2$ is in the literature. We represent R^2 by the complex plane and suppose that $\mathsf{C}[\mathfrak{S}]$ is the sector $\alpha < \arg z < \beta$. We set $G(e^{i\theta}) = g(\theta)$. Now choose three angles θ, θ_1, θ_2 such that

$$\alpha < \theta_1 < \theta < \theta_2 < \beta.$$

It is assumed that $\mathfrak{S}$ is not the whole plane, so that $\beta - \alpha \leqslant \pi$. Then

$$\mathbf{u}_1 = (\cos\theta_1, \sin\theta_1), \qquad \mathbf{u}_2 = (\cos\theta_1, \sin\theta_1),$$

while $\mathbf{x}$ is taken to be $(\cos\theta, \sin\theta)$. This gives

$$\Delta = \sin(\theta_2 - \theta_1), \qquad \Delta_1 = \sin(\theta_2 - \theta), \qquad \Delta_2 = \sin(\theta - \theta_1),$$

and (13.2.22) becomes

$$\sin(\theta_2 - \theta_1)\, g(\theta) \leqslant \sin(\theta_2 - \theta)\, g(\theta_1) + \sin(\theta - \theta_1)\, g(\theta_2). \tag{13.2.23}$$

This is the functional inequality studied by Pólya in 1929. It governs the radial growth indicator of $\log |f(re^{i\theta})|$ if f is holomorphic and of exponential growth in a sector. We have now seen that it holds for the radial indicator of a function sub-additive in an angular semi-module. Further, and this is the oldest result, it is satisfied by the function of support of a convex domain in the plane. This was first pointed out by the German mathematician Wilhelm Blaschke (1885–1962) in 1914.

Blaschke was an eminent differential geometer who also did important work on convexity and analytic function theory.

The case $m = 3$ is also of some interest. Here the coefficients in (13.2.22) are known in classical vector analysis as *box products* and the formula would be written

$$[\mathbf{u}_1\mathbf{u}_2\mathbf{u}_3]\, G(\mathbf{u}) \leqslant [\mathbf{u}\,\mathbf{u}_1\mathbf{u}_2]\, G(\mathbf{u}_1) + [\mathbf{u}_1\mathbf{u}\,\mathbf{u}_2]\, G(\mathbf{u}_2) + [\mathbf{u}_1\mathbf{u}_2\mathbf{u}]\, G(\mathbf{u}_3). \quad (13.2.24)$$

The box product is obviously a scalar. Expressed in terms of the traditional cross and dot products, we have $[\mathbf{uvw}] = (\mathbf{u} \times \mathbf{v}) \cdot \mathbf{w}$.

The functional equation

$$H(\mathbf{x}) = \sum_{j=1}^{m} a_j H(\mathbf{u}_j), \qquad \mathbf{x} = \sum_{j=1}^{m} a_j \mathbf{u}_j \quad (13.2.25)$$

is satisfied by any linear functional on R^m, i.e. we can take

$$H(\mathbf{x}) = \sum_{j=1}^{m} b_j x_j, \quad \text{if} \quad \mathbf{x} = (x_1, x_2, \ldots, x_m), \quad (13.2.26)$$

where the b's are arbitrary real numbers. Cf. Problem 7, Exercise 1.2, where it is stated that any linear functional of C^n is an inner product. The same obviously holds in R^m.

This fact has a bearing on the family of solutions of the functional inequality (13.2.25). It shows that if G is a member of the family, so is $G + H$ for any H of the form (13.2.26). This fact may be used to improve the local properties of an indicator function. For (13.2.23) the corresponding "sinusoid function"

$$h(\theta) \equiv a_1 \cos\theta + b_1 \sin\theta \quad (13.2.27)$$

has been used for such purposes.

The time has come to elucidate the connection between indicators $G(\mathbf{x})$ on the one hand and convex solids on the other. As a matter of fact, the connection is twofold. Given a convex solid K in R^m and a pole P, $P \in \mathsf{K}$, we can construct two distinct solutions of the inequality (13.2.15) each of which describes K uniquely but in different language. One of these solutions is the *gauge function* of K in the sense of Minkowski, the other its *function of support* in the sense of Blaschke–Minkowski. The first of these gives the equation of $\partial\mathsf{K}$, the boundary of K, in terms of point coordinates with P as the origin. Here

$$p(\mathbf{x}) = 1 \quad (13.2.28)$$

is the equation of $\partial\mathsf{K}$ where $p(\mathbf{x})$ is the gauge function of K which satisfies (13.2.15). The other describes K in terms of "tangential coordinates". If $\mathbf{u}$ is a unit vector from P and $h(\mathbf{u})$ is the (oriented) length of the foot point perpendicular from the pole to the tangent plane of K with normal $\mathbf{u}$, then K comes out as enveloped by the plane bundle

$$(\mathbf{x}, \mathbf{u}) = h(\mathbf{u}). \quad (13.2.29)$$

Here $h(\mathbf{u})$, the function of support of K, also satisfies (13.2.15).

Conversely, given a finite-valued measurable solution G of (13.2.15), we can construct a convex solid $\tilde{\mathsf{K}}$ having G as its gauge function and another convex solid K having G as its function of support, in both cases with respect to a pole which is taken as the origin of coordinates. These two solids are said to be *polar* to each other. The language used here (poles, polar, point coordinates, tangential coordinates) goes back to the German mathematician and experimental physicist Julius Plücker (1801–68), creator in both the fields of mathematics and of physics (algebraic geometry, line geometry, and spectral analysis).

Gauge functions occurred already in Section 1.2 and played a basic role in the discussion of linear functionals in Section 10.3. We recall the definition and basic properties of such functions, here specialized to Euclidean space R^m.

Definition 13.2.4. *Let K be a convex solid in R^m containing the origin as an interior point. The gauge function $p(\mathbf{x})$ of K is defined as*

$$p(\mathbf{x}) = \sup \{a;\ \mathbf{x} \in R^m,\ a\mathbf{x} \in \mathsf{K}\}. \tag{13.2.30}$$

Theorem 13.2.9. *The gauge function has the following properties*: (1) $p(\mathbf{0}) = 0$, (2) $p(\alpha\mathbf{x}) = \alpha p(\mathbf{x})$, $0 < \alpha$, $\forall\, \mathbf{x}$, (3) $p(\mathbf{x} + \mathbf{y}) \leqslant p(\mathbf{x}) + p(\mathbf{y})$, (4)

$$\{\mathbf{x};\ p(\mathbf{x}) < 1\} \subset \mathsf{K} \subset \{\mathbf{x};\ p(\mathbf{x}) \leqslant 1\}. \tag{13.2.31}$$

The proof is left to the reader. Cf. Theorem 10.3.1.

For an ellipsoid with center at the origin and semi-axes a, b, c, the gauge function is simply the square root of the left member of the point equation of the ellipsoid, i.e.

$$p(\mathbf{x}) = \left\{\frac{x_1{}^2}{a^2} + \frac{x_2{}^2}{b^2} + \frac{x_3{}^2}{c^2}\right\}^{\frac{1}{2}}.$$

Definition 13.2.5. *Let K be a convex solid in R^m containing the origin as an interior point. Consider the family of parallel hyperplanes*

$$H(\mathbf{u}, b)\colon (\mathbf{x}, \mathbf{u}) = b, \mathbf{u} \text{ fixed unit vector},\ 0 \leqslant b. \tag{13.2.32}$$

Here $\mathbf{x} \in R^m$ and $(\mathbf{x}, \mathbf{u})$ is the usual inner product. Set

$$h(\mathbf{u}) = \sup \{b;\ \mathsf{K} \cap H(\mathbf{u}, b) \neq \emptyset\}. \tag{13.2.33}$$

Then h is known as the function of support of K, with value $h(\mathbf{u})$ in the direction $\mathbf{u}$, while

$$(\mathbf{x}, \mathbf{u}) = h(\mathbf{u}) \tag{13.2.34}$$

is the corresponding plane of support.

Here the basic results are given by

Theorem 13.2.10. *If K is a bounded closed convex solid containing the origin in its interior, then h satisfies*

$$h(a_1\mathbf{u}_1 + a_2\mathbf{u}_2 + \cdots + a_m\mathbf{u}_m) \leqslant a_1 h(\mathbf{u}_1) + a_2 h(\mathbf{u}_2) + \cdots + a_m h(\mathbf{u}_m). \tag{13.2.35}$$

Here $\{\mathbf{u}_j\}$ *is any basis of unit vectors for* R^m *confined to an arbitrary half-space and the* a_j*'s are any non-negative numbers. Further,* h *is bounded and continuous If* K *is unbounded but confined to a half-space, say* $x_1 \leqslant a$, *let* C_0 *be the cone formed by the rays for which* $h(\mathbf{u}) < +\infty$. *Let the basis* $\{\mathbf{u}_j\}$ *be confined to* C_0. *Then* (13.2.35) *holds for any non-negative weights and h is bounded and continuous on any compact subset of* C_0.

Proof. From (13.2.34) it follows that we can extend the definition of h from unit vectors to arbitrary vectors by setting $\mathbf{v} = a\mathbf{u}$ and $h(\mathbf{v}) = ah(\mathbf{u})$, $0 < a$. This ensures positive homogeneity. We know from the discussion of indicators G that in order to prove (13.2.35) it is enough to show that $h(\mathbf{v})$ is convex. This follows from the convexity of K. For we have by (13.2.33) and the extension of the definition of h,

$$h(\mathbf{v}) = \sup\,(\mathbf{x}, \mathbf{v}), \qquad \mathbf{x} \in \mathsf{K}, \tag{13.2.36}$$

so that

$$(\mathbf{x}, \mathbf{v}) \leqslant h(\mathbf{v}), \qquad \forall\, \mathbf{x} \in \mathsf{K}. \tag{13.2.37}$$

If, now, $\mathbf{v}_1$, $\mathbf{v}_2$ are two vectors for which $h(\mathbf{v}_1)$ and $h(\mathbf{v}_2)$ are finite, then for all $\mathbf{x} \in \mathsf{K}$

$$(\mathbf{x}, \mathbf{v}_1) \leqslant h(\mathbf{v}_1), \qquad (\mathbf{x}, \mathbf{v}_2) \leqslant h(\mathbf{v}_2),$$

and for $0 < \alpha < 1$

$$\alpha(\mathbf{x}, \mathbf{v}_1) + (1 - \alpha)\,(\mathbf{x}, \mathbf{v}_2) \leqslant \alpha h(\mathbf{v}_1) + (1 - \alpha)\,h(\mathbf{v}_2).$$

By the additivity of the inner product the left member equals

$$[\mathbf{x}, \alpha\mathbf{v}_1 + (1 - \alpha)\,\mathbf{v}_2],$$

and if the inequality

$$(\mathbf{x}, \alpha\mathbf{v}_1 + (1 - \alpha)\,\mathbf{v}_2) \leqslant \alpha h(\mathbf{v}_1) + (1 - \alpha)\,h(\mathbf{v}_2)$$

holds for all $\mathbf{x} \in \mathsf{K}$, it also holds for the supremum, so that

$$h[\alpha\mathbf{v}_1 + (1 - \alpha)\,\mathbf{v}_2] \leqslant \alpha h(\mathbf{v}_1) + (1 - \alpha)\,h(\mathbf{v}_2). \tag{13.2.38}$$

This is the basis for (13.2.35).

The argument also shows that the vectors $\mathbf{v}$ for which $h(\mathbf{v})$ is finite is a convex set. This may be the whole space as in the case of a bounded convex set K. If this is not the case, then the vectors for which $h(\mathbf{v})$ is finite-valued form a convex cone C_0 confined to a half-space. This is the case when K is unbounded.

In the first case h is bounded and hence continuous by Theorem 13.3.7 below. In the second case the same conclusion holds on any compact subset of C_0. $\blacksquare$

Before taking up the converse problem—given an indicator G, find K and $\tilde{\mathsf{K}}$—we shall illustrate by a concrete problem.

Example 5. We take $m = 2$, $R^2 = C$ and the function

$$F(z) = F(x + iy) = ax + \tfrac{1}{4}y^2x^{-1}, \qquad x > 0. \tag{13.2.39}$$

Here a is a fixed positive number. The function F is closely related to that of formula (13.2.8). Thus the reader will know that F is subadditive in the right half-plane. Moreover, it is its own indicator, so that $G(z) = F(z)$.

The first problem is to interpret G as a gauge function. We have clearly

$$p(z) = G(z) = ax + \tfrac{1}{4} y^2 x^{-1} \tag{13.2.40}$$

and the closed convex region $\tilde{\mathsf{K}}$ with this function as its gauge measure is the interior and boundary of the ellipse

$$4ax^2 - 4x + y^2 = 0 \tag{13.2.41}$$

which passes through the origin.

To get the polar region, we set $z = e^{i\theta}$ and get

$$h(e^{i\theta}) \equiv g(\theta) = a \cos\theta + \tfrac{1}{4} \sin^2\theta \sec\theta. \tag{13.2.42}$$

It is a simple matter to show that the straight lines

$$x \cos\theta + y \sin\theta = a \cos\theta + \tfrac{1}{4} \sin^2\theta \sec\theta \tag{13.2.43}$$

for $|\theta| < \frac{1}{2}\pi$ have the parabola

$$y^2 = a - x \tag{13.2.44}$$

as their envelope. The interior and boundary of the parabola is the unbounded convex polar region K.

We have now

Theorem 13.2.11. *Let G be a finite-valued continuous function, defined convex and positive-homogeneous in some open cone C_0 of R^m. Here C_0 is either all of R^m or confined to a half-space. Then there exist two convex solids, $\tilde{\mathsf{K}}$ and K, such that G is gauge function of $\tilde{\mathsf{K}}$ and support function of K.*

Proof. I. We define $\tilde{\mathsf{K}}$ by

$$\tilde{\mathsf{K}}: \{\mathbf{x};\, \mathbf{x} \in \mathsf{C}_0,\, G(\mathbf{x}) \leqslant 1\}. \tag{13.2.45}$$

(1) The set $\tilde{\mathsf{K}}$ is convex. If $\mathbf{x}_1, \mathbf{x}_2 \in \tilde{\mathsf{K}}$ and $0 < \alpha < 1$, then

$$G[\alpha\mathbf{x}_1 + (1 - \alpha)\,\mathbf{x}_2] \leqslant \alpha G(\mathbf{x}_1) + (1 - \alpha)\, G(\mathbf{x}_2) \leqslant \alpha \cdot 1 + (1 - \alpha) \cdot 1 = 1,$$

so $\alpha\mathbf{x}_1 + (1 - \alpha)\,\mathbf{x}_2 \in \tilde{\mathsf{K}}$.

(2) $\tilde{\mathsf{K}}$ is closed relative to C_0. For if $\{\mathbf{x}_n\} \subset \mathsf{C}_0$ is a Cauchy sequence with a limit point $\mathbf{x}_0$ in C_0, then $G(\mathbf{x}_n) \leqslant 1$ together with the continuity of G in C_0 implies $G(\mathbf{x}_0) \leqslant 1$.

Thus $\tilde{\mathsf{K}}$ is convex, closed relative to C_0 and has G as its gauge function by virtue of (13.2.45) and the properties of G.

II. K is defined by

$$\mathsf{K}: \{\mathbf{x};\, \mathbf{x} \in R^m,\, (\mathbf{x}, \mathbf{u}) \leqslant G(\mathbf{u}),\, \mathbf{u} \in \mathsf{C}_0\}. \tag{13.2.46}$$

(i) K is convex. For if $\mathbf{x}_1, \mathbf{x}_2 \in \mathsf{K}$ and $0 < \alpha < 1$, then

$$\alpha(\mathbf{x}_1, \mathbf{u}) + (1 - \alpha)\,(\mathbf{x}_2, \mathbf{u}) \leqslant \alpha G(\mathbf{u}) + (1 - \alpha)\, G(\mathbf{u}) = G(\mathbf{u})$$

for all $\mathbf{u} \in \mathsf{C}_0$, i.e.

$$(\alpha \mathbf{x}_1 + (1 - \alpha)\,\mathbf{x}_2, \mathbf{u}) \leqslant G(\mathbf{u}), \qquad \forall\, \mathbf{u} \in \mathsf{C}_0,$$

and $\alpha \mathbf{x}_1 + (1 - \alpha)\,\mathbf{x}_2 \in \mathsf{K}$.

(ii) K is closed. If $\{\mathbf{x}_n\} \subset \mathsf{K}$ and is a Cauchy sequence in R^m with limit $\mathbf{x}_0$, then for each n and all $\mathbf{u} \in \mathsf{C}_0$

$$(\mathbf{x}_n, \mathbf{u}) \leqslant G(\mathbf{u})$$

and by the continuity of the inner product

$$\lim_{n \to \infty} (\mathbf{x}_n, \mathbf{u}) = (\mathbf{x}_0, \mathbf{u}) \leqslant G(\mathbf{u}).$$

(iii) K is unbounded, if C_0 is confined to a half-space. To fix the ideas, suppose that if $\mathbf{u} = (u_1, u_2, \ldots, u_m) \in \mathsf{C}_0$ (Cartesian coordinates!), then $u_1 > 0$. Consider the vector $\mathbf{x} = (-a, 0, 0, \ldots, 0)$, $a > 0$. Then

$$(\mathbf{x}, \mathbf{u}) = -au_1 < 0, \qquad \mathbf{u} \in \mathsf{C}_0.$$

It follows that K contains a ray, the negative x_1-axis, and is consequently unbounded.

If C_0 is all of R^m, then $|G(\mathbf{u})| \leqslant M$, some finite number, for all unit vectors $\mathbf{u}$. This gives $|(\mathbf{x}, \mathbf{u})| \leqslant M$ for $\mathbf{x} \in \mathsf{K}$. For each $\mathbf{x}_0$ on $\partial \mathsf{K}$, there is a finite positive a_0 and a unit vector $\mathbf{u}_0$ such that $\mathbf{x}_0 = a_0 \mathbf{u}_0$ and $(\mathbf{x}_0, \mathbf{u}_0) \leqslant M$ gives $a_0 \leqslant M$, so that all of K is confined to the sphere $\|\mathbf{x}\| \leqslant M$.

(iv) G is support function of K. This follows from (13.2.46). ∎

It may be shown that through every point on $\partial \mathsf{K}$, the boundary of K, passes at least one support hyperplane—that is, a plane of the form

$$(\mathbf{x}, \mathbf{u}) = G(\mathbf{u}). \tag{13.2.47}$$

EXERCISE 13.2

1. Prove Lemma 13.2.1.
2. Verify Lemma 13.2.5 and Theorem 13.2.1 when $\mathfrak{S}$ is in R^3.
3. Verify Lemma 13.2.6 in R^3.
4. Verify (13.2.9).
5. Verify the assertions made in Example 2.
6. What is the envelope of the family of straight lines

$$x \cos \theta + y \sin \theta = \sin^2 \theta \sec \theta, |\theta| < \tfrac{1}{2}\pi?$$

7. Let k, s, t be positive numbers, k fixed. Show that for all s, t

$$(1 + s)^{k+1} \leqslant (1 + t)^k (1 + s^{k+1} t^{-k}).$$

[*Hint*: What is the maximum of the right member for fixed s?]

8. Prove that $y^{k+1}x^{-k}$ is subadditive in the right half-plane and determine the indicator G.
9. Find the convex domain K for which $h(\theta) = \sec\theta$, $|\theta| < \frac{1}{2}\pi$, is the support function.
10. Rotate the set $\{(x, y); 0 \leqslant y < x^2, 0 < x\}$ about the x-axis. Show that the open solid of revolution obtained in this manner is an angular semi-module $\mathfrak{S}$ and find $\mathsf{C}[\mathfrak{S}]$, its convex hull.
11. Let $\mathbf{x} = (x_1, x_2, x_3)$ be a vector in R^3 and show that the mapping $\mathbf{x} \to F(\mathbf{x}) = (x_1{}^2 + x_2{}^2)(x_3)^{-1}$ is subadditive in the half-space $x_3 > 0$.
12. If $\mathbf{u} = (\cos\alpha, \cos\beta, \cos\gamma)$ is a unit vector in R^3, find the indicator G of F with F as in Problem 11. This is a function of γ alone, say $h(\gamma)$.
13. With $h(\gamma)$ as defined, show that the family of planes
$$x_1 \cos\alpha + x_2 \cos\beta + x_3 \cos\gamma = h(\gamma), |\gamma| < \tfrac{1}{2}\pi,$$
envelops a paraboloid of revolution. This is the surface of the convex solid K with G as support function.
14. Determine the convex solid $\tilde{\mathsf{K}}$ having the same G as gauge function. Show that it is a sphere.
15. Verify (13.2.21).
16. Verify (13.2.23).
17. Prove Theorem 13.2.9.
18. Verify $[\mathbf{uvw}] = (\mathbf{u} \times \mathbf{v}) \cdot \mathbf{w}$.
19. Show that (13.2.27) satisfies (13.2.23) with equality.
20. Verify that (13.2.26) is a solution of (13.2.25). Determine a subadditive function $F(\mathbf{x})$ which has $H(\mathbf{x})$ as radial growth indicator. Is F unique? Determine the corresponding convex solids, K and $\tilde{\mathsf{K}}$.
21. Verify the assertions made in Example 5.
22. How would you prove part II (iv) of Theorem 13.2.11?
23. Try to prove that there is a support plane passing through every point of the boundary $\partial\mathsf{K}$ of a convex solid K.

13.3 CONVEX FUNCTIONS

In formula (13.1.1) we take
$$x \circ y = \tfrac{1}{2}(x + y), \qquad G(u, v) = \tfrac{1}{2}(u + v), \tag{13.3.1}$$
so the inequality reads
$$f[\tfrac{1}{2}(x + y)] \leqslant \tfrac{1}{2}[f(x) + f(y)]. \tag{13.3.2}$$
This inequality, first considered by Otto Hölder in 1889 and in more detail by the Danish mathematician and telephone engineer J. L. W. V. Jensen (1859–1925) in 1906, characterizes *convex functions*. If $-f$ is convex, then f is said to be *concave*.

The geometric meaning of (13.3.2) is that the midpoint of any chord of the curve $t = f(s)$ lies above or on the curve. Here "curve" means any, not necessarily rectifiable or continuous, graph.

Actually a convex function is either continuous or very irregular, unbounded in every interval. Thus a measurable convex function is actually continuous. We shall ultimately prove this under the additional assumption that f is finite-valued. This will be postponed until we have proved some implications of (13.3.2) with or without assuming continuity.

Theorem 13.3.1. *Let f be defined in (a, b) and satisfy (13.3.2) whenever x and y belong to (a, b). Let $x_1, x_2, \ldots, x_n$ be n points in (a, b), not necessarily distinct. Let $r_1, r_2, \ldots, r_n$ be n non-negative rational numbers of sum unity. Then*

$$f\left(\sum_{j=1}^{n} r_j x_j\right) \leqslant \sum_{j=1}^{n} r_j f(x_j). \tag{13.3.3}$$

The inequality remains valid if some or all of the weights r_j are irrational, provided f is continuous.

Proof. The first step is to verify (13.3.3) if $n = 2^m$ and all weights are equal, $r_j = 2^{-m}$. From (13.3.2) we get for $m = 2$

$$\begin{aligned} 4f[\tfrac{1}{4}(x_1 + x_2 + x_3 + x_4)] &\leqslant 2f[\tfrac{1}{2}(x_1 + x_2)] + 2f[\tfrac{1}{2}(x_3 + x_4)] \\ &\leqslant f(x_1) + f(x_2) + f(x_3) + f(x_4). \end{aligned}$$

Complete induction shows then that

$$f\left(\frac{1}{n}\sum_{j=1}^{n} x_j\right) \leqslant \frac{1}{n}\sum_{j=1}^{n} f(x_j) \tag{13.3.4}$$

holds for all n of the form $n = 2^m$. To extend the formula to arbitrary positive integers we adapt an argument due to Cauchy (1821) as follows.

Suppose that (13.3.4) holds for some value of n and for all choices of n points in (a, b). It will be shown that the formula holds also for the preceding integer $n - 1$ (retrogressive induction). Let $x_1, x_2, \ldots, x_{n-1}$ be given and set

$$x_n = \frac{1}{n-1}(x_1 + x_2 + \cdots + x_{n-1}).$$

We have then

$$\begin{aligned} f(x_n) = f\left(\frac{1}{n}[(n-1)x_n + x_n]\right) &= f\left[\frac{1}{n}(x_1 + x_2 + \cdots + x_{n-1}) + \frac{1}{n}x_n\right] \\ &\leqslant \frac{1}{n}[f(x_1) + f(x_2) + \cdots + f(x_{n-1})] + \frac{1}{n}f(x_n), \end{aligned}$$

which is (13.3.4) with n replaced by $n - 1$. Since (13.3.4) holds for all powers of 2, this artifice enables us to extend the validity of the formula to all positive values of n.

If now $r_1, r_2, \ldots, r_n$ are n given non-negative rational numbers of sum unity, we can write

$$r_1 = \frac{p_1}{q}, \qquad r_2 = \frac{p_2}{q}, \qquad \ldots, \qquad r_n = \frac{p_n}{q},$$

where q is the least common denominator of the given fractions and the p's are positive integers $<q$ or 0. We then apply formula (13.3.4) where $n = q$ and each x_j is repeated p_j times. This gives (13.3.3).

If f is continuous, the assumption that the weights r_j are rational may be dropped, and we can allow arbitrary non-negative real numbers. Such numbers can be approximated arbitrarily closely by rationals for which (13.3.3) is known to be valid. Passing to the limit with the r's on both sides of the inequality and using continuity, we obtain the desired result. ∎

In particular, we see that the continuity of f permits us to generalize (13.3.2) to

$$f[\alpha x + (1 - \alpha) y] \leqslant \alpha f(x) + (1 - \alpha) f(y) \tag{13.3.5}$$

valid for any α between 0 and 1.

We proceed to a further discussion of the geometric implications of (13.3.2) and (13.3.5). Take four values of x

$$a < x_1 < x_2 < x_3 < x_4 < b$$

and set $P_j = [x_j, f(x_j)]$. Form the polygon $[P_1, P_2, P_3, P_4, P_1]$. It is a convex closed polygon, i.e. if two points P and Q lie inside or on the polygon, so does the line segment joining P and Q. Since f is a convex function by assumption, any one of the arcs of the curve $t = f(s)$ bounded by two points P_j and P_k, $1 \leqslant j < k \leqslant 4$, lies below or on the corresponding chord P_jP_k. Here the reader is advised to draw a figure. If m_{jk} denotes the slope of the straight line through P_j and P_k, there are a number of inequalities between the slopes implied by the geometry of the situation. The following is basic:

$$m_{12} \leqslant m_{23} \leqslant m_{34}, \tag{13.3.6}$$

but we have also

$$\begin{aligned} &m_{12} \leqslant m_{13} \leqslant m_{14}, \qquad m_{12} \leqslant m_{23} \leqslant m_{24}, \\ &m_{14} \leqslant m_{24} \leqslant m_{34}, \qquad m_{13} \leqslant m_{23} \leqslant m_{34}. \end{aligned} \tag{13.3.7}$$

If we now express the slopes in terms of the coordinates of the endpoints of the chords, a number of inequalities result of which we shall list only those given by (13.3.6), namely,

$$\frac{f(x_2) - f(x_1)}{x_2 - x_1} \leqslant \frac{f(x_3) - f(x_2)}{x_3 - x_2} \leqslant \frac{f(x_4) - f(x_3)}{x_4 - x_3}. \tag{13.3.8}$$

The inequalities have a number of implications. Set

$$x_2 = x_1 + h, \qquad x_4 = x_3 + h, \qquad 0 < h,$$

to obtain

$$\frac{1}{h}[f(x_1 + h) - f(x_1)] \leqslant \frac{1}{h}[f(x_3 + h) - f(x_3)]. \tag{13.3.9}$$

This states that the difference quotient

$$\frac{1}{h}[f(x + h) - f(x)] \tag{13.3.10}$$

is, for fixed $h > 0$, a non-decreasing function of x. In the same manner one shows that, for fixed x, the difference quotient does not increase as h decreases toward zero.

Similarly, setting

$$x_1 = x - h_2, \qquad x_2 = x - h_1, \qquad x_3 = x, \qquad 0 < h_1 < h_2,$$

we obtain an inequality which reduces to

$$\frac{1}{h_2}[f(x) - f(x - h_2)] \leqslant \frac{1}{h_1}[f(x) - f(x - h_1)]. \tag{13.3.11}$$

This shows that the difference quotient

$$\frac{1}{h}[f(x) - f(x - h)] \tag{13.3.12}$$

does not decrease as h decreases to zero, x being fixed. Similarly, we show that for any choice of positive h and h_1

$$\frac{1}{h}[f(x) - f(x - h)] \leqslant \frac{1}{h_1}[f(x + h_1) - f(x)]. \tag{13.3.13}$$

In particular, for $h_1 = h$ this gives

$$f(x + h) - 2f(x) + f(x - h) \geqslant 0. \tag{13.3.14}$$

Thus the second difference of a convex function is non-negative while the first difference of a continuous convex function is increasing as a function of x for fixed $h > 0$. ∎

We can now prove

Theorem 13.3.2. *A function f which is continuous and convex in an interval (a, b) has a left hand derivative $D^-f(x)$ as well as a right hand derivative $D^+f(x)$ at each point x of (a, b) and*

$$D^-f(x) \leqslant D^+f(x). \tag{13.3.15}$$

If f has a unique derivative in some subinterval (c, d), then that derivative $Df(x)$ is a non-decreasing function of x in (c, d).

Proof. This follows from the properties of the difference quotients (13.3.10) and (13.3.12). Let us consider the latter. For fixed x, the quotient does not decrease as h

decreases to zero and by (13.3.13) it is bounded above. Hence it has a finite limit $D^-f(x)$. Similarly, the quotient (13.3.10) is for fixed x a non-decreasing function of h as h decreases to zero and, again by (13.3.13), it is bounded below. Hence $D^+f(x)$ exists and (13.3.15) shows that the limit of the left hand member cannot exceed the limit of the right, so that (13.3.15) must hold.

Suppose now that $D^-f(x) = D^+f(x) = Df(x)$ for x in some interval (c, d). If now $c < x_1 < x_3 < d$, we conclude from (13.3.9) that

$$Df(x_1) = \lim_{h\downarrow 0} \frac{1}{h}[f(x_1 + h) - f(x_1)] \leqslant \lim_{h\downarrow 0} \frac{1}{h}[f(x_3 + h) - f(x_3)] = Df(x_3)$$

as asserted. ∎

We have

$$\lim_{h\to 0} h^{-2}[f(x + h) - 2f(x) + f(x - h)] = f''(x) \tag{13.3.16}$$

at all points where the second derivative exists.

These observations lead to

Theorem 13.3.3. *If f is continuous and convex in (a, b) and if f is twice differentiable, then*

$$f''(x) \geqslant 0, \qquad \forall\, x. \tag{13.3.17}$$

Conversely, if f has a non-negative second derivative in (a, b), then f is convex there.

Proof. The first assertion follows from (13.3.14) and (13.3.16). For the second assertion the elementary calculus argument goes as follows. If f'' exists and satisfies (13.3.17), then $f'(x)$ exists, is continuous and is non-decreasing. There is a unique curve tangent, the curve lies above or on its local tangent and the graph is concave upwards; i.e. f is convex. For an alternate proof, see Problem 3 of Exercise 13.3. ∎

For the following theorem cf. page 401.

Theorem 13.3.4. *Let f be convex in the open interval (a, b) and bounded above in some closed subinterval $[c, d]$; then f is continuous in (a, b).*

This is a special case of corresponding results for the m-dimensional case which is next on the agenda.

Let D be a convex domain in R^m, i.e. an open connected set such that $\mathbf{x}_1, \mathbf{x}_2 \in D$, and $0 < \alpha < 1$ implies that $\alpha\mathbf{x}_1 + (1 - \alpha)\,\mathbf{x}_2 \in D$. Let f be a mapping from R^m to R^1 defined on D. Then f is said to be convex in D if

$$f[\tfrac{1}{2}(\mathbf{x}_1 + \mathbf{x}_2)] \leqslant \tfrac{1}{2}[f(\mathbf{x}_1) + f(\mathbf{x}_2)] \tag{13.3.18}$$

for all $\mathbf{x}_1$, $\mathbf{x}_2$ in D.

Theorem 13.3.5. *If f is defined in D and convex in the sense of* (13.3.18), *then for any choice of n vectors $\mathbf{x}_j$ in D and any non-negative rational numbers r_j of sum unity we have*

$$f\left[\sum_{j=1}^{n} r_j \mathbf{x}_j\right] \leqslant \sum_{j=1}^{n} r_j f(\mathbf{x}_j). \tag{13.3.19}$$

If f is continuous, irrational weights may be allowed.

Proof. The proof given for (13.3.3) shows that formula (13.3.19) holds for any n collinear points in D. To extend to higher dimensions we use induction on the dimensionality of the closed convex hull of the points $\mathbf{x}_j$. Suppose that the inequality is known to be valid provided the points $\mathbf{x}_1, \mathbf{x}_2, \ldots, \mathbf{x}_{n-1}$ belong to a convex subset of D of dimension k. Then for any choice of $\mathbf{x}_n \in D$ and any choice of non-negative weights of sum unity

$$\begin{aligned} f[r_1\mathbf{x}_1 + r_2\mathbf{x}_2 + \cdots + r_{n-1}\mathbf{x}_{n-1} + r_n\mathbf{x}_n] \\ &= f\left\{(1 - r_n)\frac{r_1\mathbf{x}_1 + r_2\mathbf{x}_2 + \cdots + r_{n-1}\mathbf{x}_{n-1}}{r_1 + r_2 + \cdots + r_{n-1}} + r_n\mathbf{x}_n\right\} \\ &\leqslant (1 - r_n) f\left\{\frac{r_1\mathbf{x}_1 + r_2\mathbf{x}_2 + \cdots + r_{n-1}\mathbf{x}_{n-1}}{r_1 + r_2 + \cdots + r_{n-1}}\right\} + r_n f(\mathbf{x}_n) \\ &\leqslant (1 - r_n)\sum_{j=1}^{n-1} \frac{r_j}{1 - r_n} f(\mathbf{x}_j) + r_n f(\mathbf{x}_n). \end{aligned}$$

This is (13.3.19) for an arbitrary number of points spanned by a convex hull of dimension $k+1$. In deriving this inequality we have used (13.3.18) and the induction hypothesis. Thus (13.3.19) is valid for all n and all points $\mathbf{x}_j$ in D. ▮

There are various analogues of Theorems 13.3.2 and 13.3.3 for the m-dimensional case. They are left to the reader. On the other hand, we have to examine questions of boundedness and continuity.

Theorem 13.3.6. *Let $\mathbf{x} \to f(\mathbf{x})$ be defined in a convex domain D of R^m where it is finite-valued, measurable, and convex. Then f is bounded on compact subsets of D.*

Proof. Let D_0 be a compact convex subset of D. Suppose that f is not bounded above in D_0. Then there exists a sequence $\{\mathbf{x}_n\}$ such that (i) $\mathbf{x}_n \in D_0$, $\forall n$, (ii) $\lim_{n\to\infty} \mathbf{x}_n = \mathbf{x}_0$ exists and $\mathbf{x}_0 \in \bar{D}_0$, (iii) $f(\mathbf{x}_n) > n$. Let d be the distance of $\mathbf{x}_0$ from ∂D, the boundary of D, and let $\mathbf{t}$ be any vector such that $\|\mathbf{t}\| \leqslant \frac{1}{2}d$. For such values of $\mathbf{t}$ and for large numbers n it is seen that $\mathbf{x}_n + \mathbf{t}$ and $\mathbf{x}_n - \mathbf{t}$ belong to D. Hence

$$n < f(\mathbf{x}_n) = f[\tfrac{1}{2}(\mathbf{x}_n + \mathbf{t}) + \tfrac{1}{2}(\mathbf{x}_n - \mathbf{t})] \leqslant \tfrac{1}{2}[f(\mathbf{x}_n + \mathbf{t}) + f(\mathbf{x}_n - \mathbf{t})]$$

and either $f(\mathbf{x}_n + \mathbf{t})$ or $f(\mathbf{x}_n - \mathbf{t})$ must exceed n. Set

$$S_n = [\mathbf{x};\ \|\mathbf{x} - \mathbf{x}_n\| \leqslant \tfrac{1}{2}d, \qquad f(\mathbf{x}) > n].$$

It is a measurable set and its measure is at least

$$\tfrac{1}{2}\,(\tfrac{1}{2}\,d)^m\, V_m \tag{13.3.20}$$

where V_m is the measure of the unit sphere in R^m. Now set

$$T_n = [\mathbf{x};\ \|\mathbf{x} - \mathbf{x}_0\| \leqslant \tfrac{1}{2}d, \qquad f(\mathbf{x}) > n].$$

This is also a measurable set and for large values of n its measure differs arbitrarily little from that of S_n since the measure of the symmetric difference of the two spheres

$$[\mathbf{x};\ \|\mathbf{x} - \mathbf{x}_n\| \leqslant \tfrac{1}{2}d] \qquad \text{and} \qquad [\mathbf{x};\ \|\mathbf{x} - \mathbf{x}_0\| \leqslant \tfrac{1}{2}d]$$

goes to zero as $n \to \infty$. We now note that

$$T_n \supset T_{n+1} \supset \cdots \supset \cap\, T_k \equiv T,$$

so that T is a measurable subset of D of measure at least equal to (13.3.20). But in T all the inequalities $f(\mathbf{x}) > n$ are valid so that $f(\mathbf{x}) = +\infty$, for all $x \in T$. This contradicts the assumption that f is finite-valued in D. It follows that there exists an $M = M(D_0)$ such that

$$f(\mathbf{x}) \leqslant M(D_0), \qquad \forall\, \mathbf{x} \in D_0. \tag{13.3.21}$$

Next, suppose that f is not bounded below in D_0. Then there would exist a sequence $\{\mathbf{s}_n\}$ such that (i) $\mathbf{s}_n \in D_0$, for all n, (ii) $\lim_{n\to\infty} \mathbf{s}_n = \mathbf{s}_0$ exists and is in $\bar{D}_0$, (iii) $f(\mathbf{s}_n) < -2n$. Then for $\mathbf{t} \in D_0$

$$f[\tfrac{1}{2}(\mathbf{s}_n + \mathbf{t})] \leqslant \tfrac{1}{2}f(\mathbf{s}_n) + \tfrac{1}{2}f(\mathbf{t}) < -n + M(D_0).$$

Set

$$U_n = [\mathbf{x};\ \mathbf{x} = \tfrac{1}{2}(\mathbf{s}_n + \mathbf{t}),\ \mathbf{t} \in D_0],$$
$$U_0 = [\mathbf{x};\ \mathbf{x} = \tfrac{1}{2}(\mathbf{s}_0 + \mathbf{t}),\ \mathbf{t} \in D_0].$$

Let $V_n = \bigcup_{k \geqslant n} U_k$. Then $V_n \supset V_{n+1} \supset \cdots \supset \cap\, V_n \equiv V$. Now V coincides with the set U_0 which is of positive measure. Further, in V all inequalities $f(\mathbf{x}) < -n$ are satisfied. This implies that $f(\mathbf{x}) = -\infty$ for all $\mathbf{x}$ in V and again this is a contradiction, since f can nowhere take on the value $-\infty$. Hence f is bounded below as well as above in D_0. ∎

Theorem 13.3.7. *A measurable finite-valued function convex in the convex domain D is continuous on compact subsets of D.*

Proof. Let D_0 be a convex compact subset of D. By the preceding theorem there is a finite $B = B(D_0)$ such that $|f(\mathbf{x})| \leqslant B$, for all $\mathbf{x} \in D_0$. Set for $\mathbf{x} \in D_0$, $\mathbf{x} + \mathbf{h} \in D_0$

$$\delta^+(\mathbf{x}) = \limsup_{\|\mathbf{h}\| \downarrow 0}\, [f(\mathbf{x} + \mathbf{h}) - f(\mathbf{x})],$$

$$\delta^-(\mathbf{x}) = \liminf_{\|\mathbf{h}\| \downarrow 0}\, [f(\mathbf{x} + \mathbf{h}) - f(\mathbf{x})].$$

These quantities are uniformly bounded in D_0. We have then

$$f(\mathbf{x} + \tfrac{1}{2}\mathbf{h}) = f[\tfrac{1}{2}(\mathbf{x} + \mathbf{h}) + \tfrac{1}{2}\mathbf{x}] \leqslant \tfrac{1}{2}f(\mathbf{x} + \mathbf{h}) + \tfrac{1}{2}f(\mathbf{x}),$$

so that

$$f(\mathbf{x} + \tfrac{1}{2}\mathbf{h}) - f(\mathbf{x}) \leqslant \tfrac{1}{2}[f(\mathbf{x} + \mathbf{h}) - f(\mathbf{x})].$$

Hence

$$\delta^+(\mathbf{x}) \leqslant \tfrac{1}{2}\delta^+(\mathbf{x}) \qquad \text{or} \qquad \delta^+(\mathbf{x}) \leqslant 0.$$

We have also $\delta^-(\mathbf{x}) \leqslant \delta^+(\mathbf{x}) \leqslant 0$. Since f is convex

$$f(\mathbf{x} + \mathbf{h}) - f(\mathbf{x}) \geqslant -[f(\mathbf{x} - \mathbf{h}) - f(\mathbf{x})],$$

so that

$$\begin{aligned}\liminf_{\|\mathbf{h}\|\downarrow 0}[f(\mathbf{x} + \mathbf{h}) - f(\mathbf{x})] &\geqslant \liminf_{\|\mathbf{h}\|\downarrow 0}\{-[f(\mathbf{x} - \mathbf{h}) - f(\mathbf{x})]\} \\ &= -\limsup_{\|\mathbf{h}\|\downarrow 0}[f(\mathbf{x} - \mathbf{h}) - f(\mathbf{x})]\end{aligned}$$

or

$$\delta^-(\mathbf{x}) \geqslant -\delta^+(\mathbf{x}) \geqslant 0.$$

Hence $\delta^-(\mathbf{x}) = 0$ and the same holds for $\delta^+(\mathbf{x})$. It follows that

$$\lim_{\|\mathbf{h}\|\to 0} |f(\mathbf{x} + \mathbf{h}) - f(\mathbf{x})| = 0$$

for all $\mathbf{x}$ in D_0. Thus local boundedness implies local continuity. ∎

Corollary 1. *Theorem 13.2.4.*

Corollary 2. *Theorem 13.2.8.*

EXERCISE 13.3

1. Let f and F be continuous, real-valued, twice differentiable functions, and let F be defined on the range of f. If f is convex, find sufficient conditions for $F[f(x)]$ to be convex.
2. If f is positive, find sufficient conditions for $\log f$ to be convex.
3. Prove the second part of Theorem 13.3.3 by showing that such a function satisfies (13.3.3). Set $u = \Sigma r_j f(x_j)$ and expand $f(x)$ in powers of $u - x_j$ by Taylor's theorem with remainder.
4. If f is a continuous function, convex in (a, b), derive the inequalities (13.3.8) directly from (13.3.5).
5. If f is continuous and convex, show that

$$f[\tfrac{1}{2}(a + b)] \leqslant \frac{1}{b - a}\int_a^b f(t)\,dt.$$

6. Let g be non-negative and continuous and let $p \geqslant 1$. Show that

$$\left\{\frac{1}{b-a}\int_a^b g(t)\,dt\right\}^p \leqslant \frac{1}{b-a}\int_a^b [g(t)]^p\,dt.$$

Use properties of convex functions and the definition of the integral by Riemann sums. The inequality is simply (4.4.17).

7. Let g be real and continuous. Show that

$$\exp\left\{\frac{1}{b-a}\int_a^b g(t)\,dt\right\} \leqslant \frac{1}{b-a}\int_a^b \exp[g(t)]\,dt.$$

8. Let g be positive and continuous. Show that

$$\frac{1}{b-a}\int_a^b \log[g(t)]\,dt \leqslant \log\left\{\frac{1}{b-a}\int_a^b g(t)\,dt\right\}$$

9. If f is continuous and convex in (a, b) and if $a + h < x < b - h$, show that

$$f(x) \leqslant \frac{1}{2h}\int_{x-h}^{x+h} f(t)\,dt.$$

Actually this condition is sufficient as well as necessary for convexity of a continuous function.

10. Denote the right member of the preceding inequality by $T[f](x)$. Use the sup-norm in $C^+[a, b]$ and show that $\|T\| = 1$. Show that $f(x) = T[f](x)$ is satisfied by $\alpha x + \beta$ for any non-negative numbers α and β.

11. Let D be a convex domain in the plane, f a continuous and convex function defined in D. Formulate and prove some generalizations of Theorem 13.3.2 for this case.

12. Suppose that f has continuous second order partials. How does Theorem 13.3.3 generalize?

13. Can the surface $z = f(x, y)$ have saddle points?

14. Let f have continuous derivatives of order up to and including the third. Expand $f(x, y)$ in powers of $(x - x_0)$ and $(y - y_0)$ for $(x_0, y_0) \in D$. In the Taylor expansion the second order terms form a quadratic form. What can be said about this form?

15. Check the proof of Theorem 13.3.5.

The next four problems are accredited to E. G. Eggleston (1966).

16. If f is continuous and convex in R^m and $f(\mathbf{0}) = 0$, prove that $r^{-1}f(r\mathbf{x})$ increases with r for any fixed $\mathbf{x}$.

17. Prove the same is true for $r^{-1}[f(\mathbf{x}_0 + r\mathbf{x}) - f(\mathbf{x}_0)]$ for fixed $\mathbf{x}$ and $\mathbf{x}_0$.

18. Show that the first differential

$$\delta f(\mathbf{x}_0; \mathbf{h}) = \lim_{\alpha \downarrow 0} \alpha^{-1}[f(\mathbf{x}_0 + \alpha\mathbf{h}) - f(\mathbf{x}_0)]$$

exists for all $\mathbf{x}_0$. Note that α is restricted to positive values.

19. If f is a gauge function, show that $\delta f(\mathbf{x}_0; \mathbf{h}) \leqslant f(\mathbf{h})$.

13.4 NON-ARCHIMEDEAN VALUATIONS

This originally purely algebraic concept has some aspects involving functional inequalities on a product space. The algebraists are concerned with certain mappings of a field or of a ring into R. For our purposes we may restrict ourselves to the case where the domain of definition of the mapping is R, R^+ or Z^+. In the notation of Section 13.1

$$\circ = +, \qquad G(u, v) = \max(u, v) \tag{13.4.1}$$

and the inequality is

$$f(x + y) \leqslant \max[f(x), f(y)]. \tag{13.4.2}$$

With what N. Bourbaki would call "un abus de langage" we refer to such a function as a *valuation* of the domain.

If $D = R^+$ or Z^+, then any decreasing function of x satisfies (13.4.2). This means that in R^+ a valuation may tend arbitrarily fast to $+\infty$ as x decreases to 0 and/or arbitrarily fast to $-\infty$ as x increases to $+\infty$.

Theorem 13.4.1. *A measurable valuation defined on R^+ which assumes the value $+\infty$ at most in a set of measure zero is bounded above on compact subsets.*

Proof. We use the same argument as in the proof of Theorem 13.1.1. If the assertion is false for a particular valuation f, then there is an interval $[a, b]$, $0 < a < b < \infty$ and a point set $\{s_n\}$ such that (1) $a < s_n < b$, (2) $\lim s_n = s_0 \geqslant a$, and (3) $f(s_n) > n$ for all n. Then for $0 < s < s_n$

$$n < f(s_n) = f(s_n - s + s) \leqslant \max[f(s_n - s), f(s)]. \tag{13.4.3}$$

It follows that either $f(s) > n$ or $f(s_n - s) > n$ and the measurable set

$$S_n = \{t;\, n < f(t),\, 0 < t < s_n\}$$

has measure

$$m[S_n] \geqslant \tfrac{1}{2} s_n \geqslant \tfrac{1}{2} a.$$

From this point on the argument follows that given in the proof of Theorem 13.1.1 without modification. The reader should convince himself that this is indeed the case. ∎

Actually a stronger statement than Theorem 13.4.1 may be made.

Theorem 13.4.2. *A measurable valuation on R^+ which equals $+\infty$ at most in a set of measure zero is bounded above on every interval $[a, \infty)$ where $0 < a$.*

Proof. Suppose that

$$f(x) \leqslant M \qquad \text{for} \qquad a \leqslant x < 2a. \tag{13.4.4}$$

Such an M exists by the preceding theorem. If now $2a \leqslant x < 4a$, then $a \leqslant \tfrac{1}{2}x < 2a$ and

$$f(x) = f(\tfrac{1}{2}x + \tfrac{1}{2}x) \leqslant \max[f(\tfrac{1}{2}x), f(\tfrac{1}{2}x)] \leqslant M.$$

The proof is completed by induction. ∎

Though $f(x)$ is bounded above and conceivably also below, it need not tend to any limit as $x \to \infty$. Thus the Dirichlet function $f(x)$ which is 0 or 1 according as x is rational or irrational is a valuation on R and does not tend to any limit.

In the discussion of Čebyšev constants in Section 15.6 we shall consider valuations on Z^+. Here the condition

$$a_{m+n} \leqslant \max(a_m, a_n), \qquad 0 < m, n, \tag{13.4.5}$$

implies that a_n is bounded above but implies neither boundedness from below nor the existence of a limit of a_n. The convergence properties of valuation sequences show some peculiar features some of which are illustrated in Exercise 13.4.

The generalization to

$$f(x \circ y) \leqslant \max[f(x), f(y)] \tag{13.4.6}$$

may possibly be of some interest.

EXERCISE 13.4

1. Check the proof of Theorem 13.4.1.
2. What is the analogue of Theorem 13.1.2?
3. Extend Theorems 13.4.1 and 13.4.2 to valuations on R.
4. Suppose that f is a non-negative valuation on R and that $\lim_{h \to 0} f(h) = f(0) = 0$. Show that f is continuous everywhere.
5. If $f(n) = a_n$ is a valuation on Z^+, show that $a_n \leqslant a_1$ for all n. Is the sequence necessarily bounded below?
6. A sequence $\{a_n\}$ consists of the block 1, 1, 0 repeated infinitely often. The cluster points of the sequence are obviously 0 and 1. Show that it is a valuation sequence.
7. Show that if a valuation sequence $\{a_n\}$ has a unique limit b, then $b \leqslant a_n$ for all n.
8. Show that every integer $\geqslant j^2$ admits of a representation of the form $mj + n(j + 1)$ where m and n are non-negative integers. Is the representation unique?
9. Show that if two consecutive entries in a valuation sequence $\{a_n\}$ satisfy $a_j \leqslant a$, $a_{j+1} \leqslant a$, then $\limsup_{m \to \infty} a_n \leqslant a$.
10. The following is the most common algebraic valuation, the p-adic case. It is defined on Q, the field of rational.numbers. Let p be a given prime. If $x = m/n \in Q$, set $x = p^t a/b$ where a, b, m, n, t are integers and a and b prime to p. Define $f(x) = -t$ and show that f is a valuation on Q. In higher algebra it is customary to demand that a valuation satisfy $f(xy) = f(x) + f(y)$ as well as (13.4.2). This condition is evidently satisfied by $f(x) = -t$.

COLLATERAL READING

For subadditive and suboperative functions on the line, see:

HILLE, E. and R. S. PHILLIPS, *Functional Analysis and Semi-groups,* American Mathematical Society Colloquium Publications, Providence, R.I. (1957), Chapter 7. For the theory of semi-modules see *ibid.,* Chapter 8.

Various extensions, in particular to R^m, are treated by:

IONESCU TULČEA, C. T., Suboperative functions and semi-groups of operators, *Arkiv Mat.* **4** (7) (1960) 55–61.

ROSENBAUM, R. A., Sub-additive functions, *Duke Math. J.* **17** (1950) 227–47.

Minkowski's work on convex solids goes back at least to 1891 and may be consulted in his *Gasammelte Werke* (Chelsea Reprint, 1967). See, in particular, the posthumous memoir:

Theorie der konvexen Körper, insbesondere Begründung ihres Oberflächenbegriffs, *Gesammelte Werke,* Vol. II, pp. 131–229.

Among later papers see:

BLASCHKE, W., Beweise zu Sätzen von Brunn und Minkowski über die Minimaleigenschaft des Kreises, *Jahresber. Deut. Math. Ver.* **23** (1914) 210–34.

RADEMACHER, H., Zur Theorie der Minkowskischen Stützebenenfunktion, *Sitzungsber. Berliner Math. Ges.* **20** (1921) 14–19.

Modern treatment of the theory of convexity and convex functions is to be found in many places. The author has found the following monograph particularly useful:

EGGLESTON, E. G., *Convexity,* Cambridge Tracts in Mathematics and Mathematical Physics, No. 47, Cambridge University Press, London (1966).

For the classical theory of convex functions see:

HARDY, G. H., J. E. LITTLEWOOD, and G. PÓLYA, *Inequalities,* Cambridge University Press, London (1934).

HÖLDER, O., Über einen Mittelwertsatz, *Göttinger Nachrichten,* 1889, pp. 38–47.

JENSEN, J. L. W. V., Sur les fonctions convexes et les inégalités entre les valeurs moyennes, *Acta Math.* **30** (1906) 175–93.

For the Lindelöf–Phragmén–Pólya range of ideas see:

CARTWRIGHT, M. L., *Integral Functions,* Cambridge Tracts in Mathematics and Mathematical Physics, No. 44, Cambridge University Press, London (1962), Chapters III and VIII.

HILLE, E., *Analytic Function Theory,* Vol. II, Ginn, Boston, (1962), sections 11.3 and 19.3.

PHRAGMÉN, E. and E. LINDELÖF, Sur une extension d'un principe classique de l'analyse et sur quelques propriétés des fonctions monogènes dans le voisinage d'un point singulier, *Acta Math.* **31** (1908) 381–406.

PÓLYA, G., Untersuchungen über Lücken und Singularitäten von Potenzreihen, *Math. Z.* **29** (1929) 549–640.

For algebraic valuation theory see:

ALBERT, A. A., *Modern Higher Algebra,* University of Chicago Press, Chicago (1937).

14 FUNCTIONAL EQUATIONS

Numerous functional equations have been encountered in earlier chapters of this treatise. Thus in the discussion of fixed point theorems we were concerned with a mapping $f \to T[f]$ of a metric space $\mathfrak{X}$ into itself. The invariant elements are solutions of the functional equation

$$f = T[f].$$

When T or some power of T is a contraction, existence and uniqueness of a solution could be proved. Examples were discussed in Chapter 12.

To any functional inequality in a product space corresponds a functional equation. Thus to (13.1.2) corresponds

$$f(x \circ y) = G[f(x), f(y)].$$

Some instances of this and of the more general equation

$$f[g(x, y)] = H[f(x), f(y), x, y]$$

will be discussed below. Here g and H are given and f is to be found. We have here a much greater freedom of choice in the mappings than in Chapter 13. Not merely the domain but also the range is at our disposal.

Differential and integral equations are usually excluded from consideration by workers on functional equations as an economy of effort. We shall not do this. The field is vast and our selection of material may appear haphazard. Each functional equation creates its own problems and there is little of general theory available. We have tried to emphasize the *a priori* aspects: what information about the solutions is hidden in the equation and obtainable without solving the equation?

There are three sections: "Cryptoanalysis"; Cauchy's equations and generalizations; and Uniqueness theorems.

14.1 "CRYPTOANALYSIS"

Mathematics is full of codes which have to be deciphered. Thus any functional equation contains a message more or less well hidden. All properties of the solutions are built into the equation and our problem is to bring them out by asking appropriate questions. Note that this information is available before an explicit solution is obtained, if it is obtainable. In the following an effort will be made to show what type of *a priori* information is obtainable and how to obtain it. This type of "cryptoanalysis" will be illustrated by special examples. There is no general theory at our disposal.

First let us remind the reader that it is necessary to specify what mappings are to be considered. It is usually necessary to state the domain of f and also the range. Sometimes even limitations on the graph $\{x, f(x)\}$ have to be taken into account and may seriously affect the existence and nature of solutions. Equations look formally the same but in one case the variables may be real numbers, in another matrices, quaternions or vectors.

Secondly, even after such basic decisions have been made, we note that properties of solutions may be of two types: *categorical* and *conditional.* The first type pertains to all solutions in the given range, the second only to solutions satisfying some restrictive condition such as boundedness or measurability, etc. The first type occurs fairly seldom, the second type is what one normally operates with in this theory.

Example 1. Take the simple first order differential equation

$$w'(z) = w(z). \tag{14.1.1}$$

Here the natural domain and range are the complex plane C. The following is an incomplete list of categorical properties:

(1) *If $w(z)$ is a solution, so is $w(z + a)$ for any fixed a.*

(2) $w(0)\, w(z + a) = w(a)\, w(z)$ *for any fixed a.*

(3) *Either $w(z) \neq 0$ for all z or $w(z) \equiv 0$.*

(4) *$w^{(n)}(z)$ exists for all n and equals $w(z)$.*

(5) *$w(z)$ is an entire function of z.*

Verifications are left to the reader.

Example 2. For a contrast, take one of Cauchy's equations

$$f(x + y) = f(x) + f(y), \tag{14.1.2}$$

which will be studied in detail in the next section. We consider mappings f of reals into reals defined for all real x. Among the categorical properties we note that $f(0) = 0$ and $f(nx) = nf(x)$, more generally, $f(qx) = qf(x)$ for any rational q. On the other hand, measurability or boundedness on compact subsets are not categorical, nor is the relation $f(rx) = rf(x)$ for any real number r true for all solutions.

Example 3. The functional equation

$$R(s) - R(t) = (t - s)\, R(s)\, R(t) \tag{14.1.3}$$

is basic in resolvent theory where it is known as the *first resolvent equation.* See Sections 1.5 and 9.2 where explicit solutions are to be found first for matrix algebras and then for a general non-commutative Banach algebra $\mathfrak{B}$ with unit element e. Let us consider the latter case. Here s is a complex variable; $R(s)$, a $\mathfrak{B}$-valued function, is defined and satisfies (14.1.3) in some domain D of the complex plane.

One categorical property is discernible: $R(s)$ and $R(t)$ must commute since the equation is unchanged if s and t are interchanged. Other properties appear to be conditional. We shall show under mild restrictions on R that this function will be $\mathfrak{B}$-holomorphic in the terminology of Section 8.1. More precisely we shall show that

local boundedness $\Rightarrow$ continuity $\Rightarrow$ differentiability $\Rightarrow$ analyticity.

Suppose that in a disk $|s - a| < r$ contained in D we have

$$\|R(s)\| \leqslant M(a). \tag{14.1.4}$$

This implies

$$\|R(s) - R(t)\| \leqslant [M(a)]^2 \, |s - t|,$$

a Lipschitz condition. Hence a locally bounded solution is continuous. Further, a continuous solution is differentiable since

$$\frac{R(t) - R(s)}{t - s} = - R(s)\, R(t) \to - [R(s)]^2 \tag{14.1.5}$$

as $t \to s$, so that $R'(s)$ exists and

$$R'(s) = - [R(s)]^2.$$

This implies the existence of derivatives of all orders and

$$R^{(n)}(s) = (-1)^n n! [R(s)]^{n+1}. \tag{14.1.6}$$

Cf. Section 9.2. Thus locally the solution of (14.1.3) is given by

$$R(s) = \sum_{n=0}^{\infty} A^{n+1}(a - s)^n, \qquad A = R(a). \tag{14.1.7}$$

The power series converges at least for $|s - a| < [\|A\|]^{-1}$ and we know conversely that any such power series satisfies (13.1.3) for s and t in the circle of convergence.

The following examples are concerned with *a priori* estimates of rates of growth and nature of infinitudes of solutions of ordinary differential equations. A special case of such a problem occurs in Theorem 12.4.2. This type of discussion does not seem to have made much headway for other types of functional equation. Normally it appears to be irrelevant, but it should be meaningful for difference equations and for functions defined by addition theorems.

Example 4. We start with the non-linear first order differential equation

$$w'(z) = 1 + [w(z)]^2. \tag{14.1.8}$$

Domain and range are the complex plane. Since the equation is *autonomous*, i.e. unchanged under the shift $z \to z + a$, we can exhibit one categorical property right away: if $w(z)$ is a solution, so is $w(z + a)$. Another categorical property is involved in the following trichotomy: either $w(z) \neq \pm i$ for all z or $w(z) \equiv i$ or $w(z) \equiv -i$. This follows from the fact that there is one and only one solution which takes on the value $+i$ at $z = a$ and the same holds for $-i$.

Another categorical property may be formulated as follows: if $w(z)$ is a non-constant solution of (14.1.8), then $w(z)$ necessarily becomes infinite for z tending to some finite value b. Moreover, the infinitudes are simple poles of residue -1 and for any given value b there is one (and only one) solution which admits of $z = b$ as a pole. This naturally follows from the fact that any non-constant solution of the equation is of the form $\tan(z - a)$. But this is such a basic fact that it should be obtainable directly from the equation without the detour via the explicit solutions. There are at least two ways of accomplishing this.

One of the methods leads to another categorical property of the solutions. There is a one-parameter family of linear fractional transformations

$$v = \frac{w + c}{1 - cw} \tag{14.1.9}$$

which leaves the equation invariant. Here c is any real or complex number. Thus if $w(z)$ is a solution of (14.1.8) so is

$$\frac{w(z) + c}{1 - cw(z)}. \tag{14.1.10}$$

The connection of this result with the addition theorem of the tangent function is obvious. See Problem 5, Exercise 14.1. For our purposes the transformation

$$v = -\frac{1}{w} \tag{14.1.11}$$

is more useful. Actually it may be obtained as a limiting case of (14.1.9) by letting $c \to \infty$. This change of dependent variable gives

$$v'(z) = 1 + [v(z)]^2, \tag{14.1.12}$$

i.e. reproduces the original equation except for notation. Now it is clear that this equation has a solution which assumes the value 0 at $z = b$, say $v(z; b, 0)$. The equation says that the derivative takes the value 1 at $z = b$. This means that (14.1.8) has the solution $-[v(z; b, 0)]^{-1}$ with a simple pole at $z = b$ where the residue is -1. This is an *a priori* verification of the existence of solutions with a simple pole and residue -1 at a preassigned point.

The second method is quicker but less precise. We suspect the existence of a solution of (14.1.8) which becomes infinite as z approaches some point b. At such a point $w'(z)$ and $[w(z)]^2$ both become infinite and, comparing orders of infinity, we should get some inkling of the facts. If α and β are positive numbers, suppose that

$$w(z) \sim \beta(z-b)^{-\alpha} \tag{14.1.13}$$

up to terms of lower order. Then, formally,

$$w'(z) \sim -\alpha\beta(z-b)^{-\alpha-1}, \qquad [w(z)]^2 \sim \beta^2(z-b)^{-2\alpha},$$

again up to terms of lower order. Writing down the condition that leading terms on both sides of equation (14.1.8) must cancel, we get $-\alpha-1=-2\alpha$, $-\alpha\beta=\beta^2$, or $\alpha=1$, $\beta=-1$. Thus we are justified in expecting simple poles of residue -1 at the singular points. We know that this is correct. Since b is arbitrary, we say that equation (14.1.8) has *movable singularities which are simple poles*, another categorical property.

Our last example is more sophisticated, and here the last word has not been said. See P. J. Rijnierse (1968) and E. Hille (1969).

Example 5. The equation

$$y''(x) = x^{-\frac{1}{2}}[y(x)]^{3/2} \tag{14.1.14}$$

was introduced in nuclear physics by L. M. Thomas and Enrico Fermi independently of each other in 1927. It is known as the *Thomas–Fermi equation.* We are not concerned with electrical fields in an atom; just *a priori* properties of solutions of the equation regardless of physical significance, if any.

It is natural to restrict oneself to real positive values of x and to solutions with positive values in some interval (c, d). The equation then shows that $y''(x) > 0$ in (c, d), so the graph of y is concave upward and there can be at most one (positive) minimum in the interval. The equation has movable singularities (categorical property of the equation) and actually of two different kinds according as y tends to zero or to infinity as x approaches the point $x = b$ in question. The existence of points of the first kind is obvious since the initial values $y(b) = 0$, $y'(b) = c \neq 0$ are admissible and determine a solution which is positive to the left of $x = b$ if $c < 0$ and to the right of b if $c > 0$. This turns out to be an *algebraic branch point.* The second type of singularity where y becomes infinite is less obvious but is equally common. As a matter of fact, as soon as a solution starts to grow it is doomed to become infinite for some finite value of x. This observation seems to go back to L. Brillouin (1934). What happens to the solution beyond the critical value appears to be a moot question.

Let us see what information the second method under Example 4 can give in this case. Suppose that as x increases toward some number b

$$y(x) \sim \beta(b-x)^{-\alpha}, \qquad \alpha > 0, \qquad \beta > 0, \tag{14.1.15}$$

up to terms of lower order. Again, proceeding formally, we have

$$x^{-1/2} \sim b^{-1/2}, \qquad [y(x)]^{3/2} \sim \beta^{3/2}\,(b-x)^{-3\alpha/2},$$
$$y''(x) \sim \alpha\beta(\alpha+1)\,(b-x)^{-\alpha-2}$$

up to terms of lower order. Comparison of dominant terms gives

$$\alpha = 4, \qquad \beta = 400b. \tag{14.1.16}$$

So far so good, but there is no Laurent expansion at $x = b$, so the singularity cannot be a pole.

Nevertheless, there do exist solutions which become infinite as x increases to a preassigned point $x = b > 0$ and which satisfy the inequality

$$y(x) < 400b(b-x)^{-4}. \tag{14.1.17}$$

This also follows from *a prioristic* considerations but it would take us too far afield to develop these more sophisticated methods.

EXERCISE 14.1

1. Verify the assertions under Example 1 above.
2. Discuss the equation
$$(s-t)\,S(a)\,S(s)\,S(t) = (s-a)\,S(t) - (t-a)\,S(a)$$
along the lines of Example 3 above. Assume $S(s) \in \mathfrak{B}$ and to be defined and bounded in a domain D containing the point a.
3. Discuss the dissolvent equation
$$sD(s) - tD(t) = (s-t)\,D(s)\,D(t)$$
along similar lines. $\mathfrak{B}$ need not have a unit element and the domain D should not contain the origin.
4. Verify that (14.1.9) leaves (14.1.8) invariant. What happens when $c = i$ or $-i$?
5. From this fact derive the addition theorem for a non-constant solution of (14.1.8).
6. Determine the probable nature of the movable singularities in the case of the equation $w'(z) = 1 + [w(z)]^4$.
7. Same question for $[w'(z)]^2 = 1 + [w(z)]^4$.
8. Consider the equation $R'(z) = -[R(z)]^2$ in the setting of Example 3 where R is $\mathfrak{B}$-valued in some domain of the complex plane. The *a priori* methods of Example 4 suggest that a movable singularity should be a simple pole with an idempotent as a residue. The discussion in Section 1.5 shows that this hypothesis is wrong already for matrices. Where is the error in the reasoning?
9. Could a solution of this equation be algebraically singular wherever it exists?
10. In Example 2 it is stated that $f(qx) = qf(x)$ for all rational numbers q when f is a solution of (14.1.2). Prove this.

14.2 CAUCHY'S EQUATIONS AND GENERALIZATIONS

Cauchy in the 1820's studied the four equations

$$f(x+y) = f(x) + f(y), \tag{14.2.1}$$

$$f(x+y) = f(x)f(y), \tag{14.2.2}$$

$$f(xy) = f(x)f(y), \tag{14.2.3}$$

$$f(xy) = f(x) + f(y). \tag{14.2.4}$$

They are closely related. For the time being x and y are real and f real or complex valued. Cauchy proved that if f is a continuous solution of the first equation, then there exists a constant a such that

$$f(x) = ax, \qquad \forall\, x. \tag{14.2.5}$$

Now actually, if f is continuous at a single point, then it is continuous everywhere and (14.2.5) holds. Suppose that f is continuous at $x = b$. Then

$$f(b+h) - f(b) = f(h) \to 0 \quad \text{with} \quad h.$$

Hence

$$f(s+h) - f(s) = f(h) \to 0 \quad \text{with} \quad h,$$

so that f is continuous also at $x = s$, i.e. everywhere. This observation is due to the French mathematician Gaston Darboux (1842–1917) in 1875. The introduction of measurability and the Lebesgue integral led to further results.

Theorem 14.2.1. *If a solution f of* (14.2.1) *is Lebesgue integrable over finite intervals, then* (14.2.5) *holds.*

Proof. We have

$$\begin{aligned} f(x) &= \int_0^1 f(x+y)\,dy - \int_0^1 f(y)\,dy \\ &= \int_x^{x+1} f(s)\,ds - \int_0^1 f(s)\,ds. \end{aligned}$$

Now a definite integral is a continuous function of its limits because

$$\int_S |f(s)|\,ds \to 0 \quad \text{with} \quad m[S]. \tag{14.2.6}$$

This makes f continuous, and the integral of a continuous function is differentiable with respect to variable limits of integration. Hence f is differentiable, and the usual formulas of the calculus give

$$f'(x) = f(x+1) - f(x) = f(1) \quad \text{or} \quad f(x) = f(1)x, \tag{14.2.7}$$

since $f(0) = 0$.

For real-valued solutions one-sided boundedness on a set of positive measure or, as an alternative, measurability on such a set is sufficient to ensure continuity. The first result stated is due to Alexander Ostrowski (1929), who also proved that a non-measurable solution must be unbounded on every set of positive measure.

In the meantime, non-measurable solutions had been constructed in 1905 by Georg Hamel (1877–1954). The gist of Hamel's construction is the following. On the basis of the axiom of choice one proves that the reals can be well ordered and this implies the existence of a non-denumerable basis $\{x_\alpha\}$ for R^1 in terms of which each real number x has a unique representation

$$x = \sum_\alpha q(\alpha)\, x_\alpha, \tag{14.2.8}$$

where the coefficients $q(\alpha)$ are rational numbers and the sum involves only a finite number of summands. We choose now an arbitrary non-denumerable set of real numbers $\{z_\alpha\}$ and define f on the basis elements by

$$f(x_\alpha) = z_\alpha, \qquad \forall\, \alpha, \tag{14.2.9}$$

and set

$$f(x) = \sum_\alpha q(\alpha)\, z_\alpha. \tag{14.2.10}$$

This defines $f(x)$ uniquely for all x. Moreover, f satisfies (14.2.1). For if x is given by (14.2.8) and

$$y = \sum_\alpha r(\alpha)\, x_\alpha,$$

then

$$x + y = \sum_\alpha [q(\alpha) + r(\alpha)]\, x_\alpha, \qquad f(x+y) = \sum_\alpha [q(\alpha) + r(\alpha)]\, z_\alpha,$$

so that f is a solution. Note that in the expression for y, members belonging to other α's than in x might be different from 0 but there is still only a finite number of coefficients different from 0. Moreover, if not all z_α's are proportional to the corresponding x_α's, then this solution is not of the form (14.2.5), so it cannot be continuous or measurable.

Solutions of (14.2.1) obviously have the property

$$f(nx) = nf(x) \tag{14.2.11}$$

for all positive integers n. This implies that

$$f(rx) = rf(x) \tag{14.2.12}$$

for all rational numbers r. On the other hand, the relation need not hold for irrational values of r if f is non-measurable. This striking result is due to Z. Daróczy (1961) and L. Losonczi (1964). The following construction of such solutions was communicated to the author by J. H. B. Kemperman (1969).

Let ρ and σ be two distinct real transcendental numbers and let $Q(\rho)$ and $Q(\sigma)$ be the field extensions of the rational field Q obtained by adjunction of ρ and σ respectively. The elements of $Q(\rho)$ are simply rational functions with rational coefficients in the mark ρ and $Q(\sigma)$ is obtained by replacing ρ by σ. This defines an isomorphic mapping of $Q(\rho)$ onto $Q(\sigma)$ which we denote by Z. Next we "construct" a Hamel basis for R^1 over $Q(\rho)$, say the set $\{x_\alpha\}$. Then every $x \in R^1$ has a unique representation of the form

$$x = \sum_\alpha q_\rho(\alpha)\, x_\alpha, \tag{14.2.13}$$

where each $q_\rho(\alpha) \in Q(\rho)$ and the number of summands is finite. If now

$$y = \sum_\alpha r_\rho(\alpha)\, x_\alpha, \quad \text{then} \quad x + y = \sum_\alpha [q_\rho(\alpha) + r_\rho(\alpha)]\, x_\alpha.$$

Since the representation of a real number in terms of a Hamel basis is unique, the last formula is the Hamel representation of $x + y$. We now define

$$f(x) = \sum_\alpha \mathsf{Z}[q_\rho(\alpha)]\, x_\alpha = \sum_\alpha q_\sigma(\alpha)\, x_\alpha. \tag{14.2.14}$$

Here Z is the isomorphic mapping of $Q(\rho)$ onto $Q(\sigma)$ and $q_\sigma(\alpha)$ is the image of $q_\rho(\alpha)$. Since sums go into sums under Z, we have

$$\begin{aligned} f(x+y) &= \sum_\alpha \mathsf{Z}[q_\rho(\alpha) + r_\rho(\alpha)]\, x_\alpha \\ &= \sum_\alpha \mathsf{Z}[q_\rho(\alpha)]\, x_\alpha + \sum_\alpha \mathsf{Z}[r_\rho(\alpha)]\, x_\alpha \\ &= \sum_\alpha q_\sigma(\alpha)\, x_\alpha + \sum_\alpha r_\sigma(\alpha)\, x_\alpha = f(x) + f(y). \end{aligned}$$

Thus f satisfies (14.2.1). We shall now prove that

$$f(\rho x) = \sigma f(x), \qquad \forall\, x. \tag{14.2.15}$$

Here we need that Z acting on $Q(\rho)$ takes products into products as well as sums into sums. Thus

$$\begin{aligned} f(\rho x) &= \sum_\alpha \mathsf{Z}[\rho q_\rho(\alpha)]\, x_\alpha = \sum_\alpha \mathsf{Z}(\rho)\mathsf{Z}[q_\rho(\alpha)]\, x_\alpha \\ &= \sigma \sum_\alpha q_\sigma(\alpha)\, x_\alpha = \sigma f(x) \end{aligned}$$

as asserted. Thus (14.2.12) need not hold for irrational values of r. This solution is, of course, non-measurable.

Richard C. Metzler has called my attention to the fact that the construction may be modified so that (14.2.15) holds for a finite number of distinct values of ρ and corresponding values of σ, all transcendental numbers. If ρ is algebraic, (14.2.15) requires that σ be algebraic—in fact, ρ and σ must satisfy the same irreducible equation. See, further, Exercise 14.2.

Let us make a brief excursion into the complex field. Equation (14.2.1) also makes sense for complex variables:

$$f(z_1 + z_2) = f(z_1) + f(z_2), \tag{14.2.16}$$

where, however, there are continuous solutions besides az. We can take

$$f(x + iy) = ax + by \tag{14.2.17}$$

for any constants a and b. If a and b are real, this is the general form of a linear functional on R^2. The expression defines a differentiable function of z iff $b = ai$.

We return to real variables. We may say that measurability implies continuity, which implies differentiability for solutions of (14.2.1). The same holds for the other Cauchy equations which are satisfied by

$$e^{ax}, \; x^a, \qquad a \log x, \tag{14.2.18}$$

respectively. Here a is an arbitrary real or complex number, and in the case of (14.2.4) the variables x and y are positive.

So far we have assumed f to be a mapping from reals to reals, reals to complex, or complex to complex. But there are many other possibilities. Thus letting x and y remain real numbers, we can let f define a mapping of R^1 into $\mathfrak{M}_n$, the n by n matrices over C, and ask for solutions of the various equations in this setting. In the case of (14.2.1) any continuous solution would be of the form

$$f(x) = \mathcal{A}x, \tag{14.2.19}$$

where $\mathcal{A}$ is an arbitrary constant matrix. Equation (14.2.2) leads to

$$f(x) = \mathfrak{J} \exp [\mathcal{A}x],$$

where, again, $\mathcal{A}$ is any constant matrix, the exponential function is defined by the usual series, and $\mathfrak{J}$ is an idempotent commuting with $\mathcal{A}$.

We can also turn the tables: let $f(\mathfrak{X})$ be a number and $\mathfrak{X}$ and $\mathfrak{Y}$ matrices. The equation

$$f(\mathfrak{X} + \mathfrak{Y}) = f(\mathfrak{X}) + f(\mathfrak{Y}) \tag{14.2.20}$$

now characterizes the *additive functionals* on $\mathfrak{M}_n$. Note that such a functional need be neither homogeneous nor bounded. Assuming these additional properties, a solution is given by

$$f(\mathfrak{X}) = \sum_{j=1}^{n} \sum_{k=1}^{n} \xi_{jk} x_{jk}, \quad \text{where} \quad \mathfrak{X} = (x_{jk}) \tag{14.2.21}$$

and the coefficients ξ_{jk} are arbitrary complex numbers. This formula is obtained by writing

$$\mathfrak{X} = \sum_{j=1}^{n} \sum_{k=1}^{n} x_{jk} \mathcal{E}_{jk}, \tag{14.2.22}$$

where $\mathcal{E}_{jk}$ is the matrix with a 1 in the place (j, k) and zeros elsewhere. These matrices form a basis for $\mathfrak{M}_n$ and we define f on the basis elements by

$$f(\mathcal{E}_{jk}) = \xi_{jk}$$

and proceed by linearity to obtain (14.2.21).

Equation (14.2.3) in the setting $\mathfrak{M}_n \to C$ now characterizes *multiplicative functionals.* The determinant of $\mathfrak{X}$ is clearly a solution; more generally we can take

$$f(\mathfrak{X}) = [\det \mathfrak{X}]^a \tag{14.2.23}$$

where a is a fixed number. This makes sense when a and $\det \mathfrak{X}$ are real positive. The general solution found by M. Hosszú (1959) is $f(\mathfrak{X}) = g(\det \mathfrak{X})$, where g is multiplicative $g(xy) = g(x)g(y)$ in the domain in question R^1 or C.

We may, of course, also allow mappings from $\mathfrak{M}_n$ into $\mathfrak{M}_n$. Now the problem becomes more involved. We see by inspection that

$$\mathcal{F}(\mathfrak{X}) = \mathcal{A}\mathfrak{X} + \mathfrak{X}\mathcal{B} + \mathcal{C}\mathfrak{X}\mathcal{D} \tag{14.2.24}$$

is a solution of

$$\mathcal{F}(\mathfrak{X} + \mathfrak{Y}) = \mathcal{F}(\mathfrak{X}) + \mathcal{F}(\mathfrak{Y}) \tag{14.2.25}$$

for any choice of the constant matrices $\mathcal{A}$, $\mathcal{B}$, $\mathcal{C}$, $\mathcal{D}$. This is a continuous solution and not even the most general one. The general solution is known, even of the more general problem of mapping m by n matrices into p by q matrices. Since only addition is involved, we can consider an m by n matrix as a vector with mn components and the problem is reduced to solving (14.2.1) for vectors. With obvious change of notation we are now concerned with solutions of the equation

$$\mathbf{F}(\mathbf{x} + \mathbf{y}) = \mathbf{F}(\mathbf{x}) + \mathbf{F}(\mathbf{y}), \tag{14.2.26}$$

where $\mathbf{x} \to \mathbf{F}(\mathbf{x})$ is a mapping from C^n to C^p. According to Aczél (personal communication, cf. pp. 215–16, 348 of his 1966 treatise) one proceeds as follows. Let $\mathbf{e}_1, \cdots, \mathbf{e}_n$ be a basis of C^n and $\mathbf{E}_1, \cdots, \mathbf{E}_p$ a basis of C^p. Each component $F_k(\mathbf{x})$ of $\mathbf{F}(\mathbf{x})$ is additive:

$$F_k(\mathbf{x}_1 + \mathbf{x}_2) = F_k(\mathbf{x}_1) + F_k(\mathbf{x}_2), \tag{14.2.27}$$

which extends to any finite number of summands. If, now,

$$\mathbf{x} = \sum_{j=1}^{n} x_j \mathbf{e}_j, \tag{14.2.28}$$

we have

$$F_k(\mathbf{x}) = F_k\left[\sum_{j=1}^{n} x_j \mathbf{e}_j\right] = \sum_{j=1}^{n} F_k(x_j \mathbf{e}_j) \equiv \sum_{j=1}^{n} g_{jk}(x_j), \tag{14.2.29}$$

where each scalar function g_{jk} is additive:

$$g_{jk}(x + y) = g_{jk}(x) + g_{jk}(y).$$

Combining, we get

$$\mathbf{F}(\mathbf{x}) = \sum_{k=1}^{p} \left\{ \sum_{j=1}^{n} g_{jk}(x_j) \right\} \mathbf{E}_k, \tag{14.2.30}$$

where the g_{jk} are arbitrary additive functions, i.e. solutions of (14.2.1). This is the general solution of (14.2.26). If the g_{jk} are continuous, then so is $\mathbf{F}(\mathbf{x})$, but in general this would not be the case. From the solution (14.2.30) of (14.2.26) we can get the general solution of the matrix case (14.2.25) which served as our point of departure. The details are left to the reader.

There is, of course, no reason for stopping at vectors and matrices. Thus (14.2.1) is only another way of writing

$$T(x_1 + x_2) = T(x_1) + T(x_2), \tag{14.2.31}$$

which characterizes an *additive mapping from one semi-group to another*. The transformation need not be bounded, if $\mathfrak{X}$ and $\mathfrak{Y}$ are metric spaces where boundedness makes sense.

In the same spirit, if s and t are real and if $T(s)$ is a mapping from R^1 into $\mathfrak{E}(\mathfrak{X})$, the B-algebra of linear bounded transformations from $\mathfrak{X}$ to $\mathfrak{X}$, then the equation

$$T(s + t) = T(s)T(t) \tag{14.2.32}$$

characterizes *one-parameter groups or semi-groups of transformations*.

The Cauchy equations may be generalized in a different direction. Let $\circ$ denote a binary associative operation, say from $R^1 \times R^1$ to R^1, let $G(u, v)$ be a mapping from $R^1 \times R^1$ to R^1, and consider the equation

$$f(x \circ y) = G[f(x), f(y)]. \tag{14.2.33}$$

The Cauchy equations are clearly of this type with $\circ$ being addition or multiplication and $G(u, v) = u + v$ or uv. This class of equations was studied by C. T. Ionescu Tulcea in 1960. His general result is much too complicated to give here. The following much-specialized version is due to R. C. Metzler (personal communication, 1969). It is stated without a proof.

Theorem 14.2.2. *Let $(s, t) = s \circ t$ be a composition on $R^1 \times R^1$ to R^1 defined for $(s, t) \in D \subset R^1 \times R^1$ and continuous in D. Suppose there exists a neutral element e such that for all $(e, t) \in D$ we have $e \circ t = t$. Suppose, further, that for (x, t) and (e, u) in D we can solve the equation $x \circ t = u$ for $x \equiv p_u(t)$ uniquely and in such a way that the mapping $(u, t) \to p_u(t)$ is continuous. In addition, suppose that for each t under consideration there is a compact neighborhood $N(t)$ of t and a constant $\alpha(t)$ such that for all $s \in N(t)$ and open sets $W \subset N(t)$, $\mu[p_s^{-1}(W)] \leqslant \alpha(t)\mu(W)$ where μ is Lebesgue measure. Then if G in (14.2.33) is continuous on the range of f and if f is measurable, f is necessarily continuous.*

The conditions on $\circ$ are clearly satisfied by $+$ and $\cdot$ with neutral elements 0 and 1, respectively. A further case is

$$s \circ t = \frac{s+t}{1+st}, \qquad e = 0. \tag{14.2.34}$$

This operation is related to the addition theorem of the hyperbolic tangent and to the theory of special relativity.

The case where $s \circ t = s + t$ and $G(u, v)$ is a symmetric analytic function is known as an *addition theorem*. About 100 years ago Weierstrass determined all functions which have an *algebraic addition theorem*, i.e. where G is an algebraic symmetric function. It turned out that f has to be an algebraic function of one of the three arguments, z, e^{az}, $\wp(bz)$, where $\wp(z)$ is the elliptic function of Weierstrass. This is under the assumption of complex variables. If s and t are restricted to real values, the solutions are piecewise analytic and expressible by the same functions.

The case

$$f(s+t) = G[f(s), f(t)] \tag{14.2.35}$$

with f mapping C into a Banach algebra, in particular $\mathfrak{M}_n$, was examined by N. Dunford and E. Hille in 1944. Here $G(u, v)$ is a symmetric analytic function. Under fairly general assumptions, a continuous solution will possess derivatives of all orders. The value of $f(0)$ is restricted to the roots of the equation

$$G(a, a) = a, \tag{14.2.36}$$

$f'(0)$ can be chosen practically arbitrarily, but the values of the higher derivatives are uniquely determined by a and $f'(0)$.

The case $s \circ t = st$ is also of some interest. This leads to a so-called *multiplication theorem* with

$$f(st) = G[f(s), f(t)]. \tag{14.2.37}$$

This type of multiplication theorem should not be confused with the type considered in Section 12.5. The case where f is a linear mapping $\mathbf{T}$ of a commutative algebra into itself has been subjected to a considerable amount of investigation. Such a $\mathbf{T}$ is known as a *Bourlet operator* after C. Bourlet (1897). According to G. I. Targonski (1967), if the algebra possesses a unit element and has no divisors of zero, there are essentially only three distinct types of Bourlet operators with the corresponding functional equations:

$$\mathbf{T}[uv] = B[\mathbf{T}u][\mathbf{T}v], \tag{14.2.38}$$

$$\mathbf{T}[uv] = \tfrac{1}{2}[u\mathbf{T}v + + v\mathbf{T}u], \tag{14.2.39}$$

$$\mathbf{T}[uv] = u\mathbf{T}v + v\mathbf{T}u. \tag{14.2.40}$$

The last operator is known as a *derivation*.

EXERCISE 14.2

1. Verify that the continuous solutions of (14.2.2) to (14.2.4) have the form stated in the text.
2. Construct non-measurable solutions of (14.2.2).

The next four problems are byproducts of discussions with J. H. B. Kemperman and R. C. Metzler.

3. If f is chosen as a non-measurable solution of (14.2.1) satisfying (14.2.15), show that this implies that
$$f(sx) = tf(x),$$
where s is any element of $Q(\rho)$ and $t = \mathsf{Z}(s)$ is the corresponding element of $Q(\sigma)$.
4. Show that if (14.2.15) holds for an algebraic irrational number ρ, then σ must also be algebraic and ρ and σ satisfy the same irreducible algebraic equation. [*Hint*: Note that there is a polynomial in ρ which is a rational number and use the result of the preceding problem.]
5. Carry through the construction with $\rho = \sqrt{2}$, $\sigma = -\sqrt{2}$.
6. The following is an example of a non-measurable quadratic generalized polynomial in the sense of Section 8.4. Take a Hamel basis $\{x_\alpha\}$ for the reals and form
$$Q(x) = \sum_\alpha \sum_\beta q_\alpha q_\beta x_\alpha x_\beta, \qquad x = \sum_\alpha q_\alpha x_\alpha.$$
Find $\delta Q(x, h)$ and $\delta^2 Q(x, h)$ and verify that the higher variations vanish identically.
7. With the aid of the result (14.2.30) for the vector case, find the general solution of (14.2.25) for matrices.
8. Verify (14.2.21).
9. Verify (14.2.24).
10. Fill in missing details in the derivation of (14.2.30).
11. If $\mathfrak{X} = C[0, \infty]$ and for $s > 0$, $T(s)[f](t) = f(s + t)$, show that $\{T(s)\}$ is a semi-group of linear bounded transformations and find $\|T(s)\|$.
12. Let $\mathfrak{B}$ stand for a B-algebra with unit element e and take the mapping f of C into $\mathfrak{B}$ defined by $s \to f(s) = (e + p)^s$ where p is a nilpotent of order $k > 1$. Show by operational calculus or otherwise that
$$(e + p)^s = e + \binom{s}{1} p + \binom{s}{2} p^2 + \cdots + \binom{s}{k-1} p^{k-1}.$$
Show that $f(s)f(t) = f(s + t)$ for all s, t.
13. Find the addition theorem for $(1 - z)^{\frac{1}{2}}$.
14. If $G(u, v) = u(1 - v^2)^{\frac{1}{2}} + v(1 - u^2)^{\frac{1}{2}}$, find a function with the corresponding addition theorem.
15. The function $\wp(z)$ is that solution of the differential equation
$$[w'(z)]^2 = 4[w(z)]^3 - g_2 w(z) - g_3$$

which becomes infinite at $z = 0$. Here g_2 and g_3 are arbitrary constants, not both zero. Use the method of Section 14.1 to determine the nature of the singularity at $z = 0$ and find the principal part of the Laurent expansion (i.e. the terms which become infinite at 0).

16. Find the continuous solutions of

$$f\left(\frac{x+y}{1+xy}\right) = f(x) + f(y).$$

Verify that the conditions of Theorem 14.2.2 are satisfied for $|x| < 1$, $|y| < 1$. [*Hint*: There is a transformation of the variables which reduces this equation to the form (14.2.1).]

17. If the graph of a continuous solution of

$$f[\tfrac{1}{2}(x+y)] = \tfrac{1}{2}[f(x) + f(y)]$$

passes through two given points in the plane, show that the solution is unique and determine it explicitly.

For the remaining problems, see Targonski (1967). As basic algebra take the algebra $\mathfrak{A}$ of polynomials over the complex field. Examples are given of Bourlet operators satisfying one of the equations (14.2.38) to (14.2.40). The operators $\mathbf{T}$ are defined in terms of a fixed element g of $\mathfrak{A}$.

18. Show that $\mathbf{T}[f](x) = f[g(x)]$ satisfies (14.2.38) with $B = 1$. Such a $\mathbf{T}$ is called a *substitution operator*. What is the relation between g and $\mathbf{T}$?

19. There may be an element f of $\mathfrak{A}$ such that $f[g(x)] = \lambda f(x)$. In other words, $\mathbf{T}$ has a characteristic value λ with characteristic function f. The functional equation

$$f[g(x)] = \lambda f(x)$$

is known as *Schröder's equation* after E. Schröder (1871). Show that if $g(x) = x^2$, the corresponding Schröder equation can have no solution for $\lambda \neq 1$ and only constant solutions for $\lambda = 1$ in the polynomial algebra $\mathfrak{A}$ but does have solutions in a function algebra containing $\log x$. Which solutions?

20. Show that the *multiplication operator* $\mathbf{T}[f](x) = g(x)f(x)$ satisfies (14.2.39) and express g in terms of $\mathbf{T}$.

21. Show that $\mathbf{T}[f](x) = g(x)f'(x)$ satisfies (14.2.40) and express g in terms of $\mathbf{T}$.

14.3 UNIQUENESS THEOREMS

Around 1960 the most active centers of research in the theory of functional equations were located in Hungary with J. Aczél and his pupils at the University of Debrecen and the group led by M. Hosszú and E. Vincze at the Technical University of Miscolc. To Vincze (1962) we owe a general method of solution for certain classes of functional equations based on the implications of linear independence. Hosszú will figure briefly later in this section. Here we shall merely discuss one of Aczél's many basic contributions, a uniqueness theorem (1964) which is general, easy to state, and easy to prove. It deals with transformations from $R^1 \times R^1$ to R^1 defined by a mapping $(x, y) \to F(x, y)$ where $F(x, y)$ lies strictly between x and y if $x \neq y$. Such a mapping F is called *intern* by Aczél. We recall that a mapping $u \to H(u)$ is said to be *injective* (see Section 1.3) if $H(u_1) = H(u_2)$ implies $u_1 = u_2$. After these preliminaries we can state and prove

Theorem 14.3.1. *Let a mapping F from $R^1 \times R^1$ to R^1 be defined and continuous for $(x, y) \in (A, B) \times (A, B)$ as well as intern so that*

$$x < y \quad \textit{implies} \quad x < F(x, y) < y, \qquad x < F(y, x) < y. \tag{14.3.1}$$

Let $H(u, v, x, y)$ be a function of four variables, injective either in u or in v. Then the functional equation

$$f[F(x, y)] = H[f(x), f(y), x, y] \tag{14.3.2}$$

with the initial conditions

$$f(a) = c, \qquad f(b) = d, \qquad A < a < b < B, \tag{14.3.3}$$

has at most one continuous solution.

Proof. Suppose there were two continuous solutions $f_1(x)$ and $f_2(x)$ which we may assume to be defined in all of (A, B). We want to show that the relations $f_1(a) = f_2(a)$, $f_1(b) = f_2(b)$ imply $f_1(x) = f_2(x)$ for all x. This is proved in three steps, one for each of the subintervals of (A, B) corresponding to the partition points $x = a$ and $x = b$.

I. The interval $[a, b]$. If there should exist a point x_0 with $a < x_0 < b$ and $f_1(x_0) \neq f_2(x_0)$, then we determine two points C and D, $a \leqslant C < x_0 < D \leqslant b$, by the following considerations. Let S_1 be the set of points x in the interval $[a, x_0]$ where $f_1(x) = f_2(x)$ and let C be the supremum of x for $x \in S_1$. There exists such a C and $a \leqslant C$. Similarly, if S_2 is the set of points x in $(x_0, b]$ where $f_1(x) = f_2(x)$, let D be the infimum of x in S_2. Here $D \leqslant b$. Further, by the continuity of f_1 and f_2,

$$f_1(C) = f_2(C), \qquad f_1(D) = f_2(D) \tag{14.3.4}$$

and

$$f_1(x) \neq f_2(x) \quad \text{if} \quad C < x < D. \tag{14.3.5}$$

From (14.3.2) and (14.3.4) one obtains

$$\begin{aligned} f_1[F(C, D)] &= H[f_1(C), f_1(D), C, D] \\ &= H[f_2(C), f_2(D), C, D] = f_2[F(C, D)]. \end{aligned} \tag{14.3.6}$$

This contradicts (14.3.5) since $C < F(C, D) < D$, and thus proves the equality of $f_1(x)$ and $f_2(x)$ in $[a, b]$.

II. The interval (b, B). Denote by E the supremum of the points t in (b, B) for which $f_1(x) = f_2(x)$ for all x in $[a, t]$. If $E = B$ we are through.

If $E < B$, note that there exists a sequence $\{t_n\}$ such that (i) $E < t_n < B$, $\forall\, n$, (ii) $t_n \downarrow E$, (iii) $f_1(t_n) \neq f_2(t_n)$, and (iv) $a < F(t_n, a) < E$ for infinitely many values of n. Only the last property requires some cogitation. If we should have $F(t_n, a) \geqslant E$ for all large values of n, then

$$F(E, a) = \lim_{n\to\infty} F(t_n, a) \geqslant E,$$

and this would contradict the intern property of F. Hence (iv) is valid. We have then, for t_n satisfying (iv),

$$f_1[F(t_n, a)] = H[f_1(t_n), f_1(a), t_n, a]$$
$$\neq H[f_2(t_n), f_2(a), t_n, a] = f_2[F(t_n, a)]. \tag{14.3.7}$$

Here we have assumed that the injective property of H holds with respect to the first argument. If it should hold with respect to the second argument instead, it suffices to permute t_n and a in the formulas. Here the inequality is obtained since the first arguments $f_1(t_n)$ and $f_2(t_n)$ differ by virtue of (iii), while the three other arguments are the same in both cases. On the other hand, from (iv) it follows that

$$f_1[F(t_n, a)] = f_2[F(t_n, a)] \tag{14.3.8}$$

and this contradicts (14.3.7). Hence we must have $E = B$ and equality holds in (b, B).

III. The interval (A, a). Use the same type of argument as under II. ∎

Aczél's theorem asserts the existence of at most one continuous solution of (14.3.2) satisfying a given two-point condition. It does not affirm the existence of such a solution. For the existence of a solution it would seem to be necessary to assume continuity of H in its four arguments together with rather strong supplementary conditions. If *a priori* information concerning the existence of continuous solutions is available, such a solution may be computed to any required degree of accuracy by using the intern property of F over and over again. Note that the equation (14.3.2) is such that if the values of a solution are known at two points x_1 and x_2, $x_1 < x_2$, then the value is also known at an intermediary point since

$$f[F(x_1', x_2)] = H[f(x_1), f(x_2), x_1, x_2]. \tag{14.3.9}$$

The points in $[x_1, x_2]$, which may be reached by repeated application of this device, are apt to be dense in the interval and hence determine a continuous solution everywhere.

Illustrations are furnished by *Jensen's equation* or the *equation of the arithmetic means*

$$f[\tfrac{1}{2}(s + t)] = \tfrac{1}{2}[f(s) + f(t)] \tag{14.3.10}$$

and by the *equation of the geometric means*

$$f(\sqrt{st}) = \tfrac{1}{2}[f(s) + f(t)], \qquad 0 < s, \quad 0 < t, \tag{14.3.11}$$

satisfied by

$$s \to f(s) = as + b, \tag{14.3.12}$$

$$s \to f(s) = a \log s + b, \tag{14.3.13}$$

respectively. Here a and b are arbitrary constants. These are the continuous solutions. Judging by Jensen's equation, there should be no lack of non-measurable solutions of equations of the form (14.3.2). For if f is any non-measurable solution of (14.2.1), it is also a solution of (14.3.10).

These special equations for mean values may have served as the starting point for Aczél's discussion of two-point conditions and their importance for functional equations. The reader has undoubtedly noticed that (14.3.12) is the equation of a straight line, and there is one and only one straight line through two given points.

Aczél also considered other examples from the theory of mean values. The following functional equation stems from *information theory*:

$$f\left\{h\left(\frac{xg(x) + yg(y)}{x + y}\right)\right\} = \frac{xf(x) + yf(y)}{x + y}. \tag{14.3.14}$$

Here g is a given strictly monotone continuous function and h is its inverse. The variables are restricted to

$$\{(x, y);\ 0 < x,\ 0 < y,\ x + y \leqslant 1\}.$$

In the notation of Theorem 14.3.1 we have

$$F(x, y) = h\left(\frac{xg(x) + yg(y)}{x + y}\right), \qquad H(u, v, x, y) = \frac{xu + yv}{x + y}.$$

Here H is strictly increasing both in u and in v and hence injective. To fix the ideas, suppose that g is strictly increasing and $0 < x < y$. Then

$$g(x) < \frac{xg(x) + yg(y)}{x + y} < g(y).$$

Since h is also strictly increasing,

$$x = h[g(x)] < h\left[\frac{xg(x) + yg(y)}{x + y}\right] < h[g(y)] = y.$$

Since continuity of F is obvious, all conditions of the theorem are satisfied, so a continuous solution satisfying a two-point condition is unique.

Now solutions can be read off by inspection. It is clear that $f(x) \equiv 1$ is a solution and direct substitution shows that $f(x) \equiv g(x)$ is also a solution. Hence the general continuous solution is

$$f(x) = \alpha + \beta g(x). \tag{14.3.15}$$

Note that this solution involves two arbitrary constants the values of which can be determined by two endpoint conditions.

The condition that F be intern is quite restrictive and would, for instance, exclude the consideration of an addition theorem. This defect has been remedied, at least in part, through later work by Aczél and Hosszú (1965). They have shown that the equation

$$f[F(x, y)] = H[f(x), f(y)] \tag{14.3.16}$$

can have at most one continuous solution satisfying a two-point condition (14.3.3), if F is continuous and if F and H are both strictly increasing (both strictly decreasing) with respect to both variables involved. They also found that if $F(x, x) \not\equiv x$ and if $a \neq F(a, a)$, then there exists at most *one continuous solution satisfying the one-point condition* $f(a) = c$, instead of the two-point condition (14.3.3). This is interesting because it explains, for instance, the different behavior of the very similar Jensen and Cauchy equations, the general solution of the first being a two-parameter, that of the second a one-parameter, manifold of functions.

In 1969 Aczél's theorem was extended to topological vector spaces by C. T. Ng. We shall give a brief account of Ng's work, but to simplify the exposition we restrict ourselves to Euclidean spaces only.

The first notion to be generalized is that of an *intern mapping*. Let R^m be the Euclidean space of m dimensions over the reals. If $\mathbf{x}$, $\mathbf{y} \in R^m$, $\mathbf{x} \neq \mathbf{y}$, denote the line through $\mathbf{x}$ and $\mathbf{y}$ by

$$L\langle \mathbf{x}, \mathbf{y} \rangle = \{\mathbf{y} + t(\mathbf{x} - \mathbf{y});\ t \in R^1\}, \tag{14.3.17}$$

and the open line segment joining $\mathbf{x}$ and $\mathbf{y}$ by

$$L(\mathbf{x}, \mathbf{y}) = \{\mathbf{y} + t(\mathbf{x} - \mathbf{y});\ 0 < t < 1\}. \tag{14.3.18}$$

Let E be a closed convex subset of R^m. A mapping $\mathbf{F}$ from $E \times E$ to E is said to be *intern* if, whenever $\mathbf{x}$, $\mathbf{y} \in E$ with $\mathbf{x} \neq \mathbf{y}$,

$$\mathbf{F}(\mathbf{x}, \mathbf{y}) \in L(\mathbf{x}, \mathbf{y}). \tag{14.3.19}$$

The injective mapping of $L\langle \mathbf{x}, \mathbf{y} \rangle$ into R^1 defined by

$$\mathbf{y} + t(\mathbf{x} - \mathbf{y}) \to t \tag{14.3.20}$$

is a *perspectivity* $p_{\mathbf{x}}^{\mathbf{y}}$, *a fortiori*, a homeomorphism. Note, further, that if $\mathbf{f}_1$ and $\mathbf{f}_2$ are continuous mappings from R^m into R^n, then the set

$$S = \{\mathbf{x};\ \mathbf{x} \in R^m,\ \mathbf{f}_1(\mathbf{x}) = \mathbf{f}_2(\mathbf{x})\} \tag{14.3.21}$$

is closed in R^m.

For the proof of Ng's main theorem two preliminary lemmas are needed which are of independent interest.

Lemma 14.3.1. *With E and $\mathbf{F}$ as above, suppose that $\mathbf{f}_1$ and $\mathbf{f}_2$ are continuous mappings of E into R^n satisfying*

$$\mathbf{f}[\mathbf{F}(\mathbf{x}, \mathbf{y})] = \mathbf{H}[\mathbf{f}(\mathbf{x}), \mathbf{f}(\mathbf{y}), \mathbf{x}, \mathbf{y}], \tag{14.3.22}$$

where $\mathbf{H}$ *is a mapping from* $R^n \times R^n \times E \times E$ *into* R^n. *Then the set*

$$S = \{\mathbf{x};\, \mathbf{x} \in E,\, \mathbf{f}_1(\mathbf{x}) = \mathbf{f}_2(\mathbf{x})\} \tag{14.3.23}$$

is necessarily closed and convex.

Proof. Since E is closed in R^m and S is closed in E, S is closed in R^m. For any two distinct points $\mathbf{x}$ and $\mathbf{y}$ of S, consider the line $L\langle \mathbf{x}, \mathbf{y}\rangle$ and the mapping $p_{\mathbf{x}}^{\mathbf{y}}$ of the line into R^1. In particular, we fix the attention on the set $L(\mathbf{x}, \mathbf{y}) \ominus S$ which is mapped into a subset of $0 < t < 1$. Since S is closed, this subset is open in $(0, 1)$, and if not void is the countable union of open disjoint intervals. Consider the latter alternative and let t_1 and t_2 be the endpoints of one of these intervals and set

$$\mathbf{x}_1 = \mathbf{y} + t_1(\mathbf{x} - \mathbf{y}), \qquad \mathbf{x}_2 = \mathbf{y} + t_2(\mathbf{x} - \mathbf{y}).$$

These points define a line segment $L(\mathbf{x}_1, \mathbf{x}_2) \subset L(\mathbf{x}, \mathbf{y})$, no points of which belong to S, while the endpoints do belong. Now $\mathbf{F}(\mathbf{x}_1, \mathbf{x}_2)$ is a well-defined point of $L(\mathbf{x}_1, \mathbf{x}_2)$ by the intern property of $\mathbf{F}$, and it belongs to E, the set of definition of $\mathbf{f}_1$ and $\mathbf{f}_2$. Now the functional equation (14.3.22) gives

$$\begin{aligned}\mathbf{f}_1[\mathbf{F}(\mathbf{x}_1, \mathbf{x}_2)] &= \mathbf{H}[\mathbf{f}_1(\mathbf{x}_1), \mathbf{f}_1(\mathbf{x}_2), \mathbf{x}_1, \mathbf{x}_2] \\ &= \mathbf{H}[\mathbf{f}_2(\mathbf{x}_1), \mathbf{f}_2(\mathbf{x}_2), \mathbf{x}_1, \mathbf{x}_2] = \mathbf{f}_2[\mathbf{F}(\mathbf{x}_1, \mathbf{x}_2)].\end{aligned}$$

Thus $\mathbf{F}(\mathbf{x}_1, \mathbf{x}_2)$ belongs to both S and $L(\mathbf{x}_1, \mathbf{x}_2)$, contrary to the assumption that $L(\mathbf{x}_1, \mathbf{x}_2) \cap S$ is void. This contradiction shows that $L(\mathbf{x}, \mathbf{y}) \ominus S$ is void, that is, $L(\mathbf{x}, \mathbf{y}) \subset S$. Now $\mathbf{x}$ and $\mathbf{y}$ being arbitrary points of S, it follows that S is convex as well as closed. ∎

Lemma 14.3.2. *Let the set* E *and the mapping* $\mathbf{F}$ *be as above with* $\mathbf{F}$ *continuous in both variables. Let* $\mathbf{f}_1$, $\mathbf{f}_2$ *be mappings of* E *into* R^n, *again satisfying* (14.3.22), *though not necessarily continuous. Let* $\mathbf{H}\colon R^n \times R^n \times E \times E \to R^n$ *be injective either in the first argument or in the second. If* $\mathbf{f}_1$ *and* $\mathbf{f}_2$ *are identical in some* E*-neighborhood of a point* $\mathbf{a} \in E$, *then* $\mathbf{f}_1$ *and* $\mathbf{f}_2$ *are identical in their entire domain of definition* E.

Proof. To fix the ideas, suppose that $\mathbf{H}$ is injective with respect to the first variable. Let $\mathbf{b} \in E$, $\mathbf{b} \neq \mathbf{a}$. Define

$$\mathbf{F}^0(\mathbf{x}, \mathbf{a}) = \mathbf{x} \tag{14.3.24}$$

and

$$\mathbf{F}^{p+1}(\mathbf{x}, \mathbf{a}) = \mathbf{F}[\mathbf{F}^p(\mathbf{x}, \mathbf{a}), \mathbf{a}] \tag{14.3.25}$$

recursively for all p and all $\mathbf{x} \in E$. Take, in particular, $\mathbf{x} = \mathbf{b}$. Then each $\mathbf{F}^p(\mathbf{b}, \mathbf{a})$, $p \neq 0$, belongs to $L(\mathbf{b}, \mathbf{a})$ by the intern property of $\mathbf{F}$. Hence there is a number t_p, $0 < t_p < 1$, such that

$$\mathbf{F}^p(\mathbf{b}, \mathbf{a}) = \mathbf{a} + t_p(\mathbf{b} - \mathbf{a}) \tag{14.3.26}$$

and, since $\mathbf{F}$ is intern, the sequence $\{t_p\}$ is strictly decreasing to a limit, say $t_0 \geqslant 0$. It is claimed that $t_0 = 0$ and $\lim_{p\to\infty} \mathbf{F}^p(\mathbf{b}, \mathbf{a})$, which clearly exists, equals $\mathbf{a}$. For $\mathbf{F}$ is continuous in its first argument and if

$$\lim_{p\to\infty} \mathbf{F}^p(\mathbf{b}, \mathbf{a}) = \mathbf{a} + t_0(\mathbf{b} - \mathbf{a}) \neq \mathbf{a},$$

then

$$\lim_{p\to\infty} \mathbf{F}^{p+1}(\mathbf{b}, \mathbf{a}) = \lim_{p\to\infty} \mathbf{F}[\mathbf{F}^p(\mathbf{b}, \mathbf{a}), \mathbf{a}] = \mathbf{F}[\lim_{p\to\infty} \mathbf{F}^p(\mathbf{b}, \mathbf{a}), \mathbf{a}]$$
$$= \mathbf{F}[\mathbf{a} + t_0(\mathbf{b} - \mathbf{a}), \mathbf{a}] \neq \mathbf{a} + t_0(\mathbf{b} - \mathbf{a}).$$

This contradiction shows that $t_0 = 0$ and $\lim \mathbf{F}^p(\mathbf{b}, \mathbf{a}) = \mathbf{a}$.

Let V denote the E-neighborhood of $\mathbf{x} = \mathbf{a}$ where $\mathbf{f}_1$ and $\mathbf{f}_2$ are identical. All points $\mathbf{F}^p(\mathbf{b}, \mathbf{a})$ are in E and hence ultimately in V. Suppose k is so large that

$$\mathbf{F}^k(\mathbf{b}, \mathbf{a}) \in V.$$

Then

$$\mathbf{f}_1[\mathbf{F}^k(\mathbf{b}, \mathbf{a})] = \mathbf{f}_2[\mathbf{F}^k(\mathbf{b}, \mathbf{a})].$$

It is desired to show that

$$\mathbf{f}_1[\mathbf{F}^p(\mathbf{b}, \mathbf{a})] = \mathbf{f}_2[\mathbf{F}^p(\mathbf{b}, \mathbf{a})], \qquad p = 0, 1, 2, \cdots, k - 1. \tag{14.3.27}$$

Here we use retrogressive induction on p passing from p to $p - 1$. The functional equation gives

$$\mathbf{H}\{\mathbf{f}_1[\mathbf{F}^{k-1}(\mathbf{b}, \mathbf{a})], \mathbf{f}_1(\mathbf{a}), \mathbf{F}^{k-1}(\mathbf{b}, \mathbf{a}), \mathbf{a}\} = \mathbf{f}_1[\mathbf{F}^k(\mathbf{b}, \mathbf{a})]$$
$$= \mathbf{f}_2[\mathbf{F}^k(\mathbf{b}, \mathbf{a})] = \mathbf{H}\{\mathbf{f}_2[\mathbf{F}^{k-1}(\mathbf{b}, \mathbf{a})], \mathbf{f}_2(\mathbf{a}), \mathbf{F}^{k-1}(\mathbf{b}, \mathbf{a}), \mathbf{a}\}.$$

Since $\mathbf{H}$ is injective with respect to the first argument, and $\mathbf{f}_1(\mathbf{a}) = \mathbf{f}_2(\mathbf{a})$, this gives

$$\mathbf{f}_1[\mathbf{F}^{k-1}(\mathbf{b}, \mathbf{a})] = \mathbf{f}_2[\mathbf{F}^{k-1}(\mathbf{b}, \mathbf{a})],$$

so that (14.3.27) holds for $p = k - 1$. We can then proceed recursively and see that (14.3.27) holds for all indicated values of p. In particular, for $p = 0$

$$\mathbf{f}_1(\mathbf{b}) = \mathbf{f}_2(\mathbf{b}).$$

Since $\mathbf{b}$ is arbitrary, the equality holds everywhere in E. ∎

Combining these two lemmas, we get Ng's main theorem.

Theorem 14.3.2. *Let E be a closed convex subset of R^m and $\mathbf{F}$ an intern mapping from $E \times E$ to E, continuous in both variables. Let $\mathbf{H}: (\mathbf{u}, \mathbf{v}, \mathbf{x}, \mathbf{y}) \to \mathbf{H}(\mathbf{u}, \mathbf{v}, \mathbf{x}, \mathbf{y})$ be a mapping from $R^n \times R^n \times E \times E$ into R^n, injective with respect to either $\mathbf{u}$ or $\mathbf{v}$. Consider the functional equation*

$$\mathbf{f}[\mathbf{F}(\mathbf{x}, \mathbf{y})] = \mathbf{H}[\mathbf{f}(\mathbf{x}), \mathbf{f}(\mathbf{y}), \mathbf{x}, \mathbf{y}]. \tag{14.3.28}$$

If $\mathbf{f}_1$ and $\mathbf{f}_2$ are two continuous mappings from E to R^n satisfying this equation *and if it be known that* $\mathbf{f}_1$ *and* $\mathbf{f}_2$ *are identical on some subset* D *of* E *whose closed convex hull* $C[D]$ *has a non-void interior as a subset of* E, *then* $\mathbf{f}_1$ *and* $\mathbf{f}_2$ *are identical on all of* E.

In particular, we have the following

Corollary. *If* D *contains* $m+1$ *points* $\mathbf{a}_j$ *such that the vectors* $\mathbf{a}_2 - \mathbf{a}_1$, $\mathbf{a}_3 - \mathbf{a}_1, \cdots, \mathbf{a}_{m+1} - \mathbf{a}_1$ *are linearly independent, then there exists at most one continuous solution of* (14.3.28) *satisfying* $m+1$ *initial conditions*

$$\mathbf{f}(\mathbf{a}_j) = \mathbf{b}_j, \qquad j = 1, 2, \cdots, m+1. \tag{14.3.29}$$

For $m = 1$ this becomes Theorem 14.3.1 for any closed subinterval of (A, B).

Aczél's definition of an intern mapping given by (14.3.1) is the natural one on the line. On the other hand, Ng's extension to higher dimensions and topological vector spaces is only one of several possibilities. J. B. Miller (personal communication) has proved an extension of Aczél's theorem to R^2 based on partial ordering where $\mathbf{x} < \mathbf{y}$, $\mathbf{x} \neq \mathbf{y}$, implies $\mathbf{x} < \mathbf{F}(\mathbf{x}, \mathbf{y}) < \mathbf{y}$. As reported by Aczél and Ng, a general uniqueness theorem has been found by Ng for "cell-intern" mappings in R^n.

One of the earliest instances of a two-point uniqueness theorem is a result due to Picard (1890). It occurs in the memoir where he brought forth the method of successive approximations. It reads:

Theorem 14.3.3. *Let* $(x, y) \to F(x, y)$ *be continuous in a rectangle* $|x| < A$, $|y| < B$, *and strictly increasing as a function of* y *for each fixed* x. *Then the differential equation*

$$y'' = F(x, y) \tag{14.3.30}$$

can have at most one solution satisfying conditions of the form

$$y(a) = c, \qquad y(b) = d, \qquad -A < a < b < A, \qquad -B < c, \qquad d < B. \tag{14.3.31}$$

Proof. Suppose there were two solutions f_1 and f_2 with these properties. Consider a maximal subinterval (a_1, b_1) of (a, b) in which

$$g(x) = f_2(x) - f_1(x)$$

keeps a constant sign, say $g(x) > 0$. Here $a \leqslant a_1 < b_1 \leqslant b$ and

$$g(a_1) = g(b_1) = 0$$

since the subinterval is maximal. Thus in (a_1, b_1)

$$g''(x) = f_2''(x) - f_1''(x) = F[x, f_2(x)] - F[x, f_1(x)] > 0,$$

since $F(x, y)$ is an increasing function of y. This means that g is convex and g' is increasing. Since $g(x) > 0$ for $a_1 < x < b_1$ and $g(a_1) = 0$ we must have $g'(a_1) \geqslant 0$, so $g' > 0$ in (a_1, b_1) and this forces $g(b_1)$ to be positive, which is a contradiction. It follows that $f_1(x) \equiv f_2(x)$ in (a_1, b_1). If (a_1, b_1) is a proper subinterval of (a, b), the same type of argument applies to any other subinterval in which g keeps a constant sign and hence ultimately to all of (a, b). Thus there is at most one solution of (14.3.30) that can satisfy the two-point condition. ∎

EXERCISE 14.3

1. Verify (14.3.13). Why the names attached to equations (14.3.10) and (14.3.11)?
2. Let p and q be arbitrary fixed positive numbers, g a given continuous and strictly monotone function, h its inverse. Solve the equation
$$f[F(x, y)] = \frac{pf(x) + qf(y)}{p + q}, \qquad F(x, y) = h\left(\frac{pg(x) + qg(y)}{p + q}\right).$$
Verify that the conditions of Theorem 14.3.1 are satisfied so that there is a unique continuous solution satisfying a given two-point condition. This equation occurs in the theory of averages.
3. If the $m + 1$ vectors $\mathbf{a}_2 - \mathbf{a}_1, \mathbf{a}_3 - \mathbf{a}_1, \ldots, \mathbf{a}_{m+1} - \mathbf{a}_1$ are linearly independent in R^m, show that so are $\mathbf{a}_2 - \mathbf{a}_1$, $\mathbf{a}_3 - \mathbf{a}_2$, $\ldots$, $\mathbf{a}_{m+1} - \mathbf{a}_m$, and vice versa. Geometrical interpretation?
4. The equation of the arithmetic mean for complex variables
$$f[\tfrac{1}{2}(z_1 + z_2)] = \tfrac{1}{2}[f(z_1) + f(z_2)]$$
has the obvious solution $z \to f(z) = \alpha + \beta z$ where α and β are arbitrary complex numbers. But there are also continuous solutions analogous to (14.2.17). Show that $x + iy \to \alpha + \beta x + \gamma y$ is a solution for arbitrary α, β, γ and that the solution is uniquely determined by a three-point condition $f(a_j) = b_j, j = 1, 2, 3$, provided the triangle formed by the three points $\{a_j\}$ in the complex plane is non-degenerate.
5. The equation of the arithmetic mean in R^3
$$\mathbf{f}[\tfrac{1}{2}(\mathbf{x} + \mathbf{y})] = \tfrac{1}{2}[\mathbf{f}(\mathbf{x}) + \mathbf{f}(\mathbf{y})]$$
with $\mathbf{x}$, $\mathbf{y}$ and the range of $\mathbf{f}$ in R^3 has continuous solutions which are uniquely determined by a four-point condition involving a non-degenerate tetrahedron. What is the form of the solution?
6. Apply Ng's theorem to the equation (13.2.25).
7. Show that the only solutions of
$$f(x + y) = [f(x) + f(y)]^2$$
are two specific constants.

Find continuous solutions of the following functional equations:

8. $f(x + y) = f(x)\,g(y)$, g given continuous, $g(0) = 1$.
9. $[f(x)]^2 = f(x + y)f(x - y)$.
10. $f(x + y) + f(x - y) = 2f(x)f(y)$. (J. d'Alembert, 1769.)
11. $|f(x + iy)|^2 = |f(x)|^2 + |f(iy)|^2$.
12. A classical method of solving functional equations in two variables is to assume differentiability of the functions involved, differentiating with respect to one of the

variables, perhaps several times. If now one of the variables is given a special value, the resulting differential equation may be solvable. Apply this method to the first Cauchy equation and to Nos. 8, 9 and 10 above.

COLLATERAL READING

For the whole field or parts thereof consult:

ACZÉL, J., *Lectures on Functional Equations and Their Applications,* Academic Press, New York (1966).

KUCZMA, M., *Functional Equations in a Single Variable.* Monografie Matematyczne, Vol. 46, PWN (Polish Scientific Publishers), Warsaw (1968).

TARGONSKI, G. I., *Seminar on Functional Operators and Equations,* Springer-Verlag, Berlin (1967).

For differential equations, see:

HILLE, E., *Lectures on Ordinary Differential Equations,* Addison-Wesley, Reading, Mass. (1969).

References made in the text involve the following papers:

ACZÉL, J., Ein Eindeutigkeitssatz in der Theorie der Funktionalgleichungen und einige ihrer Anwendungen, *Acta Math. Acad. Sci. Hung.* **15** (1964) 355–64.

ACZÉL, J. and M. HOSSZÚ, Further uniqueness theorems for functional equations, *Acta Math. Acad. Sci. Hung.* **16** (1965) 51–55.

BOURLET, C., Sur les opérations en général et les équations différentielles linéaires d'ordre infini, *Ann. Sci. École Norm. Super.* (3) **14** (1897) 133–50.

BRILLOUIN, L., *L'atome de Thomas–Fermi,* Actualités Scientifiques et Industrielles, No. 160, Hermann, Paris (1934).

CAUCHY, A. L., *Cours d'Analyse de l'École Polytechnique,* Vol. 1, *Analyse Algébrique,* Paris, 1821; *Œuvres,* (2) **3** (1897) 98–113.

DARBOUX, G., Sur la composition des forces en statique, *Bull. Sci. Math.* (1) **9** (1875) 281–8.

DARÓCZY, Z., Notwendige und hinreichende Bedingungen für die Existenz von nichtkonstanten Lösungen linearer Funktionalgleichungen, *Acta Sci. Math. Szeged* **22** (1961) 31–41.

DUNFORD, N. and E. HILLE, The differentiability and uniqueness of continuous solutions of addition formulas, *Bull. Amer. Math. Soc.* **50** (1944) 67–75.

HAMEL, G., Eine Basis aller Zahlen und die unstetigen Lösungen der Funktionalgleichung $f(x+y)=f(x)+f(y)$, *Math. Ann.* **60** (1905) 459–62.

HILLE, E., On the Thomas–Fermi equation, *Proc. Nat. Acad. Sci. U.S.* **62** (1969) 7–10.

HOSSZÚ, M., A remark on scalar valued multiplicative functions of matrices, *Publ. Math. Debrecen* **6** (1959) 288–9.

IONESCU TULČEA, C. T., Suboperative functions and semi-groups of operators, *Arkiv Mat.* **4** (1960) 55–61.

LOSONCZI, L., Bestimmung aller nichtkonstanten Lösungen von linearen Funktionalgleichungen, *Acta Sci. Math. Szeged* **25** (1964) 250–4.

NG, C. T., Uniqueness theorems for a general class of functional equations, *J. Australian Math. Soc.* **11** (1970) 362–6.

OSTROWSKI, A., Mathematische Miszellen. XIV. Über die Funktionalgleichung der Exponentialfunktion und verwandte Funktionalgleichungen, *Jahresber. Deut. Math. Ver.* **38** (1929) 54–62.

PICARD, É., Mémoire sur la théorie des équations aux derivées partielles et la méthode des approximations successives, *J. Math. Pures Appl.* (4) **6** (1890) 145–210.

RIJNIERSE, P. J., Algebraic solutions to the Thomas–Fermi equation for atoms, University of St. Andrews, Ph.D. Thesis, 1968.

SCHRODER, E., Über iterierte Funktionen, III, *Math. Ann.* **3** (1870) 296–322.

VINCZE, E., Eine allgemeine Methode in der Theorie der Funktionalgleichungen, I, *Publ. Math. Debrecen,* **9** (1962) 149–63.

15 MEAN VALUES†

Mean values such as the arithmetic, geometric, and the power means have figured in many places of this treatise. In the present chapter we shall discuss in some detail a class of mean values referred to as A-averages which includes as special cases those just mentioned. Our point of departure will be a set of postulates given, independently of each other, by A. N. Kolmogorov and M. Nagumo in 1930, and, from a different angle, by B. de Finetti in 1931.

The A-averages lead to functional equations of the "intern" type. A non-constant solution of one of these equations is either unbounded in every interval or is continuous as well as strictly monotone and defines an A-average.

This class of mean values has numerous important applications. We shall consider some geometric extremal problems and study in some detail two classes of set functions—transfinite diameters and Čebyšev constants. Both were originally defined for sets in the complex plane and geometric means, but the extensions to arbitrary complete metric spaces and A-averages do not lack interest. There are also important connections with generalized potential theory. We shall encounter many connections between the subject-matter of this chapter and those of Chapters 13 and 14.

There are eight sections: The postulates; Associated functional equations; Remarks on summability; Some geometric extremal problems; The transfinite A-diameter; The Čebyšev constants; Some examples; and Potential theories.

15.1 THE POSTULATES

We shall consider an average subject to the following conditions:

(A_1) *Let (a, b) be a given real interval. For each natural number n and for each set of n numbers $x_1, x_2 \ldots, x_n$ belonging to (a, b) there is a number $A(x_1, x_2 \ldots, x_n)$ in (a, b) called the A-average of these numbers.*

(A_2) *$A(x_1, x_2, \ldots, x_n)$ is a continuous symmetric function of its arguments, strictly increasing in each of them.*

(A_3) $A(x, x, \ldots, x) = x$.

(A_4) *For each n and each $k < n$ let $y = A(x_1, x_2, \ldots, x_k)$, then $A(x_1, x_2, \ldots, x_k, x_{k+1}, \ldots, x_n) = A(y, y, \ldots, y, x_{k+1}, \ldots, x_n)$, where y is repeated k times.*

---

† Comments by J. Aczél and C. T. Ng have been helpful in the editing of this chapter.

There are various immediate consequences of the postulates which will be stated as lemmas.

Lemma 15.1.1. *Unless all the x's are equal,*

$$\min x_j < A(x_1, x_2 \dots, x_n) < \max x_j. \tag{15.1.1}$$

Proof. Use the strictly increasing character of A together with (A_3). ∎

Lemma 15.1.2. *The average of k sets $x_1, x_2 \dots, x_n$ is the same as the average of one set:*

$$A(x_1, \dots, x_n, x_1, \dots, x_n, \dots, x_1, \dots, x_n) = A(x_1, \dots, x_n). \tag{15.1.2}$$

Proof. Let y denote the right member of the proposed equality. Using (A_4) repeatedly, we can replace each of the k aggregates $x_1, x_2, \dots, x_n$ by y repeated n times. The result is the average of y repeated kn times and this is y by (A_3). Hence the two sides of (15.1.2) are equal.

This gives a convenient method of extending and contracting averages. The same device gives *the principle of repeated averages.*

Lemma 15.1.3. *From a set E of n numbers in (a, b), say $x_1, x_2, \dots, x_n$, select k numbers, $1 < k < n$, and form their average. The average of all the averages for k fixed obtainable in this way equals $A(x_1, x_2, \dots, x_n)$.*

Proof. The basis for this fact is the preceding lemma together with the identity between binomial coefficients

$$n\binom{n-1}{k-1} = k\binom{n}{k}. \tag{15.1.3}$$

We extend the given set E so as to obtain $\binom{n-1}{k-1}$ copies of each x_j. The extended set E^* has a number of elements equal to the left side of (15.1.3). This identity shows that we can separate the elements of E^* into $\binom{n}{k}$ distinct subsets S_p, each being a selection of k elements of E. Averaging over E or over E^* gives the same result by Lemma 15.1.2. On the other hand, in the average over E^* we may replace the elements of a subset S_p by their average $y\binom{n,k}{p} = A(S_p)$ repeated k times. Thus the average over E^* equals the average of the $\binom{n}{k}$ averages $A(S_p)$ each repeated k times. By Lemma 15.1.2 this reduces to the average of the averages as stated. ∎

Corollary. *With the same notation, for each $k < n$,*

$$\min y\binom{n,k}{p} < A(x_1, x_2, \dots, x_n) < \max y\binom{n,k}{p} \tag{15.1.4}$$

unless all the x's are equal.

The last remark follows from the fact that the y's are equal iff the x's are equal.

This means that the averaging process is *oscillation reducing* in the following sense. Let S be an arbitrary set of numbers in the interval (a, b). Select n distinct numbers from S and form their A-average. The set of all such averages involving n numbers is a set S_n. For each n set

$$a_n = \inf S_n, \qquad b_n = \sup S_n. \tag{15.1.5}$$

Then for all $n > 1$

$$a_{n-1} \leqslant a_n \leqslant b_n \leqslant b_{n-1}, \tag{15.1.6}$$

so that

$$b_n - a_n \leqslant b_{n-1} - a_{n-1}. \tag{15.1.7}$$

This follows from the corollary.

EXERCISE 15.1

1. Verify that $\mathsf{M}_\alpha(x)$ is an A-average.
2. Same question for the geometric mean with
$$A(x_1, x_2, \ldots, x_n) = (x_1 x_2 \ldots x_n)^{1/n}.$$
3. Same question for the harmonic mean with
$$A(x_1, x_2, \ldots, x_n) = \left\{\frac{1}{n}\left[\frac{1}{x_1} + \frac{1}{x_2} + \cdots + \frac{1}{x_n}\right]\right\}^{-1}.$$
4. [Aczél] Let m be a positive odd integer and take $(a, b) = (-\infty, \infty)$. Define a mean value by
$$n[A(x_1, \ldots, x_n)]^m = x_1{}^m + x_2{}^m + \cdots + x_n{}^m.$$
Verify that this is an A-average.
5. [Aczél] Take $(a, b) = (-\frac{1}{2}\pi, \frac{1}{2}\pi)$ and define
$$A(x_1, \ldots, x_n) = \arcsin\left\{\frac{1}{n}\left[\sin x_1 + \cdots + \sin x_n\right]\right\},$$
where the arc sine has its principal value between $-\frac{1}{2}\pi$ and $\frac{1}{2}\pi$. Verify that this is an A-average.
6. Give a complete proof of Lemma 15.1.1.
7. Prove the corollary of Lemma 15.1.3.
8. Prove (15.1.6).

15.2 ASSOCIATED FUNCTIONAL EQUATIONS

The postulates obviously do not determine the A-averages uniquely but they do lead to simple expressions for $A(x_1, x_2, \ldots, x_n)$.

Our object is to show, for each A-average, the existence of a two-parameter family of continuous, strictly monotone functions f such that for each n and each choice of $x_1, x_2, \ldots, x_n$ in (a, b) we have

$$f[A(x_1, x_2, \ldots, x_n)] = \frac{1}{n} \sum_{j=1}^{n} f(x_j). \tag{15.2.1}$$

Conversely, if there exists a continuous strictly monotone function f which satisfies (15.2.1) for all n, all x_j in (a, b) and for some function $A(x_1, \ldots, x_n)$, then $A(x_1, \ldots, x_n)$ necessarily satisfies postulates (A_1) to (A_4). It will be shown that if f is a solution of (15.2.1), then so is $\alpha f + \beta$ for any constants α and β.

We have here a family of functional equations of which the simplest member is

$$f[A(x_1, x_2)] = \tfrac{1}{2}[f(x_1) + f(x_2)]. \tag{15.2.2}$$

This looks like Jensen's equation (14.3.10), to which it reduces in the case of the arithmetic means where

$$A(s, t) = \tfrac{1}{2}(s + t).$$

See also Eq. (14.3.11) for the geometric means. The first equation is satisfied by $f(u) = \alpha u + \beta$, the second by $f(u) = \alpha \log u + \beta$, as observed in Section 14.3.

There is a close connection between these mean-values and convex functions. Thus the technique used in proving Theorem 13.3.1 applies in proving that (15.2.2) implies (15.2.1) for all n. Further, just as in the case of convex functions, a mean-value function f, i.e. a solution of (15.2.2), is either unbounded on every interval or continuous (and strictly monotone). It is obviously the solutions of the second kind that are of interest in the theory of averages.

Since $A(s, t)$ is an intern transformation in the sense of Aczél, Theorem 14.3.1 applies to the present situation, but here we need a sharper statement. It is not enough that there is *at most* one continuous solution satisfying a given two-point condition; we have to show that there is *at least* one such solution.

Theorem 15.2.1. *Let A be an A-average and let f satisfy* (15.2.2) *for all x_1, x_2 in (a, b). Then f also satisfies* (15.2.1) *for all n and all $x_1, x_2, \ldots, x_n$ in (a, b).*

Proof. We prove first that (15.2.1) holds for $n = 2^m$ and start with $m = 2$. By (A_4) and Lemma 15.2.1

$$A[x_1, x_2, x_3, x_4] = A[A(x_1, x_2), A(x_3, x_4)],$$

so that

$$\begin{aligned} f[A(x_1, x_2, x_3, x_4)] &= \tfrac{1}{2}\{f[A(x_1, x_2)] + f[A(x_3, x_4)]\} \\ &= \tfrac{1}{4}[f(x_1) + f(x_2) + f(x_3) + f(x_4)], \end{aligned}$$

which is (15.2.1) for $n = 4$. Complete induction takes care of the case $n = 2^m$ for $m > 2$.

Next, suppose that (15.2.1) holds for a particular value of n. We shall then show that it holds for $n - 1$. Set

$$y = A(x_1, x_2, \ldots, x_{n-1})$$

and note that

$$f[A(x_1, x_2, \dots, x_{n-1}, y)] = \frac{1}{n}[f(x_1) + f(x_2) + \cdots + f(x_{n-1}) + f(y)].$$

Now by (A_4)

$$A(x_1, x_2, \dots, x_{n-1}, y) = A(y, y, \dots, y, y) = y.$$

Hence

$$f(y) = \frac{1}{n}[f(x_1) + f(x_2) + \cdots + f(x_{n-1}) + f(y)]$$

or

$$f[A(x_1, x_2, \dots, x_{n-1})] = \frac{1}{n-1}[f(x_1) + f(x_2) + \cdots + f(x_{n-1})]$$

as asserted. Since (15.2.1) is known to be true for n equal to a power of 2, it follows that the formula is true for all n. This type of argument goes back to Cauchy (*Analyse algébrique*, Paris, 1821). ∎

This means that we can disregard (15.2.1) and concentrate on the functional equation (15.2.2).

Theorem 15.2.2. *Let* A *be an A-average and let f satisfy* (15.2.2) *for all x_1, x_2 in (a, b). If f is bounded in some neighborhood of $s = s_0$, $a < s_0 < b$, then f is continuous at $s = s_0$.*

Proof. Suppose that $s_0 + h$ also belongs to the neighborhood where $|f(s)| < M$ for all h with $|h| < \eta$. Then by (A_2) and Lemma 15.1.1

$$A(s_0, s_0 + h) = s_0 + p(s_0, h). \tag{15.2.3}$$

Here $h \to p(s_0, h)$ is a continuous map and

$$\operatorname{sgn} p(s_0, h) = \operatorname{sgn} h, 0 < |p(s_0, h)| < |h|. \tag{15.2.4}$$

Then by (15.2.2)

$$f[A(s_0, s_0 + h)] - f(s_0) = \tfrac{1}{2}[f(s_0 + h) - f(s_0)].$$

We take absolute values and set

$$\limsup_{h \to 0} |f(s_0 + h) - f(s_0)| = \delta(s_0). \tag{15.2.5}$$

By the continuity of A together with (15.2.3) and (15.2.4) we have also

$$\limsup_{h \to 0} |f[A(s_0, s_0 + h)] - f(s_0)| = \delta(s_0),$$

so that

$$\delta(s_0) = \tfrac{1}{2}\delta(s_0) \quad \text{or} \quad \delta(s_0) = 0$$

since $\delta(s_0)$ is finite. Thus $f(s)$ is continuous at $s = s_0$ as asserted. ∎

Thus we see that *local boundedness implies local continuity*. We shall see that *local unboundedness implies global unboundedness* under mild restrictions.

Consider the mapping defined by $t \to A(s, t)$ when s is fixed in (a, b). Since A is a strictly increasing function of t and $A(s, s) = s$, it is seen that A maps (a, b) onto an interval (s^-, s^+) where

$$a \leqslant s^- < s < s^+ \leqslant b. \tag{15.2.6}$$

We have now

Theorem 15.2.3. *If f satisfies* (15.2.2) *in (a, b), if $f(s) \neq -\infty$ for all s in (a, b), and if f is not bounded above in some interval $(s_0 - \eta, s_0 + \eta)$ with $a < s_0 - \eta < s_0 + \eta < b$, then for any choice of $s_1 \neq s_0$ with $a < s_1 - \delta < s_1 + \delta < b$, f cannot be bounded above in $(s_1 - \delta, s_1 + \delta)$. Similarly for lower unboundedness if $f(s) \neq +\infty$ for all s.*

Proof. Consider first the interval (s_0^-, s_0^+) obtained by setting $s = s_0$ in (15.2.6). Choose a point t in this interval. By assumption there exists a sequence $\{x_n\}$ such that (1) $s_0 - \eta < x_n < s_0 + \eta$, (2) $x_n \to s_0$, (3) $f(x_n) > 2n$. Then

$$f[A(x_n, t)] = \tfrac{1}{2}f(x_n) + \tfrac{1}{2}f(t) > n + \tfrac{1}{2}f(t),$$

which goes to infinity with n since $f(t) \neq -\infty$. Hence

$$\lim_{n \to \infty} f[A(x_n, t)] = +\infty$$

while

$$\lim_{n \to \infty} A(x_n, t) = A(s_0, t).$$

This shows that f is unbounded above everywhere in (s_0^-, s_0^+). If this is the interval (a, b), we are through. If not, let $b_0 < b$ be the least upper bound of the values of s for which f is locally unbounded above. Now for any choice of a sequence $\{\delta_n\}$ with $\delta_n \downarrow 0$, the function f is not bounded above in any one of the intervals $(b_0 - \delta_n, b_0 - \delta_{n+1})$ by assumption. Consider

$$f\{A[b_0 - \delta, \tfrac{1}{2}(b_0 + b)]\} = \tfrac{1}{2}f(b_0 - \delta) + \tfrac{1}{2}f(b_0 + b).$$

If b should happen to be $+\infty$, we replace $\frac{1}{2}(b_0 + b)$ by some large positive number, before proceeding to the next step. On the one hand, there exists a sequence $\{\delta_n\}$ with $\delta_n \downarrow 0$, such that $f(b_0 - \delta_n) > 2n$; on the other hand, A is continuous, so that

$$A[b_0 - \delta_n, \tfrac{1}{2}(b_0 + b)] \to A[b_0, \tfrac{1}{2}(b_0 + b)] > b_0.$$

This shows that a $b_0 < b$ cannot be the least upper bound for the values of s in any neighborhood of which f is not bounded above. Thus $b_0 = b$. In the other direction it is seen that the interval of local unboundedness above extends all the way to $s = a$. In the same way it is proved that local unboundedness below together with $f(s) \neq +\infty$ implies global unboundedness below. ∎

Thus it is clear that a solution f of (15.2.2) must be bounded and hence continuous on compact subsets of (a, b) if it is going to be of any use for the theory of A-averages.

Theorem 15.2.4. *A continuous solution of* (15.2.2) *is either a constant or strictly monotone.*

Proof. Aczél's uniqueness theorem applies to the present situation. Suppose that f is a continuous solution of (15.2.2) and that there exist s_1 and s_2, $s_1 \neq s_2$, such that $f(s_1) = f(s_2) = y_1$. Then $f(s) = y_1$ is a solution satisfying the given two-point condition. Since the solution is unique we are dealing with a constant solution of (15.2.2). If no such s_1, s_2 exist, then f, being continuous, is strictly monotone. ∎

We come now to the existence theorem.

Theorem 15.2.5. *Given four real numbers* s_0, s_1, y_0, y_1 *with* $a < s_0 < s_1 < b$, $y_0 < y_1$. *Then there exists a unique continuous, strictly increasing solution of* (15.2.2) *defined for* $a < s < b$ *such that*

$$f(s_0) = y_0, \qquad f(s_1) = y_1. \tag{15.2.7}$$

Proof. The emphasis here is on "there exists"; since Aczél's uniqueness theorem applies, we know in advance that a continuous solution is unique, if it satisfies (15.2.7), and $y_0 < y_1$ implies that it is strictly increasing. In order to obtain a strictly decreasing solution instead, we must assume $y_0 > y_1$.

In this discussion the functional equation (15.2.2) will be understood in the following sense. If s and t are any two points in (a, b), if two real numbers $f(s)$ and $f(t)$ are uniquely defined, then $f(u)$ exists also for

$$u = A(s, t) \quad \text{and} \quad f(u) = \tfrac{1}{2}[f(s) + f(t)]. \tag{15.2.8}$$

By repeated application of this principle it is seen that, if f is defined for $u = s_0$ and $u \equiv s_1$, then $f(u)$ exists and is well defined in a countable subset S of $[s_0, s_1]$ and this set is mapped in a 1–1 manner onto a subset Y of $[y_0, y_1]$. We proceed to a description of these two sets.

This is most conveniently given in terms of an auxiliary set D consisting of all the dyadic rationals in $[0, 1]$:

$$d = a_0 + a_1 2^{-1} + a_2 2^{-2} + \cdots + a_n 2^{-n} \tag{15.2.9}$$

where a_j is 0 or 1 and $a_0 = 1$ iff all other a's are zero. To each such number d we assign a number s_d in $[s_0, s_1]$ where the indexing is determined by the composition rule

$$s_\gamma = A(s_\alpha, s_\beta) \Leftrightarrow \gamma = \tfrac{1}{2}(\alpha + \beta) \tag{15.2.10}$$

and α, β, $\gamma \in D$. Here we begin with $\alpha = 0$, $\beta = 1$ and proceed by successive averaging. At the nth stage of the process we have labeled all those points of S for which the index is of the form $d = k2^{-n}$ with $k = 0, 1, 2, \ldots, 2^n$. If here k is even, the point in question has already been labeled but it is clear that the index is unchanged. We shall state or prove various properties of the three sets D, S, Y. We start with Y.

Lemma 15.2.1. *If $d \in D$, then*

$$f(s_d) = dy_1 + (1-d)y_0. \tag{15.2.11}$$

Proof. We use induction on n in the expression (15.2.9) for d. The formula is clearly true for $d = 0, \frac{1}{2}, 1$, i.e. for $n = 1$. Suppose it is true for $n = m$ and consider a point s_d which was indexed at the $(m+1)$th stage. Such a d is of the form

$$d = (2p+1)\,2^{-m-1} = \tfrac{1}{2}[p\,2^{-m} + (p+1)\,2^{-m}] \equiv \tfrac{1}{2}(\alpha + \beta)$$

with obvious notation. Since (15.2.11) is true for $n = m$ by assumption, we have

$$\begin{aligned} f(s_d) &= \tfrac{1}{2}[f(s_\alpha) + f(s_\beta)] \\ &= \tfrac{1}{2}\{p\,2^{-m}y_1 + [1 - 2^{-m}p]y_0 + (p+1)\,2^{-m}y_1 + [1-(p+1)2^{-m}]y_0\} \\ &= (2p+1)\,2^{-m-1}y_1 + [1-(p+1)\,2^{-m-1}]y_0 = dy_1 + (1-d)y_0. \end{aligned}$$

Thus the formula holds also for $n \equiv m+1$ and hence for all $d \in D$. ∎

An obvious corollary is that Y is dense in $[y_0, y_1]$. It is clear that D is dense in $[0, 1]$.

Lemma 15.2.2. *S is dense in $[s_0, s_1]$.*

Proof. If S were not dense in $[s_0, s_1]$ there would exist an interval (α_0, β_0) no point of which belongs to S, while the endpoints are at least in the closure of S. If α_0 and β_0 are both in S, then so is $A(\alpha_0, \beta_0)$, which lies in (α_0, β_0). This is a contradiction. If $\alpha_0 \in S$, $\beta_0 \in \bar{S}$, the continuity of A would give

$$\alpha_0 < A(\alpha_0, \beta_0) < \beta_0$$

and there is still a contradiction. The remaining possibilities are disposed of in the same manner. It follows that S is dense in $[s_0, s_1]$. ∎

End of Proof of Theorem 15.2.5. At this stage we have f defined for all s in the dense set S. This definition is to be completed in such a way that f is shown to exist as a continuous strictly increasing function, first for all s in $[s_0, s_1]$ and then in the rest of the interval (a, b). The procedure in the first case is obvious. If s_0 is not in S, it belongs to $\bar{S}$ and we can find a sequence $\{s_k\}$ which converges to s_0 and such that $s_k \in S$, for all k. Here each s_k is an s_{d_k} where $d_k \in D$ and the sequence $\{d_k\}$ converges to a limit $d_0 \in \bar{D}$. Here we have used the fact that the mapping of D onto S is (1, 1), continuous, and monotone. But now the mapping of S onto Y is also continuous and equi-monotone. To $s = s_k$ corresponds $y = y_k$ where

$$y_k = f(s_k) = d_k y_1 + (1 - d_k)y_0. \tag{15.2.12}$$

Hence we can define

$$f(s_0) = \lim_{k\to\infty} f(s_k) = d_0 y_1 + (1 - d_0) y_0. \qquad (15.2.13)$$

Since the mapping of D onto Y is (1, 1), continuous, and monotone, it follows that d_0 is uniquely determined by s_0 and does not depend upon the choice of the particular sequence $\{s_k\}$ as long as it belongs to S and converges to s_0. The obvious equality

$$f(s') - f(s'') = (d' - d'')(y_1 - y_0) \qquad (15.2.14)$$

will help the reader to clarify this important point. In this manner f is defined everywhere in $[s_0, s_1]$ as a continuous, strictly increasing solution of the functional equation (15.2.2) which satisfies the given two-point condition and it is the only solution with these properties.

We still have to extend the definition of f to the rest of the interval (a, b). It is enough to indicate how this is done for the interval (s_1, b). The only tool at our disposal is the functional equation

$$f[A(s, t)] = \tfrac{1}{2}[f(s) + f(t)],$$

which the proposed solution is bound to satisfy. The equation links the values of f for three values of u, namely $u = s$, $u = t$, $u = A(s, t)$. If two of these values are known, then the third is uniquely determined. We now know the value $f(u)$ for any u in $[s_0, s_1]$. We have to choose two values of u in $[s_0, s_1]$ such that the third associated value lies in (s_1, b). An advantageous choice would be t and s such that

$$s_0 < t < s_1 < s < b, \qquad A(s, t) = s_1. \qquad (15.2.15)$$

Such a choice, however, is not always possible. What is desired is to solve the equation

$$A(s, t) = s_1 \qquad (15.2.16)$$

for s, given the value of t in $[s_0, s_1]$. It is a case of the implicit function theorem but under weaker assumptions than those of Theorem 6.1.1. On the other hand, $A(s, t)$ has rather special properties which guarantee the existence of a unique solution for t close to s_1.

We return to (15.2.6), interchanging the roles of s and t, and note that the mapping $s \to A(s, t)$, for fixed t, of (t, b) onto (t, t^+) varies continuously with t. For $t = s_1$ we have $t^+ = s_1^+ > s_1$. This means that for a small $\delta > 0$ and $0 < s_1 - t < \delta$ there is an $\varepsilon = \varepsilon(\delta)$ such that $0 < s_1^+ - t^+ < \varepsilon$. This says that for such a t the function $A(s, t)$ maps the interval $s_1 < s < b$ onto an interval containing the point $u = s_1^+ - \varepsilon$ and, *a fortiori*, the point $u = s_1$. For $A(s, t)$ increases from $A(s_1, t) < s_1$ to $t^+ > s_1^+ - \varepsilon > s_1$ as s goes from t to b. It follows that, for $s_1 - t$ small positive, Eq. (15.2.16) has a unique solution $s = s_1(t)$ such that

$$A[s_1(t), t] = s_1. \qquad (15.2.17)$$

We can then define

$$f[s_1(t)] = 2f(s_1) - f(t). \tag{15.2.18}$$

If $f(t) = dy_1 + (1-d)y_0$

$$f[s_1(t)] = (2-d)y_1 - (1-d)y_0 \equiv d_1 y_1 + (1-d_1)y_0 \tag{15.2.19}$$

where $1 < d_1 < 2$. The function $s_1(t)$ is again continuous and strictly monotone on its interval of definition, say $[t_1, s_1]$. Here t_1 is the least value of $t > s_0$ such that $s_1(t) \leqslant b$. Thus f has been defined on some interval $(s_1, s_1(t_1))$. If this is not all of (s_1, b), the argument is repeated. Instead of (15.2.16) we use an equation

$$A(s, t) = s_2 \tag{15.2.20}$$

where s_2 equals $s_1(t_1)$ if f is defined for this value, otherwise some smaller value. Actually, if $s_1(t_1) < b$ we can take $s_2 = s_1(t_1)$, for f is necessarily bounded to the left of this point, since otherwise it would be unbounded in the whole interval of definition, as we see by an adaptation of the argument used in proving Theorem 15.2.3. It follows that f is at least right-continuous at this point and the right hand limit can be taken as the definition of f for $u = s_1(t_1)$. We can then carry through the same argument as above and obtain f defined in a larger interval. In the same way we extend to the left into (a, s_0).

Suppose that $b_0 < b$ is the least upper bound for the values of s for which $f(s)$ may be defined in this manner. Again we see that f must be bounded to the left of $s = b_0$ and tend to a finite limit as s increases to b_0. But then we can solve the equation

$$A(s, t) = b_0 \tag{15.2.21}$$

for s in terms of t for $t < b_0$ but close to b_0. Here $s(t) > b_0$ and we can set

$$f[s(t)] = 2f(b_0) - f(t).$$

This shows that we must have $b_0 = b$ and in the same manner it is seen that the greatest lower bound for the values of s for which f is definable equals a. ∎

There is a converse of this result.

Theorem 15.2.6. *Let $s \to f(s)$ be a continuous, strictly monotone mapping defined on an interval (a, b). Let g be the inverse function of f so that*

$$g[f(s)] = f[g(s)] = s, \qquad \forall\, s \in (a, b).$$

Then

$$A(s_1, s_2, \ldots, s_n) \equiv g\left\{\frac{1}{n}\sum_{j=1}^{n} f(s_j)\right\} \tag{15.2.22}$$

is defined for all n and all s_j in (a, b). It is, moreover, a mean value satisfying postulates (A_1) to (A_4).

Proof. To fix the ideas, suppose that f is strictly increasing. Then, for a given

n and a choice of the s_j as indicated, the quantity

$$\frac{1}{n}\sum_{j=1}^{n} f(s_j) \tag{15.2.23}$$

belongs to the range of f and hence to the domain of g. Thus $A(\dots)$ is well defined. Moreover, the sum is an increasing continuous function of each of its arguments since f has these properties. Since g is continuous and strictly increasing, $A(\dots, s_j, \dots)$ is a continuous strictly increasing function of s_j. Thus (A_1) and (A_2) are satisfied. Postulate (A_3) is obviously true and (A_4) is a trivial consequence of the identity

$$g\left\{\frac{1}{n}\left(k\frac{1}{k}[f(s_1) + \cdots + f(s_k)] + f(s_{k+1}) + \cdots + f(s_n)\right)\right\}$$

$$= g\left\{\frac{1}{n}\sum_{j=1}^{n} f(s_j)\right\}$$

for we have

$$f[A(y, \dots, y, x_{k+1}, \dots, x_n)] = \frac{1}{n}\left[kf(y) + \sum_{j=k+1}^{n} f(x_j)\right], \quad kf(y) = \sum_{j=1}^{k} f(x_j),$$

whence (A_4) follows. ∎

Among the possible choices for f we note the following:

(1) $f(s) = s^\alpha$, $\alpha \neq 0$, which gives the power means

$$\mathsf{M}_\alpha(x) = \left\{\frac{1}{n}\sum_{j=1}^{n} x_j^{\alpha}\right\}^{1/\alpha}. \tag{15.2.24}$$

Normally we have to take $(a, b) = (0, \infty)$ but if α is an odd positive integer, $(-\infty, +\infty)$ is admissible.

(2) $\alpha = -1$, $f(s) = 1/s$, $s > 0$, gives the harmonic mean.

(3) The case $\alpha = 0$ is clearly excluded in (15.2.24). Since f is not uniquely determined by the mean we may replace f by $cf + d$, choosing c and d in such a manner that a limit exists for $\alpha \to 0$. The desired choice is

$$\frac{1}{\alpha}[s^\alpha - 1].$$

The limit function $\log s$ defines the geometric mean.

(4) A choice which has not figured above is $f(s) = \exp(\alpha s)$ with $\alpha \neq 0$. This gives

$$A(s, t) = \frac{1}{\alpha}\log \tfrac{1}{2}[\exp(\alpha s) + \exp(\alpha t)], \tag{15.2.25}$$

which is basic in information theory.

Mean values involving the sine, the tangent, and their inverses figure in various exercises above and below. Since there is an ample supply of continuous strictly monotone functions, the reader can construct mean values *ad lib.*

EXERCISE 15.2

1. Some mean values are *translation-invariant*; i.e. if each s_j is replaced by $s_j + a$ for a fixed a, then the average is changed by the same amount:
$$A(s_1 + a, s_2 + a, \ldots, s_n + a) = A(s_1, s_2, \ldots, s_n) + a.$$
Show that the arithmetic mean, $f(s) = s$, and the means defined by (15.2.25) have this property.

2. According to M. Nagumo, these are the only translation invariant means. Prove. [*Hint*: It is suggested to show that $f(s + a) = g(a)f(s) + h(a)$ with $g(a) \neq 0, \forall a$. From this conclude that f and g have continuous derivatives and $f'(s) \equiv \alpha f(s) + \beta$ with constant α and β. Integration proves the assertion.]

3. Some mean values are *homogeneous*; i.e.
$$A(as_1, as_2, \ldots, as_n) = aA(s_1, s_2, \ldots, s_n).$$
Show that the power means and the geometric mean have this property.

4. [Nagumo] Show that these are the only homogeneous means satisfying the postulates. [*Hint*: Find a direct proof or reduce to Problem 2 by setting $s = e^t$.]

5. The method used in proving Theorem 15.2.2 can be used to throw some light on existence questions for some equations of the Aczél–Hosszú type (14.3.16). Suppose that $A(s, t)$ is a mean value in the sense used here and consider the equation
$$f[A(s, t)] = \tfrac{1}{2}[f(s) + f(t)] + \sum_{k=1}^{\infty} c_k[f(s) - f(t)]^{2k}$$
where the c's are non-negative and the series $\sum c_k u^{2k}$ defines an entire function. Let f be a locally bounded solution whose oscillation
$$\limsup_{h \to 0} |f(s + h) - f(s)| = \omega(s)$$
satisfies the (highly restrictive) condition
$$\sum_{k=1}^{\infty} c_k[\omega(s)]^{2k-1} < \tfrac{1}{2}.$$
Show that f is necessarily continuous at all points where this condition holds.

6. Determine $s^-(t)$ and $s^+(t)$ when $f(s) = \arctan s$. How do they vary with t?

7. If $A(s, t)$ is an admissible mean value and if
$$s^-(t) = \lim_{s \downarrow a} A(s, t), \qquad s^+(t) = \lim_{s \uparrow b} A(s, t),$$
show that these mappings, if finite, are continuous, strictly monotone functions of t.

8. Same question for $s(t, s^0)$, the solution of

$$A(s, t) = s^0 \quad \text{with} \quad s(s^0, s^0) = s^0.$$

9. In the proof of Theorem 15.2.5 carry out the extension of f to the interval (a, s_0).

10. Suppose that $s \to f(s)$ defines an admissible mean value and that f is analytic, holomorphic at every point $s = s_0$ of (a, b). Prove or disprove the assertion that $A(s, t)$ is also analytic.

11. Suppose instead that $A(s, t)$ is given and is analytic; i.e. for every point (s_0, t_0) with real coordinates in the interval (a, b) there is an absolutely convergent double series expansion in terms of powers of $s - s_0$ and $t - t_0$. Discuss the analyticity of the corresponding function f.

12. In information theory the mapping $(s, t) \to \inf(s, t)$ plays an important role. Set $A(s_1, s_2, \ldots, s_n) = \inf s_j$. This is not a mean value in the sense used here but becomes acceptable if (A_2) is slightly weakened. How should this be done? Verify that the other postulates hold.

13. If $A(s, t)$ is defined as in the preceding problem, what is the corresponding function f? What two-point conditions are admissible? Is $(s, t) \to A(s, t)$ an intern transformation?

14. What would be the answer to Problem 10 in the present case?

15.3 REMARKS ON SUMMABILITY

Averaging processes have two important applications: to the smoothing of statistical data and to the summation of non-convergent infinite series or, equivalently, to the limitation of infinite sequences. We shall make some remarks on the second application. Here the preservation of existing limits is essential.

Theorem 15.3.1. *An averaging process* A *satisfying* (A_1) *to* (A_4) *is limit preserving, i.e. if* $\{x_n\}$ *is a convergent sequence of numbers in the interval* (a, b), $\lim_{n\to\infty} x_n = y_0$, *where* $y_0 \neq a$ *and* b, *then*

$$\lim_{n\to\infty} A(x_1, x_2, \ldots, x_n) = y_0. \tag{15.3.1}$$

For the proof we need the following.

Lemma 15.3.1. *The average of* k *numbers* c *and* n *numbers* d *converges to* d *if* n *goes to infinity in such a manner that* $k/n \to 0$.

Proof. This is an immediate consequence of the existence of a continuous strictly monotone function f satisfying (15.2.1). If $A_{k,n}$ is the average of k entries c and n entries d we have

$$f(A_{k,n}) = \frac{k}{k+n} f(c) + \frac{n}{k+n} f(d). \tag{15.3.2}$$

As $n \to \infty$ while $k/n \to 0$, the right hand side goes to $f(d)$. If g is the existing continuous strictly monotone inverse of f, then

$$\lim_{n\to\infty} A_{k,n} = \lim g\left[\frac{k}{k+n} f(c) + \frac{n}{k+n} f(d)\right]$$

$$= g[\lim f(A_{k,n})] = g[f(d)] = d$$

as asserted. ∎

Proof of Theorem 15.3.1. Since f does not necessarily exist or be continuous at either end of the interval we have assumed that $a < y_0 < b$. Suppose now that for all j we have $a < \alpha < x_j < \beta < b$ and that for a given $\varepsilon > 0$ it is known that $\alpha < y_0 - \varepsilon \leqslant x_j \leqslant y_0 + \varepsilon < 3$ for $j > k$. Then by (A_2) and Lemma 15.1.1 $A(x_1, x_2, \ldots, x_n)$ lies between

$$A(\alpha, \alpha, \ldots, \alpha, y_0 - \varepsilon, \ldots, y_0 - \varepsilon)$$

and

$$A(\beta, \beta, \ldots, \beta, y_0 + \varepsilon, \ldots, y_0 + \varepsilon).$$

Here α and β occur k times each while $y_0 - \varepsilon$ and $y_0 + \varepsilon$ occur $n - k$ times each. By the lemma, the first of these expressions converges to $y_0 - \varepsilon$, the second to $y_0 + \varepsilon$ as $n \to \infty$. Since ε is arbitrary, the assertion follows. ∎

Repeated smoothing of data is a well-known device in statistics and repeated averaging is also used in the theory of summability. This device was introduced by Otto Hölder (of the Hölder inequality) in 1882 for the arithmetic means, $f(s) = s$. More generally, let A_1 and A_2 be two averages, distinct or not, and form

$$[A_1 \cdot A_2](x_1, x_2, \ldots, x_n) = A_1[A_2(x_1), A_2(x_1, x_2), \ldots, A_2(x_1, x_2, \ldots, x_n)]. \tag{15.3.3}$$

We have then what is known as a *consistency theorem*:

Theorem 15.3.2. *If* A_1 *and* A_2 *satisfy the postulates and if* $\{x_n\}$ *is a sequence such that*

$$\lim_{n\to\infty} A_2(x_1, x_2, \ldots, x_n) = y_0, \tag{15.3.4}$$

then

$$\lim [A_1 \cdot A_2](x_1, x_2, \ldots, x_n) = y_0. \tag{15.3.5}$$

Proof. Combine (15.3.1) and (15.3.3). ∎

Since, in general, $\mathsf{A}_1 \cdot \mathsf{A}_2 \neq \mathsf{A}_2 \cdot \mathsf{A}_1$, a sequence $\{x_n\}$ which has an $[\mathsf{A}_1 \cdot \mathsf{A}_2]$-limit need not have an $[\mathsf{A}_2 \cdot \mathsf{A}_1]$-limit, and, if both exist, they need not be equal, that is, A_1 and A_2 need not be consistent in their common domain of applicability.

There is no reason to stop with two "factors" in the composition of averages. Except for Hölder means, not much use has been made of this device. The composition is a fairly clumsy device and the restriction to a common interval (a, b) for the two processes may cause difficulties.

EXERCISE 15.3

1. Give some instances where Theorem 15.3.2 remains valid for $y_0 = a$ or b.
2. Suppose it is known that $x_{2k} \to c$, $x_{2k+1} \to d$ where $c \neq d$. Show that, nevertheless, the sequence $\{x_n\}$ has an A-limit the value of which depends upon the choice of A.
3. Suppose that A_1 and A_2 are two processes satisfying the postulates and that $a < s < A_1(s, t) < A_2(s, t) < t$ for all s, t in (a, b). Given two numbers c and d, $a < c < d < b$, form the sequences $\{c_n\}$ and $\{d_n\}$ where $c_0 = c$, $d_0 = d$ and
$$c_n = A_1(c_{n-1}, d_{n-1}), d_n = A_2(c_{n-1}, d_{n-1}), n > 0.$$
Show that the first sequence is increasing, the second decreasing and they have the same limit. For $A_1(s, t) = \sqrt{st}$, $A_2(s, t) = \frac{1}{2}(s + t)$, the limit is the *arithmetico-geometric mean of Gauss.*
4. Give an example of a sequence of positive numbers for which the geometric and the arithmetic means have different limits.
5. It was shown by T. Carleman (1923) that
$$\sum_{n=1}^{\infty} (a_1 a_2 \dots a_n)^{1/n} \leqslant e \sum_{n=1}^{\infty} a_n$$
if the right member is a convergent series with positive terms. Is the series
$$\sum_{n=1}^{\infty} \frac{1}{n} (a_1 + a_2 + \cdots + a_n)$$
convergent under such circumstances?

15.4 SOME GEOMETRIC EXTREMAL PROBLEMS

We shall study a geometric extremal problem whose solution is essential for the discussion in Section 15.7. The point of departure is an innocent-looking problem in maxima and minima with side conditions. A triangle is inscribed in the unit circle, the lengths of the sides being s_1, s_2, s_3. What triangle will maximize a given symmetric function of s_1, s_2, s_3? The solution is often, but not always, given by the equilateral triangle. There may exist improper solutions formed by degenerate triangles, i.e. double diameters. Here the methods of the calculus carry us a long way. Now consider the corresponding problem in three dimensions: a tetrahedron is inscribed in the unit sphere, what configuration will maximize a given symmetric function of the lengths of the edges? Here the methods of the calculus are apt to fail right away. The problem generalizes to higher dimensions. In three space we could, of course, also pose the problem for the faces of the tetrahedron instead of for the edges.

To return to edges and higher dimensions, let n points $P_1, P_2, \ldots, P_n$ be given on the unit sphere in R^{n-1}

$$\sum_{j=1}^{n-1} x_j^2 = 1. \tag{15.4.1}$$

Join the points by line segments P_jP_k for $1 \leqslant j < k \leqslant n$. These are the edges of an *n-simplex*. It is said to be *regular* if all the edges have the same length. We may ask: For what n-simplex is the sum of the lengths of the edges a maximum? Instead of maximizing the sum of the lengths, we could maximize the arithmetic mean of the lengths and, for that matter, any average of the type described in this chapter. Here the outcome may be expected to depend upon what averaging process is used. The answer is not obvious and already in the plane unexpected things will happen. For the equilateral triangle inscribed in the unit circle the length of the edge is $\sqrt{3}$ and any averaging process satisfying (A_3) would give the same value. If any other triangle should give a larger average, considerations of symmetry show that the triangle must be degenerate. For a double diameter we obtain $A(2, 2, 0)$, if this expression makes sense, and the problem reduces to the question of for what A-averages is

$$A(2, 2, 0) > \sqrt{3}? \tag{15.4.2}$$

Take, in particular, $\mathsf{A} = \mathsf{M}_p$, $p > 0$, the pth power mean. Here the left member of (15.4.2) is

$$A(2, 2, 0) = 2(\tfrac{2}{3})^{1/p}$$

and for large values of p this is arbitrarily close to $2 > \sqrt{3}$.

We shall give a solution of this *simplicial problem* in any Euclidean space for a large class of averages.

Theorem 15.4.1. *The regular n-simplex in the unit sphere in R^{n-1} maximizes the A-average of the lengths of the edges of the simplex provided the function f defining* A *satisfies one of the following four conditions*:

(1) $f(s) = s^p$, $p \leqslant 2$, $p \neq 0$;

(2) $f(s) = \log s$;

(3) $f(s)$ *is strictly convex and decreasing*;

(4) $f(s)$ *is strictly concave and increasing.*

The solution is unique except for $p = 2$. If $f(s) = s^p$ and $p > 2$, then the regular simplex does not give a maximum, and if n is even, the maximum is furnished by a multiple diameter.

The proof will be given in several stages. The point of departure is the generalized Parallelogram Law, formula (2.5.7). This gives the solution for $f(s) = s^2$.

We recall that if $\mathfrak{X}$ is any product space with the usual conventions and notation, then for any choice of n vectors $\mathbf{x}_1, \mathbf{x}_2, \ldots, \mathbf{x}_n$, distinct or not, we have

$$\sum_{1 \leqslant j \leqslant k \leqslant n} \|\mathbf{x}_j - \mathbf{x}_k\|^2 + \left\| \sum_1^n \mathbf{x}_j \right\|^2 = n \sum_1^n \|\mathbf{x}_j\|^2. \tag{15.4.3}$$

In particular, if the $\mathbf{x}_j$'s are unit vectors, we get the inequality

$$\sum_{1 \leqslant j \leqslant k \leqslant n} \|\mathbf{x}_j - \mathbf{x}_k\|^2 \leqslant n^2 \tag{15.4.4}$$

with equality iff

$$\sum_1^n \mathbf{x}_j = 0, \tag{15.4.5}$$

i.e. the centroid of the endpoints of the vectors falls at the origin.

At this stage there is no relation between the number n of the vectors and the dimension of the space. Suppose now that the $\mathbf{x}_j$'s are unit vectors in R^{n-1}. Then the summands on the left, $\frac{1}{2}n(n-1)$ in number, are the squares of the lengths of the edges of the n-simplex and we thus obtain

Lemma 15.4.1. *If $f(s) = s^2$, the regular n-simplex maximizes the M_2-average of the lengths of the sides of the inscribed simplex. The maximizing configuration is unique iff $n = 3$.*

To see the truth of the last remark we observe that if three unit vectors $\mathbf{x}_1, \mathbf{x}_2, \mathbf{x}_3$ are located in the plane and subjected to the condition

$$\mathbf{x}_1 + \mathbf{x}_2 + \mathbf{x}_3 = 0,$$

then they determine the vertices of an equilateral triangle inscribed in the unit circle. On the other hand, if $n > 3$, then the condition $\Sigma \mathbf{x}_j = 0$ does not determine the $\mathbf{x}$'s up to a rotation. Thus besides the regular simplex, there are infinitely many simplices for which the sum of the squares of the lengths of the edges reaches the maximum value n^2. For future reference we note that the length of one of the edges in a regular n-simplex is

$$\sqrt{2}\left(\frac{n}{n-1}\right)^{\frac{1}{2}} \tag{15.4.6}$$

if the simplex is inscribed in the unit sphere in R^{n-1}. This expression tends to the limit $\sqrt{2}$ as $n \to \infty$, a fact important for the following.

This settles the mean square case. The remaining cases of Theorem 15.4.1 are essentially an exercise in the use of Hölder's inequality and the properties of convex functions.

We start with the power means M_p with $0 < p < 2$. Let n points $p_1, p_2, \ldots, p_n$ be given on the unit sphere (15.4.1) in R^{n-1} and let d_{jk} denote the length of the line segment joining p_j and p_k. These are the edges of the corresponding n-simplex and our problem is to maximize $\mathsf{M}_p(d_{jk})$. Now for fixed entries the power mean is an increasing function of the order p. Hence

$$\mathsf{M}_p(d_{jk}) \leqslant \mathsf{M}_2(d_{jk}) \leqslant \max \mathsf{M}_2(d_{jk}) = \sqrt{2}\left(\frac{n}{n-1}\right)^{\frac{1}{2}}. \tag{15.4.7}$$

Here equality holds in the first place iff all d_{jk} are equal, in the second place iff the centroid of the vertices is at the origin. Both conditions hold simultaneously iff the simplex is regular. Thus we have proved

Lemma 15.4.2. *For $0 < p < 2$ the p-th mean of the lengths of the edges of an n-simplex inscribed in the unit sphere in R^{n-1} is a maximum iff the simplex is regular.*

Again the maximum is given by (15.4.6) and it is now reached for a configuration which is unique up to a rotation about the origin.

Next we take the power means with $2 < p$. Here the discussion is based upon the observation that $s^p \leqslant s^2$ for $0 \leqslant s \leqslant 1$ with equality iff s is either 0 or 1. This leads to the following sequence of inequalities:

$$\sum_{j<k} (d_{jk})^p = 2^p \sum_{j<k} (\tfrac{1}{2}\, d_{jk})^p \leqslant 2^p \sum_{j<k} (\tfrac{1}{2}\, d_{jk})^2 = 2^{p-2} \sum_{j<k} (d_{jk})^2 \leqslant 2^{p-2} n^2. \tag{15.4.8}$$

Note that the edges of the simplex have a length at most equal to 2. Thus dividing by 2 in the first step ensures that $0 \leqslant \frac{1}{2}\, d_{jk} \leqslant 1$, so that the inequality $s^p \leqslant s^2$ can be used. Now equality holds in the first doubtful place iff each d_{jk} is either 0 or 2 and in the second place iff the centroid is at the origin. If n is even, $n = 2m$, both conditions may be satisfied by letting m vertices coincide at one point of the sphere and the remaining m vertices coincide at the antipode. In other words, choose m vectors $\mathbf{x}_j$ equal to $\mathbf{x}$ and the other m vectors equal to $-\mathbf{x}$ where $\mathbf{x}$ is a unit vector. If we move the $\mathbf{x}_j$'s, one pair at a time, one to $\mathbf{x}$ and the other to $-\mathbf{x}$, we see that we end up with m^2 distances d_{jk} equal to 2 and the remaining $m^2 - m$ distances equal to 0. For this choice we have equality all the way through in (15.2.8). It follows that the maximum value of the pth mean, $p > 2$, of the lengths of the edges of the inscribed n-simplex is

$$2^{1-1/p}\left(\frac{n}{n-1}\right)^{1/p}, \tag{15.4.9}$$

and this is reached for n even by a degenerate n-simplex, a multiple diameter. Since this supremum exceeds (15.4.6) for $p > 2$, the regular simplex cannot maximize the M_p-power means. In particular, this holds for the tetrahedron which served as our point of departure.

We turn now to case (3): f strictly convex and decreasing. Now if f is strictly convex,

$$f\left(\frac{1}{m}\sum_{j=1}^{m} s_j\right) < \frac{1}{m}\sum_{j=1}^{m} f(s_j) \tag{15.4.10}$$

unless all the s_j's are equal. As above, let d_{jk} denote the length of the edge P_jP_k and let d_N stand for the average of the d_{jk} as defined by f. Here $N = \frac{1}{2}n(n-1)$ and

$$f(d_N) = \frac{1}{N}\sum_{j<k} f(d_{jk}) \geqslant f\left(\frac{1}{N}\sum_{j<k} d_{jk}\right) \geqslant f\left[\left(\frac{2n}{n-1}\right)^{\frac{1}{2}}\right]. \tag{15.4.11}$$

To get equality throughout, the centroid of the vertices must be at the origin and all d_{jk} must be equal. Note that f is both strictly convex and strictly decreasing. We have also used the known value of the maximum for M_1 which enters in the last inequality in (15.4.11) as well as in (15.4.6). It follows that for each $n > 3$ the maximizing configuration is given by the regular n-simplex, which is unique up to a rotation.

In particular, this result applies to $f(s) = s^p$ for $p < 0$. It also applies to the A-average defined by $f(s) = \cot(\frac{1}{4}\pi s)$ where $a = 0$ and $b = 2$.

In the same manner we prove case (4). This includes $f(s) = \log s$ and $f(s) = s^p$ for $0 < p < 1$. Here $(a, b) = (0, \infty)$. Another possibility is $f(s) = \sin(\frac{1}{4}\pi s)$, $(a, b) = (0, 2)$.

EXERCISE 15.4

1. Justify the discussion of (15.4.2) for the pth power means. Determine the critical value of p, $p = p_0$, beyond which the equilateral triangle fails to maximize the average.
2. The calculus problem of finding maxima and minima for the pth means of the lengths of the sides in a triangle inscribed in the unit circle reduces to finding maxima and minima of
$$F(x, y) = [\sin x]^p + [\sin y]^p + [\sin(x+y)]^p,$$
where x, y, $x + y$ are in $[0, \pi]$. Verify! Show that $x = y = \frac{1}{3}\pi$ (i.e. the equilateral triangle) gives a local maximum of $F(x, y)$ for $0 < p < 4$, a local minimum for $4 < p$ and neither a maximum nor a minimum for $p = 4$.
3. Verify (15.4.3).
4. Write out a proof for case (4) of Theorem 15.4.1.
5. Let f be strictly increasing and strictly convex. Let d_n denote the maximum of the corresponding A-average for the n-simplex inscribed in the unit sphere. It is supposed that f is defined for $s = 0$ and $s = 2$ with $0 < f(0) < f(2) < \infty$. Show that
$$\sqrt{2} < \liminf_{n\to\infty} d_n \leqslant \limsup_{n\to\infty} d_n < 2.$$
[*Hint*: Use the fact that the graph of f for $0 < s < 2$ lies below the chord joining $(0, f(0))$ with $(2, f(2))$. Actually $\lim d_n$ exists, as will be shown later.]
6. Take $f(s) = (1/s)\tan cs$ with $0 < c < \frac{1}{4}\pi$. Show that f defines an A-average which satisfies the conditions of the preceding problem. Make plausible that for $n = 2m$ the completely degenerate n-simplex consisting of just a multiple diameter gives the

maximum value for the average. Show also that if $\lim d_{2m} = d_0$, then d_0 is the root of the equation

$$\tan cx = \tfrac{1}{2}[c + \tfrac{1}{2}\tan 2c]\, x$$

in the interval $(0, 2)$. For $c = \frac{1}{6}\pi$, $d_0 = 1.766$ approximately, i.e. nearer to 2 than to $\sqrt{2}$, the limits given in Problem 5.

7. The function $s \to f(s) = \tan(\frac{1}{4}\pi s)$ defines an A-average in $(-2, 2)$. In this case the A-average of the lengths of the edges of the n-simplex has no proper maximum, but if the supremum is denoted by d_n, show that $d_n = 2$ for all n. This is also the upper bound for the topological diameter.

15.5 THE TRANSFINITE A-DIAMETER

A beautiful application of mean values is furnished by the notion of the *transfinite diameter* of a set. The original concept was introduced in 1923 by Mihály Fekete (born in Hungary 1886, died in Jerusalem 1957) in connection with a question involving algebraic numbers. The precise nature of this problem has no bearing on the following discussion.

Given a bounded closed set E in the complex plane, Fekete wanted a measure of how far apart n points of E could get on the average. He was working with discriminants, i.e. expressions of the form

$$\prod_{1 \leqslant j < k \leqslant n} (z_j - z_k),$$

so it was natural for him to use the geometric mean for averaging, that is,

$$\left\{ \prod_{1 \leqslant j < k \leqslant n} |z_j - z_k| \right\}^{1/N}, \qquad N = \tfrac{1}{2} n(n-1). \tag{15.5.1}$$

Denote the maximum of this expression for any choice of n points in E by $d_n(E)$. Fekete could show that the sequence $\{d_n(E)\}$ is decreasing, so that

$$d_0(E) = \lim_{n \to \infty} d_n(E) \tag{15.5.2}$$

exists. He called this number the *transfinite diameter* of E. It is obviously at most equal to the topological diameter $d(E)$ and usually considerably less. Thus for a circular disk it equals the radius, for a line segment it is a quarter of the length. Fekete found that this concept gave him the solution of his algebraic problem, but he soon realized that he had got hold of something much more significant, and he spent the rest of his active life as a mathematician on investigations of the transfinite diameter. Thus this concept has a bearing on the so-called *Čebyšev constant* (see next section), on the *logarithmic equilibrium potential* of E and, if E is simply connected, on the *exterior conformal mapping radius* of E. Here work by G. Szegö also played an important role. For the latter concepts, see Section 15.8.

In 1931 Pólya and Szegö considered the corresponding problem in three dimensions. Since they wanted to preserve the contact with potential theory

(Newtonian in this case), they found that the averaging should be based not on the geometric but rather on the harmonic mean. The former corresponds to $f(s) = \log s$, the latter to s^{-1}. These are the simplest logarithmic and Newtonian potential functions respectively, s being the distance from a fixed to a variable point in the space. Pólya and Szegö also considered $f(s) = s^p$ in two and three dimensions.

It is clear that a transfinite diameter is definable for an arbitrary metric space and arbitrary A-averages. Much work along such lines has been done by F. Leja (Krakow). We shall consider this problem. Actually the connection with potential theory of a generalized kind is preserved also in this case.

Let $\mathfrak{X}$ be a complete metric space, let E be a bounded infinite point set in $\mathfrak{X}$, and let A be an averaging process, satisfying the postulates, defined by a continuous strictly monotone function $f(s)$.

We take n points $P_1, P_2, \ldots, P_n$ of E, note the distances

$$d(P_j, P_k) = d_{jk}, \qquad j \neq k, \tag{15.5.3}$$

and take the A-average of the d_{jk} for $1 \leqslant j < k \leqslant n$, which is denoted by $A(d_{jk})$ for short. This is a positive number, not exceeding the topological diameter $d(E)$. Set

$$d_n(E) = \sup A(d_{jk}) \tag{15.5.4}$$

when the n points P_j range over E. Here and in the following we restrict ourselves to A-averages for which the formulas make sense; a sufficient but not necessary condition is that $(a, b) = (0, \infty)$.

Lemma 15.5.1. *The sequence $\{d_n(E)\}$ is non-increasing. We set*

$$\lim_{n \to \infty} d_n(E) = d_0(E; \mathsf{A}). \tag{15.5.5}$$

Proof. The properties of the average A will be used. For a given n and a given $\varepsilon > 0$ we can find $n + 1$ points Q_j in E with

$$d(Q_j, Q_k) = \delta_{jk}$$

such that

$$A(\delta_{jk}) > d_{n+1}(E) - \varepsilon \tag{15.5.6}$$

by the definition of the supremum. Here there are $\frac{1}{2}n(n + 1)$ distances δ_{jk}. We can take $n - 1$ sets of these distances and average, obtaining the same result by Lemma 15.1.2. These $\frac{1}{2}(n - 1)n(n + 1)$ distances are now grouped into $n + 1$ sets with $\frac{1}{2}n(n - 1)$ distances in each set. This can be done in such a manner that in the jth set no distance involving Q_j occurs. Let η_j be the average of the distances in the jth set. Here

$$\eta_j \leqslant d_n(E) \tag{15.5.7}$$

for each j since we are dealing with distances between n points in E. Then by (A_4) and Lemma 15.1.3

$$A(\delta_{jk}) = A(\eta_1, \eta_2, \ldots, \eta_{n+1}) \leqslant d_n(E). \tag{15.5.8}$$

Hence

$$d_{n+1}(E) - \varepsilon \leqslant d_n(E) \tag{15.5.9}$$

for all $\varepsilon > 0$. Thus the sequence is non-increasing and the limit (15.5.5) exists. ∎

This set function $d_0(E; \mathsf{A})$ has certain properties of continuity and monotony. Thus we get

Lemma 15.5.2. *If $E_1 \subset E_2$, then*

$$d_0(E_1; \mathsf{A}) \leqslant d_0(E_2; \mathsf{A}). \tag{15.5.10}$$

Proof. Any choice of n points of E_1 is also a choice of n points in E_2. Thus the supremum in the second case must be at least as large as in the first case. Since this holds for all n, it holds also for the limits. ∎

This is as far as we can go. Even if E_1 is quite a small subset of E_2, the two transfinite diameters may be equal. Thus in Fekete's case the circular disk and its perimeter have the same transfinite diameters. In fact, this is quite natural and may be expected in much more general cases, for the points that maximize the average for n points will usually find the boundary as the best possible location.

We have also

Lemma 15.5.3. *Let E_ε be the set of points having a distance from E not exceeding ε, then*

$$\lim_{\varepsilon \downarrow 0} d_0(E_\varepsilon) = d_0(E). \tag{15.5.11}$$

Proof. This is essentially an exercise in carrying out repeated limiting processes in the appropriate order. We choose an integer n, a positive number η, and n points $Q_1, Q_2, \ldots, Q_n$ in E_ε so that their average distance $A(Q) > d_n(E_\varepsilon) - \eta$. Now for each Q_j there is a point P_j in E such that

$$d(P_j, Q_j) \leqslant \varepsilon,$$

and we may assume that $P_j \neq P_k$ if $j \neq k$. Then

$$d(Q_j, Q_k) \leqslant d(P_j, P_k) + 2\varepsilon$$

and

$$d_n(E_\varepsilon) - \eta \leqslant A[\ldots, d(P_j, P_k) + 2\,\varepsilon, \ldots]. \tag{15.5.12}$$

Now A is a continuous function in each of its $N = \frac{1}{2}n(n-1)$ arguments at the point in R^N whose coordinate at the place (j, k) is $d(P_j, P_k)$. Hence we can find a $\sigma(\varepsilon)$ which goes to zero with ε such that

$$A[\ldots, d(P_j, P_k) + 2\,\varepsilon, \ldots] < A[\ldots, d(P_j, P_k), \ldots] + \sigma(\varepsilon) < d_n(E) + \sigma(\varepsilon).$$

Hence

$$d_n(E_\varepsilon) \leqslant d_n(E) + \sigma(\varepsilon) + \eta. \tag{15.5.13}$$

Here the left member is $\geqslant d_0(E_\varepsilon)$ so that

$$d_0(E_\varepsilon) \leqslant d_n(E) + \sigma(\varepsilon) + \eta.$$

The left member is a non-decreasing function of ε, so it tends to a limit as $\varepsilon \downarrow 0$ and $\sigma(\varepsilon) \to 0$. Hence

$$\lim_{\varepsilon \downarrow 0} d_0(E_\varepsilon) \leqslant d_n(E) + \eta$$

for every n. Now let $n \to \infty$ to obtain

$$\lim_{\varepsilon \downarrow 0} d_0(E_\varepsilon) \leqslant d_0(E) + \eta$$

for every $\eta > 0$ and hence also for $\eta = 0$. On the other hand, the limit in the left member is at least $d_0(E)$ and (15.5.11) holds. ∎

EXERCISE 15.5

1. Verify that if E is a line segment and A is the geometric mean, then $d_0(E; A)$ is a quarter of the length of the segment.
2. What is the answer if A is the arithmetic mean instead?
3. Find $d_0(E; \mathsf{A})$ for the unit disk and the geometric mean. [*Hint*: Assume that for each n the nth roots of unity define a maximizing configuration by reasons of symmetry. The value of the integral $\int_0^\pi \log \sin t\, dt = -\pi \log 2$ may be required.]
4. Find $d_0(E; \mathsf{M}_2)$ for the unit ball in R^m.

15.6 THE ČEBYŠEV CONSTANTS

This is another class of set functions obtained by an averaging process. The original definition dealt with sets in the complex plane and the use of geometric means. Given a bounded closed set E in the complex plane, consider the set $\{P_n\} = \mathsf{P}_n$ of all polynomials of degree n in the complex variable z with leading coefficient 1. The absolute value of each such polynomial attains a maximum in E. For what polynomial in the set P_n is the maximum as small as possible? This is a *minimax problem*. There exists a unique polynomial $T_n(z; E)$ for which the minimax is assumed. This is known as the nth Čebyšev polynomial for the set E after the Russian mathematician Pafnuti Livovič Čebyšev (1821–94), who first considered such questions. The transliteration of Russian names varies from one language to another. Here we are using the Czech spelling, which is that used by *Mathematical Reviews*. The "T" in $T_n(z)$ is a relic of older spellings starting with "Tch" or "Tsch".

We now set

$$M_n = M_n(E) = \left[\max_{z \in E} |T_n(z; E)|\right]^{1/n}. \tag{15.6.1}$$

It may be shown that the sequence $\{M_n\}$ converges to a limit known as the Čebyšev constant of E

$$\check{C}(E) = \lim_{n\to\infty} M_n(E). \tag{15.6.2}$$

It was shown by Fekete that his transfinite diameter $d_0(E)$ coincides with $\check{C}(E)$.

Now the absolute value of a polynomial of degree n and leading coefficient 1 is simply the product of the distances from a variable point z to n fixed points, $z_1, z_2, \ldots, z_n$, the roots of the polynomial P_n under consideration. If we extract the nth root of the absolute value, we are simply taking the geometric mean of the distances.

This is a familiar situation which invites generalizations. We take an arbitrary complete metric space $\mathfrak{X}$ and a bounded infinite set E in $\mathfrak{X}$. Take an averaging process A satisfying the postulates and such that the interval of definition (a, b) contains $(0, d(E)]$ where $d(E)$ is the topological diameter of E. Take now n points $P_1, P_2, \ldots, P_n$ in $\mathfrak{X}$ but not necessarily in E and form the A-average of the distances from a point P to $P_1, P_2, \ldots, P_n$,

$$A[d(P, P_1), d(P, P_2), \ldots, d(P, P_n)] \equiv g(P). \tag{15.6.3}$$

This function $g(P)$ has a supremum for P in E. We now ask: What is the infimum of the set of suprema for a given integer n and arbitrary choice of the points P_1 to P_n? We set

$$\inf_g \sup_{P\in E} g(P) = M_n(E). \tag{15.6.4}$$

The sequence $\{M_n(E)\}$ is bounded, for if all points P_j as well as P are in E, then $d(P, P_j)$ cannot exceed $d(E)$ for any j, and the same bound must then apply to $M_n(E)$ for all n. Actually the sequence is convergent, but to prove this we need some further inequalities.

Lemma 15.6.1. *We have*

$$M_{m+n}(E) \leqslant \max [M_m(E), M_n(E)] \tag{15.6.5}$$

for all m and n. In particular,

$$M_n(E) \leqslant M_1(E), \qquad \forall\, n. \tag{15.6.6}$$

More generally,

$$M_{kn}(E) \leqslant M_k(E), \qquad \forall\, k, n. \tag{15.6.7}$$

Proof. If $\varepsilon > 0$ is given, we can find two functions of P:

$$g_1(P) = A[d(P, P_1), d(P, P_2), \ldots, d(P, P_m)],$$
$$g_2(P) = A[d(P, Q_1), d(P, Q_2), \ldots, d(P, Q_n)],$$

whose suprema on E are within ε of M_m and M_n, respectively. We then form the function

$$g(P) = A[d(P, P_1), \ldots, d(P, P_m), d(P, Q_1), \ldots, d(P, Q_n)]$$

whose supremum on E is at least $M_{m+n}(E)$. By (A_4) we have for all P

$$g(P) = A[g_1(P), \ldots, g_1(P), g_2(P), \ldots, g_2(P)]$$

with m entries $g_1(P)$ and n entries $g_2(P)$. This gives

$$g(P) \leqslant \max [g_1(P), g_2(P)],$$

and on E the right member does not exceed

$$\max [M_m(E) + \varepsilon, M_n(E) + \varepsilon] = \max [M_m(E), M_n(E)] + \varepsilon.$$

This implies (15.6.5) since $\varepsilon > 0$ is arbitrary. The two remaining inequalities are simply special cases of (15.6.5) combined with (A_4). ∎

For a fixed $E \subset \mathfrak{X}$ the sequence $\{M_n(E)\}$ is a valuation sequence in the sense of Section 13.4 by virtue of (15.6.5). We refer to Exercise 13.4 for various ways in which such a sequence may behave. The main point here is that a valuation sequence need not converge. This means that additional information is required to ensure convergence.

Lemma 15.6.2. *For any number $b \geqslant d(E)$*

$$M_{n+1}(E) \leqslant A[M_n, M_n, \ldots, M_n, b) \tag{15.6.8}$$

where M_n figures n times.

Proof. We use the function $g_2(P)$ defined above and take an arbitrary point Q in E and form with n entries $g_2(P)$,

$$g(P) = A[g_2(P), g_2(P), \ldots, g_2(P), d(P, Q)]. \tag{15.6.9}$$

The supremum of $g(P)$ on E is at least $M_{n+1}(E)$. On the other hand, on E the right member is at most

$$A[M_n(E) + \varepsilon, M_n(E) + \varepsilon, \ldots, M_n(E) + \varepsilon, d(E)].$$

Since ε is arbitrary (15.6.8) results. ∎

With the aid of these inequalities we shall now prove the convergence of $\{M_n(E)\}$.

Theorem 15.6.1. *The sequence $\{M_n E\}$ converges to a limit*

$$\check{C}(E; \mathsf{A}) = \inf M_n(E), \tag{15.6.10}$$

called the Čebyšev constant of the set E with respect to the average A.

Proof. Problems 7, 8, and 9 under Exercise 13.4 suggest the following arrangement of the proof. Set

$$\beta = \limsup_{n \to \infty} M_n(E). \tag{15.6.11}$$

We shall show that $\beta \leqslant M_n$ for all n, so that a unique limit must exist and equal β. Let us first assume that there is a number γ and an integer j such that

$$M_j(E) < \gamma, \qquad M_{j+1}(E) < \gamma. \tag{15.6.12}$$

By Problem 8, Exercise 13.4, every integer $n > j^2$ admits of a representation of the form

$$n = pj + q(j+1), \tag{15.6.13}$$

where p and q are non-negative integers. This combined with Lemma 15.6.1 shows that for such values of n

$$M_n(E) \leqslant \max\,[M_j(E), M_{j+1}(E)] < \gamma. \tag{15.6.14}$$

That is, if j and γ exist for which (15.6.12) holds, then necessarily $\beta < \gamma$. The proof proceeds by showing that if

$$M_n < \beta$$

for some n, then j and γ exist with $\beta > \gamma$ and this contradiction establishes that $\beta \leqslant M_n(E)$ for all n. Here is where the second lemma comes in handy.

Suppose that for some m we have $M_m(E) = \alpha < \beta$. Then for all n of the form

$$n = 2^k m$$

we would also have $M_n \leqslant \alpha$ from (15.6.7). For such an n we use (15.6.8) to obtain

$$M_{n+1}(E) \leqslant A[M_n(E), M_n(E), \ldots, M_n(E), d(E)] \leqslant A[\alpha, \alpha, \ldots, \alpha, d(E)].$$

As n goes to infinity with k, the last member converges to α by Lemma 15.3.1. It follows that (15.6.12) holds for some large value of j, say $j = 2^k m$, if we take for γ some number greater than α. Since $\alpha < \beta$, we can take $\gamma < \beta$ and this gives the contradiction. Hence we must have

$$M_n(E) \geqslant \beta = \limsup_{p\to\infty} M_p(E) \tag{15.6.15}$$

for all n, and this proves the existence of $\lim M_n(E)$. ∎

Here we cannot claim equality between the transfinite diameter $d_0(E; \mathsf{A})$ and the Čebyšev constant $\check{C}(E; \mathsf{A})$ in general, even though equality holds for the geometric mean and a complex plane set. Already Pólya and Szegö found counter examples. On the other hand, we do have an inequality.

Theorem 15.6.2. *For every admissible choice of the average* A

$$\check{C}(E; \mathsf{A}) \leqslant d_0(E; \mathsf{A}). \tag{15.6.16}$$

Proof. In E we choose $n+1$ points $P_1, P_2, \ldots, P_n, P_{n+1}$ in such a manner that the average distance between the first n points exceeds $d_n(E) - \varepsilon$ for a given $\varepsilon > 0$. Thus

$$A[d(P_j, P_k);\, 1 \leqslant j < k \leqslant n] \geqslant d_n(E) - \varepsilon.$$

On the other hand, if we average over all $n+1$ points, the result is at most $d_{n+1}(E)$, so that

$$d_{n+1}(E) \geqslant A[d(P_j, P_k)] = A[d(P_{n+1}, P_1), d(P_{n+1}, P_2), \ldots, d(P_{n+1}, P_n), \\ d(P_1, P_2), d(P_1, P_3), \ldots, d(P_{n+1}, P_n)].$$

Here P_{n+1} is at our disposal and may be chosen so that the average of the n distances from P_{n+1} is arbitrarily close to the supremum of

$$g(P) = A[d(P, P_1), d(P, P_2), \ldots, d(P, P_n)], \qquad P \in E,$$

which in its turn is at least equal to $M_n(E)$. The average of the remaining $\frac{1}{2}n(n-1)$ distances is at least equal to $d_n(E) - \varepsilon$ by the choice of the first n points. Hence

$$d_{n+1}(E) \geqslant A[M_n(E) - \eta, \ldots, M_n(E) - \eta, d_n(E) - \varepsilon, \ldots, d_n(E) - \varepsilon].$$

Here η is small positive, there are n entries of the first kind and $\frac{1}{2}n(n-1)$ of the second. Since ε and η are arbitrarily small, we can let these numbers tend to zero and appeal to the continuity of A to obtain

$$d_{n+1}(E) \geqslant A[M_n(E), \ldots, M_n(E), d_n(E), \ldots, d_n(E)].$$

Here we note that $d_n(E) \geqslant d_{n+1}(E)$ and that A is strictly increasing. Hence

$$d_{n+1}(E) \geqslant A[M_n(E), \ldots, M_n(E), d_{n+1}(E), \ldots, d_{n+1}(E)]. \qquad (15.6.17)$$

At this point we fall back on the generating function of A and, to fix the ideas, we assume that f is strictly increasing. This gives

$$f[d_{n+1}(E)] \geqslant f[A(\ldots)] = \frac{2}{n(n+1)}\{nf[M_n(E)] + \tfrac{1}{2}n(n-1)f[d_{n+1}(E)]\}.$$

This simplifies to

$$f[d_{n+1}(E)] \geqslant f[M_n(E)] \qquad (15.6.18)$$

and

$$d_{n+1}(E) \geqslant M_n(E) \geqslant \check{C}(E; \mathsf{A}) \qquad (15.6.19)$$

by (15.6.10). Passing to the limit with n we get (15.6.16). ∎

EXERCISE 15.6

In the first nine problems $\mathfrak{X}$ is the complex plane and A is the geometric mean.

1. Show that a polynomial $P_n(z) \in \mathsf{P}_n$ which has zeros outside the closed convex hull H of E cannot be a minimizing polynomial. The closed convex hull is the least closed convex set containing E.
2. Prove the existence of $T_n(z; E)$. [*Hint*: The maxima of $|P_n(z)|$ in E form a bounded set, so the infimum exists and is a positive number (why positive ?). There exists a sequence of elements in P_n for which the maxima converge to the infimum. Select subsequences of polynomials whose zeros converge to limits in H.]

3. Show that the minimizing polynomial $T_n(z; E)$ is unique if the equation $\max |T_n(z; E)| = [M_n(E)]^n$ has $n + 1$ distinct roots in E. [*Hint*: For a competing polynomial, $U_n(z)$, consider the set $cT_n(z) + (1 - c)\, U_n(z)$, $0 \leqslant c \leqslant 1$, and discuss maxima.]
4. The original Čebyšev polynomial (of the first kind) of degree n is most conveniently defined by
$$T_n(z) = 2^{-n}\{[z + (z^2 - 1)^{\frac{1}{2}}]^n + [z - (z^2 - 1)^{\frac{1}{2}}]^n\} = 2^{1-n} \cos (n \arccos z).$$
Show that these two expressions are equal and are actually polynomials in z of degree n and leading coefficient 1. Determine zeros and maxima and minima.
5. Use the criterion of No. 3 or otherwise to show that $T_n(z)$ is the unique Čebyšev polynomial of degree n for the interval $[-1, 1]$.
6. If $S_n(z) = 2^{n-1} T_n(z)$, verify the composition property
$$(S_m \circ S_n)(z) = S_m[S_n(z)] = S_{mn}(z).$$
7. Solve the Schröder equation (see Problem 19, Exercise 14.2)
$$f[S_n(z)] = \lambda f(z).$$
8. Find the Čebyšev constant for the interval $[-1, 1]$.
9. Show that the Čebyšev constant for the unit disk equals 1.

In the remaining problems $\mathfrak{X}$ and A are arbitrary unless otherwise restricted.

10. If $E_1 \subseteq E_2$, show that $C(E_1; \mathsf{A}) \leqslant C(E_2; \mathsf{A})$.
11. Compare the Čebyšev constants for the power means
$$C(E; \mathsf{M}_\alpha) \quad \text{and} \quad C(E; \mathsf{M}_\beta), \ \alpha < \beta.$$
12. Show that Problem No. 1 generalizes to any B-space $\mathfrak{X}$ and any arbitrary A. Only those functions $g(P)$ of formula (15.6.3) need be considered for which the points P_j lie in the convex hull of E.

15.7 SOME EXAMPLES

It is of some interest to examine the unit ball U of a Banach space $\mathfrak{X}$ from the point of view of transfinite diameters. It is clear that no matter what average we use (with domain containing $[0, 2]$) the transfinite diameter $d_0(U; \mathsf{A})$ cannot exceed 2, which is the topological diameter. But in a surprisingly large number of cases the value 2 is actually attained.

Theorem 15.7.1. *$d_0(U; \mathsf{A}) = 2$ for all admissible A if $\mathfrak{X}$ is one of the spaces $C[a, b]$, $L_1(a, b)$, $L_\infty(a, b)$, l_1, m.*

Proof. The idea of the proof is the same in all these cases. For any natural number n we may exhibit n elements which are unit vectors two units apart. We consider the two cases $C[a, b]$ and l_1, leaving the others to the reader.

In the case of $C[a, b]$, divide the interval into n equal parts. On each subinterval we erect an isosceles triangle of height 1. For the function f_j the jth triangle points upwards, all other triangles pointing downwards. Here $\|f_j\| = 1$ and $\|f_j - f_k\| = 2$ for $j \neq k$. The average distance between these n elements f_j is obviously 2, so $d_n(U) = 2$ for each n and this gives $d_0(U; \mathsf{A}) = 2$.

A similar construction works in l_1. Take the sequence of unit vectors $\{\mathbf{x}_k\}$ where $\mathbf{x}_k$ has a one in the kth place and zeros elsewhere. Then $\|\mathbf{x}_k\| = 1$ and $\|\mathbf{x}_j - \mathbf{x}_k\| = 2$ for $j \neq k$. ∎

The Čebyšev constants of the unit ball are subject to the following restriction.

Theorem 15.7.2. *The Čebyšev constant of the unit ball in a B-space cannot exceed 1.*

Proof. We have to consider

$$\inf_{\{\mathbf{x}_j\}} \sup_{\|\mathbf{x}\| \leq 1} A[\|\mathbf{x} - \mathbf{x}_j\|; j = 1, 2, \dots, n] \tag{15.7.1}$$

where the $\mathbf{x}_j$'s are n arbitrary elements of $\mathfrak{X}$. If we take all $\mathbf{x}_j = \mathbf{0}$, the supremum is the sup of $\|\mathbf{x}\|$, i.e. 1, so the inf sup can be at most 1. ∎

These two theorems show that there may very well be a wide gap between the values of the transfinite diameter and the corresponding Čebyšev constant. Thus in general the inequality (15.6.16) cannot be replaced by equality. The actual determination of Čebyšev constants for unit balls of Banach spaces of infinite dimension does not seem to have been attacked.

Let us return to the transfinite diameter. A pertinent question is the following: Is there a choice of A most natural to a given metric space $\mathfrak{X}$? Is there any sense in saying that such and such a definition of transfinite diameter is the most natural? What criteria are applicable to settle such a question? Should we give preference to a definition for which transfinite diameter and Čebyšev constant coincide? Then none of the spaces mentioned in Theorem 15.7.1 would admit of a natural definition. A criterion which is sometimes used is to require that the unit ball should have transfinite diameter unity. Again such a criterion cannot be applied to the cases mentioned above. Still another criterion is to require that the function f defining the average be an elementary potential function in the space. This gives the geometric mean with $f(s) = \log s$ in R^2, the harmonic mean with $f(s) = 1/s$ in R^3, the power mean M_{2-n} with $f(s) = s^{2-n}$ in R^n. This would make the arithmetic mean with $f(s) = s$ the natural choice in R^1 since linear functions are "harmonic" in this space. Beyond these cases the connections with potential theory become rather tenuous. As we shall see in the next section, however, it is possible to build a potential theory around almost any function f which defines a mean value. The existence of infinitely many parallel potential theories valid in the same space is a fact. The choice is merely transferred from transfinite diameters to potential theories.

The question of finding transfinite diameters of the unit ball in $L_2(a, b)$ or, equivalently, in a Hilbert space leads to problems which are solvable unless the chosen averaging process creates trouble. Actually the solution in a large number of cases is given by Theorem 15.4.1.

Theorem 15.7.3. *Let U be the unit ball in $L_2(-\pi, \pi)$ or any isometric and isomorphic image thereof. Then $d_0(U; \mathsf{A}) \geqslant \sqrt{2}$ for all A. We have $d_0(U; \mathsf{A}) = \sqrt{2}$, if the generating function f of A satisfies one of the conditions* (1) *to* (4) *of Theorem 15.4.1. If $f(s) = s^p$, $2 < p$, then $d_0(U; \mathsf{A}) = 2^{1-1/p}$.*

Proof. The set of functions

$$t \to (2\pi)^{-\frac{1}{2}} e^{kit}, k = 0, \pm 1, \pm 2, \ldots \tag{15.7.2}$$

is an orthonormal system for the space $L_2(-\pi, \pi)$ and the distance between distinct elements of the set is constantly equal to $\sqrt{2}$. Hence for every n we can find n elements of the space of norm 1 whose distance from each other equals $\sqrt{2}$. It follows that $d_n(U; \mathsf{A}) \geqslant \sqrt{2}$ for every choice of A and all n and this gives $d_0(U; \mathsf{A}) \geqslant \sqrt{2}$ as asserted.

To prove the remaining assertions we have to analyze the results obtained in Theorem 15.4.1. This theorem states that under the listed conditions the regular n-simplex maximizes the average of the lengths of the edges of any n-simplex inscribed in the unit sphere. The space is supposed to be an inner product space of sufficiently high dimension, at least $n - 1$. There is no need, however, to assume that the dimension is exactly $n - 1$. This assumption, even when explicitly made, is used nowhere in the proof. Thus there is nothing to prevent us from taking the sphere to be the unit ball of $L_2(-\pi, \pi)$. It then follows that

$$d_n(U; \mathsf{A}) = \sqrt{2}\left(\frac{n}{n-1}\right)^{\frac{1}{2}} \tag{15.7.3}$$

and thus $d_0(U; \mathsf{A}) = \sqrt{2}$.

Finally, if $\mathsf{A} = \mathsf{M}_p$ with $2 < p$, we conclude from (15.4.8) that

$$d_{2k}(U; \mathsf{A}) = 2\left(\frac{k}{2k-1}\right)^{1/p} \tag{15.7.4}$$

and this gives $d_0(U; \mathsf{A}) = 2^{1-1/p}$ as asserted. ∎

For the spaces l_p and L_p definitive results are lacking.

Theorem 15.7.4. *If U is the unit ball in l_p or in $L_p(a, b)$ with $1 < p$, then*

$$d_0(U; \mathsf{A}) \geqslant 2^{1/p}. \tag{15.7.5}$$

Proof. Consider l_p and the unit vectors (δ_{jm}). Here the distance between (δ_{jm}) and (δ_{km}) is $2^{1/p}$ for all $k \neq j$. This gives $d_n(U; \mathsf{A}) \geqslant 2^{1/p}$ for all A and all n. This implies (15.7.5). The case L_p is left to the reader. ∎

The inequality (15.7.5) is possibly the best of its kind for $1 \leqslant p \leqslant 2$. For $2 < p$

better inequalities are known; both the derivation and the result are complicated and not very likely to be definitive. At any rate, it appears that for $2 < p$ the transfinite diameter of U is an increasing function of p which converges to 2 as $p \to \infty$.

EXERCISE 15.7

1. Prove Theorem 15.7.1 for $\mathfrak{X} = m$.
2. Same question for $\mathfrak{X} = L_1(a, b)$ and $L_\infty(a, b)$.
3. Prove that Theorem 15.7.1 holds also for the space of functions f holomorphic and bounded in the unit disk of the complex plane. The metric is defined by the sup norm. [*Hint*: The integral powers of z may be worthy of consideration.]
4. Let P_0 be a fixed point in R^n, $n > 2$, P a variable point and r the Euclidean distance from P_0 to P. Consider the generalized Laplacian operator

$$\sum_{k=1}^{n} \frac{\partial^2}{\partial x_k^2},$$

   and show that r^{2-n} is annihilated by this operator and thus is a harmonic function in R^n.
5. Prove Theorem 15.7.4 for $L_p(a, b)$.
6. Let $\omega_n = \exp(2\pi i/n)$. Then $d_{jk} = |\omega_n^j - \omega_n^k| = 2\sin[\pi(k-j)/n]$, $1 \leqslant j < k < n$. Define $d_n(\Omega)$ as the A-average of the numbers $d_{jk}, j < k$. Further, prove that $d_0(\Omega) = \lim d_n(\Omega)$ exists and, if A is generated by $f(s)$, then

$$f(d_0) = 2\int_0^1 (1 - s) f[2 \sin \pi s]\, ds.$$

   In particular, for the arithmetic means, $d_0(\Omega) = 4/\pi$. [*Hint*: Use (15.2.3), the definition of the Riemann integral and the properties of f.]
7. If $\mathfrak{X}$ is a B-space over the complex field and if $\mathbf{x} \in U$, then all the points $\omega_n^j \mathbf{x}$ also belong to U. Use this to show that if A is taken as the arithmetic mean, then

$$d_0(U; \mathsf{A}) \geqslant \frac{4}{\pi}.$$

15.8 POTENTIAL THEORIES

In this last section of the chapter and of the treatise we shall give a sketch of the type of potential theory that can be developed starting from the generating function of an A-average. The discussion will be informal and expository, terms will be left somewhat vague and proofs will be mostly lacking. It is desired to whet the reader's appetite, not to satiate it.

Classical potential theories are of two kinds: the logarithmic and the Newtonian. The first theory is built around the basic harmonic function and the two-dimensional Laplacian

$$\log r \quad \text{and} \quad \frac{\partial^2 U}{\partial x^2} + \frac{\partial^2 U}{\partial y^2} = 0 \tag{15.8.1}$$

of which it is a solution; in the second,

$$\frac{1}{r} \quad \text{and} \quad \frac{\partial^2 U}{\partial x^2} + \frac{\partial^2 U}{\partial y^2} + \frac{\partial^2 U}{\partial z^2} = 0 \tag{15.8.2}$$

play a similar role. Here r is the distance between a fixed point and a variable point in the space. In fourth and higher dimensions we generalize via Laplace's equation and the corresponding basic solution

$$r^{2-n} \quad \text{and} \quad \sum_{j=1}^{n} \frac{\partial^2 U}{\partial x^2} = 0. \tag{15.8.3}$$

Further generalization may be given along two different lines. We may generalize the Laplacian or the basic function. The first road leads to a study of solutions of linear partial differential equations of the elliptic type. We shall avoid this path. Secondly, we note that the basic functions in the cases listed are generating functions of A-averages: f defined for $s > 0$, continuous and strictly monotone. Given such a function, we can carry over the classical formulas of potential theory and construct something that may be considered a potential theory.

The use of such more general kernels goes back to the early 1930's with Pólya and Szegö, Marcel Riesz (1886–1969) and Otto Frostman as pathfinders. The general notion of capacity (more general than below) has been developed by G. Choquet and Lennart Carleson.

In order to get our bearings, consider the Newtonian case. This begins with the notion of *Borel sets* (there are other alternatives, however, but we should not get lost in subtleties). In any topological space $\mathfrak{X}$, the σ-algebra generated by the open sets is called the class of Borel sets (Émile Borel, 1871–1956), here denoted by Σ. Cf. Problem 2, Exercise 4.1. Thus Σ contains all open sets, all closed sets, countable unions, and finite intersections. On Σ shall be defined a class of *finite, non-negative, countably additive set functions* $\mu(S)$. Here $S \in \Sigma$ and $\mu(S)$ is the measure of S in the σ-algebra $(\mathfrak{X}, \Sigma, \mu)$. In applications to potential theory "mass" or "charge" is often used instead of "measure". "Countably additive" means that

$$\mu(\textstyle\bigcup_1^\infty S_k) = \sum_{k=1}^{\infty} \mu(S_k) \quad \text{if} \quad S_j \cap S_k = \emptyset \quad \text{for} \quad j \neq k, \tag{15.8.4}$$

the series being convergent. See Definition 4.1.2.

If $f(\mathbf{x})$ is real-valued and Σ-measurable and $E \in \Sigma$, one can define the so-called

Radon–Stieltjes integral of f over E with respect to μ (cf. p. 237 *et seq.*)

$$\int_E f(\mathbf{x})\, d\mu(\mathbf{x}) \tag{15.8.5}$$

by the usual discussion of approximating sums of the Riemann–Stieltjes type. The integral is named after the Austrian mathematician Johann Radon (1887–1956). We mention without further explications the possibility of forming *product measures* and establishing the extension of the *Fubini theorem for double integrals.*

If $\mathfrak{X} = R^3$, the Newtonian potential of μ on E is by definition

$$U(\mathbf{x};\mu) = \int_E \|\mathbf{x} - \mathbf{y}\|^{-1}\, d\mu(\mathbf{y}). \tag{15.8.6}$$

Here $\mathbf{x}$ and $\mathbf{y}$ are vectors in R^3 and $\mathbf{y}$ is restricted to E. Further, $E \in \Sigma$ and $\mu(E) = 1$. The corresponding energy integral is

$$I(\mu; E) = \int_E \int_E \|\mathbf{x} - \mathbf{y}\|^{-1}\, d\mu(\mathbf{x})\, d\mu(\mathbf{y}). \tag{15.8.7}$$

This integral always exists but may be $+\infty$. In order to form these integrals, it is not necessary to assume that $\mathfrak{X} = R^3$, but they are physically meaningful iff $\mathfrak{X} = R^3$.

With Σ as above and $\mathfrak{X} = R^3$ consider now the set $\Gamma(E)$ of all countably additive measures, finite and non-negative, defined on Σ and with $\mu(E) = 1$. To each such measure μ corresponds an energy integral $I(\mu; E)$. Here it is possible that all energy integrals equal $+\infty$. Then E is said to be of capacity 0, Newtonian capacity to be more precise.

If E is not of capacity zero, then we set

$$\inf I(\mu; E) = V(E), \tag{15.8.8}$$

$$C(E) = [V(E)]^{-1}. \tag{15.8.9}$$

Here $C(E)$ is by definition the *Newtonian capacity* of E. If the boundary ∂E of E is sufficiently smooth, there is a measure $\mu_0 \in \Gamma(E)$ with

$$I(\mu_0; E) = V(E). \tag{15.8.10}$$

If μ_0 is unique, it is known as the *equilibrium distribution* and the corresponding potential function

$$\int_E \|\mathbf{x} - \mathbf{y}\|^{-1}\, d\mu_0(\mathbf{y}) \tag{15.8.11}$$

is known as the *equilibrium potential.*

These terms hail from electrostatics. If a conductor E is given and furnished with a unit charge, then the charge will distribute itself on the surface of the conductor. The distribution and the resulting potential are then known as the equilibrium distribution and the equilibrium potential. $V(E)$ is known as *Robin's constant* [Gustave Robin (1855–97)].

These are the notions which should be generalized. The kernel r^{-1} is replaced by an arbitrary kernel $K(r)$ defined for $r > 0$, continuous and strictly monotone. Thus K is the generating function of an A-average. $K(0)$ may be finite or infinite. It is convenient to assume $K(r) > 0$. True, this excludes the logarithmic potential, but for the following development it is preferable to deal with positive energies and positive potentials. We consider the case $\mathfrak{X} = R^m$ and the set Σ of all Borel sets there. Again let $\Gamma(E)$ be the set of all countably additive non-negative measures μ defined on a closed bounded set $E \in \Sigma$ and its subsets in Σ with $\mu(E) = 1$. Define the *K-potential* by

$$U(\mathbf{x}; \mu, K) = \int_E K(\|\mathbf{x} - \mathbf{y}\|)\, d\mu(\mathbf{y}) \tag{15.8.12}$$

and the corresponding energy integral

$$I(\mu; E, K) = \int_E \int_E K(\|\mathbf{x} - \mathbf{y}\|)\, d\mu(\mathbf{x})\, d\mu(\mathbf{y}). \tag{15.8.13}$$

It may happen that the energy integral is infinite for all $\mu \in \Gamma(E)$. In this case E is said to be of *K-capacity* 0. This presupposes that $K(0) = +\infty$. In any case, consider

$$V_K(E) = \inf I(\mu; E, K) \tag{15.8.14}$$

and determine the K-capacity from the equation

$$K[C_K(E)] = V_K(E). \tag{15.8.15}$$

The solution is unique since K is strictly monotone and continuous.

Suppose now that $C_K(E) > 0$. We can then find a sequence of mass distributions $\mu_n \in \Gamma(E)$ such that

(1) $I(\mu_n; E, K) \to V_K(E)$.

(2) There is a $\mu_0 \in \Gamma(E)$ such that

$$\int_E f(\mathbf{x})\, d\mu_n(\mathbf{x}) \to \int_E f(\mathbf{x})\, d\mu_0(\mathbf{x}) \tag{15.8.16}$$

for every f continuous on E.

(3) $I(\mu_0; E, K) = V_K(E)$.

If (15.8.16) holds, we say that μ_n *converges weakly* to μ_0. If now μ_0 is unique, it can be obtained by the following construction. We recall that K is the generating function of an A-average, say A_K. Since E is compact in R^m, we can find, for any n, a set of n points $\mathbf{x}_1, \mathbf{x}_2, \ldots, \mathbf{x}_n$ in E such that

$$A_K(\|\mathbf{x}_j - \mathbf{x}_k\|; 1 \leqslant j < k \leqslant n) = d_n(E), \tag{15.8.17}$$

so that

$$\binom{n}{2} K[d_n(E)] = \sum_{1 \leqslant j < k \leqslant n} K(\|x_j - x_k\|). \tag{15.8.18}$$

At each point $\mathbf{x}_j$ we place a mass of $1/n$. This defines $\mu_n \in \Gamma(E)$ and a simple calculation shows that

$$I(\mu_n; E, K) = K[d_n(E)] \to K[d_0(E)], \qquad n \to \infty, \tag{15.8.19}$$

whence

$$d_0(E; \mathsf{A}_K) = C_K(E). \tag{15.8.20}$$

This relation reveals another application of transfinite diameters. For $K(r) = \log r$ (which does not satisfy our assumptions!) this connection was established by Fekete and for $K(r) = 1/r$ by Pólya and Szegö.

In this discussion we have assumed that $C_K(E) > 0$. If this is not true, then the sequence of potential functions corresponding to μ_n need not converge weakly to a limit.

Among the properties of the equilibrium potential, the following should be noted:

(1) $U(\mathbf{x}; \mu_0, K) \geqslant V_K(E)$ for all $\mathbf{x} \in E$, excepting at most a set of K-capacity 0.

(2) $U(\mathbf{x}) \leqslant V_K(E)$ everywhere on the support of μ_0.

(3) There is a constant B depending only upon the dimension m of the space such that $U(\mathbf{x}) \leqslant BV_K(E)$ for all $\mathbf{x}$.

The notion of a Green's function may also admit of an extension to K-potentials. Green's function will then be of the form

$$G(\mathbf{x}, \mathbf{z}; E) = K(\|\mathbf{x} - \mathbf{z}\|) - \int_E K(\|\mathbf{x} - \mathbf{y}\|)\, d_{\mathbf{y}}\nu(\mathbf{y}, \mathbf{z}). \tag{15.8.21}$$

It is a symmetric function of $\mathbf{x}$ and $\mathbf{z}$ and leads to the representation

$$u(\mathbf{z}) = \int_E u(\mathbf{y})\, d_{\mathbf{y}}\nu(\mathbf{y}, \mathbf{z}) \tag{15.8.22}$$

for any K-potential corresponding to a distribution of mass on E.

As the last topic to be mentioned briefly here we consider the relation between the transfinite diameter and conformal mapping discovered by Fekete and mentioned in Section 15.5.

Let E be a bounded continuum in the complex plane whose complement is connected. E has a transfinite diameter with respect to the geometric mean and for each n there is a set of n points $\{z_{jn}; j = 1, 2, \ldots, n\}$ in E such that

$$\prod_{1 \leqslant j < k \leqslant n} |z_{jn} - z_{kn}| = [d_n(E)]^N, \qquad N = \tfrac{1}{2}n(n-1). \tag{15.8.23}$$

These points are in general not uniquely determined but this does not affect what follows. For each n there exists a set $\{z_{jn}\}$ with the aid of which we can form an nth *Fekete polynomial* for E,

$$F_n(z; E) = F_n(z) = \prod_{j=1}^{n} (z - z_{jn}). \tag{15.8.24}$$

Set

$$f_n(z) = [F_n(z)]^{1/n}, \tag{15.8.25}$$

where the nth root is so chosen that

$$\lim_{z\to\infty} z^{-1} f_n(z) = 1.$$

The functions f_n are holomorphic in the complement of E. A basic property of the Fekete polynomials (here stated without proof) is that

$$\lim_{n\to\infty} \max_{z\in E} |f_n(z)| = d_0(E), \tag{15.8.26}$$

the geometric transfinite diameter of E. This property ensures the existence of a subsequence of $\{f_n\}$ which converges everywhere on the complement of E to a limit function f, uniformly on compact sets. As a matter of fact, the whole sequence converges to f. This function f is holomorphic on the complement and it is *univalent*, i.e. $z_1 \neq z_2$ implies $f(z_1) \neq f(z_2)$. Furthermore, it maps the complement of E *conformally* on the exterior of the circle

$$|w| = d_0(E). \tag{15.8.27}$$

Since f has for large values of $|z|$ an expansion of the form

$$f(z) = z + \sum_{k=0}^{\infty} A_k z^{-k}, \tag{15.8.28}$$

it follows that f is the function which maps the complement of E on the exterior of the circle with center at the origin and radius r_e, the *exterior conformal mapping radius* of E. Thus

$$d_0(E) = r_e(E), \tag{15.8.29}$$

which is the identity proved by Fekete.

On the other hand, if μ_n is the distribution of mass on E where each point z_{jn} carries the mass $1/n$, then

$$-\log|f_n(z)| = \int_E \log\frac{1}{|z-t|}\, d\mu_n(t). \tag{15.8.30}$$

As $n \to \infty$, the left member converges to $-\log|f(z)|$ and μ_n converges weakly to μ_0, the equilibrium distribution on E, so that

$$-\log|f(z)| = \int_E \log\frac{1}{|z-t|}\, d\mu_0(t), \tag{15.8.31}$$

which is the equilibrium potential $U(z;\mu_0, E)$. All three functions are harmonic in the complement of E. $U(z;\mu_0, E)$ approaches $V(E)$ on ∂E, excepting at most a set of logarithmic capacity 0. The first member of (15.8.31) approaches $-\log d_0(E)$

almost everywhere on ∂E, the exceptional set again being of logarithmic capacity 0. Hence

$$-\log d_0(E) = V(E) = -\log C(E) \tag{15.8.32}$$

or

$$d_0(E) = C(E), \tag{15.8.33}$$

so that *the transfinite diameter also equals the logarithmic capacity of E* as observed by Szegö.

The coefficients A_k in (15.8.28) can be expressed in terms of the *moments of the equilibrium distribution*

$$M_k = \int_E t^k d\mu_0(t) \tag{15.8.34}$$

via the series for the logarithmic derivative

$$\frac{f'(z)}{f(z)} = \sum_{k=0}^{\infty} M_k z^{-k-1}. \tag{15.8.35}$$

The actual formulas for A_k in terms of the M_j are fairly complicated, see the exercise below. The expressions coincide with the Newton–Waring formulas for the kth symmetric function of n variables in terms of the corresponding power sums.

EXERCISE 15.8

1. Prove that $I(\mu; E, K)$ is finite when $K(0)$ is.
2. Verify (15.8.19).
3. Verify (15.8.29).
4. Show that the exterior mapping function satisfies

$$-\log f(z) = \int_E \log \frac{1}{z-t}\, d\mu_0(t), \qquad -\pi < \arg z \leqslant \pi,$$

where the logarithm has its principal value. Use this to verify (15.8.35). [*Hint*: The harmonic conjugate of the left member of (15.8.30) is obtained by replacing the kernel by its harmonic conjugate.]

5. Compute the coefficients A_0, A_1, A_2 in (15.8.28).
6. Show that

$$A_k = \sum \frac{(-1)^{p_1+p_2+\cdots+p_k}}{(p_1)!\,(p_2)!\dots(p_k)!}\left(\frac{M_1}{1}\right)^{p_1}\left(\frac{M_2}{2}\right)^{p_2}\cdots\left(\frac{M_k}{k}\right)^{p_k},$$

where the summation extends over those non-negative integers p_j such that

$$p_1 + 2p_2 + \cdots + kp_k = k.$$

7. The potentials of M. Riesz are based on the kernel

$$R_\alpha(\mathbf{x}) = \frac{\Gamma[(\frac{1}{2}(m-\alpha))]}{\pi^{\frac{1}{2}m}\Gamma(\frac{1}{2}\alpha)} \|\mathbf{x}\|^{\alpha - m}, \qquad 0 < \alpha < m.$$

If α, β, $\alpha + \beta$ are all in $(0, m)$, the kernel has the elegant composition property

$$R_{\alpha+\beta}(\mathbf{x}) = \int_{R^m} R_\alpha(\mathbf{s})\, R_\beta(\mathbf{x} - \mathbf{s})\, d\mathbf{s}.$$

Verify this for $m = 3$. It is enough to consider vectors of the form $\mathbf{x} = (t, 0, 0)$. Introduce polar coordinates.

COLLATERAL READING

For the whole chapter, see the survey article:

HILLE, E., *Topics in Classical Analysis. Lectures on Modern Mathematics,* Vol. III (ed. T. L. Saaty), Wiley, New York (1965).

Sections 3, 4, 5, and the extensive bibliography bear on the present chapter. The reader is referred to this survey for further details. For Section 15.4 see:

HILLE, E., Some geometric extremal problems, *J. Australian Math. Soc.* **6** (1966) 122–8.

For Section 15.8 see also:

CARLESON, L., *Selected Problems on Exceptional Sets,* Van Nostrand, Princeton, N.J. (1967). [Bibliography with 1049 entries.]

CHOQUET, G., Theory of capacities, *Ann. Inst. Fourier, Grenoble* **5** (1953–4) 131–295.

HILLE, E., *Analytic Function Theory,* Vol. 2, Ginn, Boston (1962). Sections 16.3 to 16.5 and 17.3.

INDEX

ABCDE7987654321